Nuclidic Masses

Proceedings of the Second International Conference
on Nuclidic Masses, Vienna, Austria

July 15—19, 1963

Sponsored by the International Union of Pure and Applied Physics and
the Nuclear Science Committee of the National Academy of Sciences —
National Research Council (USA) with financial support from the
United Nations Educational, Scientific and Cultural Organization

Edited by

Walter H. Johnson, Jr.

University of Minnesota, School of Physics
Minneapolis, Minnesota, USA

With 188 Figures

1964

Springer-Verlag Wien GmbH

ISBN 978-3-7091-5558-5   ISBN 978-3-7091-5556-1 (eBook)
DOI 10.1007/978-3-7091-5556-1

© 1964 by Springer-Verlag Wien

Originally published by Springer-Verlag/Wien in 1964

Softcover reprint of the hardcover 1st edition 1964

Titel Nr. 9122

Editor's Preface

The Second International Conference on Nuclidic Masses was held in Vienna, Austria, July 15–19, 1963, using facilities of the International Atomic Energy Agency. This was the third conference in the general area of nuclidic masses in recent years. The first, a symposium held at the Max Planck Institut für Chemie in 1956, was international in character but not in name. The First International Conference on Nuclidic Masses was held at McMaster University in September of 1960 in conjunction with and shortly after the meeting of the General Assembly of the International Union of Pure and Applied Physics and the Kingston Conference on Nuclear Structure.

The Second International Conference on Nuclidic Masses was held under the sponsorship of the International Union of Pure and Applied Physics and the Nuclear Science Committee of the National Academy of Sciences—National Research Council of the United States. Financial support for the conference came from the United Nations Educational, Scientific, and Cultural Organization. The conference committee was made up of the following individuals:

Chairman: J. H. E. MATTAUCH

General Secretary: H. E. DUCKWORTH

Local Secretary: F. P. VIEHBÖCK

W. W. BUECHNER	B. GROSS
E. R. COHEN	M. J. HIGATSBERGER
A. DE SHALIT	A. O. C. NIER
J. W. M. DuMOND	H. H. STAUB
B. S. DZHELEPOV	D. M. VAN PATTER

A. H. WAPSTRA

The conference committee requested that an acknowledgment be made of the aid received by the committee from the International Atomic Energy Agency, especially Dr. H. SELIGMAN, Dr. B. GROSS and Mr. HEROLD.

Acknowledgment of the efficient conference service staff should also be made. This staff included the following persons:

Frau Dr. Bohrn	Hr. Kaltseis
Frau Chambalu	Hr. Lugmair
Hr. Edl	Hr. Magerle
Hr. Fiedler	Frl. Radl

The conference committee is grateful to the above individuals for their contributions to the conference.

A tape recording of the proceedings of the conference aided in the preparation of the discussions following the papers. In addition, persons who participated in the discussion were asked to put in writing their contributions immediately after each discussion. These written contributions aided the editor greatly in reconstructing the discussion. Each author has corrected the proofs of his paper and also the proofs of the discussion related to his paper. Except for this, the discussions have not been checked by the participants. They are the editor's version of what was said and should be considered in that light.

The editor expresses his thanks to the authors of papers presented at the conference for their cooperation. The editor wishes to especially acknowledge the help of Fr. Dr. Bohrn during the conference. The thanks of the editor is also expressed to Dr. E. R. Cohen and Prof. J. W. M. DuMond who translated the paper by Prof. Batuecas from French to English and also to Mr. Petroff of the IAEA who translated the paper by Dr. Kravtsov and the paper by Demirkhanov, Dorokhov and Dzkuya from Russian to English. The aid of Miss Ardis Hovland, Mrs. Rose Mary Haji-Sheikh, Miss Gladys Knutson, and Matthew Cunningham here at the University of Minnesota is also gratefully acknowledged.

Minneapolis, Minnesota, June 1964.

Walter H. Johnson

Contents

Session I. Theoretical
Chairman: A. DE SHALIT

Session II. Fundamental Constants
Chairman: J. W. M. DuMOND

Session III. Energy Differences Between Nuclear States
Chairman: D. M. van Patter

Session IV. Calibration Energies for Nuclear Measurements
Chairman: H. H. Staub

Session V. Nuclear Q-Value Determinations
Chairman: W. W. Buechner

Session VI. Mass Values from Mass Spectroscopy
Chairman: A. O. C. Nier

Nuclear Masses and Nucleosynthesis

By

J. A. Wheeler

Princeton University, Princeton, New Jersey, USA

The manuscript for this paper was not available at the time of publication of the proceedings.

Mass Laws and Nucleosynthesis*

By

P. A. Seeger

California Institute of Technology, Pasadena, California, USA

With 6 Figures

I. Introduction

The purpose of this paper is to show that there is a considerable body of experimental evidence from geophysics and astrophysics which can be used to test and compare various semi-empirical mass laws. As an experimentalist, I am somewhat surprised to find this paper in the theoretical session of this conference, but on the other hand there is no other place to put it. As you perhaps know, Professor WILLIAM A. FOWLER of the California Institute of Technology has for many years been an instigator of mass laws; my own mania for computers has ensnared me as his current pawn.

II. Processes of Nucleosynthesis

In work on the various processes involved in nucleosynthesis[1], the basic data are the nuclidic abundances in the solar system and similar stellar systems, shown schematically in Fig. 1, which is based on the data of SUESS and UREY[2]. A young star subsists on charged particle reactions

* Supported in part by the Office of Naval Research and in part by the National Aeronautics and Space Administration.

involving protons, alpha particles, and light nuclei; these reactions can in general be studied experimentally. As stars evolve and the universe grows older, material is gradually converted to elements in the neighborhood of iron, at the peak of the curve of binding energy per particle. Thus the prominent abundance peak at mass number $A = 56$ is produced. The most prominent feature of the schematic abundance plot for A greater

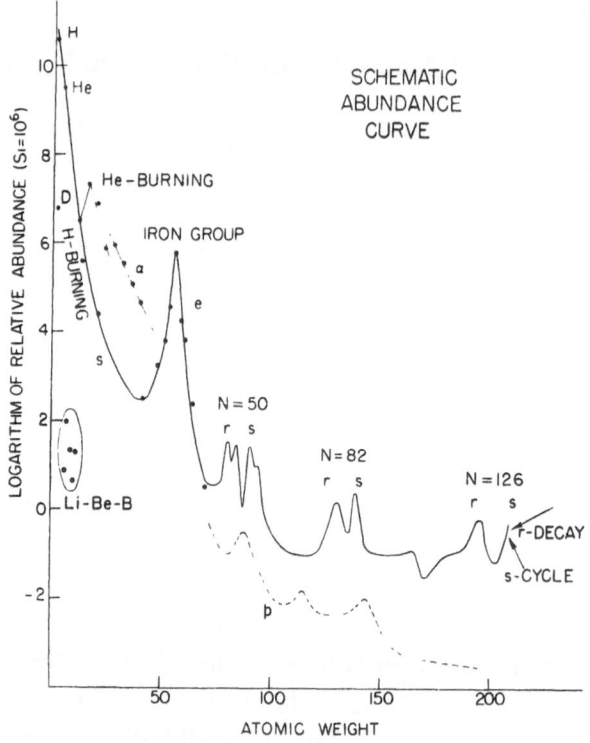

Fig. 1. Schematic abundance curve, based on the data of Suess and Urey

than 56 is the appearance of twin peaks at $A = 80$ and 90, $A = 130$ and 140, and $A = 194$ and 208. These peaks can be explained as the results of two different processes of neutron addition to the iron group. In each case the peak is attributed to the occurrence of a magic number of neutrons.

Fig. 2 illustrates the paths in the $Z-A$ plane followed by these two neutron capture processes. If the neutron captures occur at a slow rate compared to the intervening β-decays (s-process), then the path follows very nearly the line of beta stability. The nuclides involved are thus available for laboratory study, and the s-process abundances can be calculated[3] from experimentally determined neutron capture cross sections. Peaks are found where the capture cross section is low, for instance at magic neutron numbers: $N = 50$ occurs at $A = 90$, $N = 82$ at $A = 140$, and $N = 126$ at $A = 208$.

On the other hand, if the neutron flux and temperature are such that all the (n, γ) and (γ, n) reactions occur at a much more rapid rate than any β-decays (r-process), the result will be that for material of a given Z a statistical equilibrium will be reached between the various isotopes by (n, γ) and (γ, n) reactions. The average number of neutrons may be 30 more than the β-stable isotope. Such super-n-uated nuclei will of course eventually

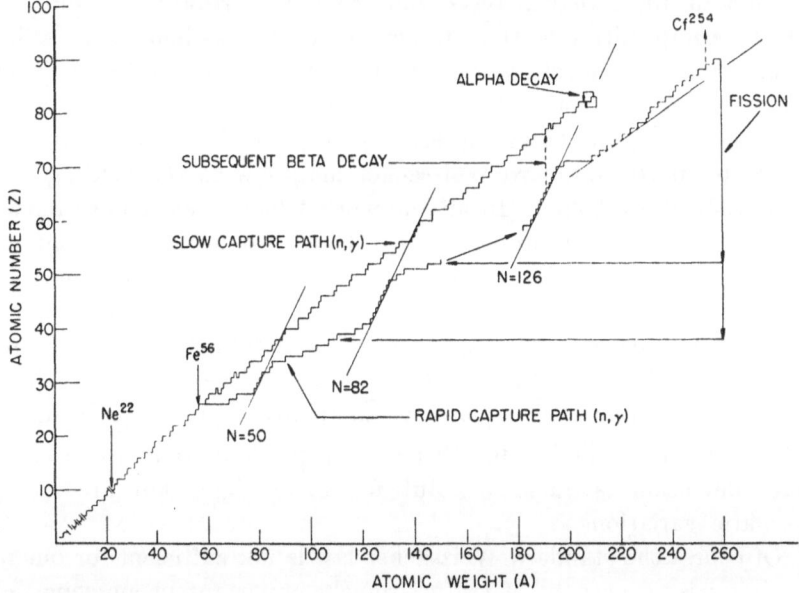

Fig. 2. The neutron capture paths of the s-process and the r-process (schematic)

β-decay to $Z + 1$, where statistical equilibrium is again quickly established between the nuclei, neutrons, and photons. Thus a path is followed in the $Z - A$ plane which is quite far from the region of experimentally available nuclides; only after neutron addition ceases can the material β-decay to the region of stability. Thus for instance nuclei of Cd^{130}, with 48 protons and 82 neutrons, decay to Te^{130}, and the existence of $N = 82$ as a magic number in the progenitors produces the observed abundance peak at $A = 130$, ten units lower in A than the $N = 82$ abundance peak formed in the s-process.

In order to calculate the path of the r-process for a given neutron flux and temperature, it is necessary to know the neutron binding energy to every one of the nuclei involved. Here we come to a major difference between r-process calculations and the other synthesis mechanisms: there are only rare exceptions in which experimental measurements of the basic

1*

nuclear quantities used in the calculations can be made—in general the neutron binding energies must be calculated on the basis of nuclear systematics. In other words, we need a good mass law, which can be extrapolated to extremely neutron rich nuclei. Conversely, we submit that one test of a mass law is its ability to "predict" the observed r-process isotopic abundance data. I would now like to discuss our requirements for a "good" mass law.

III. Requirements for a Mass Law

First of all statistical tests come to mind. However, I must admit that the complexities of the problem leave me without any feeling for the meanings of the statistical tests. I hope that Professor BREITEN-BERGER will clarify the statistical problems this afternoon. To illustrate my dilemma, Fig. 3 shows the error plot (in MeV) of a recent attempt of mine to fit the standard Weizsäcker mass law to the binding energies in the 1961 Mass Table[4]. In all the work I have done I have outwardly ignored the statistical problems, and have instructed the computer to proceed as if it were doing an ordinary least squares problem. I used all the data, weighted inversely as the square of the published error. The result was in this case a χ^2 greater than 2 million, which should be sufficient warning that the problem is outside the range of validity of statistical tests, and least squares is not really a proper procedure. The least squares test number for this case is $\sqrt{\chi^2/r} = 51.1$; the root-mean-square deviation is 2.0 MeV. But far more significant are the great systematic variations.

Of course the standard Weizsäcker law is not sufficient for our needs, since it takes no account of two extremely important phenomena—closed shell effects and nuclear deformations—and only empirically treats a third—pairing correlation energy. These three effects are essential to the r-process: pairing, because the neutron binding energy always involves a difference between an even N and an odd N; shell effects, to produce the observed abundance peaks; and deformation, to account for a broad peak in the rare earth region. Fig. 4 is the error plot (again in MeV) for a mass law which includes theoretical functions for these correction terms, added to the standard mass law. The fit is of course better—standard deviation 0.69 MeV. The least squares test number is also considerably reduced, to 18.03; however, since I have found without exception that the test number decreases when new terms are added to the law, I don't believe the test number to be a sufficient test of mass laws with different numbers of terms.

In order to have any confidence that a mass law can be extrapolated, I feel that it must have the smallest possible number of terms, since almost all terms are strongly correlated and the probability of obtaining physically

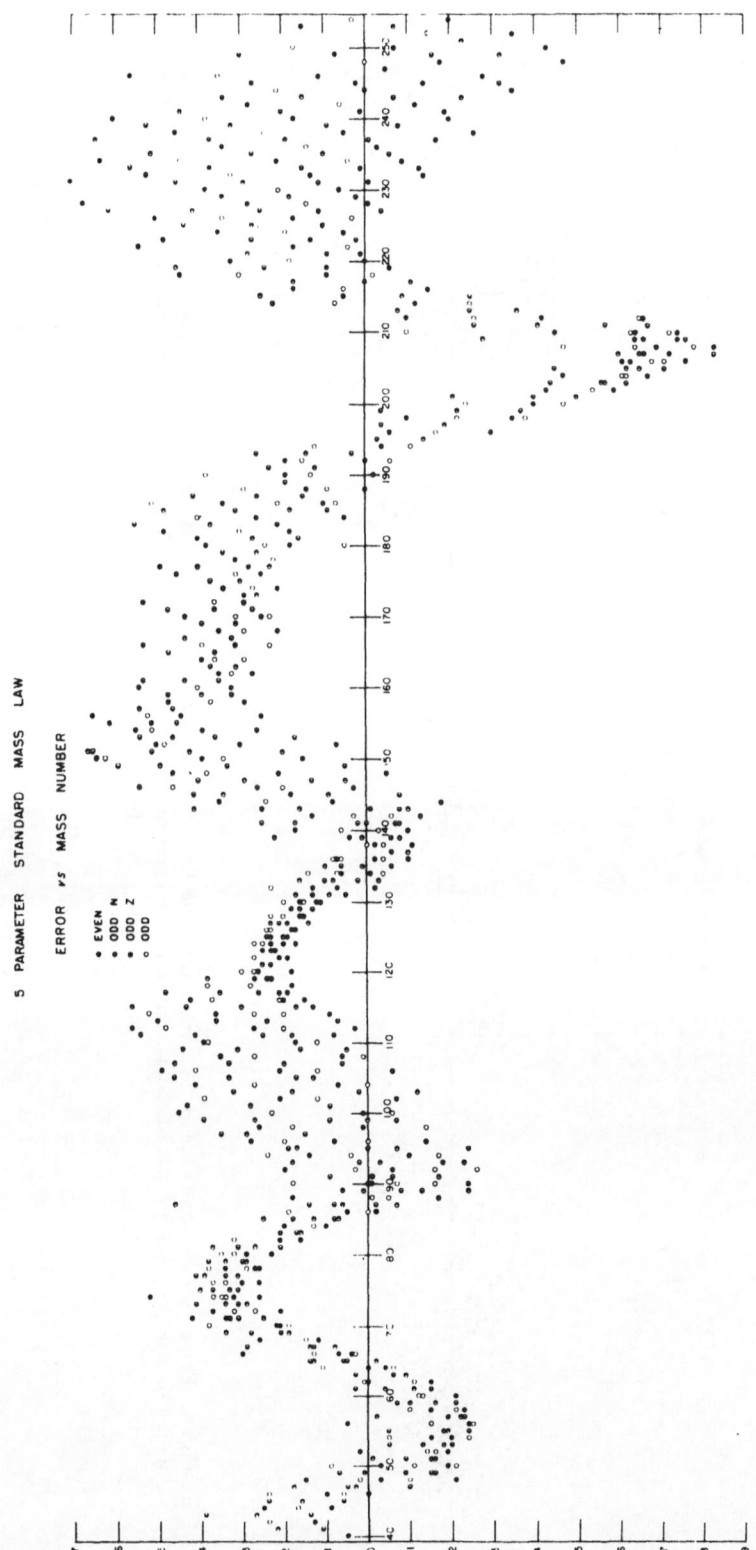

Fig. 3. Error plot (in MeV) for the standard Weizsäcker mass law, including volume term, surface term, isotopic term, composition dependent surface term, and a pairing term of the form $\pm A^{-1/2}$. Coulomb energy is calculated using constants derived from electron scattering experiments

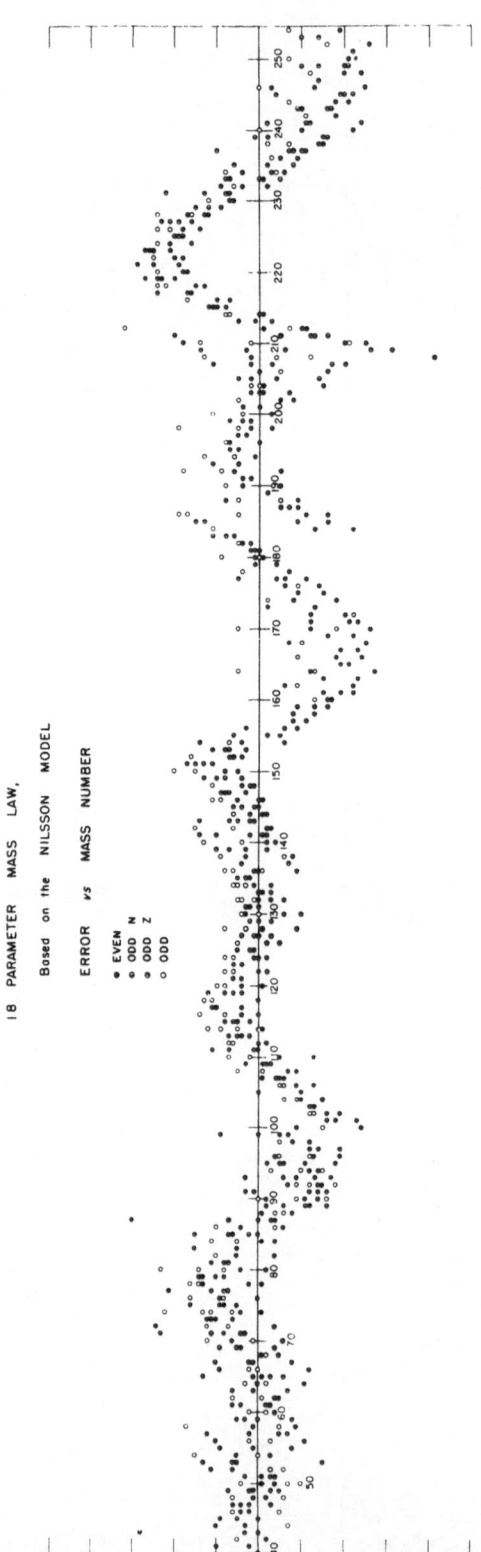

Fig. 4. Error plot (in MeV) for an 18 parameter mass law. The terms are the first 4 of the standard law, plus one for each neutron shell (5), one for each proton shell (4), and one for each neutron or proton shell (5) to account for shell effects, deformation, and pairing energy

meaningful coefficients decreases as the number of coefficients increases. Furthermore, for a coefficient to have "physical meaning", the function it multiplies must be based on theory well supported by independent experimental evidence—such as the Nilsson model and the Bardeen equation pairing model which are used in this mass law. I won't go into any more detail on this mass law, except to say that it is work in progress, and that it is used as the example in the r-process calculations to follow.

IV. Calculation of the r-process Path

When one has a mass law, he may proceed to the calculation of the path of the r-process. Fig. 5 shows a portion of the results for one such

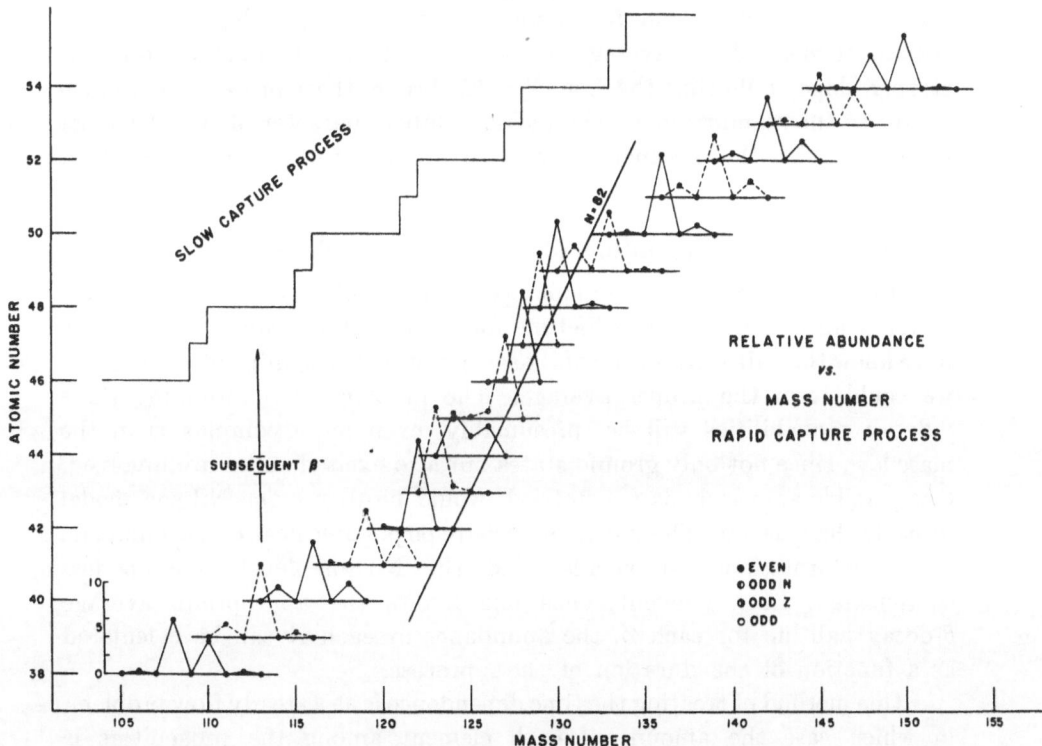

Fig. 5. Relative isotopic abundances of each of several elements during the r-process

calculation. At each value of Z, the ratio of the abundance of isotope $A + 1$ to isotope A may be expressed[1] by Boltzmann statistics as

$$\log \frac{n(A+1)}{n(A)} = \log \frac{\omega(A+1)}{\omega(A)} + \log \frac{A+1}{A} + \log n_n - 34.07 - \frac{3}{2} \log T_9 +$$

$$+ \frac{5.04}{T_9} Q_n(A),$$

where $\omega(A)$ is the statistical weight of isotope A, n_n is the neutron density in cm^{-3}, T_9 is the temperature in units of 10^9 °K, $Q_n(A)$ is the neutron binding to nucleus (Z, A) in MeV, and all logarithms are base 10. One approximation has been made in the above equation: that β-decay is so slow as not to affect the equilibrium. In these calculations, however, the statistical weights have also been omitted; they will eventually be calculated on the basis of the Nilsson model. The graph for each Z in Fig. 5 is a separate calculation, and is normalized to unity. These calculations are for $n_n = 10^{24}$ and $T_9 = 1.8$; the temperature was chosen to place the $N = 82$ abundance peak at $A = 130$.

Note that in this region the r-process path moves quickly to $N = 82$, then stays at $N = 82$ while moving through several values of Z, and finally breaks away from $N = 82$ and moves quickly to higher N. Thus several elements have average masses near $A = 130$. Furthermore, the steeper slope, following the line $N = 82$, brings the r-process path much nearer to the stability line. The magic number character of $N = 82$ must be apparent in this manner in order to produce the abundance peak at $A = 130$.

V. Calculation of r-process Abundances

In order to calculate the amount of material at each A, it is necessary to know an average β-decay lifetime for each Z. For instance, at $Z = 40$, if we knew the half-life for each of the seven isotopes of significant abundance, we could form the proper average. The problem of determining these β-decay probabilities will be, presumably, even more complex than the mass law, since not only ground states but also excited states are involved. One possible approach, as yet untried, would be to use the Nilsson model to study the β-decay. This should especially be possible near magic numbers, where deformations are smaller and the Nilsson level diagram less complicated. In any event, once one knows the appropriate average β-decay half-life for each Z, the abundance at each A can be calculated as a function of the duration of the r-process.

One method of treating the time dependence is as a steady flow problem, in which case the amount of each element among the progenitors is proportional to the half-life for that Z. Steady flow is reasonable, based on the estimated half-lives by BURBIDGE et al.[1] in which the very first β-decay, iron, is about 25% of the total cycle time, and thus controls the flow rate. Admittedly these half-life estimates are very crude; nevertheless I have used them in Fig. 6 to show what the final result of an r-process calculation would be. The solid line is the calculated abundance curve, drawn to a convenient scale, and the dots are the experimentally observed abundances[5]. I might point out that this region of A is one of the worst as far as accurate knowledge of abundances is concerned. All the xenon points, for example, might be low by a factor of two; however, the relative amounts of the Xe isotopes are fairly certain. The same is true for Te. Thus the existence of this abundance peak is quite clear. The calculation gives a somewhat narrower peak, which is a very pleasing result because it will surely be broadened by the existence of a range of temperatures rather than the fixed temperature of $1.8 \times 10^9 \,^\circ$K used here, and also by a smoothing which must have occurred during the "freezing" of the

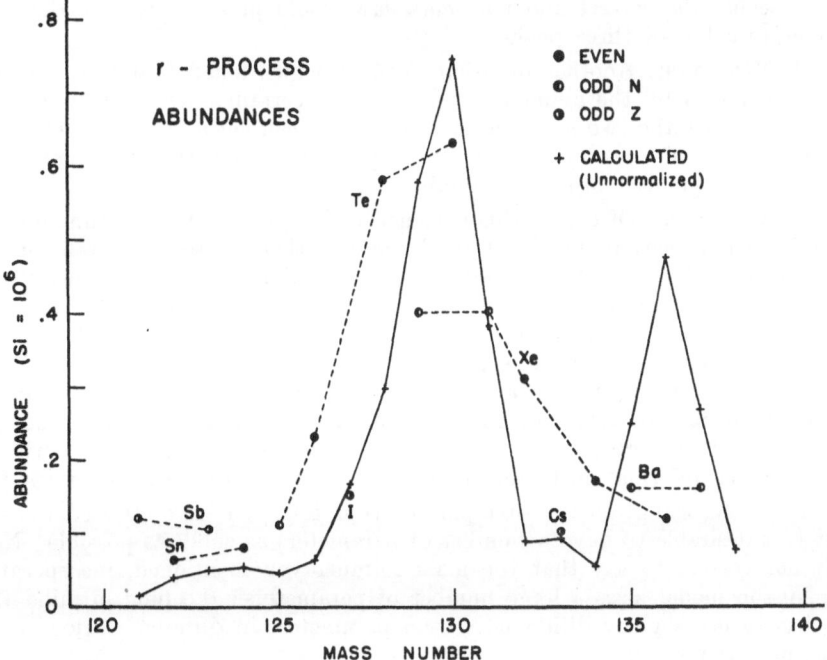

Fig. 6. Comparison of experimental and calculated isotopic abundances attributed to the r-process, in the region of the observed peak at $A = 130$

r-process, as temperature and neutron density decreased. The calculation also gives an auxiliary peak at $A = 136$, corresponding to the *proton* magic number $Z = 50$ among the progenitors. This peak does not appear so strongly in the observed abundances, but I would like to point out that the plateau at Ba is a factor of four more abundant than the r-process isotopes above $A = 140$.

VI. Conclusion

It shouldn't be necessary to point out that this work is very preliminary. At this stage of the game it is difficult to say whether the r-process would "prove" a mass law, or a mass law would "prove" the r-process. Continued work will hopefully bring the two problems into fruitful liaison.

Discussion

J. Mattauch: You mentioned that for the first peak you needed a temperature of about $1.8 \times 10^{9\circ}$. We tried to do similar calculations with our mass law but we saw that for the three double peaks we need quite different temperatures, and as far as I could gather from your paper Burbidge et al. you also required different temperatures.

P. A. Seeger: That's true. I haven't looked too carefully yet at the other peaks with this particular mass law. I do find I need about a 20% different temperature to fit the $N = 126$ peak. This, of course, is another test of the mass laws, because the correct and true mass law would presumably give the same temperature for all three peaks.

J. Mattauch: Another question: You also mentioned that the mass law might be tested by the abundance curve. How certain is this? How certain, for instance, are the two maxima near lead? We can get a very good mass law within the stability valley, but we don't get a second maximum before lead. However, we get the other two double maxima.

P. A. Seeger: Of course this is something to be tested. I presumably will be able to take various mass laws and compare them to see how well they fit these points. The point I took here, $A = 130$, is about the worst-known region in the abundance data. The existence of the peak there is quite clear from the data, and I think it is equally clear that there is actually an abundance peak at the other points also. There is a further feature in the rare-earth region which also must fit, and there the abundance data, the relative abundances among the group itself, are very well known. This is something else I have not yet tried to fit. Presumably this would tell us if a mass law is no good but will not give us much information to tell us if a mass law is good just because it fits.

H. G. Kümmel: In the first part of your talk you have stressed the fact that it is desirable to have a number of parameters as small as possible. Now, is it not correct to say that the mass formula you mentioned, incorporating the Nilsson model, uses a large number of parameters? (keeping in mind that one has to use slightly different Nilsson parameters in different regions to fit experimental results).

P. A. Seeger: Yes. In fact there are two parameters for each proton shell and two for each neutron shell—a total of 30. However, I have allowed myself no freedom with these, but have taken them from the papers of Mottelson and Nilsson. I do not use different values in different sub-shells. The values used are presumably those which give the proper level spacings and deformations, and were not adjusted to fit binding energies. I would like to point out that by using such a model, I could also calculate deformations, spins, and single particle excited states.

References

[1] E. M. Burbidge, G. R. Burbidge, W. A. Fowler, and F. Hoyle, Synthesis of the Elements in Stars. Revs. Modern Phys. 29, 547 (1957).

[2] H. E. Suess and H. C. Urey, Abundances of the Elements. Revs. Modern Phys. 28, 53 (1956).

[3] D. D. Clayton, W. A. Fowler, T. E. Hull, and B. A. Zimmerman, Neutron Capture Chains in Heavy Element Synthesis. Ann. Phys. 12, 331 (1961).

[4] L. A. König, J. H. E. Mattauch, and A. H. Wapstra, 1961 Nuclidic Mass Table. Nuclear Phys. 31, 18 (1962).

[5] Compiled by D. D. Clayton, private communication.

Semi-Empirical Analysis of Nuclear Masses

By

Nissan Zeldes

Theoretical Physics, Hebrew University of Jerusalem, Israel

With 15 Figures

My talk will be divided into three parts. First the empirical features of the nuclear mass surface will be briefly reviewed. Then we shall discuss theoretical interpretations, starting with mass formulas based on the shell model of the nucleus. Lastly, necessary corrections to mass formulas based on the liquid drop model will be considered.

Part I

The Empirical Features of Nuclear Masses

A. One of the first facts known about nuclei was the concentration of known nuclei within a narrow strip in the (N, Z) plane, the so-called line of beta stability[1]. Fig. 1 shows the beta-stable nuclei. The known unstable ones occupy a strip somewhat wider, including the stable nuclei along its central line.

A second early known fact was the approximately constant increase of nuclear binding energy when nucleons are added to build heavier and heavier stable nuclei[1]; approximately, because the increase is somewhat steeper at the beginning, and somewhat less steep in heavier nuclei. Fig. 2 shows this behavior, and the resulting course of the B/A line, reaching a maximum around Fe or Ni, and then decreasing slowly. To keep the numbers manageable, one subtracts the binding energy from a linear function of N and Z, $Z \, \Delta M_H + N \, \Delta M_n$, and calls the difference $\Delta M = Z \, \Delta M_H + N \, \Delta M_n - B$, the mass defect. This is also shown in Fig. 2.

Thirdly, the section of the mass surface in a direction of constant A, approximately perpendicularly to the line of beta stability, is parabolic to a high degree of approximation. These three facts were known in the early thirties[1], and formed the basis of the liquid drop picture of the nucleus.

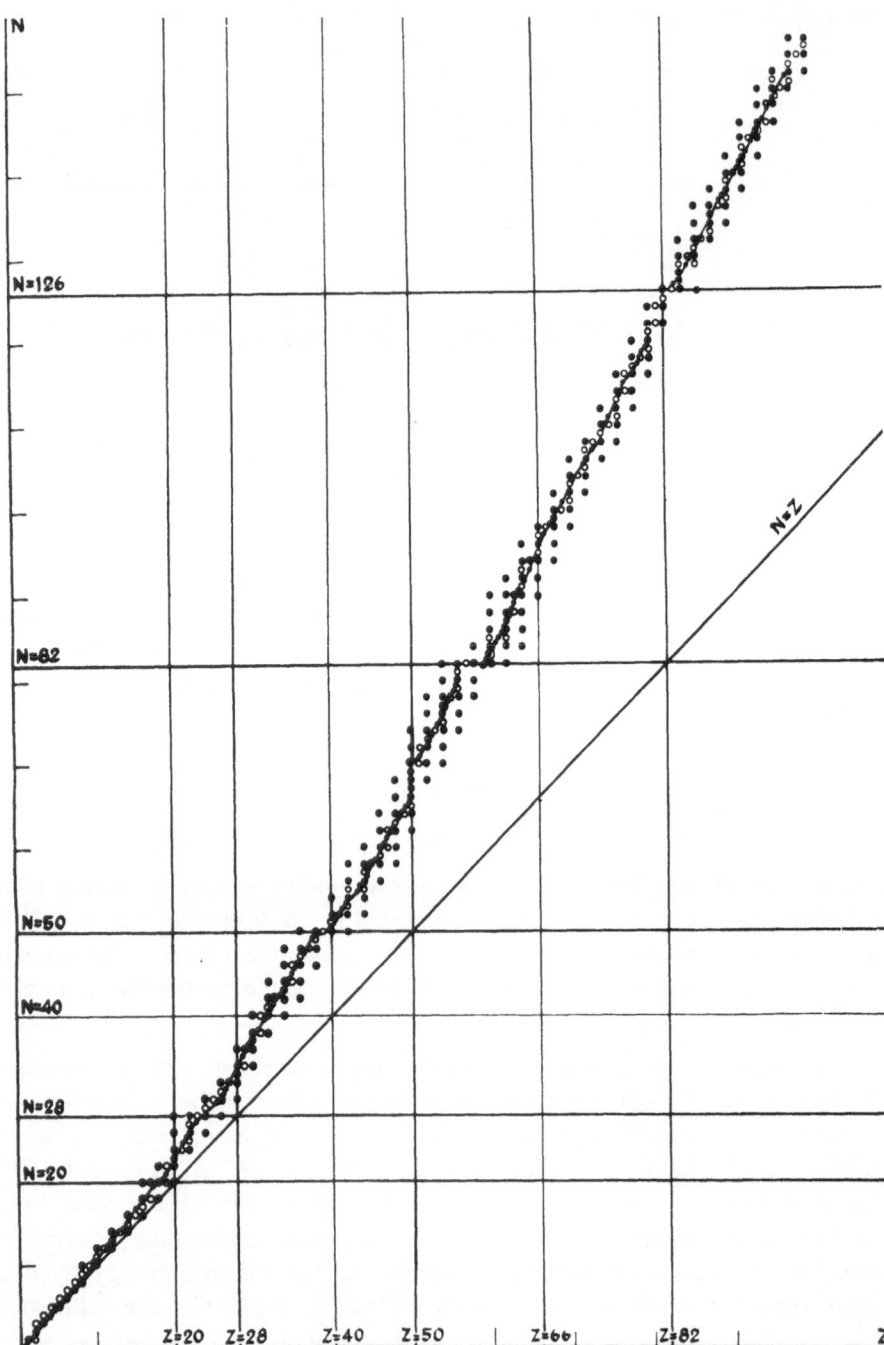

Fig. 1. Beta-stable nuclei. Full circles represent even A nuclei, empty circles odd A nuclei. Smaller full circles with a thin line running through them are the minima of isobaric sections, determined as explained in the text (cf. the discussion in connection with Fig. 6)

B. Actually, the statement concerning the parabolic section was inaccurate. It soon became evident, that there is more than one smooth mass surface. Fig. 3 shows isobaric sections of the mass surface for both

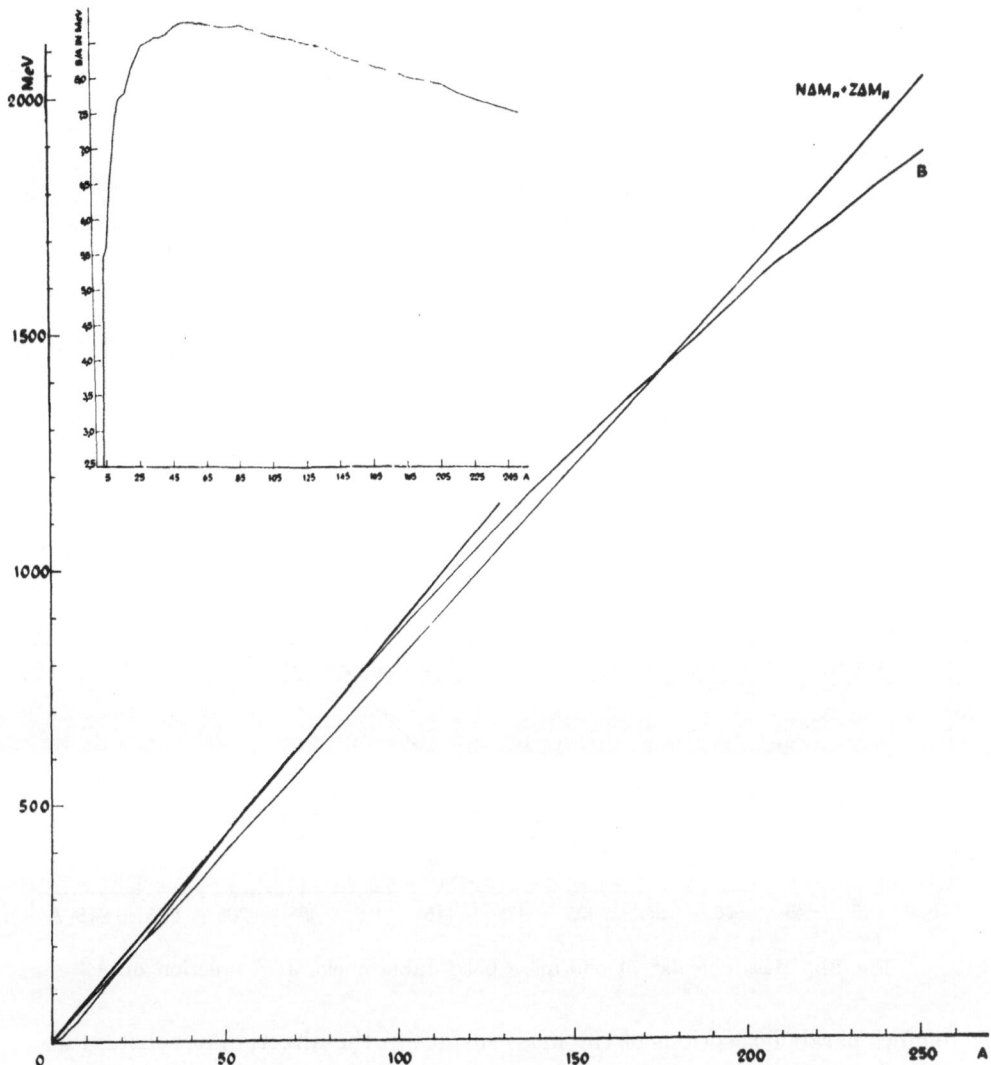

Fig. 2 a. Binding energy of odd-mass beta-stable nuclei as a function of mass number A. Binding energy per particle of odd-A beta-stable nuclei as a function of mass number A

even and odd A, and also the beta decay energies representing the mass differences between neighboring nuclei. From the zig-zag behavior of the Q_β lines it is evident that there is not one, but actually two parallel parabolas for each section. For even A the separation is large, of the order of 2 MeV, with the even-even parabola always below. This was also already known in the thirties[1]. For odd A the separation is much smaller, of changeable sign, and was discovered much later[2, 3].

The two neighboring surfaces for odd-A nuclei lie between the two even-A surfaces, usually somewhat nearer to the odd-odd surface. Fig. 4

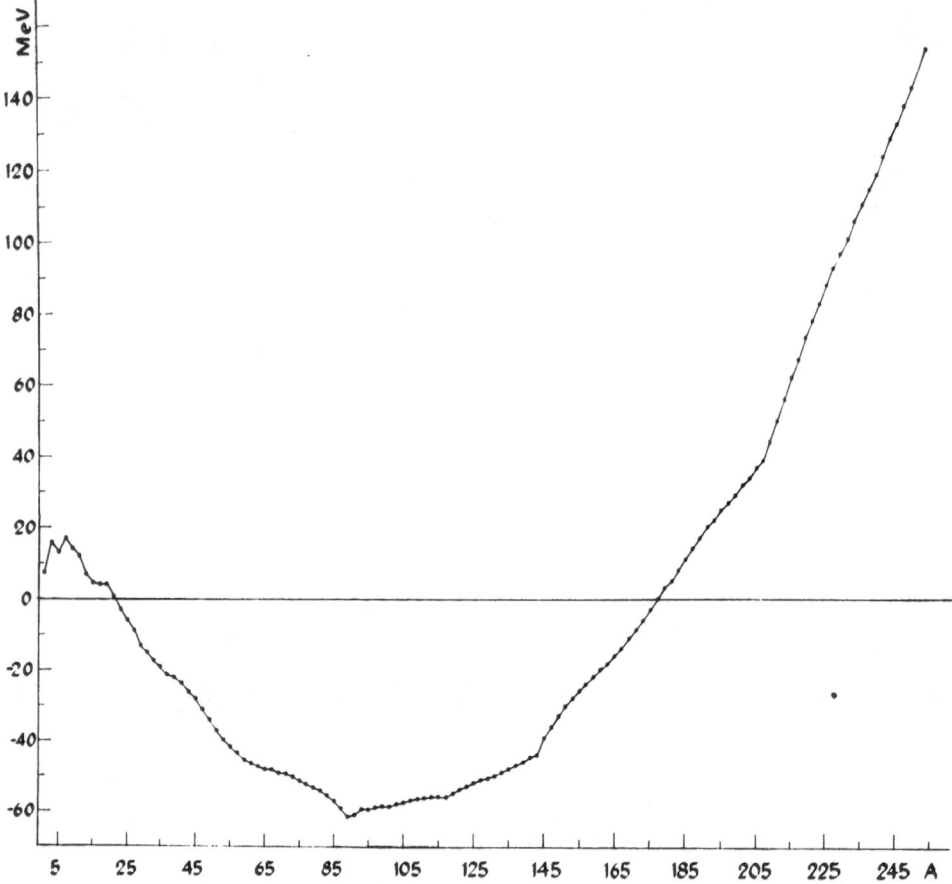

Fig. 2b. Mass defects* of odd-mass beta-stable nuclei as a function of A

shows parabolic sections of the mass surface in the direction of constant Z, for the even-Z element Zn, and for the odd-Z element Ga, together with their neutron separation energies, which are determined by the mass differences between neighboring isotopes. From the zig-zag behavior of the S_n lines, one again infers the existence of two parallel parabolas for each element, with the even-N parabola always higher. The smaller width of the zig-zag for Ga indicates that the vertical distance between the

---

* In Figs. 2, 13 and 14, the mass unit is $O^{16} = 16$; in all the other figures the mass unit is $C^{12} = 12$.

two Ga parabolas is smaller than for Zn, which is the statement made at the beginning of this section. However, this is not strictly universal, but only almost always so. It was also discovered relatively late[4, 5].

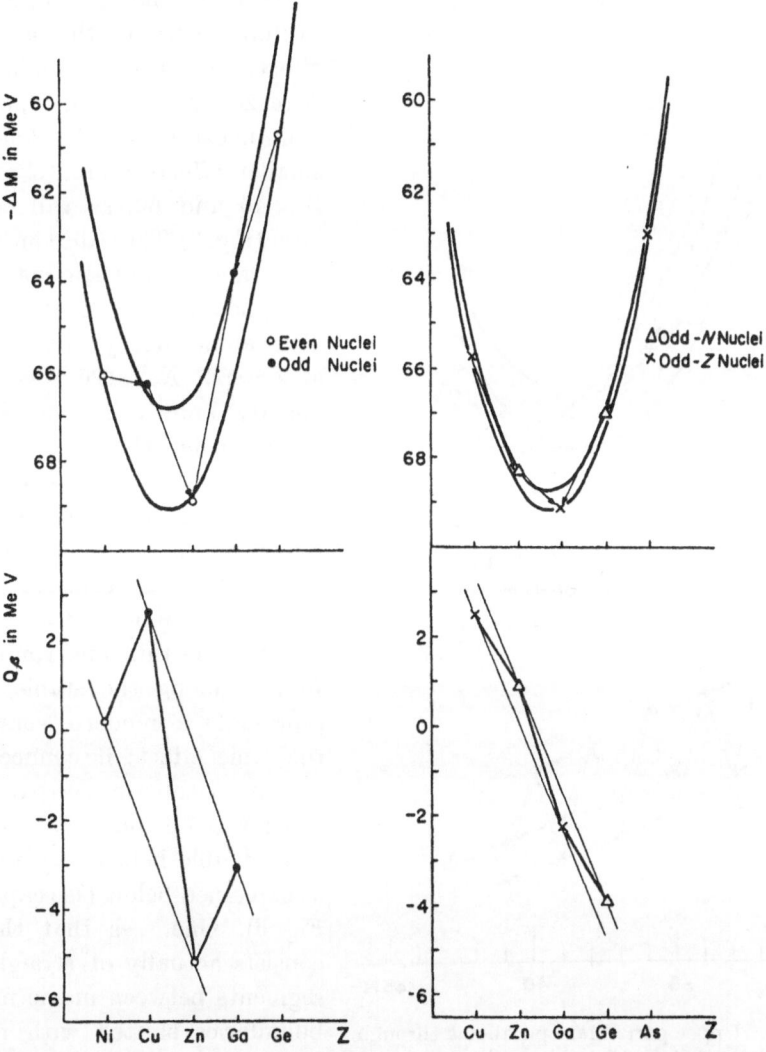

Fig. 3. Left part: Mass parabolas and beta decay energies for the $A = 66$ isobars. Right part: Mass parabolas and beta decay energies for the $A = 69$ isobars. Taken from ref. 14

Having established the existence of the four mass surfaces, let us turn to shell effects[6]. Their most direct manifestation is the sudden change in nucleon separation energies at certain nucleon numbers, the so-called magic numbers. Fig. 5 shows the sudden increase of S_{2n}* when

* We have taken double separation energies to avoid the necessity of considering separately even N and odd N.

neutron pairs are successively taken out of the nucleus and a neutron magic number is reached (numbers 20 and 126 in the figure). In the masses themselves there is a sudden increase of slope at these numbers, manifesting the same effect. For protons there is the analogous effect. The magic numbers are N or $Z = 2, 8, 20, 28, 50, 82,$ and 126 (neutrons only). Somewhat smaller effects seem to exist at $Z = 40$ and 66, as will be seen later (Fig. 9). These discontinuities in S_{2n} and S_{2p} are also manifested in alpha and beta decay energies. The alpha decay discontinuity at $Z = 82$, $N = 126$ was indeed already known in the thirties[7] much earlier than any nucleon separation energy discontinuity.

A related phenomenon which suggested the existence of shells in the thirties[8] consists of the periodic windings of the stability valley. The thin line running in Fig. 1 among the stable nuclei represents a modern version of the same effect: it connects the minima of isobaric mass parabolas, when these could be determined from double beta decay energies, as explained below (see caption of Fig. 6). One sees that the line consists actually of straight line segments between magic number boundaries shifted with respect to each other at the latter. This was first shown by CORYELL[9]. The resulting of such shifts from the changes of slopes of the mass surface at magic numbers shown in Fig. 5 is, of course, geometrically obvious. The numbers established beyond any doubt in this way are $Z = 50$ and $N = 50, 82$. At $Z = 82$ and $N = 126$ both shifts happen at the same place, compensating each other to allow the line of beta stability to continue its course uninterrupted through the doubly magic Pb[208].

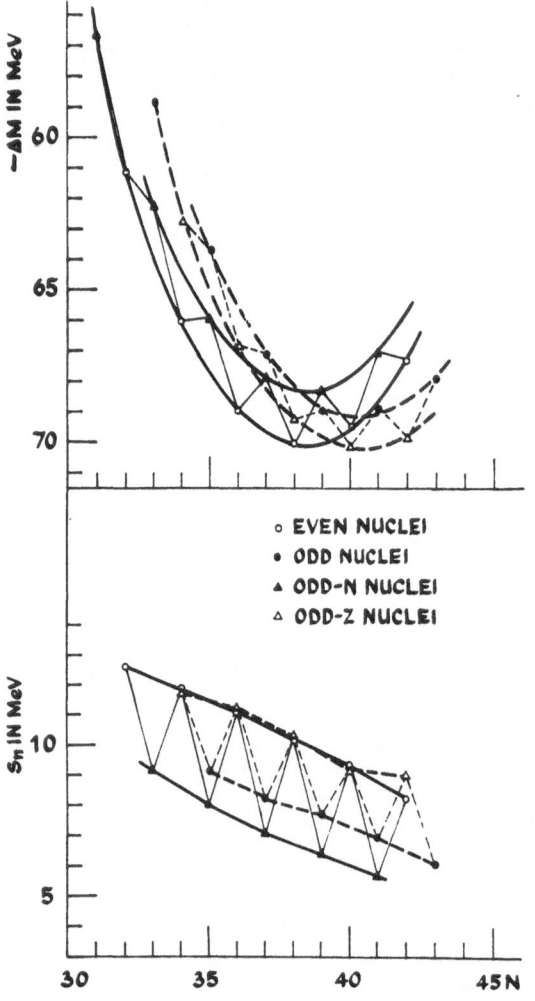

○ EVEN NUCLEI
● ODD NUCLEI
▲ ODD-N NUCLEI
△ ODD-Z NUCLEI

Fig. 4. Upper part: Mass parabolas through the zinc (full lines) and gallium (broken lines) isotopes. Lower part: Neutron Separation Energies of the zinc (full lines) and gallium (broken lines) isotopes

The last phenomenon we want to mention under heading B is the linearity of various mass differences with respect to Z and N between magic

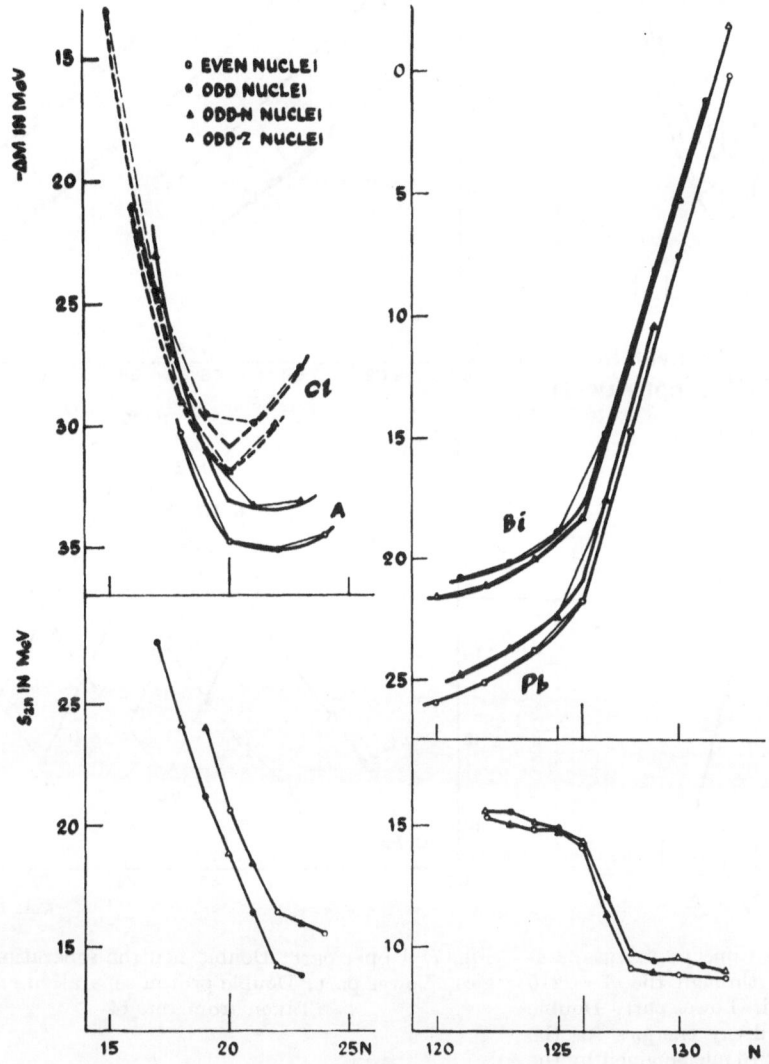

Fig. 5. Left part: Mass parabolas through the argon (full lines) and chlorine (broken lines) isotopes, and the corresponding double neutron separation energies (below). Right part: Mass parabolas through the lead and bismuth isotopes, and the corresponding double neutron separation energies (below). One notices the sudden changes associated with the magic neutron numbers 20 and 126, respectively

numbers. This became clear gradually, as more precise mass data became available. The first mentioned approximately linear systematics was that of alpha decay energies[10] and then, in particular, the more linear beta decay energies[11]. Fig. 7 shows the linear networks obtained by connecting the

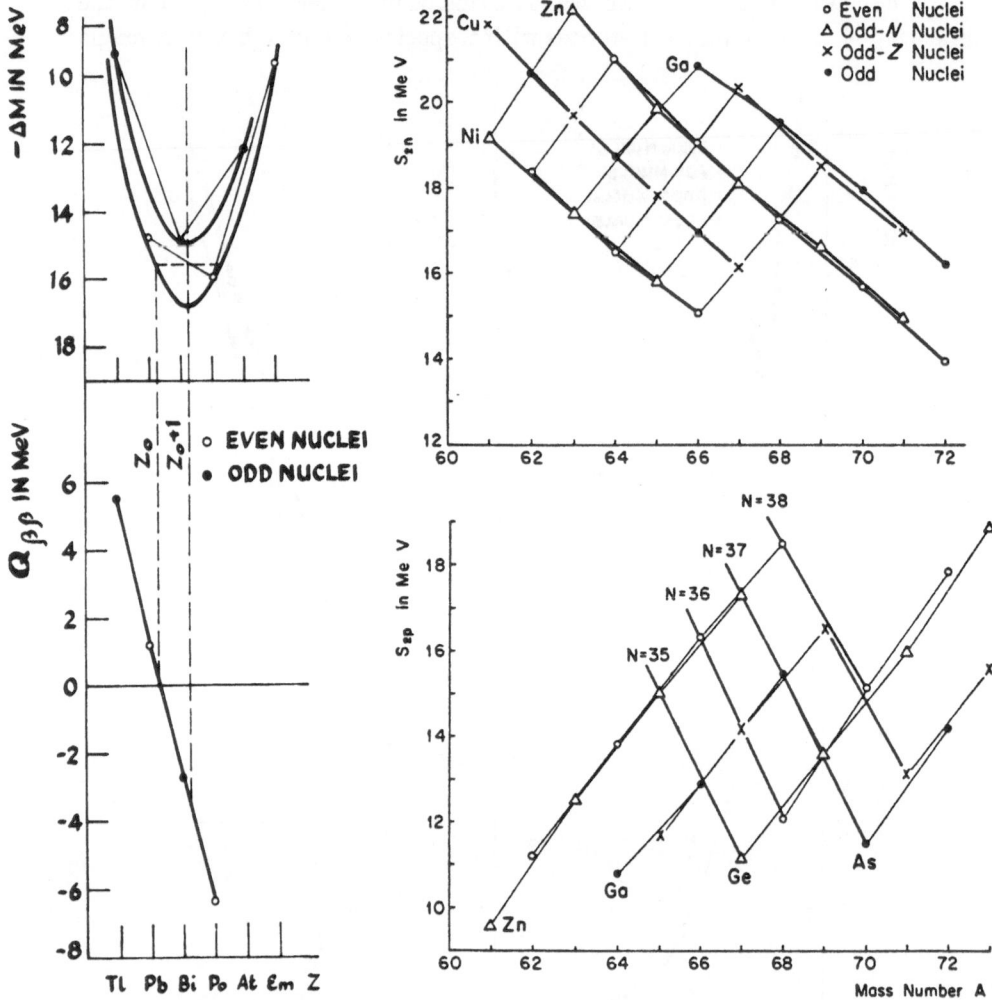

Fig. 6. Upper part: Mass parabolas through the $A = 210$ nuclei. Lower part: Double beta decay energies for the $A = 210$ nuclei, given by the mass difference between two neighboring nuclei on the same parabola. It is obvious from the figure that the Z value of the minimum of the mass parabolas is $Z_0 + 1$, where Z_0 is the Z value (not necessarily an integer) for which $Q_{\beta\beta}$ vanishes. This might be determined very easily from a graph of $Q_{\beta\beta}$ on a squared paper

Fig. 7. Upper part: Double neutron separation energies. Lower part: Double proton separation energies. Taken from ref. 14

S_{2n} and S_{2p} values of neighboring nuclei. The three last mentioned groups of phenomena, namely the splitting of the mass surface according to parity type, the shell effects, and the linear systematics, form the basis of the independent particle picture of the nucleus, as will be described later (Part II).

C. I would now like to describe results of a study carried out at Jerusalem recently[12] concerning the question of the deviation from linearity of the above mentioned systematics.

The second differences

$$\Delta_{n\,n}(Z, N) = S_{2n}(Z, N) - S_{2n}(Z, N - 2), \tag{1a}$$

$$\Delta_{p\,p}(Z, N) = S_{2p}(Z, N) - S_{2p}(Z - 2, N), \text{ and} \tag{1b}$$

$$\Delta_{n\,p}(Z, N) = S_{2p}(Z, N) - S_{2p}(Z, N - 2) \tag{1c}$$

giving the slopes of S_{2n} and S_{2p} lines like those of Fig. 7 were calculated for all known nuclei. Likewise, the third differences

$$\delta_n(Z, N) = |S_n(Z, N + 1) - 2\,S_n(Z, N) + S_n(Z, N - 1)| \; (Z \text{ even}) \tag{2a}$$

$$\delta_p(Z, N) = |S_p(Z + 1, N) - 2\,S_p(Z, N) + S_p(Z - 1, N)| \; (N \text{ even}) \tag{2b}$$

giving the splitting between the odd and even nucleon surfaces and

$$\delta_{n\,p}(Z, N) = |S_p(Z, N + 1) - 2\,S_p(Z, N) + S_p(Z, N - 1)| \tag{2c}$$

giving the extra lowering of the odd-odd surface towards the odd-mass surfaces, were calculated. These second and third differences were plotted as functions of Z and N, and some of the results are shown in Figs. 8–12.

Now, if the systematics of S_{2n} and S_{2p} were exactly linear, the second differences should have been constant between magic numbers, which would themselves be characterized by sharp discontinuities in such graphs. However, this is far from what one sees. Let us first concentrate on $\Delta_{n\,n}$ (Fig. 8). One sees sharp peaks at neutron numbers 28, 50, 82, and 126, and something similar might be seen at 20*. Below 20 the changes in S_{2n} within the shell when neutrons are added are comparable to the change associated with the crossing of magic numbers, and therefore the latter are not clearly seen on the graph, and one has to look at plots of S_{2n} themselves to find shell effects. There this difference between "light" and "heavy" magic numbers is very clearly seen as in Fig. 5, and is physically due to the much stronger interaction of two neutrons in the much smaller light nuclei. No submagic numbers can be clearly inferred from the figure.

Between magic numbers the $\Delta_{n\,n}$ values are not constant. For the very light nuclei below $N = 20$, there is a linear network of $\Delta_{n\,n}$ with respect to both N and Z (only the lines connecting isotopes were drawn in the figure). Then from $N = 28$ up to $N = 82$ there are irregular fluctuations superposed on a very clear (more or less linear) decrease of the absolute value of $\Delta_{n\,n}$. After $N = 82$, in the rare earth region, this regular decrease stops, and instead there is a hump, symmetric about the middle of the neutron shell between $N = 82$ and $N = 126$. Similarly there

---

* Unfortunately, the magic number 20 with the mark pointing at it was by mistake omitted from Fig. 8, as well as from Fig. 9–12.

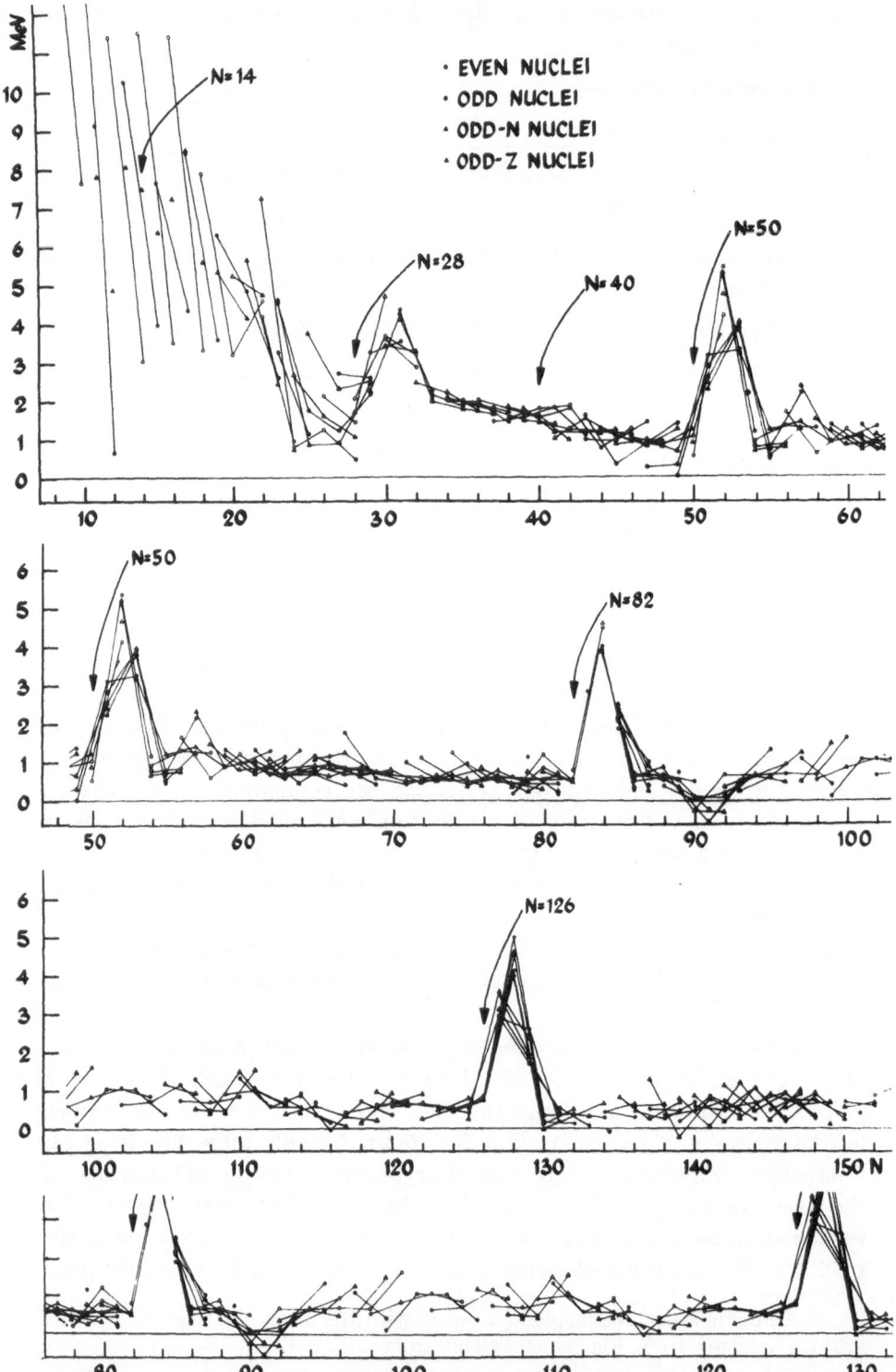

Fig. 8. — Δ_{nn} as a function of N. Isotopes of the same parity type are connected by a line. The part $82 < N < 126$ is repeated below

is the beginning of such a hump after $N = 126$, towards the middle of the next major neutron shell, completed at $N = 184$.

The proton second difference $\varDelta_{pp}$ (Fig. 9) behaves similarly: There are magic peaks at $Z = 28$, 50 and 82. But here there is also a very clear although

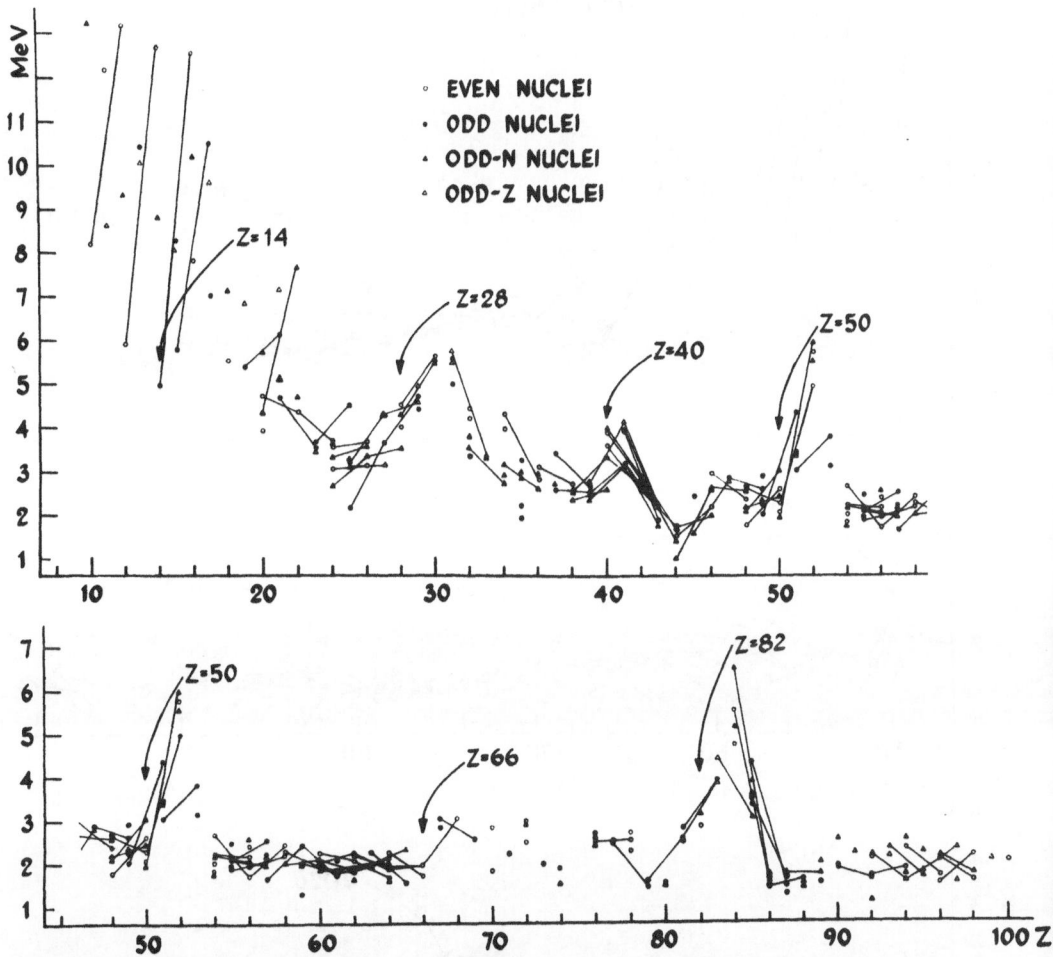

Fig. 9. — $\varDelta_{pp}$ as a function of Z. Isotones of the same parity type are connected by a line

smaller sharp peak at $Z = 40$, and perhaps some small peak at $Z = 66$ too. Thus for the protons there seem to exist also submagic numbers, corresponding to the completion of the $p_{3/2}$, $f_{5/2}$, and $p_{1/2}$ subshells ($Z = 40$), and to the completion of the $d_{5/2}$, $g_{7/2}$, and $s_{1/2}$ subshells ($Z = 66$), which seems quite consistent with the spins and energy systematics of single proton levels in these regions. Between magic numbers there is again the linear network for very light nuclei (this time only the lines connecting

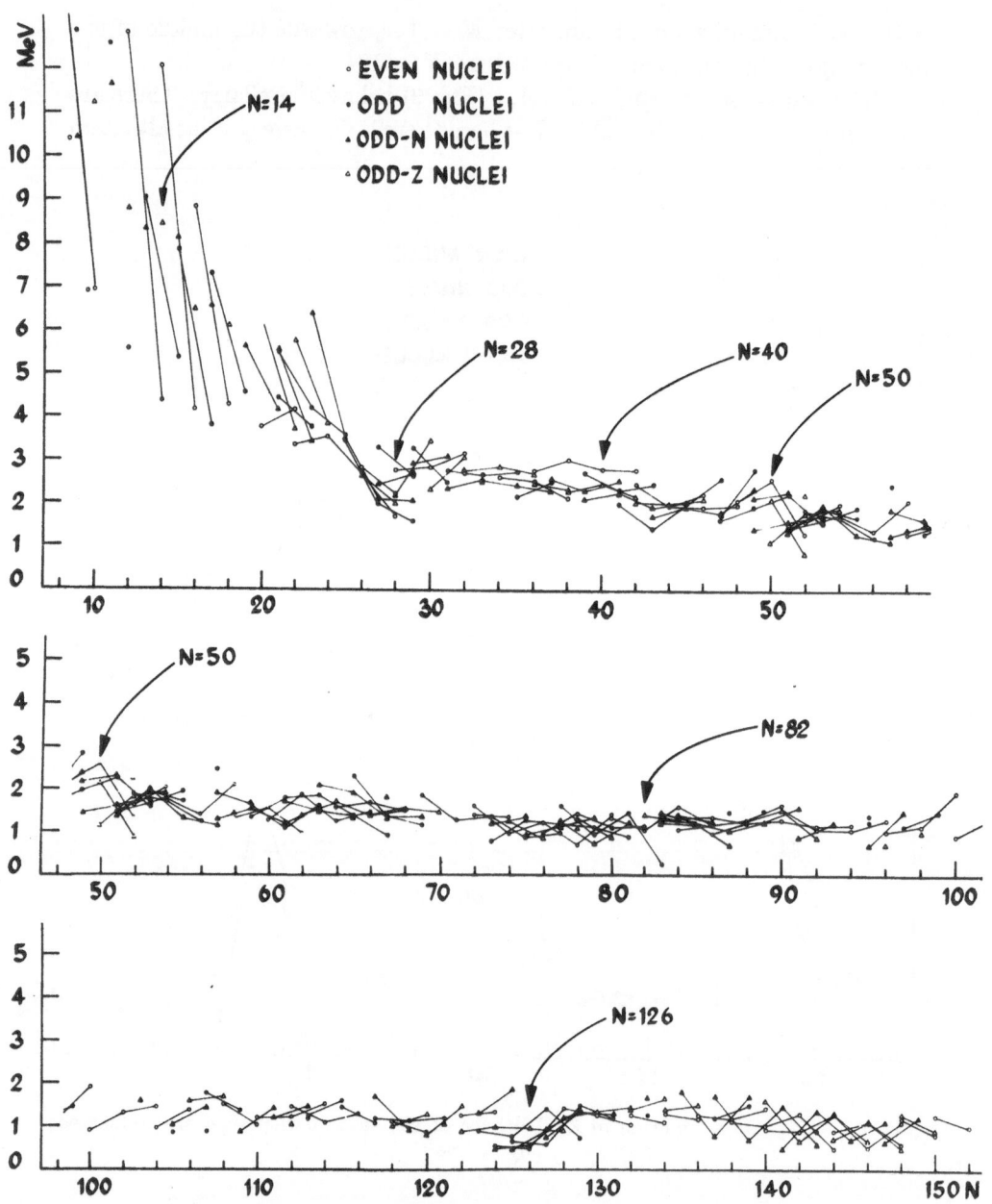

Fig. 10. Δ_{np} as a function of N. Isotopes of the same parity type are connected by a line

isotones were drawn), then regular overall decrease beyond $Z = 28$ up to $Z = 66$ (with an interruption for $40 < Z < 50$). Then the points are scarce up to $Z = 82$, and beyond $Z = 82$ there is again the beginning of a hump, as was for $N > 82$.

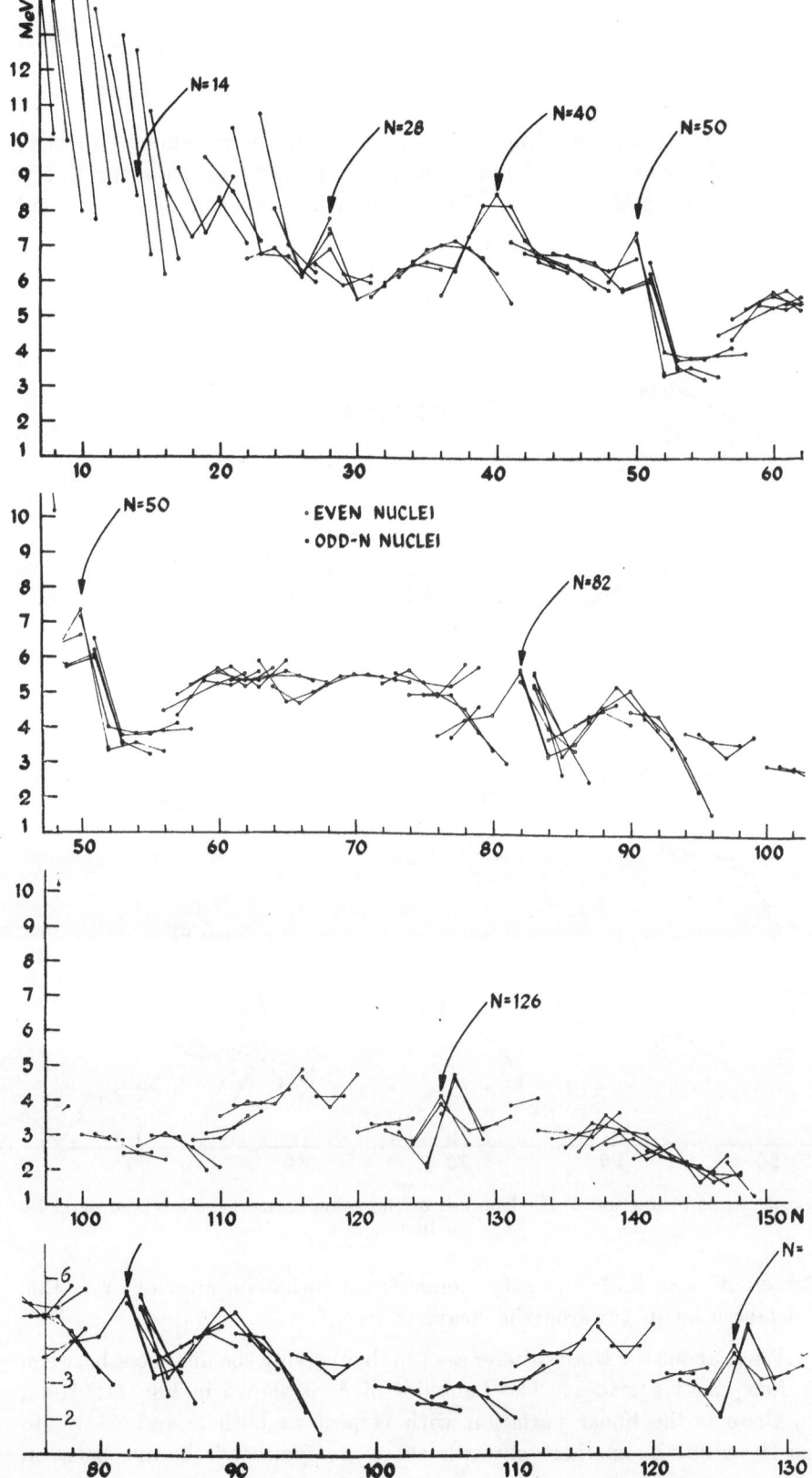

Fig. 11. δ_n as a function of N. Isotopes of the same parity type are connected by a line. The part $82 < N < 126$ is repeated below

The mixed second difference Δ_{np} (Fig. 10) shows similar linearity and overall decrease for light and for medium weight nuclei, respectively, up to the rare earths region. Then it remains approximately constant

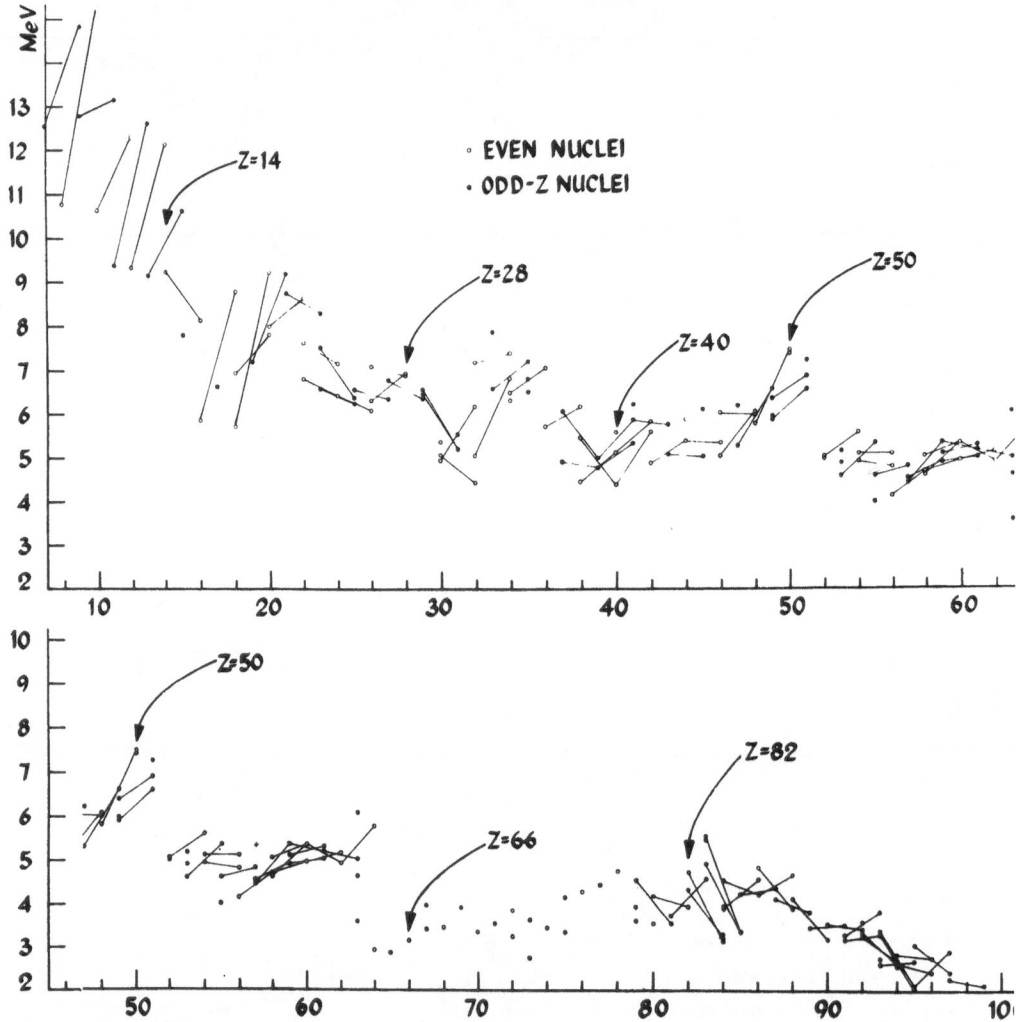

Fig. 12. δ_p as a function of Z. Isotones of the same parity type are connected by a line

between $N = 82$ and $N = 126$, increases somewhat on crossing $N = 126$, to decrease again towards the heaviest nuclei.

What about the third differences (2a, b, c) giving the distances between the four mass surfaces? The behavior of δ_n is shown in Fig. 11. Here, too, there is the linear variation with respect to both Z and N up to $N = 20$ or 28. From there onwards there are symmetric humps between magic numbers, single between 28 and 50, double with a small separation between the two peaks from $Z = 50$ to $Z = 82$, and double with a large separation between the two peaks from $Z = 82$ to $Z = 126$. From $N = 126$ onwards there is only the first half of the double hump, up to the middle of the major shell, beyond which no data exist.

With the proton separation distance, given by δ_p, the situation is similar, at least below Ca and beyond Pb (Fig. 12). In between it is possible to imagine a behavior which is symmetric with respect to the middle of the major shells, but this is not very clear.

The extra separation δ_{np} is considerably smaller than δ_n and δ_p, and will not be discussed here.

In conclusion, let us summarize the properties we have mentioned:

A.

1. Valley of stability.
2. $B/A \sim$ constant.
3. Parabolic sections.

B.

4. Splitting to four surfaces.
5. Shell effects.
6. Linear systematics.

C.

7. Deviations from linearity of the systematics:

a. Linearity of the second differences for light nuclei.

b. Symmetric behavior of Δ_{nn} and Δ_{pp} with respect to the middle of major shells for heavy nuclei.

8. Variation of the distances between the four surfaces:

a. Linearity of δ_n and δ_p for light nuclei.

b. Symmetric behavior for heavy nuclei.

A mass formula has to reproduce at least these regularities in order not to show regular deviations from the experimental masses.

Part II
Semi-Theoretical Interpretations

Roughly speaking, semi-empirical mass formulas can be divided into two broad classes. The first class is based on the picture of the nucleus as a charged liquid drop. This picture emerges naturally from the empirical properties (1–3) in group A at the end of Part I. Historically, these properties were discovered first, and consequently the first mass formulas, in fact the majority of mass formulas up to the present time, belong to this category.

The second class consists of formulas based on the other extreme picture, describing the nucleus as an ensemble of nucleons, each of which moves independently of the others, in the average field of force created by these other nucleons. This picture comes out in a natural way from the properties (4–6) under heading B, and it also gives a very simple qualitative understanding of properties (7–8) of group C. The properties classified under B and C became known much later than those of A, and consequently the corresponding shell model mass formulas are relatively less developed. Nevertheless, having first expounded all the properties (1–8) listed above, we shall start the discussion with shell model formulas. This will also help us in Part III, discussing corrections to the liquid drop model formulas.

A. The first fact suggesting the existence of nucleonic shells is the magic numbers. These are obtained in the most natural way by assuming the nucleons to move independently in a common average potential well. In this approximation the energy of the nucleus is given by

$$E = \sum T_i + \sum V_i, \tag{3a}$$

where T_i is the kinetic energy of the ith nucleon, and V_i its potential energy, or, in a self-consistent treatment, by

$$E = \sum T_i + {}^1/_2 \sum V_i. \tag{3b}$$

If one takes into account the Pauli principle in the building up of nuclei by adding up nucleons, the magic numbers, property (5) of the list at the end of Part I, are naturally obtained. In this approximation, there is no interaction among the nucleons except through the intermediary of the central field, and both the splitting, property (4), and the linear systematics, property (6), must not appear.

B. In the first approximation just mentioned, there is tremendous degeneracy. In order to calculate the next approximation, one must assume something about the wave function, that is, which linear combination of degenerate wave functions of the first approximation describes the actual state. A guide in the choice of the proper linear combination is furnished by the existence of pairing energy, which makes it always more difficult to remove an even nucleon from the nucleus than its odd neighbors (Fig. 4). Also, it is a well-known fact[6] that the total spin J and the total magnetic moment μ of an odd-A nucleus are usually those of a single nucleon in the subshell which is being filled, as if contributed only by the last odd nucleon. Likewise, the structure of the spectrum of low lying levels of an odd-A nucleus does not change when pairs of identical nucleons are added.

These facts bring one to the suggestion that there is a strong binding interaction between two identical nucleons whose spins are antiparallel, so that the resultant spin of the pair vanishes, $J_{12} = 0$. Identical nucleons will then couple in such saturated pairs as much as is consistent with the Pauli principle. Saturated pairs of this kind do not contribute to J or μ and can only induce regular shifts of the energy levels, thus preserving the structure of the spectrum as a whole. This is the single-particle approximation[6]. The reason for its validity is the short range of the nuclear forces, which tends to draw the nucleons as much as possible towards each other. This is achieved best by forming such saturated pairs.

In this approximation the energy of a nucleus with n neutrons in a shell j_n and p protons in a shell j_p outside closed shells is given by[13]

$$E = E_0 + n\,c + p\,C + \frac{n(n-1)}{2}\,d + \frac{p(p-1)}{2}\,D + n\,p\,I^0 +$$

$$+ \left[\frac{n}{2}\right]\pi + \left[\frac{p}{2}\right]\Pi + \frac{1-(-1)^n}{2} \times \frac{1-(-1)^p}{2} \times I', \qquad (4)$$

where E_0 is the energy of the closed shells core, c and C the energies of a single neutron and a single proton, respectively, in the nuclear central field, and the other terms give the interaction of the particles in the open shells: π and Π are the pairing energies of a neutron and of a proton saturated pair, and $[n/2]$ and $[p/2]$ the number of such neutron and proton pairs, respectively. d, D and I^0 describe the residual average interaction between any two neutrons, any two protons, and any neutron-proton pair, respectively, which still remains after the saturated pairs formation. I' describes some extra interaction between the odd neutron and the odd proton in an odd-odd nucleus.

When several subshells fill simultaneously, each of them contributes to the total energy according to a formula like (4). To the extent that the interaction parameters are the same in all these subshells, their contribution can be summed up to a formula of the same type, where n is now the total number of neutrons outside the core, and p the total number of such protons. In the single-particle approximation, the energy is thus a quadratic function of N and Z in a given shell region, with constant distances between the four quadratic parity-type surfaces, given by the jump discontinuities of the functions $[n/2]$ and $[p/2]$ when both n and p alternate between even and odd values as the shell fills. It can also be shown[15] that for the ground states of odd-odd nuclei, $I' > 0$ always, from which follows, that the odd-odd surface is nearer to the odd-mass surfaces than the even-even surface (Fig. 4). The quadratic dependence implies the linear systematics, property (6) at the end of Part I (Fig. 7). Thus one obtains all the properties under heading B of Part I.

The quadratic mass equation resulting from the single particle approximation, with parameters determined by a Least-Squares fit to the empirical masses, separately for each shell region, is Levy's Empirical

Mass Equation[16]. It reproduces nuclear masses with mean deviations of
between 200 to 600 keV in the various shell regions, except the very light
ones. According to Levy's quadratic formula, the second differences
Δ_{nn}, Δ_{pp} and Δ_{np} (1a, b, c) as well as the splitting distances between
the surfaces given by δ_n, δ_p and δ_{np} (2a, b, c), should be constant within
a given shell region. We have seen above (Figs. 8–12) that this is not so.

C. What have we neglected thus far? Formula (4) was calculated for
a given nucleus. It is a quadratic formula in a group of nuclei only provided
the parameters are the same throughout the group. These parameters
are defined by integrals of the type $\int \psi\, V\, \psi$ over the nuclear volume,
where ψ is a two-nucleon wave function, and V the effective two-nucleon
interaction. Therefore the ψ's should be the same throughout the group,
which will be the case, if the nuclear potential well is the same. It is
somewhat difficult to say a priori whether the nuclear field is the same
or not. It is certainly much more constant than in atoms, where the nuclear
charge has a decisive influence on the atomic field, increasing it considerably
and drawing the electrons towards the center when an electronic shell is
being filled. The nucleus lacks such a central attractive center, and the
nuclear forces show saturation, which results in a constant nuclear density.
Still, the gradual increase of the nuclear radius when nucleons are added
might change gradually the radius of the nuclear potential well, which
will cause corresponding slight changes of the ψ's and a slight variation
of the nuclear parameters*. With parameters varying linearly with respect
to n and p the nuclear energy, and with it the nuclear mass, becomes a
cubic function of N and Z: an expression of the form

$$\delta_1\, n^3 + \delta_2\, n^2 p + \delta_3\, n\, p^2 + \delta_4\, n\, p^3 \tag{5}$$

will be added to (4), and this will contribute a linear expression to Δ_{nn},
Δ_{pp} and Δ_{np}. Asking ourselves where this addition is likely to be important
we see immediately that it will be more important in the light nuclei,
where the addition of a nucleon changes conditions relatively much more
than in heavier nuclei. Thus the effect (7a) at the end of Part I is understood.

What else has been neglected? Equation (4) was calculated for a
pure shell model state, with a definite distribution of the nucleons in the
various subshells between two magic numbers. However, if the various
subshells have approximately the same energy, various such distributions
will have similar energies, and the actual nuclear state will be a superposition of various pure shell model states. It is immediately seen, that

---

* THIEBERGER[17] found it necessary to introduce such variations in order
to improve the agreement with the experimental data of a formula analogous
to (4) and applicable to light nuclei. The suggestion that such a variation is
due to increase of the nuclear radius is due to Professor PEIERLS[18].

if the parameters are the same in the various subshells, the contribution of the diagonal elements of the energy matrix to the nuclear energy will again be of the form (4). However, the extra binding coming from the non-diagonal elements is not accounted for.

What can be qualitatively said about this configuration interaction? Take the case of an even-even nucleus first. Here the interaction certainly vanishes when there are no particles at all outside the closed shells, and also when all the subshells are completely filled. Also, the number of available interacting configurations is the same for particles and holes, and is therefore symmetric with respect to the middle of the major shell between the magic numbers. Therefore, if the nuclear parameters are also the same in the various subshells, the contribution of this interaction will be symmetric with respect to the middle of the shell, and as such its neutronic part can be expanded in the form

$$\alpha\, n(\delta - n) + \beta\, n^2(\delta - n)^2 + \gamma\, n^3(\delta - n)^3 + \ldots \qquad (6\,\mathrm{a})$$

and similarly, the contribution from the interacting proton configurations can be expanded as

$$A\, p(\varDelta - p) + B\, p^2(\varDelta - p)^2 + C\, p^3(\varDelta - p)^3 + \ldots \qquad (6\,\mathrm{b})$$

where δ and $\varDelta$ are the number of places available for neutrons and for protons, respectively, in all the subshells between their respective magic numbers.

The terms with α and A cannot be distinguished experimentally from the previous terms in (4), being of the same degree as the latter. However, the term with β, for example, is of the fourth degree in n, symmetric with respect to the middle of the shell, and its second difference will contribute to $\varDelta_{nn}$ an expression of the second degree, symmetric too. Similarly, the term with γ will contribute to $\varDelta_{nn}$ a symmetric term of the fourth degree.

Where will the terms (6a, b) be important? Where there are many subshells with similar energies. Therefore they must not be found empirically in the light nuclei, where there is but one or two subshells in a major shell. Many subshells start to fill simultaneously only in heavier nuclei, in particular in the rare earths and the heavy radioactive nuclei, whose large quadrupole moments also point to strong mixtures of the essentially spherical pure shell model states. Thus the effect (7b) of Part I is also qualitatively understood.

What about the trends of the separation between the four surfaces, the effects (8a) and (8b) of Part I, manifested in Figs. 11 and 12? Clearly, if the field varies slowly in light nuclei, not only the parameters d, D and I^0, common to all four parity types, will vary linearly, but also the pairing energies. Thus, instead of constant π, Π and I' one is lead to introduce into Levy's formula expressions like

$$\pi_0 + \pi_n n + \pi_p p, \tag{7a}$$

$$\Pi_0 + \Pi_n n + \Pi_p p, \tag{7b}$$

$$I_0' + I_n' n + I_p' p \tag{7c}$$

and in the same way as (5), this correction will be more important for light nuclei.

Finally, in order to understand the humps in δ_n and δ_p in the heavier nuclei, one has to consider, that for an odd-N nucleus, for example, the symmetric neutron configuration interaction vanishes for one neutron and one hole instead of for no neutrons and no holes in the major shell. Besides, its magnitude might be considerably different than for an even-N nucleus[12]. Thus its expansion will look like

$$\alpha'(n-1)(\delta-1-n) + \beta'(n-1)^2(\delta-1-n)^2 + \ldots \tag{8a}$$

which can be transformed by simplifying brackets into the form

$$\alpha\, n(\delta-n) + \beta\, n^2(\delta-n)^2 + \ldots$$

$$+ \frac{1-(-1)^n}{2}\, [\text{const.} + \bar\alpha\, n(\delta-n) + \bar\beta\, n^2(\delta-n)^2 + \ldots]. \tag{8b}$$

The first line is similar to (6a), which is thus common to all four types of nuclei. The second line is an extra term valid for odd-N nuclei only, and will thus combine with the neutron pairing energy to give the separation distance δ_n between the odd-N and even-N surfaces. Its form is just the form needed to account for the empirical humps of Fig. 11. The proton humps of Fig. 12 can be accounted for in the same way.

To sum up, the addition of terms of the forms (5), (6), (7), and (8) to Levy's quadratic formula (4) enable it to reproduce all the empirical trends summarized ·in Part I. In the shell model picture these terms originate from two causes: (5) and (7) are due to a gradual variation of the nuclear field in light nuclei, while (6) and (8) come from configuration interaction prevailing in heavier ones.

In Jerusalem we have recently been fitting this generalized Levy's formula to experimental masses. Preliminary trials give mean deviations of between 150 keV and 250 keV in the various shell regions, except for the very light nuclei, where the linear variation of the parameters giving rise to (5) and (7) might be insufficient.

Part III
Corrections to Liquid Drop Formulas

In this last part we shall briefly consider mass formulas based on the liquid drop picture of the nucleus, concentrating on ways to correct them so that they might reproduce nuclear masses with an accuracy like that obtained with shell model formulas. It is well known that the properties under heading A at the end of Part I lead to the picture of the nucleus as a charged liquid drop, whose energy consists of three terms:

$$E = a A + b A^{2/3} + \frac{Z^2}{r_0 A^{1/3}} \qquad (9a)$$

giving, in the above order, the volume energy, the surface energy, and the Coulomb energy. Use has been made of the experimental fact that the nuclear density is approximately constant, and that nuclei have approximately a spherical shape. This form can give us properties (1) and (2).

To obtain the parabolic section, one has to assume, with WEIZSÄCKER[19] that the parameters a and b, arising from the purely nuclear forces, will be composition dependent. Demanding further that they will be symmetric about the line $N = Z$ to conform with the charge symmetry of nuclear forces, expanding in powers of $(N - Z)/(N + Z)$, and retaining the first two terms of the expansion, one has

$$E = a \left[1 + \frac{p}{a} \left(\frac{N - Z}{A} \right)^2 \right] A + b \left[1 + \frac{q}{b} \left(\frac{N - Z}{A} \right)^2 \right] A^{2/3} + \frac{Z^2}{r_0 A^{1/3}} =$$

$$= a A + b A^{2/3} + \frac{Z^2}{r_0 A^{1/3}} + p \frac{(N - Z)^2}{A} + q \frac{(N - Z)^2}{A^{4/3}}. \qquad (9b)$$

The first four terms are the classical Bethe-Weizsäcker formula[1]. The last term was put into use recently, in the mass formulas of CAMERON[20], SEEGER[21] and JOHANSSON[22]. We have heard of its far reaching implications from Professor WHEELER at the Hamilton Conference[23].

The Coulomb energy as written above is the classical Coulomb energy of a uniformly charged spherical liquid drop, with a radius proportional to $A^{1/3}$. The newer formulas of CAMERON[20] and of SEEGER[21] use a trapezoidal charge distribution with parameters as obtained from the Stanford electron scattering experiments, and the Coulomb energy is calculated for this charge distribution, leaving no free Coulomb energy parameter in the formula. The exchange Coulomb energy is also included.

Another recent development concerns the symmetry energy. SEEGER[21] and JOHANSSON[22] introduce in addition to the $(N - Z)^2/A^2$ term, a term with $|N - Z|/A^2$ into their formulas. Such a term was obtained theoretically in the supermultiplet approximation by WIGNER[24], assuming spin- and charge-independent forces between the nucleons. With present day knowledge about nuclear forces, there is no theoretical justification for introducing a term of this kind. However, it is said[22] to improve considerably the agreement of the formula with the course of the line of beta stability.

We did not study the relative merits of the different corrections proposed. However, as far as the agreement of the formula with the empirical masses is concerned, it seems from the literature that the new formulas are not significantly superior to the Bethe-Weizsäcker formula. Thus, Green[25] quotes a mean deviation of 2.7 MeV using the classical Bethe-Weizsäcker equation with new constants determined by him, while Seeger[21] using all the new improvements, quotes a mean deviation of

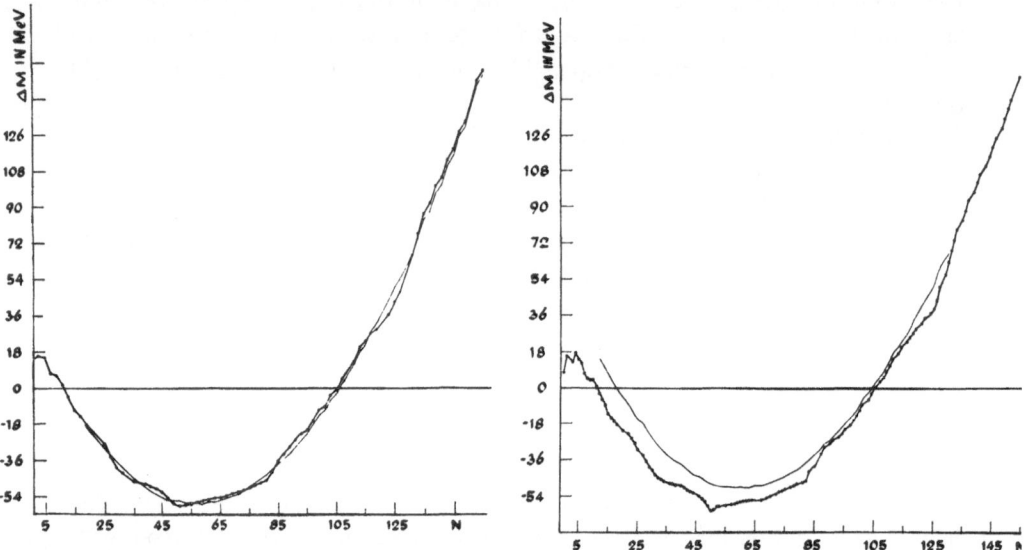

Fig. 13. Left part: The liquid-drop model part of Cameron's Mass formula (thin line) running smoothly through the empirical mass defects of odd-odd nuclei (thick line) from which this smooth part was determined. Right part: The liquid-drop model part of Seeger's mass formula (thin line) running above the actual mass defects of odd-A stable nuclei (thick line)

2.65 MeV. Thus there must be an inherent inability of such formulas to agree with empirical data to better than that order of magnitude. This we shall now examine in more detail.

As is well known, the main reason for these large discrepancies is due to shell effects. Fig. 13 shows the mass defects of odd-odd nuclei and of odd-A nuclei along the stability valley, together with liquid drop type mass surfaces. It is clear that the liquid-drop surfaces are much too smooth to account for the cusps connected with the magic numbers. The left part shows the liquid-drop model part of Cameron's formula, which was calculated so as to give the best fit to the experimental odd-odd masses shown in the figure. It is seen that between the places where the calculated line crosses the experimental masses, near the magic numbers, there is a large disagreement between the experimental and the theoretical masses, of the other of several MeV. Thus, this is the main source of error which has to be corrected.

Before starting to see how this can be done, let us remark that there are two ways to introduce corrections: One can fit the liquid-drop formula as well as one can, like CAMERON did, before introducing additional corrections. In this case, the liquid-drop formula will cut through the empirical surface many times to have closest approach to it, as shown in the left part of Fig. 13. However, one can also try to fit a corrected formula, from the beginning with correction of any desired form, in which case it is unnecessary that the liquid-drop part of the formula will make the closest possible approach to the experimental masses. The right part of Fig. 13 shows the liquid-drop model part of Seeger's formula, the corrections to which were chosen of such a form that it always lies above the actual masses. We will carry the discussion in terms of the first possibility because the exposition is somewhat clearer in this way, and also because it was much easier to calculate the liquid drop part of Cameron's formula, who gave his corrections in the form of a numerical table,

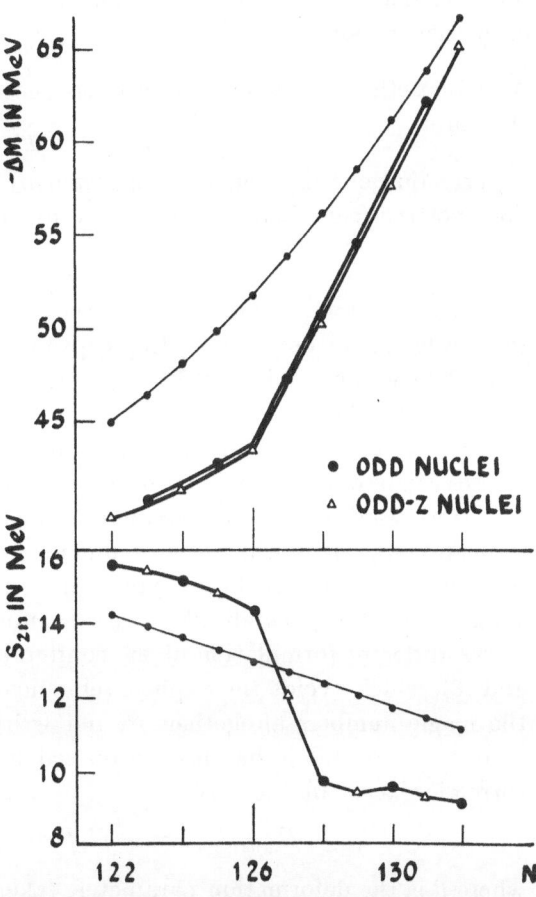

Fig. 14. Upper part: Empirical masses of the bismuth isotopes (connected by thick lines) and the values predicted for them by the liquid-drop model part of Cameron's formula (thin line). Lower part: Empirical double neutron separation energies of the bismuth isotopes (connected by a thick line) and the values predicted for them by the liquid-drop model part of Cameron's formula (thin line)

than that of Seeger's formula, whose correction terms were given as linear combinations of trigonometrical functions which have first to be looked up in tables and then pass through several more manipulations before being available for use. However, the conclusions reached are valid for both types of approximation.

How can one correct for shell effects? Fig. 14 shows the masses and the double-neutron separation energies for the Bi-isotopes, and also the values predicted for them by Cameron's formula. To the extent that the discrepancy between theory and experiment for the S_{2n} values varies linearly, one has to add to the masses a quadratic correction term for each shell region,

$$\beta_1 n + \gamma_1 n^2. \tag{10a}$$

A similar effect exists for the protons, necessitating a correction term of the form

$$\beta_2 p + \gamma_3 p^2. \tag{10b}$$

A correction of this form was suggested by Green and Edwards[26]. One does not see from Fig. 14 a necessity for a mixed quadratic term

$$\gamma_2 n p. \tag{10c}$$

However, a closer comparison of the experimental Δ_{np} (Fig. 10) with those calculated from a liquid-drop type formula shows that this is necessary too. Thus we must add to (9b) a complete quadratic expression (10a, b, c). Then we shall have the quadratic terms necessary to account for the linear systematics and the magic numbers, and the agreement with experimental masses will be of the same order of magnitude as achieved in Levy's formula.

If we want a better approximation, we have to add the extra correction terms (5), (6), (7), (8) which were found necessary to improve Levy's formula. Adding them, one will obtain agreement of the same order of magnitude as achieved by the generalized Levy's formula in Part II.

A different formulation of the configuration-interaction corrections (6) and (8), which avoids the explicit reference to the shell model picture and the magic numbers altogether, by not writing these corrections explicitly in terms of n and p, has been proposed by Johansson[22]. He writes his correction term in the form

$$E_{\text{deformation}} = - C_1 \beta + C_2 \beta^2 + C_3 \beta^3, \tag{11}$$

where β is the deformation parameter, taken directly or extrapolated from experimental deformations. We can see the equivalence of this form with (6) by looking at the deformation as function of N or Z. Fig. 15 shows how the nuclear deformation varies as a function of N. It is seen that between magic numbers, in particular in the rare earths and beyond Pb, it has the appearance of a hump of a form similar to the one discussed above in connection with Figs. 8, 9, 11, 12 and represented algebraically by a formula like (6). In fact, from the point of view of the shell model the deformation must behave like (6), since a large deviation from spherical symmetry can only be achieved within the shell model framework by a large configuration mixing.

Johansson's formula agrees with experimental masses to about 200 keV in the cases checked[22].

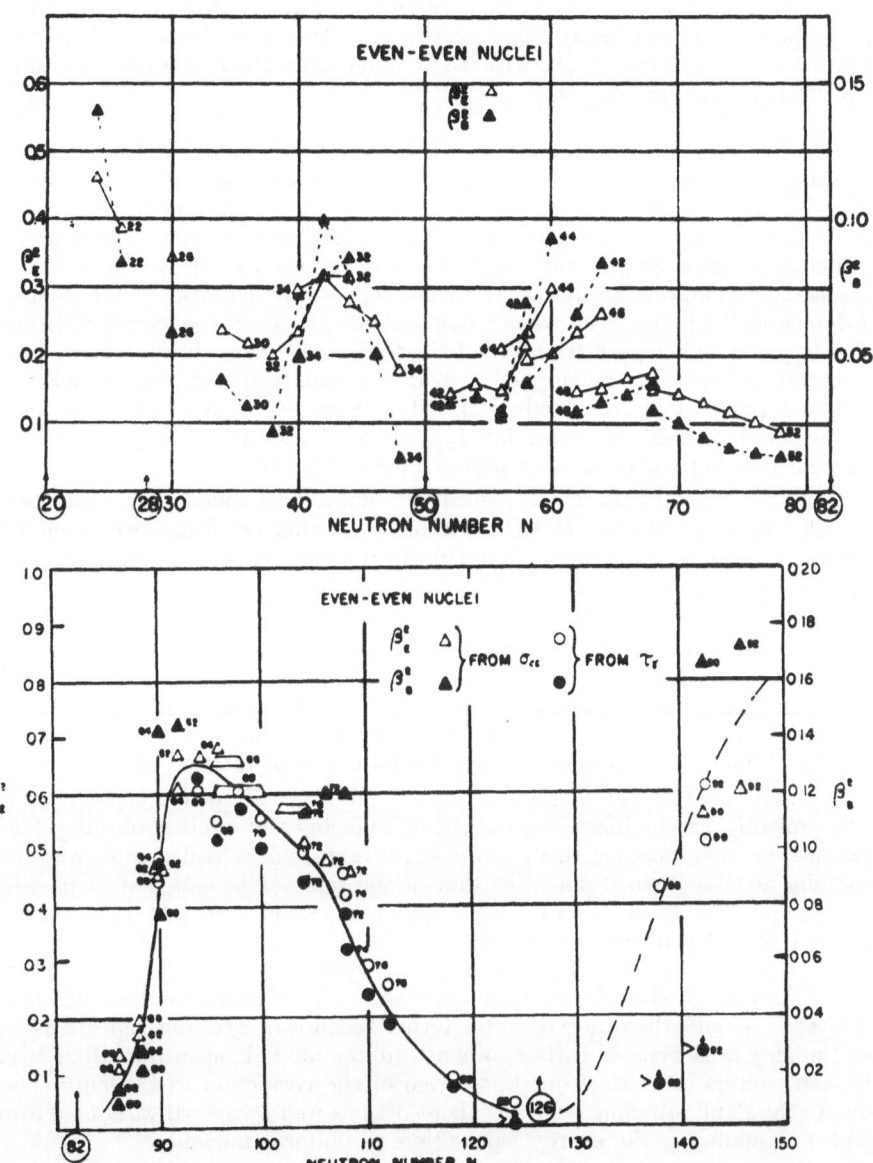

Fig. 15. Nuclear deformation as function of neutron number (taken from ref. [27])

Discussion

H. G. KÜMMEL: I have several questions:

Firstly: I always did not understand the following point in the Levy formula. The term E_0 is suggested to be the energy of the filled shells. This, however, means that it is not an independent parameter. E_0 is included by LEVY, however, as an independent quantity. What is your meaning of this?

Secondly: I do not believe that we have here the best one can do with the large number of parameters Levy or you use. One does not need all of them. The precision of our formula with 35 parameters is about the same as Levy's with many more parameters.

N. Zeldes: Both questions are strongly related, and I will answer them together. You are perfectly right in saying that one can reduce the number of independent parameters in our formulas without worsening the goodness of fit. With E_0, its use as an independent parameter is more distressing in principle, because as you have said, it is determined by the parameters of the preceding region, whereas the other parameters are by definition independent, in different shell regions. Still, when looking at the graphs of the second differences Δ_{nn}, Δ_{pp}, and Δ_{np} (Figs. 8, 9, and 10 of the text), these differences continue rather smoothly between the magic peaks, so it is certainly unnecessary to introduce different parameters to reproduce Δ_{nn}, for example, when you cross a proton magic number, and vice versa for Δ_{pp}. Also, Δ_{np} seems to vary continuously disregarding both neutron and proton magic numbers.

On the other hand, if one reduces the number of independent parameters in this way one will have to treat one big regression problem, with about 1000 nuclei as given data, instead of having to do with smaller separate problems, with about 150 data at most, and you know better than I do how this can complicate things in practice.

The question is whether this extra price is worthwhile. If the primary objective of a mass formula is to provide reliable parameters for purposes of extrapolation, then as the number of degrees of freedom in the calculation is large enough and the parameters are significant enough, the answer will be negative. This is our approach. But I admit it is a question of taste.

Our formula is still in status nascendi. In the final version smaller regions will probably be combined together, thus reducing the number of independent parameters and making them more significant and reliable. But we won't combine together two regions like that of the rare earths and that of post-lead nuclei, the first including 133 nuclei and the second 150, just to reduce the number of independent parameters.

F. Malik: Are Levy's formulas consistent with the most general single particle orbitals? Would there not be extra terms because of exchange effects as well as the Majorana type of basic two-body interaction? I mean that there would be extra terms in view of our knowledge of the correction to Coulomb energy due to the Pauli principle (which is about 5%) as well as the old work of Wigner and Feenberg on the energy calculation in uniform model.

N. Zeldes: All exchange terms are included in the formula. The most general two-body interaction will lead to it in the single-particle approximation. Wigner and Feenberg worked in the super-multiplet approximation, but I don't remember exactly what they obtained.

F. Malik: I am very surprised. Because the inclusion of exchange terms, exchange between all particles, core as well as extra core particles, will not give such a smooth A dependence. From the work on Coulomb energy which includes exchange by say Seugupta or Bethe, we find already a different A dependence for the Coulomb energy part.

References

[1] See, e. g., BETHE and BACHER, Nuclear Phys. A, Revs. Modern Phys. 8, 82 (1936).

[2] GLUECKAUF, Proc. Phys. Soc., A 61, 21 (1948).

[3] SUESS and JENSEN, Arkiv for Fysik 3, 577 (1951).

[4] GHOSHAL and SAXENA, Proc. Phys. Soc., A 69, 293 (1955).

[5] WAY, Proceeding of the Mainz Conference on Nuclidic Masses. London: Pergamon Press. 1957.

[6] See, e. g., MAYER and JENSEN, Elementary Theory of Nuclear Shell Structure. New York: John Wiley and Sons. 1955.

[7] ELSASSER, J. de Phys. et Rad. 5, 635 (1934).

[8] See, e. g. GAMOW, Structure of Atomic Nuclei. Oxford: Oxford University Press. 1937.

[9] CORYELL, Ann. Rev. Nucl. Sci. 2, 305 (1953).

[10] PERLMAN, GHIORSO, and SEABORG, Phys. Rev. 77, 26 (1950).

[11] WAY and WOOD, Phys. Rev. 94, 119 (1954).

[12] ZELDES, GRONAU, and LEV, in preparation.

[13] ZELDES, Nuclear Phys. 7, 27 (1958). A qualitative derivation of this formula was given at the Hamilton Conference [14].

[14] ZELDES, Proceedings of the Hamilton International Conference on Nuclidic Masses. Toronto: University of Toronto Press. 1961.

[15] DE-SHALIT, Phys. Rev. 105, 1528 (1957).

[16] LEVY, Phys. Rev. 106, 1265 (1957).

[17] THIEBERGER, Jerusalem Thesis, 1958 (unpublished).

[18] PEIERLS, A critical remark quoted in ref. [13].

[19] See, e. g., WEIZSÄCKER, Die Atomkerne. Leipzig: Akademische Verlagsgesellschaft m. b. H. 1937.

[20] CAMERON, Canad. Journ. Physics 35, 1021 (1957).

[21] SEEGER, Nuclear Phys. 25, 1 (1961).

[22] JOHANSSON, UCRL-10624, p. 92 (1962).

[23] WHEELER, Proceeding of the Hamilton International Conference on Nuclidic Masses. Toronto: University of Toronto Press. 1961.

[24] See, e. g., WIGNER and FEENBERG in Reports on Progress in Physics 8. London: The Physical Society. 1941.

[25] GREEN, Revs. Modern Phys. 30, 569 (1958).

[26] GREEN and EDWARDS, Phys. Rev. 91, 46 (1953).

[27] HEYDENBURG and TEMMER, Ann. Rev. Nucl. Sci. 6, 77 (1956).

Empirical Separation Energies and the BCS Pairing Theory

By

M. Beiner, K. Bleuler, P. Huguenin, and **R. de Tourreil**

University of Bonn, W-Germany

(Presented by K. BLEULER)

With 1 Figure

Separation energies as determined from nuclidic masses exhibit in a striking way the characteristic features of shell structure if plotted in a suitable three-dimensional scheme[1]. Our aim is to check this experimental behaviour numerically with the help of explicit shell-model calculations. It is shown that a rather good agreement is actually obtained if the independent particle levels of shell-structure are corrected with the help of the BCS pairing theory. This means that the direct interactions between individual nucleons are taken into account in a suitable approximation.

1. The Average Potential of Spherical Nuclei

The first part of our calculation constitutes the determination of the average potentials for protons and neutrons for all nuclei which can be assumed to have mainly spherical structure. We use in order to simplify this problem, Saxon type potentials with spin-orbit coupling which depend on four parameters (radius, depth, surface thickness, and coupling constant). These parameters are then to be considered as continuous functions of N and Z (neutron and proton number) and our first step constitutes the determination of these three functions directly from suitable empirical data. (We choose linear functions of N and Z and the Hofstadter expression for the radius-law.)

For this purpose we assume that a certain number of nuclear properties in which the bulk of nucleons contribute is already represented with a reasonable accuracy by ideal shell structure. (We consider charge distribution, average separation energy, optical properties etc.) We therefore calculate explicitly all single particle levels as functions of our parameters and we calculate the nuclear properties in question. Comparing these values with the corresponding experimental data we are able to determine the parameters as corresponding functions of N and Z. By this method the average potential is determined in a semiphenomenological way. It is, first of all, an interesting fact that the empirical data do allow the determination of a reasonable average potential. But it is, in addition, of great importance that the self-consistency of such a phenomenological potential can now be tested in the frame work of the Hartree-Fock method if explicit effective nuclear forces as determined from nucleon scattering and properties of light elements are used. In fact we obtain a quite reasonable agreement, if we use the well-known Serber two-body forces. This is true especially with respect to the characteristic dependence on $(N-Z)$ of the potential depth and the difference of proton and neutron potential, corresponding to the Coulomb energy. On the other hand, it seems that nuclear forces are, at least for the moment, not known accurately enough in order to make a straight forward Hartree-Fock approximation without the help of empirical nuclear data. In addition, the existence of the hard core makes, as is well-known, considerable mathematical difficulties.

2. Independent Particle Model and Empirical Properties

After the determination of our average potential we have well defined independent particle states for all nucleons and we can make a more detailed comparison between experimental data and theoretical results from this simplified model. In doing so, it is immediately seen that although certain general properties are in perfect agreement, very characteristic differences or discrepancies arise: if we follow for a suitable choosen series of nuclei the more detailed behaviour of separation energy as function of N or as function of Z, we realize that the empirical values are rather continuous functions which exhibit characteristic irregularities (steps) only at the so called magic numbers whereas the calculated separation energies are much more irregular, exhibiting steps of the order of $^1/_2$ to 1 MeV, after filling up each individual single particle level[2].

In the same time, the calculated steps at the magic numbers are about twice as high as the empirical ones, and the characteristic energy difference between even and odd nuclei does not occur at all in these simplified calculations.

On the other hand, as is well known, a large number of other nuclear properties, position of excited levels, spin assignments and so on, are in rather good agreement with empirical facts. Therefore the question arises how this smoothing effect of separation energies, described above, can be understood from the theoretical view-point.

3. Direct Interaction and BCS Pairing Theory

It is clear that in addition to the average potential, direct interaction between individual nucleons must be taken into account; and the question

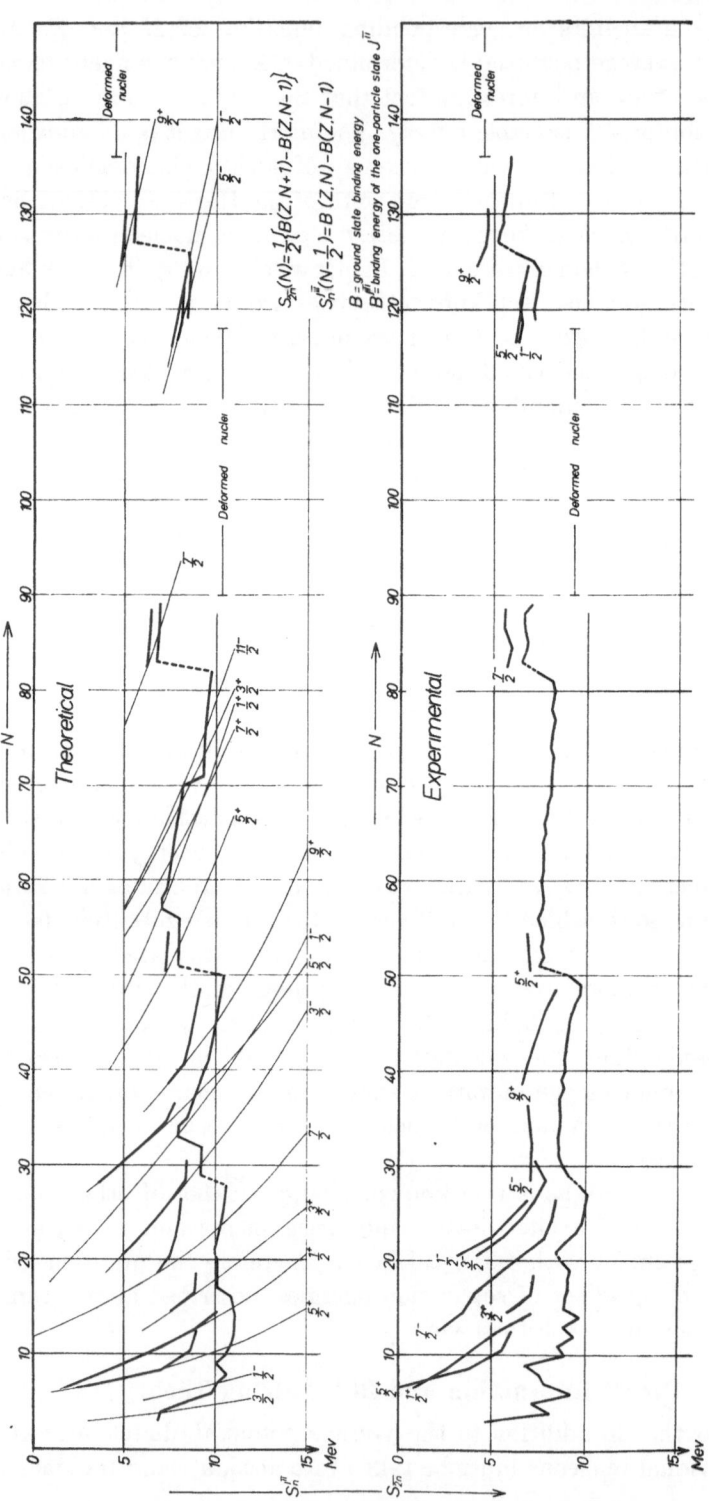

Fig. 1. Nuclear Properties as Functions of the Neutron-Number *N*. Upper part: *Calculated values.* Fine lines: Position of single-particle levels in the neighborhood of the Fermi-level as function of *N* with indication of the corresponding assignments (absolute position in MeV with respect to separation represented by O). Heavy lines: Calculated BCS-correction. *Lower line*: Ground-states of even-even nuclei. *Upper, interrupted lines*: Ground-states and first excited states of even-odd nuclei. *Lower part: Experimental values plotted in the same way*. *Lower line*: Groundstates of even-even nuclei from experimental mass values. *Upper system of lines*: Ground-states and first excited states with assignments from experimental masses, excitation energies and determinations of nuclear spins

arises how this could be done in a suitable way or a suitable approximation, which allows explicit calculations of the position of the Fermi level directly as a function of N and Z. It was realized that this program could be worked out relatively easily using the formulas of Bardeen-Cooper-Schrieffer pairing theory of superconductivity, specially arranged for nuclei[3]. There, the pairs are now constituted by opposite single particle angular momentum j instead of opposite ordinary momentum in the case of superconductivity. In addition some matrix elements must be explicitly calculated, using actual nuclear forces. We have so far used the Serber force already introduced in the calculation of the average potential and it is of some importance that the pairing theory constitutes, so to speak, a refinement or next approximation of the Hartree-Fock method. (So far, direct interaction was taken into account only in the last subshells.) In the upper part of our figure representing calculated separation energies, we have first plotted with thin lines the position of the single particle levels as a function of the neutron number N, having used an average value for the corresponding values from the BCS calculations of all levels which occur in the neighborhood of the Fermi level. There are two separate lines, the lower one representing the ground states of even nuclei, whereas the upper broken line represents the calculated ground-states and first excited states of the odd nuclei. The system of these lines has now to be compared with the corresponding systems determined from the experimental data which are shown in the lower part of our figure. It is immediately realized that there is now a rather good agreement between these two systems for the characteristic parts of the periodic table (N between 15 and 88 and between 115 and 135) in which spherical structure can be assumed. Especially the smoothing effect and the pairing separation is in accordance with experimental facts. This constitutes a characteristic success of the application of the pairing theory in which no free parameter occurs (the unperturbed wave functions and the direct interactions are fixed from the outset). The comparison of the two systems also show the great importance of a systematic and accurate knowledge of nuclear binding energies, the empirical diagram being determined by these values. A more detailed publication of these results with an additional theoretical discussion of the phenomenological spin-orbit interaction used so far is under preparation.

References

[1] Cf. the contribution on separation energies to the last conference by M. BEINER and K. BLEULER and Nuclear Phys. 22, 589 (1961).

[2] BLEULER and TERREAUX, Helv. Phys. Acta 28, 245 (1955).

[3] Cf. J. R. SCHRIEFFER, Nuclear Phys. 35, 363 (1952). — S. T. BELIAEV, Dan. Vid. Selsk. 31, No. 11, 5 (1959). — KISSLINGER and SORENSEN, Dan. Vid. Selsk. 32, No. 9 (1960).

A New Nuclidic Mass Law

By
H. Kümmel*
Oklahoma State University, Stillwater, Oklahoma, USA,
J. Mattauch and **W. Thiele**
Max-Planck-Institut für Chemie (Otto-Hahn-Institut), Mainz, Germany
and
A. H. Wapstra
Instituut voor Kernphysisch Onderzoek, Amsterdam, The Netherlands

(Presented by H. KÜMMEL)

With 7 Figures

I. Introduction

This is one of three talks about the same subject. You have heard from NISSAN ZELDES, how the study of the mass surface led him to his mass formula, and you again will learn from WLADISLAV SWIATECKI what he believes the mass formula should look like. What we have to offer in some sense is a compromise between these two extremely different models. No matter, however, which formula you like more, you will agree that the fact that there are so many different formulae is quite amusing, since physics is amusing as long as there are problems.

It is easy to criticise all formulae, including our own. But, one has to bear in mind that the relation between experimental binding energies and theory is a very loose one, since there is no theory predicting binding energies with small errors. This is a point I want to make clear presently.

The mass surface clearly exhibits the shell structure of nuclei (Fig. 1): If we plot the deviations between experimental masses and the masses as calculated from the Bethe-Weizsäcker formula, we see oscillations around the Bethe-Weizsäcker values with an amplitude of the order of 15 MeV. So, a lot has to be done to bring down these deviations—especially if we are interested in average deviations smaller than or at most equal to 0.5 MeV. This was our aim, indeed, since this is a condition *sine qua*

---

* Supported in part by the Oklahoma State University Research Foundation.

non for the extrapolation of masses in regions well outside the valley of stable nuclei. Although it certainly is well known to the majority of you, I have shown this picture to make clear that it does not suffice, say, to

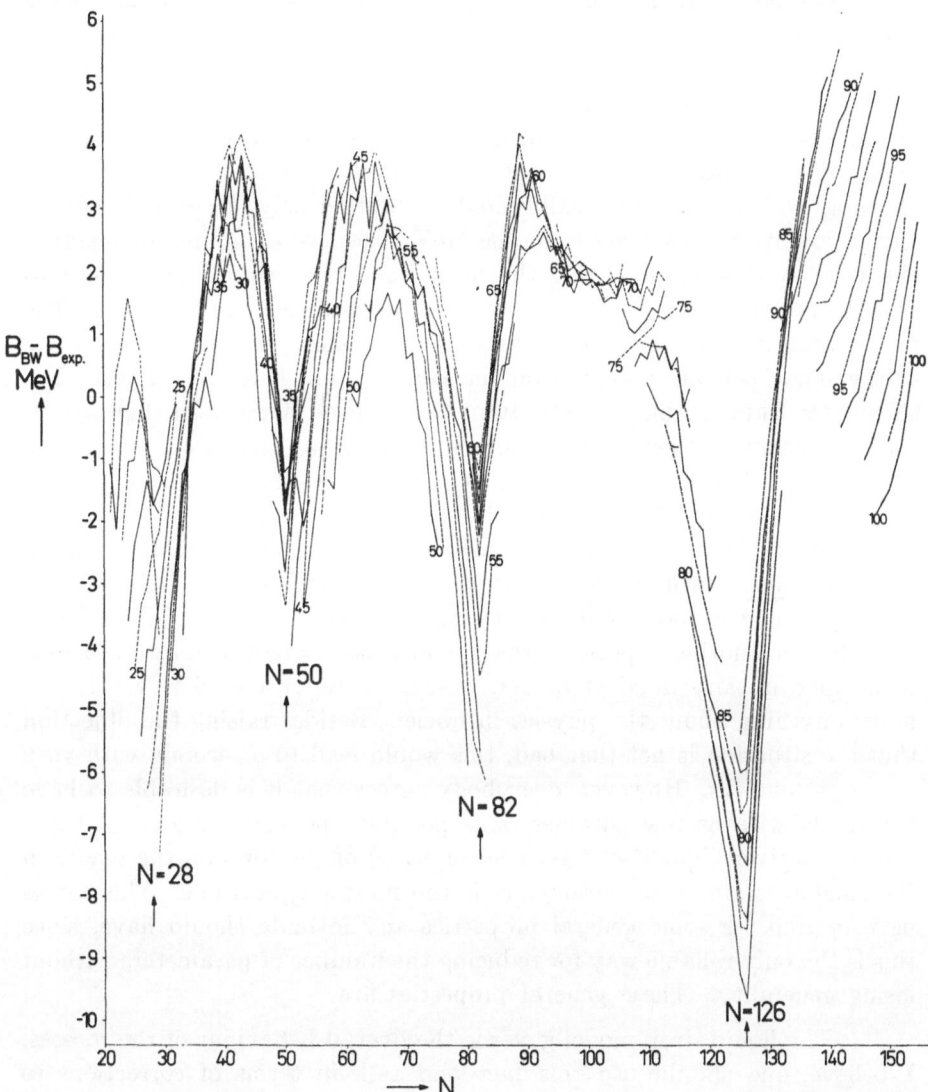

Fig. 1. Differences between liquid-drop binding energies B_{BW} and experimental values B_{exp}

account only for the bulk effects of the shell structure: The remaining errors have to be at most of the order of magnitude of 5% of the original errors. This is what I mean if I am talking about "small errors".

The consequences of the requirement of small errors are quite serious. Let us look into the shell model more closely; I should say "into the shell models" since there are many different models one may consider. I think there is no question about the fact that the shell model one really uses by no means is the consequence of a Hartree-Fock variational principle (with residual interactions). Instead, it is made to fit some experiments which have to do with levels of excited states more than with anything else. The problems involved have been discussed very clearly by INGLIS[1]. He shows that one may introduce velocity dependence as well as re-arrangement energy into the shell model. In fact, any combination of both will work, and there is no sufficiently accurate experimental evidence showing which situation really holds. "Sufficiently accurate" means: Errors involved are *smaller* than one MeV. Besides, assuming for instance the velocity dependence being the most important aspect, one runs into trouble not knowing how this velocity dependence really looks. The fashionable reduced mass approximation assumes quadratic dependence of the single particle potential on the momentum. However, as has been shown by BRUECKNER and GAMMEL[2] the reduced mass itself depends on the momentum. Using his numerical values, we find that the errors involved are of the order of 2 MeV, i. e. by no means small if one is looking for a shell model with high precision.

All these facts have made us believe that it does not make sense to look for a mass formula employing too many details of a model—at least as long as one is looking for such small errors.

There is another aspect, however, which has to be taken into account. If we give up any detailed model, we say more or less that we do not know anything about the physics in nuclei. Besides raising the objection that the situation is not that bad, this would lead to a formula with very many parameters. However, everybody agrees that it is desirable to have a formula with as few parameters as possible and there is general belief that for a given "quality" (given mean error) of the formula the one with the smallest number of parameters is the most physical one. This forces us to search for some general properties any formula should have, since this is the only reliable way for reducing the number of parameters without losing generality. These general properties are:

1. The liquid drop model governs the overall behaviour of the masses. I believe, one should use this fact and talk in terms of corrections to this model.

2. Shell structures are present. So, the shell model must be made compatible with the liquid drop model. This means that any term occuring in one model must have some meaning in the other one.

3. The binding energy surface is continuous (as far as one can formulate this for the discrete variables N and Z). This requires some justification, since not all published mass formulae have this property. If we assume N

and Z for the moment as continuous variables, the separation energy for neutrons resp. protons is given by

$$S_N = \frac{\partial B(N,Z)}{\partial N}, \qquad S_P = \frac{\partial B(N,Z)}{\partial N} \qquad (1)$$

through the binding energy $B(N,Z)$. Although there are no arguments to assume that S_P and S_N are continuous functions, one safely can say that they should not contain δ-functions: This would mean, that a certain nucleon number has very much different properties than any other number in the neighborhood. In the more realistic case that N and Z are integers and the critical lines are the lines N or $Z =$ magic numbers, the first

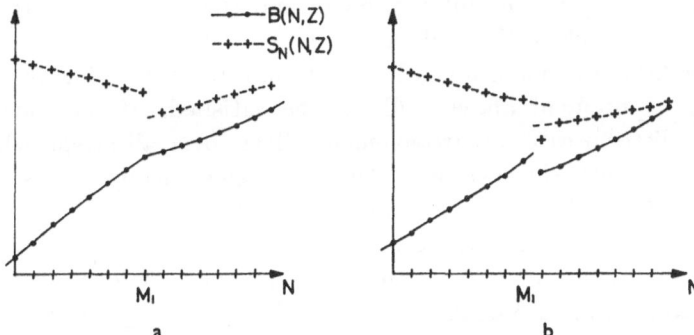

Fig. 2. Behaviour of energy surface near magic number

nucleon in a new shell would have a separation energy very much different from the second one, whereas it should behave similarly to it (differently only from nucleons in the lower filled shell). In other words: $B(N,Z)$ has to be "continuous" and only the behaviour of the energy surface according to Fig. 2a is permitted, 2b is forbidden.

4. Sometimes it is more convenient to start with the separation energies S_P and S_N of protons and neutrons instead of the total binding energy B. Apparently

$$B(N,Z) = \sum S_P(N',Z') + \sum S_N(N',Z') \qquad (2)$$

and the knowledge of either B or S_P and S_N is equivalent. Now, the binding energy must be independent of the way and the order in which the neutrons and protons have been put together to yield the nucleus. Or: eq. (2) must be independent of the path of the summation. This is a severe restriction, which in terms of continuous N and Z means

$$\frac{\partial S_P}{\partial N} = \frac{\partial S_N}{\partial Z}. \qquad (3)$$

As long as one has a continuous $B(N,Z)$ and defines S_P and S_N by eq. (1) with discontinuities only along certain lines, eq. (3) will hold. But as soon as one does not start with B, one has to consider (3) or the path independence of (2) which one may replace by

$$B(N, Z) = \int_{(0,0)}^{(N, Z)} (S_P dZ + S_N dN) \tag{4}$$

(independent of the path).

5. We have stated above that one will not be able to construct a complete shell model which is accurate enough for our purpose. Some minimum requirements, however, seem to be generally accepted: The single particle energy of the respectively last particle is equal to (minus) the separation energy. This is not as trivial as it may seem: The removal of a nucleon changes the interaction in and the geometry of the nucleus such that *all* levels will be changed, if one defines a sequence of levels as eigenvalues of a potential well (which may be quite arbitrary). In this case, the requirement above will not be satisfied. If one does it this way, one distributes the rearrangement effects over all levels, whereas we include them into the last one. This discussion merely shows that one, on the one hand, has to go beyond the simple single particle model, and that, on the other hand, there are several different ways to do this.

A second requirement says, that the single particle levels or separation energies are grouped together to form shells which yield the standard magic numbers etc.

We shall see that these 5 requirements almost automatically will lead to our mass formula. Merely some parameters have to be determined by a least squares fit: Their dependence on the shells will offset our lack of more knowledge.

II. The Mass Formula

Apparently, we need only the set of all single particle levels or negative separation energies, including of course all their dependence on N and Z. Now, already the liquid drop model binding energy (without pairing energy)

$$B_{\text{l. d.}} = u_V A - u_S A^{2/3} - u_C \frac{Z(Z-1)}{A^{1/3}} - u_\tau \frac{(N-Z)^2}{A}, \tag{1}$$

predicts some separation energies through

$$S_{P \text{ l. d.}} = \frac{\partial B_{\text{l. d.}}}{\partial Z}, \qquad S_{N \text{ l. d.}} = \frac{\partial B_{\text{l. d.}}}{\partial N} \tag{2}$$

(by the way, it predicts also some $(N-Z)$-dependence of the single particle potential, if there is such a thing at all). Of course, in this way, we can get only the smooth (not shell dependent) part of it. To obtain the shell influence, we use the requirements in 5 of the last section.

Some notations: We call M_i the magic number of the ith shell (the number closing this shell), assuming equal magic numbers for protons and neutrons. The number of levels in the ith shell may be denoted by N_i, such that $M_i - M_{i-1} = N_i$. The number of protons and neutrons in the ith shell shall be denoted by z_i and n_i respectively, the number of holes by $\hat{z}_i$ and $\hat{n}_i$.

$$\hat{z}_i = N_i - z_i, \qquad \hat{n}_i = N_i - n_i. \tag{3}$$

The single particle levels are grouped together into shells with N_i levels. Each such level depends on the numbers N and Z, respectively or n_i and z_i. One has to keep in mind, however, that only points or small sections of

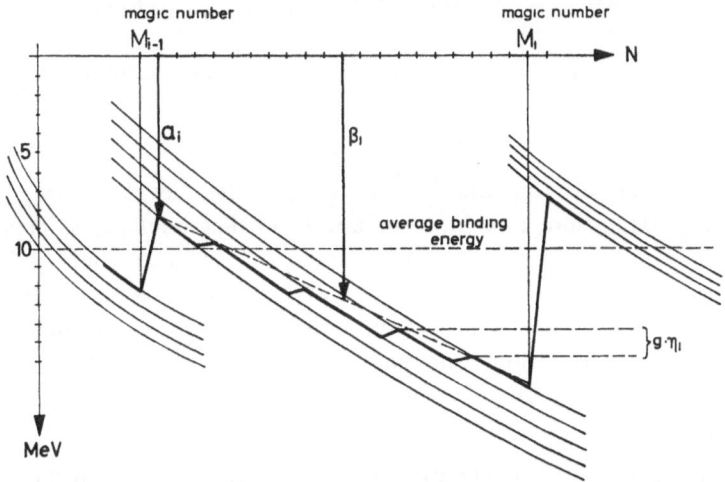

Fig. 3. Schematic representation of single particle levels

them show up in the binding energy: The νth level from the bottom of a shell occurs only for $\nu = n_i$. Nevertheless, we have drawn the levels as hypothetical quantities also where they do not show up (Fig. 3), to give an idea of what is going on. This figure is an idealized section of a figure one may find in the work of BLEULER and TERREAUX[3] who calculate such single particle levels. The signs have been chosen in agreement with experience. Also, we have admitted degeneracy, allowing levels to be occupied by more than one particle.

It is easy to formulate quantitatively, what we have made clear qualitatively: The νth level (from the bottom) occuring for the occupation number ν, may be described by an polynomial expansion

$$\varepsilon_i(\nu) = a_i + b_i(\nu - 1) + c_i(\nu - 1)^2 + \ldots \tag{4}$$

(breaking this series after the second order term), where a_i is the bottom of the shell, if there is only one nucleon in the shell, and b_i the "level distance" in a linear approximation (with $c_i = 0$), denoted by $g \cdot \eta_i$ in the figure (g = degeneracy). It is not what one usually would call level distance, since it includes shell filling and occupation number dependence.

Two things have to be remarked concerning eq. (4): First of all, it does not include the smooth and relatively slowly varying contributions from eq. (2). This has to be added later. Secondly, except for (2), there is no dependence on the occupation of the respectively "other" kind of nucleons: There is no interaction between neutrons and protons. This, too, will be included later.

The total contribution of (4) to the binding energy is

$$E(n_i) = \sum_{\nu=1}^{n_i} \varepsilon_i(\nu) = \beta_i n_i - \frac{1}{2} (\eta_i + \zeta_i n_i) n_i \hat{n}_i \qquad (5)$$

(for n_i neutrons in the ith shell). Here, we have assumed $a_i \gg b_i \gg c_i$.

$$\beta_i = a_i + \frac{1}{2} \eta_i N_i, \qquad (6)$$

is the "center of the shell" and $\eta_i \approx b_i$ the "level distance" (in the linear approximation, i. e. for constant level distance). ζ_i is proportional to c_i. The figure reflects the experimental evidence that $\eta_i > 0$: The levels go down faster than one goes up by filling the next level. This makes sure that one has minimum binding energy contribution for half filled shells. In addition, one has a similar expression for protons. It turned out that it suffices to use the same parameters for protons and neutrons.

Comparing (5) with (1) we see that we have to replace the volume energy $u_V A$ by $\sum_t \beta_i(n_i + z_i)$. We may say that u_V is the average value of the shell centers.

After this experience one would expect that one should replace also the other terms in (1) by shell dependent quantities. This, however, is not possible, since this would violate the requirements 3 and 4 of the first section. Either one gets a discontinuous binding energy, or the binding energy depends on how the nucleus has been built up. Of course, it would be possible to replace these terms by something which is more different from the liquid drop behaviour. This would violate requirement one, however, if the difference is important, and it would not bring in any improvement if the difference is not important. Thus we consider the surface, symmetry, and Coulomb terms as corrections describing the smooth part of the (N, Z) dependence of the levels or binding energy. It may be that one needs some more terms describing the smooth dependence. We will not worry about this problem here. However, once one has chosen a sufficiently good smooth contribution, there is no loss of generality if one looks only for shell dependent corrections which vanish for closed shells both of neutrons and protons. We did it this way already, and should proceed in the same manner.

Adding some terms coupling neutrons and protons is then reasonably straightforward. The simplest terms one may imagine are

$$\sum_{ik} \theta_{ik} n_i \hat{n}_i \hat{z}_k z_k. \tag{7}$$

Several more physical things can be said in favour of such terms: One would expect that empty and filled shells have minimum interaction, as it is the case with (7). Besides, the number of interaction terms in a second order perturbation theory is proportional to the number of particles and holes of both kinds of particles and this is a shell dependent effect. The first order produces the main part of the smooth terms. We should note that eq. (7) predicts a level dependence of a neutron level on the number of protons different from the standard. They do not go down strongly and uniformly as Z increases. Instead, although it may go down (if θ_{ik} has the right sign), this effect now is strongly shell dependent: It is largest if only few neutrons or neutron holes are in the shell, and if the proton shell is half filled. This seems to be quite reasonable.

By the way, only far from the valley of stable nuclei will this difference between our formula and others come into play. We should recall the fact that we do not talk about "theoretical" levels, but define levels by the separation energy only.

Furthermore, we have to introduce the pairing energy. We found, that a $A^{-1/2}$ dependence is superior to $A^{-1/3}$ and $A^{-2/3}$, so we used this one. After some trials and discussions with colleagues we postponed a more theoretical foundation of the pairing term. Although one may use an interpolated form of the pairing theory[4] for heavy nuclei, there seems to be less hope that this will work for the light ones: Here the pairing energy shows less systematics.

Finally, we have to introduce the deformation energy. As is well known, heavy nuclei sufficiently far from closed shells have strong deformations. We have described this effect either purely phenomenologically by introducing powers of $n_i \hat{n}_i z_k \hat{z}_k$, or using the model proposed recently by SWIATECKI[5], to whom we are very much obliged for fruitful discussions. As is well known, a deformation α diminishes the Coulomb energy and increases the surface energy. In addition, the shell structure deteriorates as can be seen from the Nilsson scheme[6]. We have described this effect by replacing η_i by $\eta_i \exp\left[-\left(\frac{\alpha}{\alpha_0}\right)^2\right]$ with an attenuation parameter α_0. Then one can calculate the deformation energy for fixed α, determine the equilibrium deformation from $\frac{\partial D}{\partial \alpha} = 0$ and the deformation energy by putting in this deformation α. However, there is deformation only, if $\dfrac{\partial^2 D}{\partial \alpha^2}\Big|_{\alpha=0} > 0$ (i. e. if the solution $\alpha = 0$ is instable), which leads to the condition

$$\frac{1}{2}\left(\eta_i\, n_i\, \hat{n}_i + \eta_k\, z_k\, \hat{z}_k\right) \geqslant \alpha_0{}^2\, \frac{A^{2/3}}{5}\left(2\, u_S - u_C\, \frac{Z^2}{A}\right) \tag{8}$$

for nuclei filling the ith neutron and k^{th} proton shell. The deformation energy is (in the quadratic approximation)

$$D = \frac{\alpha_0{}^2}{5}\, A^{2/3}\left(u_C\, \frac{Z^2}{A} - 2\, u_S\right)\left|\ln \frac{{}^1\!/_2(\eta_i\, n_i\, \hat{n}_i + \eta_k\, z_k\, \hat{z}_k)}{A^{2/3}\, \alpha_0{}^2/5\,(2\, u_S - u_C\, Z^2/A)} + 1\right| +$$

$$+ \frac{1}{2}\left(\eta_i\, n_i\, \hat{n}_i + \eta_k\, z_k\, \hat{z}_k\right). \tag{9}$$

Eq. (8) has a very nice interpretation and a desirable feature. Only if protons and neutrons are "in phase" (both filling shells at about the same time) will deformation occur. For fixed n_i, the deformation occurs the earlier, the higher z_i is. Both effects are well known experimentally.

We summarize: The mass formula we use is

$$B(N, Z) = a - u_S\, A^{2/3} - u_C\, \frac{Z(Z-1)}{A^{1/3}} - u_\tau\, \frac{(N-Z)^2}{A} + \delta_{N\,Z}\, \frac{\Delta}{A^{1/2}} +$$

$$+ \sum_i\left[\beta_i\,(n_i + z_i) - \frac{1}{2}\,(\eta_i + \zeta_i n_i)\, n_i\, \hat{n}_i - \frac{1}{2}\,(\eta_i + \zeta_i z_i)\, z_i\, \hat{z}_i\right] +$$

$$+ \sum_{ik} \theta_{ik}\, n_i\, \hat{n}_i\, z_k\, \hat{z}_k + D \tag{10}$$

with either phenomenological deformation energy

$$D = \sum_{ik}\, \sum_{\nu\mu,\nu'\mu'} \tau_{ik}^{(\nu\mu,\nu'\mu')}\, n_i^\nu\, \hat{n}_i^\mu\, z_k^{\nu'}\, \hat{z}_k^{\mu'} \tag{11}$$

or D according to (9), if (8) is satisfied. Here

$$\delta_{N\,Z} = \begin{cases} 0 \text{ if } N \text{ and } Z \text{ both even,} \\ 1 \text{ if } N \text{ or } Z \text{ even,} \\ 2 \text{ if } N \text{ and } Z \text{ both odd.} \end{cases} \tag{12}$$

We have added a constant term a since our mass formula begins with N and $Z > 20$. It takes into account a part of the binding energy of the first 40 nucleons.

III. Methods

We now briefly explain, how the parameters have been determined. We used standard least squares analysis with some modifications. The masses we used have been taken from the 1961 mass list[7] with some changes where the errors were very large. We have added quadratically the value of 0.1 MeV to the errors of this mass list and used the reciprocal value of it as weight. The reason for this was the feeling that we should not overemphasize the region where the masses are determined best (in the valley of stable nuclei), since this would give less weight to the other regions: These other regions, however, are the most important ones for the extrapolation. Besides, the matrices from the least squares analysis are not very well conditioned, such that we already are very near to the limits of the computer (IBM 7090) we have used. We have, so to speak, sacrificed clean statistical methods in favour of practical and physical aspects.

The calculation is reasonably straightforward, as long as all terms contain the parameters linearly. If they do not, one either has to linearize the expression putting in a guessed initial value for the corresponding parameter, or one may try it with a net of different parameters. Such a problem arose in the deformation energy, where η_i, η_k and $\alpha_0{}^2$ occur as parameters. We have used here a combination of both methods. More details need not to be given here.

IV. Results

Unfortunately, we have not yet quite finished our calculations. So the results we present here are quite preliminary. The reason is that we did incorporate the deformation energy in the more physical form only very recently. Since the most serious objections against our mass formula will be due to the large number of parameters, we have made a least squares fit using only the simplest ideas: We assume constant level distance in each shell, no interaction terms and only three terms describing a phenomenological deformation energy. In short: Besides the liquid drop terms, only the center of each shell, the level distance and deformation energy occur in it. This formula having only 18 parameters compares very favourably with all other formulae with a similar number of parameters: We have plotted the deviations between experimental masses and predicted masses in Fig. 4. The average deviation is 0.64 MeV. The formula reads:

$$B(N,Z) = a - u_S\,A^{2/3} - u_C\,\frac{Z(Z-1)}{A^{1/3}} - u_\tau\,\frac{(N-Z)^2}{A} + \delta_{NZ}\,\frac{\Delta}{A^{1/2}} +$$

$$+ \sum_i \left[\beta_i\,(n_i + z_i) - \frac{1}{2}\,\eta_i\,(n_i\,\hat{n}_i + z_i\,\hat{z}_i)\right] + \theta_{43}\,n_4\,\hat{n}_4\,z_3\,\hat{z}_3 +$$

$$+ \theta_{54}\,n_5\,\hat{n}_5\,z_4\,\hat{z}_4 + \tau_4^{(13,00)}\,n_4\,\hat{n}_4{}^3. \tag{1}$$

Secondly, in Fig. 5 we show the results of a more complicated formula with 35 parameters, using again a phenomenological description of the deformation energy:

$$B(N,Z) \text{ equal to (II, 10) with } D \text{ according to (II, 11),}$$
$$\text{containing 9 terms.} \tag{2}$$

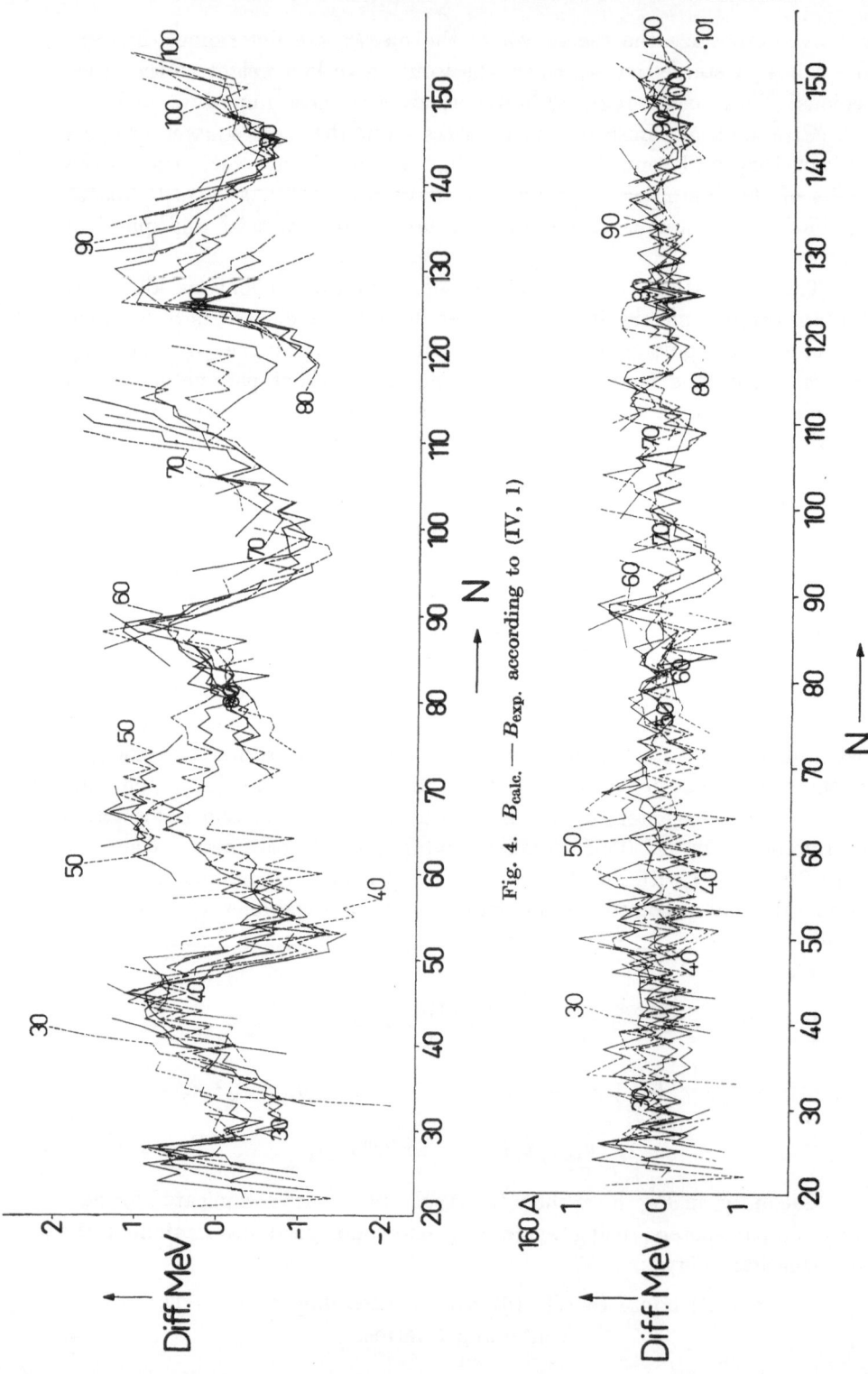

Fig. 4. $B_{calc.} - B_{exp.}$ according to (IV, 1)

Fig. 5. $B_{calc.} - B_{exp.}$ according to (IV. 2)

A list of parameters can be found in the appendix. If we study this plot in more detail, we find that there are left over some systematic deviations. It is not yet clear, however, whether they are due to systematic

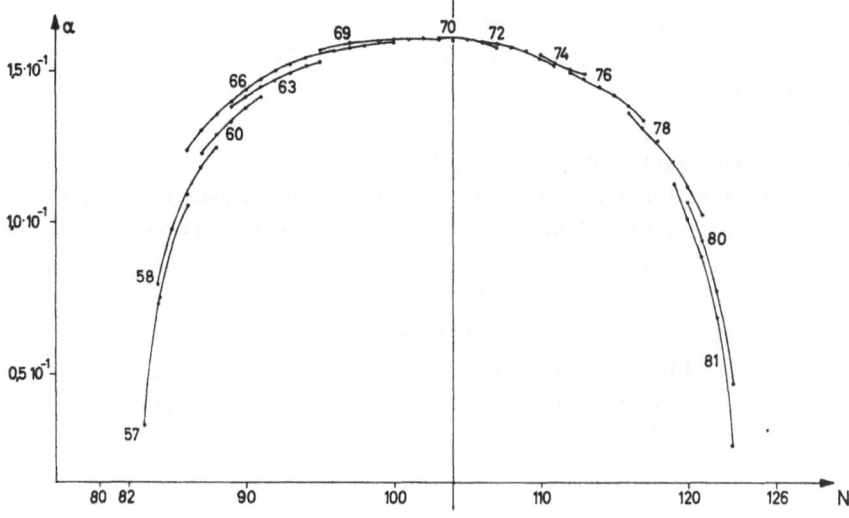

Fig. 6. Deformation parameter α for $82 < N \leqslant 126$

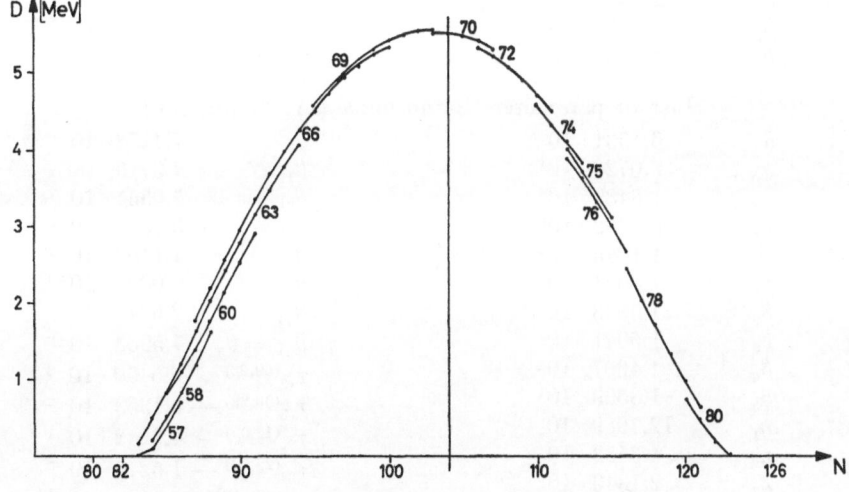

Fig. 7. Deformation energy for $82 < N \leqslant 126$

experimental errors (which may well occur, since the reaction and mass spectroscopic data build chains which determine the masses) or systematic errors of the mass formula. This will be known in the near future, if we employ the new data and new mass adjustment we will have after this conference. One has to bear in mind, however, that one cannot expect a mass formula with errors smaller than the scattering of the single particle levels, as long as one does not include subshell structures. The order of magnitude in question is several tenths of MeV, and so there is not too much to be expected without modification of the formula. Modification means: Larger, much larger number of parameters. We believe that the best one can do with major shells alone is about what has been done by us: the average deviation is of the order of 0.33 MeV.

Finally, we mention that we also have introduced the more physical deformation energy discussed above in the region $82 < N < 126$. We do not yet have a plot of it since we got these results very recently. However, deformation, deformation energy as well as the masses themselves come out quite reasonably. More details will be published as soon as we have exploited these methods more thoroughly. Some preliminary results are shown in Figs. 6 and 7.

Appendix

List of parameters for formula (IV, 1), (in keV)

a	$7.0770 \cdot 10^5$	β_5	$1.4649 \cdot 10^4$
u_S	$2.6082 \cdot 10^4$	η_1	$4.9096 \cdot 10$
u_τ	$1.8043 \cdot 10^4$	η_2	$6.2247 \cdot 10$
u_C	$5.8207 \cdot 10^2$	η_3	$4.4822 \cdot 10$
Δ	$-1.0952 \cdot 10^4$	η_1	$3.6216 \cdot 10$
β_1	$1.7079 \cdot 10^4$	η_5	$5.2325 \cdot 10$
β_2	$1.6250 \cdot 10^4$	θ_3	$7.7830 \cdot 10^{-2}$
β_3	$1.5728 \cdot 10^4$	$\theta_{5,}$	$7.1066 \cdot 10^{-2}$
β_4	$1.5232 \cdot 10^4$	$\tau_{4,}^{(13,00)}$	$-6.7458 \cdot 10^{-3}$

List of parameters for formula (IV, 2) (in keV)

a	$6.4531 \cdot 10^5$	ζ_5	$7.4172 \cdot 10^{-2}$
u_S	$2.0729 \cdot 10^4$	θ_{11}	4.3715
u_τ	$1.8489 \cdot 10^4$	θ_{21}	$7.0354 \cdot 10^{-1}$
u_C	$5.7782 \cdot 10^2$	θ_{22}	$-5.7325 \cdot 10^{-2}$
Δ	$-1.1045 \cdot 10^4$	θ_{32}	$1.1109 \cdot 10^{-1}$
β_1	$1.6021 \cdot 10^4$	$\theta_{\zeta 3}$	$9.0085 \cdot 10^{-2}$
β_2	$1.5393 \cdot 10^4$	θ_{44}	$2.6941 \cdot 10^{-2}$
β_3	$1.5041 \cdot 10^4$	$\theta_{51} = \theta_{53}$	$7.5963 \cdot 10^{-2}$
β_4	$1.4607 \cdot 10^4$	$\tau_{,2}^{(00,22)}$	$6.5450 \cdot 10^{-2}$
β_5	$1.3569 \cdot 10^4$	$\tau_{,3}^{(00,22)}$	$-6.1363 \cdot 10^{-2}$
η_1	$12.7989 \cdot 10$	$\tau_{2,}^{(12,00)}$	$-4.9242 \cdot 10^{-1}$
η_2	$4.3552 \cdot 10$	$\tau_{3,}^{(22,00)}$	$-1.5245 \cdot 10^{-2}$
η_3	$2.1340 \cdot 10$	$\tau_{4,3}^{(12,12)}$	$8.3144 \cdot 10^{-5}$
η_4	$6.7338 \cdot 10$	$\tau_{4,}^{(13,00)}$	$-5.8658 \cdot 10^{-3}$
η_5	$2.7032 \cdot 10$	$\tau_{4,}^{(23,00)}$	$-1.1699 \cdot 10^{-3}$
ζ_2	$-2.5279 \cdot 10^{-2}$	$\tau_{4,}^{(31,00)}$	$1.3964 \cdot 10^{-2}$
ζ_3	$-7.7549 \cdot 10^{-2}$	$\tau_{4,}^{(34,00)}$	$3.7429 \cdot 10^{-6}$
ζ_4	$-2.2779 \cdot 10^{-2}$		

The notation of shells is given by

No. of shell	1st	2nd	3rd	4th	5th
Nucleon number	21–28	28–50	51–82	83–126	127–184

References

[1] D. R. INGLIS, Nuclear Phys. **30**, 1 (1962).
[2] K. A. BRUECKNER and J. L. GAMMEL, Phys. Rev. **109**, 1023 (1958).
[3] K. BLEULER and CHR. TERREAUX, Helv. Phys. Acta **30**, 183 (1957).
[4] H. KÜMMEL, Z. Naturf. **16** *a*, 208 (1961).
[5] W. J. SWIATECKI, preprint (1963).
[6] S. G. NILSSON, Mat. Fys. Medd. Dan. Vid. Selsk. **29**, No. 16 (1955).
[7] L. KÖNIG, J. MATTAUCH, and A. H. WAPSTRA, Nuclear Phys. **31**, 18 (1962).

Discussion

K. BLEULER: May I just add two words to my contribution? I forgot to say that these rather long calculations we made in spherical nuclei with the pairing theory were done mainly by Dr. HUGUENIN who is with me now. I would be interested to know which value you get for the pairing energy term.

H. G. KÜMMEL: The term used was $\Delta/A^{1/2}$. So you have to know Δ. We assume the same Δ for protons and neutrons by the way. We get for Δ the values 10.952 MeV. Now I should mention that we also tried a more physical description which uses the BCS theory. But the point is that in a least squares fit you don't feel the pairing energy details; so, whether we did it this way or any other, did not make any difference.

K. BLEULER: I would like to suggest that different analytical expressions should be introduced for every shell.

H. G. KÜMMEL: We have done least square fits with different pairing terms for each shell but, we gave it up some time ago. Since ZELDES has done this, I now think that we will do it again, too.

K. BLEULER: Do you assume a pairing between protons and neutrons?

H. G. KÜMMEL: No, I don't.

K. BLEULER: Can it be seen that it really should not occur?

H. G. KÜMMEL: I would say that it is there. But it is a small contribution and if you made a least square fit of these large binding energies, apparently you don't see it.

E. BREITENBERGER: Referring to the number of parameters used, what was your criterion for acceptance or exclusion of terms?

H. G. KÜMMEL: The answer is that we simply dropped all terms which did not lead to essential improvements. What "essential" means, I cannot state very definitely. A new parameter which did not bring down the rms deviation of the masses by several tens of keV has been rejected. Of course, this also is a function of the confidence we had in the physical meaning and importance of such terms.

E. BREITENBERGER: The addition of 100 keV to the errors is open to objection. It would be ridiculous to ask, why 100 and not 99? But it is no more ridiculous to ask, why 100 rather than 50 or 200? In fact, it would seem most rational to add some 300 keV in order to make a virtue of a quandary and allow for the natural variance of the masses which is there anyway.

H. G. KÜMMEL: We have thought of all these possibilities after the discussions we had with you. But what really is the best thing we can do we do not know.

This has to be tried. The question is: if we now try 0.3 MeV, do we really get something essentially different? I do not believe we will.

E. Breitenberger: I am sure you would because there is a great difference in the experimental errors in different parts of the $N Z$ surface. In particular, since the experimental errors at the present time are fairly large in the region of deformed nuclei, you have given in your analysis particular weight to just that region because you reduced the weights in other parts where the experimental errors are very small. As I shall show afterwards in my paper, it is particularly dangerous, because it is just there that the experimental correlations between the mass values play the most important role and the result is that your deformation terms may be spurious.

H. G. Kümmel: That is very unlikely, I would say. Again my answer is to try it with different weights.

E. R. Cohen: I think Breitenberger has touched on a point that is very important. If you use the errors from the Mattauch mass table you also have to use the total error matrix. Otherwise your mass differences that you calculate are really meaningless. If you have not used the error matrix then you might to much better advantage merely weight the data equally unless you have some reason for throwing out an item of data completely. I think this is quite important because as has been mentioned I think by Zeldes also, you are not trying to determine experimental error in the data, you are trying to fit a surface to the point and the error with which you know the point you are trying to fit is much smaller than the deviations that you are trying to match to the surface. It is quite unrealistic to base your weights on the quoted errors in the Mattauch table and if you do then it really has to be in terms of the error matrix and not just the stated errors.

H. G. Kümmel: The point is, that we do not have a complete error matrix because of the limitation set up by the computer we used. So we are forced again to deviate from the ideal procedure.

N. Zeldes: I would like to make a comment about assigning weights inversely proportional to experimental errors. If one believes that the deviations of the experimental masses from the formula are due to these experimental errors this procedure would be justified by the theory of errors. But this is certainly not the case. In particular in your case of fitting a global formula, since there are regions where the errors are larger, this just makes injustice to these regions. We just rejected in each region data with uncertainties above a certain limit, and used the remaining ones with equal weight. And the limit was different for each region, being greater where the experimental errors are greater so as not to reject in these regions too large part of the data.

H. G. Kümmel: Again, all I can say is that it is to some extent a matter of taste which weight we should use. Rejecting some data completely would necessarily underemphasize regions where the masses are badly determined. Furthermore, I do not believe that the results depend very strongly on the statistical methods applied.

J. Mattauch: Let me comment on the questions by Drs. Cohen, Breitenberger, and Zeldes. You said we should have used all the points with equal weight and that experimental errors are very small compared to the deviation from any formula. I don't think that is quite true. We have in our table errors of the masses of 300, even of 500 keV. Should we throw them away? We got what you wanted, about equal weight, by adding this 100 keV quadratically to the errors. As KÜMMEL said, we have to do something not to over-emphasize those points which are known with errors of one or two keV. To accept the errors as they are in the table would mean to give no weight to any point outside the valley of stability.

E. BREITENBERGER: The essential point that I was making is that your choice of 100 keV has been made without specific reason, it carries a subjective element into the analysis. This may be unimportant, but it may also become an insidious source of self-deception. It makes me worry, what is the objective meaning of the formula. Amongst statisticians it is considered good practice to give a wide berth to such complications by taking all data with equal weights.

H. G. KÜMMEL: Certainly, there is a subjective element in the analysis. Since physical aspects cannot be forgotten and there is not yet agreement between physicists which "physics" is the correct one, subjective elements cannot be avoided. I believe that using equal weights for all nuclei would be as subjective and open to criticism as is our procedure.

(*Editor's Note.*) A comment by A. H. WAPSTRA concerning several points raised in this discussion may be found in the discussion of the paper by W. THIELE.

Semi-Empirical Interpretation
of Nuclear Masses and Deformations*

By

Wladyslaw J. Swiatecki

Lawrence Radiation Laboratory, University of California,
Berkeley, California, USA

With 3 Figures

In this talk I would like, first, to describe a semi-empirical mass formula and then to mention three items that are of interest in the interpretation of nuclear masses.

1. The Mass Formula

The semi-empirical mass formula whose general appearance is given in eq. (1) and (2) is supposed to include a rough description of shell effects on nuclear masses:

$$M(N, Z, \text{shape}) = M_{\text{Liquid Drop}}(N, Z, \text{shape}) + M_{\text{Shells}}(N, Z, \text{shape}). \quad (1)$$

It differs in one respect from similar formulae in that it aims at describing the nuclear masses, including shell effects, as a function of N, Z, *and the nuclear shape*.

The first part of eq. (1) is a conventional liquid drop formula, consisting of volume, surface and electrostatic energies and an even-odd correction. The second part is the shell correction:

$$M_{\text{Shells}}(N, Z, \text{shape}) = C \cdot S(N, Z) \cdot e^{-(\text{distortion})^2/a^2}. \quad (2)$$

It consists of an amplitude parameter C, a dimensionless shell function S of N and Z, and an "attenuating factor" exp $[-(\text{distortion})^2/a^2]$, of range a. The "distortion" in the attenuating factor is taken to be the root-mean-square deviation of the radius vector $R(\theta, \varphi)$ (specifying the nuclear surface) from the radius R_0 of the undistorted sphere. The quantity $S(N, Z)$ is an oscillating function of two variables with negative throughs at magic

---

* This work was performed under the auspices of the U. S. Atomic Energy Commission.

numbers, $N = N_{\text{magic}}$ or $Z = Z_{\text{magic}}$, and cushion-like bumps in between. In its simplest version $S(N, Z)$ is an explicit semi-empirical expression with one adjustable parameter:

$$S(N, Z) = \frac{F(N) + F(Z)}{(A/2)^{2/3}} - c A^{1/3}, \tag{3}$$

where the function F is given by

$$F(X) = \frac{3}{5} \cdot \frac{M_i^{5/3} - M_{i-1}^{5/3}}{M_i - M_{i-1}} \cdot (X - M_{i-1}) - \frac{3}{5} (X^{5/3} - M_{i-1}^{5/3}), \tag{4}$$

for X in the range between the $(i-1)$th and the ith magic numbers M_{i-1} and M_i. (Here X stands for N or Z.)

The intrinsic oscillations in the dependence of S on N and Z follow from a simple theory and the adjustable parameter c specifies, roughly speaking, the "base-line" or "zero" from which the oscillations are to be counted as positive or negative.

In all then the shell correction M_{Shells} has three adjustable parameters: the amplitude parameter C, the base-line parameter c and the range parameter a.

I cannot here go into the derivation of the formula specified by (1), (2), (3) and (4), so I will only state in words the physical ingredients on which the shell correction is based. These ingredients are two in number:

1. Shell effects on nuclear masses are due to a bunching of the single-particle levels in the nuclear potential into groups corresponding to the magic numbers. (This ingredient is common to many semi-empirical treatments of shell effects. A particularly simple assumption about the bunching leads to the simple formula given by (3) and (4).)

2. The above bunching of levels, being associated with the spherical shape of the nuclear potential, will disappear if the nucleus is sufficiently distorted. (This ingredient is specific to the present treatment. It is expressed formally with the aid of the "attenuating factor".)

I have tried to illustrate the physical content of the mass formula with the aid of a single slide (Fig. 1). This figure shows how the dependence of the mass of a nucleus on deformation is given by a liquid drop expression (the smooth curve) corrected by a shell term (the shaded negative or positive bump), which disappears for large distortions. The amplitude of the bump, in its dependence on N and Z, is given in the lower part of Fig. 1. It is an oscillating function with negative dips at magic numbers. The functional form of these wiggles follows from a simple theory of bunched levels, shown on the right.

The fact that our mass formula is a function of nuclear *shape* has several consequences. The most important one is that by minimizing the mass with respect to shape:

$$\frac{\partial M}{\partial (\text{shape})} = 0,$$

we obtain a prediction of the equilibrium deformation of a nucleus as well as its mass. From Fig. 1 we see that close to magic numbers, when the shell correction is negative, our formula will predict spherical equilibrium shapes. Between magic numbers—at least between some magic numbers, if the shell correction turns sufficiently positive—*deformed* equilibrium shapes will be predicted.

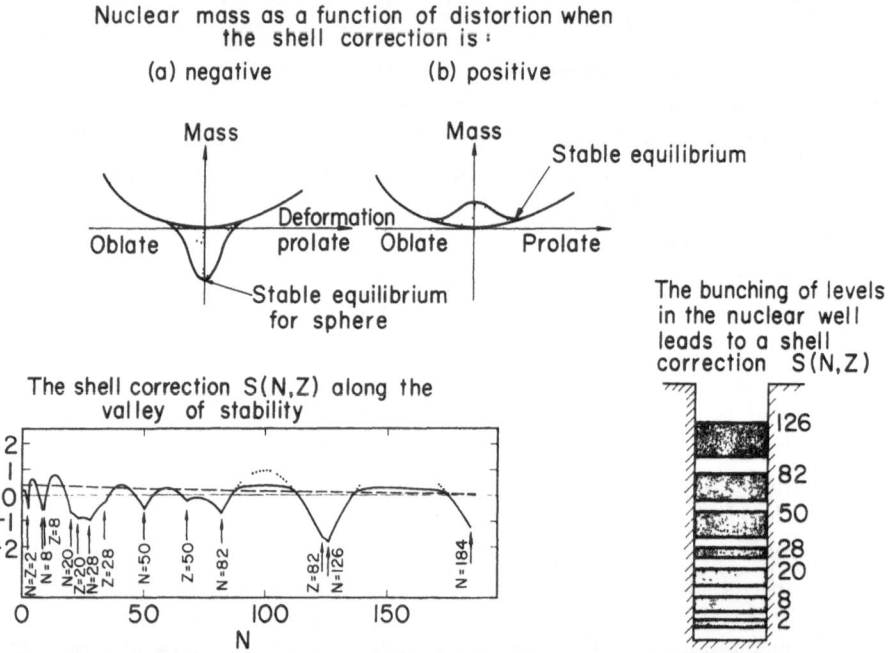

Fig. 1. The upper part shows schematically the energy of a nucleus as a function of deformation in the case when the shell correction (dotted part) is negative or positive. This corresponds to nuclei close to magic numbers or far from magic numbers, respectively. The oscillations in the shell correction are shown in the lower part of the figure, which was derived from a model of bunched levels indicated schematically on the right. The smooth dashed line is the critical amount of positive shell correction necessary to overcome the stabilizing effect of the liquid drop part of the mass formula and produce stably distorted nuclei. The masses of such nuclei, instead of following the dotted arcs, are flattened out to values a little in excess of the dashed line

The masses and deformations of nuclei, calculated with the aid of our formula, are compared with experiment in Figs. 2 and 3. Three parameters, C, c and a, were adjusted. We see that with these three adjustable parameters available to us for the description of shell effects it seems possible to give a rough account of the oscillations in the masses and, simultaneously, of the general character of nuclear deformations. The description is only rough and appreciable deviations remain. The fact that one can give even a semi-quantitative account of masses and

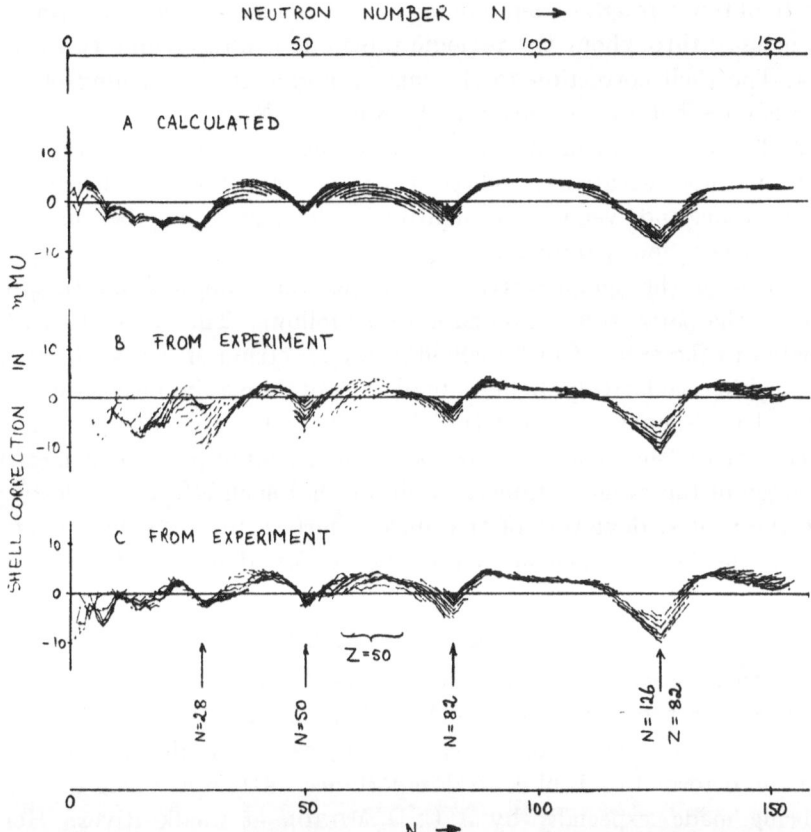

Fig. 2. The calculated shell correction M_{Shells} is plotted against the neutron number for fixed proton numbers and compared with similar plots deduced from experimental masses. The two different plots labeled "from experiment" were obtained by subtracting somewhat different smooth mass surfaces from the experimental masses. In the last plot a term with some of the properties of a Wigner term was included in the smooth mass surface. Note its effect on the lower part of the periodic table

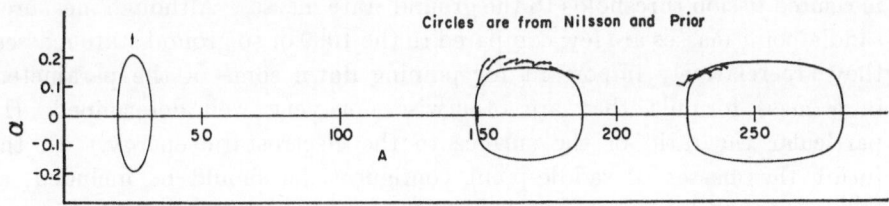

Fig. 3. The calculated ground state deformations α, given by the curves, are compared with the deformations deduced from experiment in the region of the rare earths and the actinides

deformations in such a simple way is, however, noteworthy. What it boils down to—and this is the principal result of this study—is that in order to obtain a rough understanding of the trends of nuclear deformations (and masses) throughout the periodic table one has to assume two things:

1. The shell correction to the masses is due to a bunching of levels and oscillates between negative and positive values.

2. The correction disappears for a sufficiently large distortion.

From these two assumptions the general course of nuclear deformations follows at once and even a semi-quantitative description may be constructed with relatively few parameters.

(Some of the quantitative conclusions that follow from the fitted values of the parameters C, c and a are as follows. The value of C, which is related to the amount of bunching in the spectrum of levels, shows that this bunching is relatively slight, only 10–20% of the maximum corresponding to complete degeneracy of levels. The value of the base-line parameter c is such that the shell correction is about equally often positive as negative. The value of the range parameter a shows that shell effects are destroyed when the r. m. s. deviation of the nuclear surface from a sphere is of the order of the fastest nucleon in the nucleus, or about one fermi.)

2. Saddle-Point Masses

Another consequence of the claim of our formula to give a description of masses as a function of the nuclear shape is that the formula may be applied to the discussion of nuclear fission, where the dependence of the energy on deformation is of central importance. Attempts along these lines are being made, especially by J. C. D. MILTON at Chalk River. Here I would like to make one remark concerning a group of nuclear mass determinations that is not usually included in comparisons with semi-empirical mass formulae. This group consists of the one dozen or so of measured saddle-point masses, that is, masses of nuclei not in their ground states but in elongated cylinder-like or dumb-bell-like configurations associated with fission barriers. These masses are obtained by adding measured fission thresholds to the ground state masses. Although measured saddle-point masses are few compared to the 1000 or so ground state masses, they are relatively important for pinning down some of the parameters in a mass formula that are otherwise not very well determined. (In particular the ratio of the surface to the electrostatic energy.) In the future the masses of saddle-point configurations should be included, as far as possible, in comparisons of semi-empirical formulae with experiment.

3. Electrostatic Energy of Diffuse Charge Distributions

In connection with the electrostatic energy I would like to point out a minor but useful theorem concerning the electrostatic energy of charge distributions with a diffuse surface. Although the correction for diffuseness for a *spherically symmetric* charge distribution is not too hard to work out, it seems at first sight that to generalize the correction to an arbitrarily shaped nucleus with a diffuse surface would entail a rather formidable calculation. In fact the following simplification makes the problem a trivial one. If the correction for diffuseness is worked out to lowest order in the ratio of the thickness of the surface to the size (radius) of the nucleus then, using Gauss' theorem in electrostatics, it can be shown that this lowest order correction is exactly independent of the shape of the nucleus. Thus if we denote by $\delta\varrho$ the change in the charge density required to convert a sharp distribution into a diffuse one ($\delta\varrho$, equal to $\varrho_{\text{diffuse}}(n) - \varrho_{\text{sharp}}(n)$, is assumed to be a function of the normal distance n across the surface of the distribution and is discontinuous at the location of the sharp surface) then the correction δE_c to the electrostatic energy, to lowest order in the skin thickness, is found to be

$$\delta E_c = - 4\pi \text{ (Total Charge)} \int_{-\infty}^{\infty} n\, \delta\varrho\, dn.$$

Applying this result to a diffuseness of the charge distribution $\varrho(n)$ given by a Woods-Saxon form factor:

$$\varrho(n) = \varrho_0 \cdot \frac{1}{1 + e^{n/d}},$$

we find for the electrostatic energy of an arbitrarily shaped nucleus the expression

$$E_c = \frac{3}{5} \frac{e^2}{r_0} \frac{Z^2}{A^{1/3}} g(\text{shape}) - \frac{\pi^2}{2} \frac{e^2}{r_0} \frac{Z^2}{A} \left(\frac{d}{r_0}\right)^2 + \text{higher powers of } (d/r_0 A^{1/3}),$$

where r_0 is the nuclear radius constant. The first term, a function of shape, is the electrostatic energy of a deformed *sharp* distribution. The second term is the correction for diffuseness. The fortunate circumstance that the lowest order correction is a mere constant, independent of the nuclear shape, makes the discussion of a deformed drop with a *thin* diffuse surface as easy as that of a drop with a *sharp* surface. This is a boon for the theory of deformed nuclei and in particular for the theory of fission, where the diffuseness of the charge distribution can now be included approximately with no extra labor whatever.

4. The Wigner Term $|N-Z|$

My final remark concerns the so-called Wigner term in the mass formula, characterized by a dependence of the mass on $|N - Z|$. Such a term was originally derived by WIGNER[1] under the assumption of nuclear forces whose range was very large compared to the sizes of nuclei, and later using a less well-defined hypothesis regarding the smoothness of certain interaction integrals. The Wigner term proportional to $|N - Z|$

has since been included in some semi-empirical formulae but not in others. Should such a term be included or not?

It does not seem to be widely known that some information concerning this question is contained in pre-historic studies of the interaction energy between particles of an infinitely extended Fermi gas ("nuclear matter"), for example, the work of H. A. BETHE and R. F. BACHER[2] or H. VOLZ[3]. Thus if one takes the formulae for the interaction energy given by the above authors and lets the range of the forces become large compared to the spacing between particles (in order to reproduce Wigner's assumption of long-range forces) one does indeed find an interaction energy which, in the limit, behaves like $|N - Z|$. On the other hand if one makes the range of the forces comparable with the interparticle spacing—which is, of course, the actual state of affairs—the "cusp" in the $|N - Z|$ dependence is completely smoothed out and becomes part of the familiar $(N - Z)^2$ parabola in the symmetry energy. (The physical reason for this is clear from an examination of the Fermi gas calculations. The $|N - Z|$ dependence arises from a certain interference between the "mixed densities" of the neutrons and protons in the Fermi gas, and it takes place only if many nucleon wavelengths are contained within the range of the nuclear forces.)

The above remarks would argue against the correctness of including an $|N - Z|$ term in a mass formula, although formally there is no disagreement with Wigner's original derivation of such a term in the limit of infinitely-long-range forces. The situation is, however, rendered more subtle by some tentative calculations which suggest that for a finite rather than an infinite Fermi gas there may be effects favoring, like the Wigner term, the special stability of nuclei with $N = Z$. The physical origin of these terms would appear to be different than the origin of the Wigner term and, as a result, both the dependence on A and on $|N - Z|$ would be different.

The tentative conclusion would seem to be that, although there may be experimental evidence in the measured masses for a component with some of the properties of a Wigner term, one should not include such a term in its traditional form without further analysis.

Appendix

For the sake of definiteness I shall write down explicitly one version of the new mass formula, specialized to describe the masses of nuclei possessing *ellipsoidal* shapes. The semi-axes of the ellipsoid are specified, as usual, in terms of a distortion parameter σ and a shape parameter γ (ref. [4]):

$$a = R_0 \exp\left[\sigma \cos\left(\gamma - \frac{2\pi}{3}\right)\right],$$

$$b = R_0 \exp\left[\sigma \cos\left(\gamma + \frac{2\pi}{3}\right)\right],$$

$$c = R_0 \exp\left[\sigma \cos\gamma\right].$$

The combination $\alpha^2 = \sigma^2 \left(1 - \dfrac{1}{7} \sigma \cos 3\gamma + \text{higher powers of } \sigma\right)$, proportional to the mean square deviation of an ellipsoidal surface from a sphere, is found to be useful in the present context.

The mass formula for diffuse ellipsoidal nuclei then reads as follows:

$$M(N, Z; \alpha, \gamma) = M_n N + M_H Z - c_1 A +$$

$$+ c_2 A^{2/3} \left(1 + \frac{2}{5} \alpha^2 - \frac{4}{105} \alpha^3 \cos 3\gamma + \text{higher powers of } \alpha\right) +$$

$$+ c_3 \frac{Z^2}{A^{1/3}} \left(1 - \frac{1}{5} \alpha^2 - \frac{4}{105} \alpha^3 \cos 3\gamma + \text{higher powers of } \alpha\right) -$$

$$- \frac{\pi^2}{2} \frac{e^2}{r_0} \frac{Z^2}{A} \left(\frac{d}{r_0}\right)^2 \pm \frac{12 \text{ mMU}}{A^{1/2}} + C \cdot S(N, Z) \cdot \exp\left[-\frac{1}{5} \left(\frac{r_0}{a}\right)^2 A^{2/3} \alpha^2\right].$$

In the above M_n and M_H are the masses of the neutron and of the hydrogen atom. The quantities c_1 and c_2, given by

$$c_1 = k_1 \left[1 - \varkappa \left(\frac{N - Z}{A}\right)^2\right],$$

$$c_2 = k_2 \left[1 - \varkappa \left(\frac{N - Z}{A}\right)^2\right],$$

are the volume and surface energy coefficients, taken as dependent on nuclear composition through the term $\varkappa \left(\dfrac{N - Z}{A}\right)^2$ ($\varkappa$ is a constant). The parameter c_3 is related to the nuclear radius constant by $c_3 = \dfrac{3}{5} \dfrac{e^2}{r_0}$. The diffuseness correction and the even-odd correction have been written out in full. The function $S(N, Z)$ is given in Section 1.

The values of the seven adjustable parameters giving an approximate fit to the experimental data are in the neighborhood of

$$k_1 = 17.012 \text{ mMU},$$

$$k_2 = 20.600 \text{ mMU},$$

$$c_3 = 0.78 \text{ mMU (therefore } r_0 = 1.1894 \text{ fermi)},$$

$$\varkappa = 1.803,$$

$$C = 5.7 \text{ mMU},$$

$$c = 0.27,$$

$$(a/r_0) = 0.328.$$

References

[1] E. FEENBERG and E. P. WIGNER, On the Structure of Nuclei between Helium and Oxygen. Phys. Rev. 51, 95 (1937). — E. WIGNER, On the Structure of Nuclei beyond Oxygen. Phys. Rev. 51, 947 (1937).

[2] H. A. BETHE and R. F. BACHER, Nuclear Physics. Revs. Modern Phys. 8, 82 (1936), especially equation 158.

[3] H. VOLZ, Über die Größe der Kernkräfte. Z. Physik 105, 537 (1937).

[4] B. C. CARLSON, Ellipsoidal Distributions of Charge or Mass. J. Math. Phys. 2, 441 (1961).

Acknowledgments

I would like to thank SVEN JOHANSSON and STAN THOMPSON for valuable discussions.

Discussion

N. ZELDES: SEEGER used the $|N - Z|$ term, and JOHANSSON too. They say it improves the fit to the course of the line of beta-stability. Could you please comment on this?

W. J. SWIATECKI: This is correct. From an empirical point of view a term $|N - Z|$ helps with the fit. This can also be seen from my slide 2 (Fig. 2) where the bottom line has in it a term with some of the properties of a Wigner term. One would like to know, however, whether on the basis of theory such a term should be present. The tentative conclusion is that a Wigner term in its original form is not appropriate, but something like it may be there.

N. ZELDES: Did you try a detailed comparison of the formula with experimental masses to see what is the mean deviation?

W. J. SWIATECKI: No. I was satisfied at the stage of comparing the experimental and calculated wiggles and seeing the rough correspondence. I would like to stress that the present mass formula is not meant to be compared, as regards accuracy, with the more detailed treatments you have heard described. I was interested in obraining an over-all understanding of what is going on, as regards mass and deformations, throughout the periodic table.

D. M. VAN PATTER: With regard to the correction to the Coulomb energy due to diffuseness which you discussed, there has been considerable new experimental information concerning Coulomb energy differences from the positions of the isobaric analog states discovered by ANDERSON and WONG. I wondered if you would expect this correction to be sufficient to be seen from an examination of these new experimental data.

W. J. SWIATECKI: The diffuseness correction is a few percent for heavy nuclei and perhaps 20–30% for very light nuclei. In order, however, to isolate such a term experimentally one would have to discuss simultaneously the exchange correction to the electrostatic energy, which turns out to have similar properties as the diffuseness correction.

Construction of an Extended Nuclidic Mass Table*

By

Manny Hillman

The Hot Laboratory, Brookhaven National Laboratory, Upton, N. Y.

With 4 Figures

The calculation of cross sections of high energy nuclear reactions requires a fairly accurate table of masses of a large number of unknown nuclides. The best extended nuclidic mass table[1] (at the time this work began) included quite a few with deviations between known and calculated masses greater than 1 MeV and ranging up to 6 MeV. Such discrepancies naturally cause one to be suspicious about the values calculated for unknown masses. The following report describes an attempt to construct an extended mass table with very small deviations among the known masses. It is based on a re-examination[2-4] of "parabolic systematics".

The Weiszäcker[5] mass formula is parabolic in Z at constant A. Changing the Coulomb term[1] to trapezoidal charge distribution introduces a $Z^{4/3}$ term; however, this latter term will be ignored for the time being. When expressed in the form

$$\Delta M = a + bZ + cZ^2,$$

the determination of the parameters a, b, and c *versus* A would allow the use of this formula for the calculation of unknown masses. The parameters, and their products with powers of Z are very much larger than the excess masses themselves, requiring their knowledge to a large number (~ 5) of significant figures to obtain meaningful results. Translation of the axis to the symmetry line transforms the equation to

$$\Delta M = Ao + B(Z - Zo)^2 + \delta,$$

where Ao is the minimum excess mass, Zo is the charge at the minimum, and B is the curvature of the parabola. The δ term is introduced to accomodate pairing corrections. The formula in this form has been

---

* This work was performed under the auspices of the U. S. Atomic Energy Commission.

5*

discussed previously[3, 4]. For use, the parameters do not require a large number of significant figures.

There is sufficient data in the 1961 Mass Table[6] to determine the parameters for 149 of the 382 parabolas (two for each even A) for isobars 1

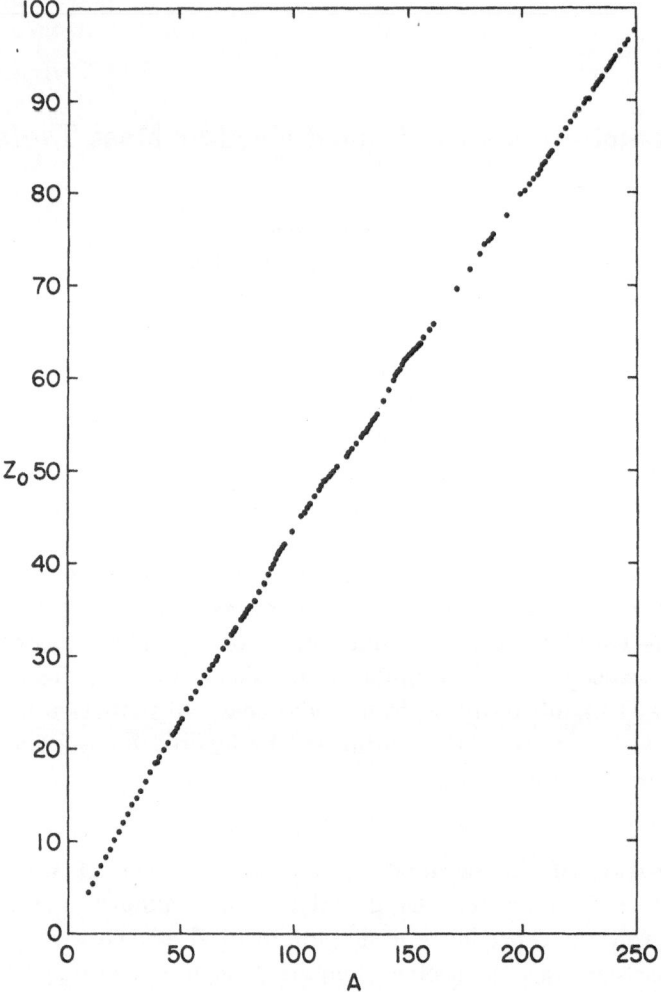

Fig. 1. Zo vs. A. Data obtained from odd-A isobars with more than three masses available and from even-A isobars with more than three masses of even-even and/or more than three masses of odd-odd nuclides available

through 255. By means of further considerations this may be improved considerably. The values for Zo and Ao fall as illustrated in Figs. 1 and 2. Only a very insignificant difference could be discerned for Zo for parabolas of odd-odd and even-even nuclides of the same isobar, and the values were accepted as identical. For even-A isobars, values of Ao

are plotted separately for the parabolas of odd-odd (crosses) and even-even (circles) nuclides.

No attempt was made to fit these curves to any formula. Instead a parabolic interpolation procedure was used to determine the missing

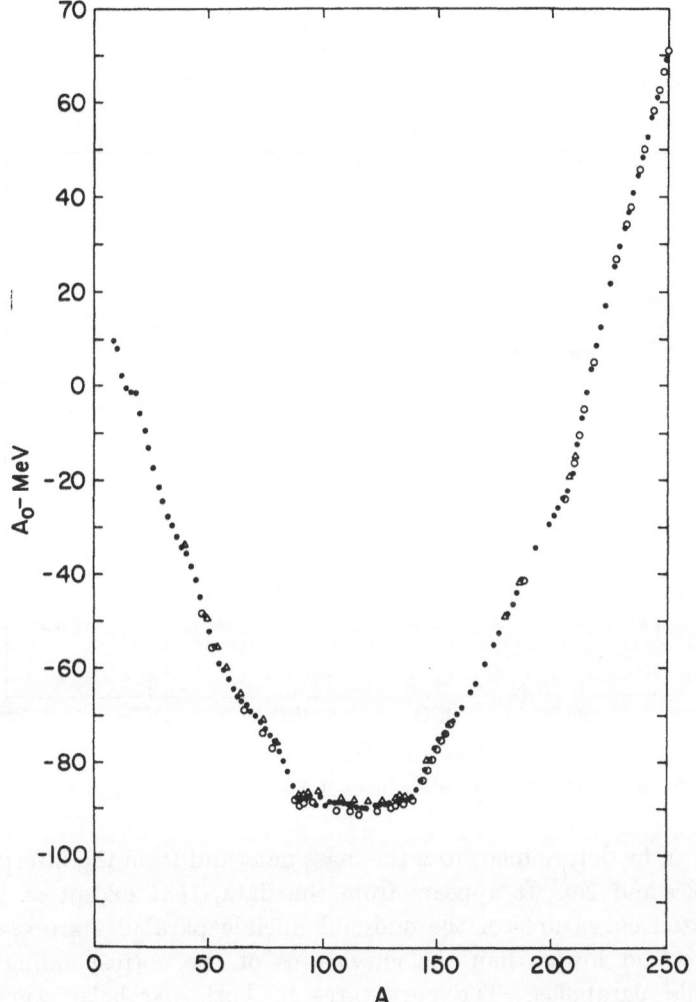

Fig. 2.   *Ao* vs. *A*. Data obtained as in Fig. 1.   ● = odd-*A* isobars;   ○ = even-*A* isobars, even-even nuclides: × = even-*A* isobars, odd-odd nuclides

points. The interpolation procedure was evaluated by determining known values by the same method. The check was quite good. The δ's were then determined as the difference of known Ao's for the even-*A* isobars and the values of the Ao's for even-*A* isobars as determined by interpolation of the Ao's for odd-*A* isobars (Fig. 3). Fitting the δ's to an apparent straight line allowed a reasonable estimation of the remaining even-A Ao's.

The entire problem can be simplified by an elementary consideration of the two parabolas at even-A isobars. *They may not cross*. For if they did, this would result in positron emission on the neutron excess side and negatron emission on the proton-rich side. Consequently the curvature, B, for the odd-odd parabola must be either greater than or equal to the curvature of the even-even parabola. The curvatures of most of the

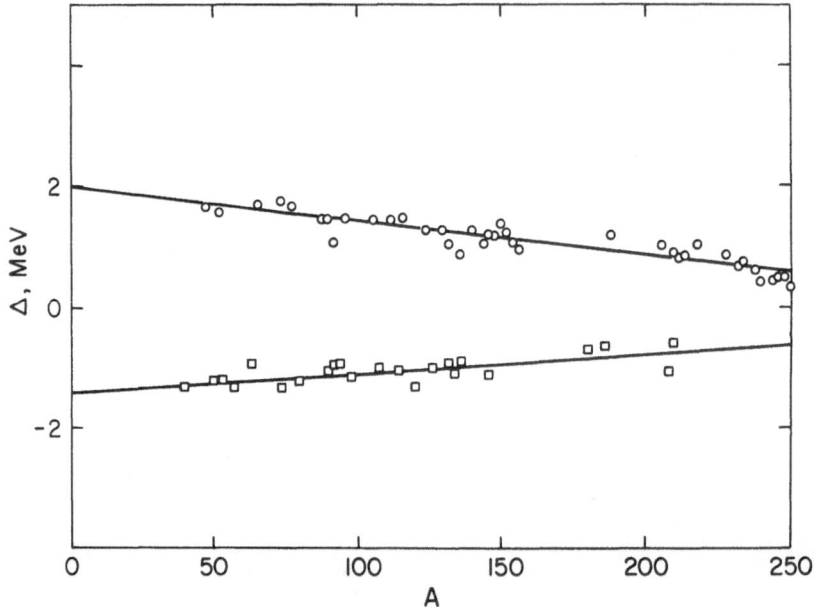

Fig. 3. δ vs. A. $\bigcirc$ = even-even nuclides; $\boxdot$ = odd-odd nuclides

parabolas can be determined from the mass data and from the interpolated values of Zo and Ao. It appears from the data, that except at low A, the calculated curvatures of the odd-odd nuclide parabolas are scattered both higher and lower than the curvatures of the corresponding even-even nuclide parabolas. The curvatures of both parabolas were then assumed to be equal. This consideration, therefore, reduces the number of parabolas to 255 and increases the number for which parameters are directly determinable to 229. The rest were determined by means of the interpolation procedure for missing Zo's and Ao's and using the estimated linear relationship for the δ's. No interpolation procedure was possible for B since the values were too scattered (Fig. 4). A few still indeterminable in this way required the use of additional data from the Nuclear Data Sheets.

The mass table constructed from these parameters is, on the surface, excellent. The standard deviation is 0.255 and there are only 8 deviations greater than 1 MeV. (For comparison with other tables only isobars of A greater than 21 were considered.) There is, however, a very serious flaw.

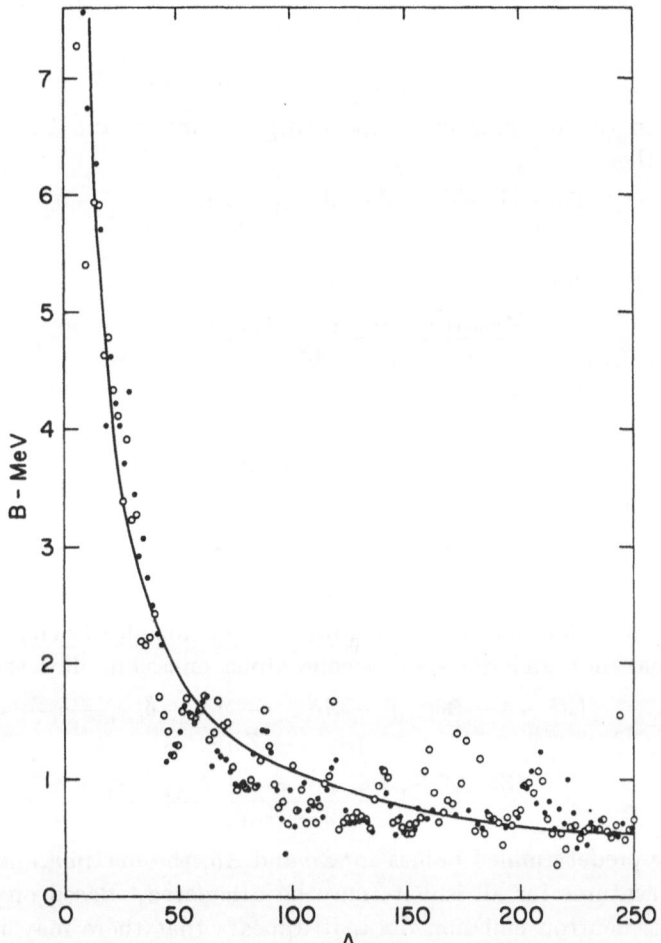

Fig. 4. B vs. A. $\bullet$ = odd-A isobars, $\bigcirc$ = even-A isobars. Solid curve represents the fit of the data to $B = a\, A^{-1} + b\, A^{-1/3}$

For many isobars, the Q for neutron emission is positive on the proton-rich side and/or negative on the neutron-rich side. The case is similar for proton emission. In order to alleviate this fault one must invoke some of the properties of the systematics of the binding energies of protons, neutrons, and alpha particles.

When considering proton and/or neutron decay systematics, one ordinarily expects that a nuclide of all protons or all neutrons should spontaneously emit protons or neutrons respectively, while nuclides containing only one proton or neutron should not spontaneously emit those respective particles. Generally, for spontaneous particle emission

$$\Delta M_i > \Delta M_f + \Delta M_p.$$

For non-emission the sign is reversed. Thus, for example, for non-neutron emission

$$Ao_i + B_i(Z_j - Zo_i)^2 + \delta_i < Ao_{i-1} + B_{i-1}(Z_j - Zo_{i-1})^2 + \delta_{i-1} + n,$$

and, since in the extreme cases the charge of initial and final nuclides is $A_i - 1$, then

$$\frac{B_i}{B_{i-1}} < \frac{(A_i - 1 - Zo_{i-1})^2 + Ao_{i-1} - Ao_i + \delta_{i-1} - \delta_i + n}{(A_i - 1 - Zo_i)^2}.$$

Similarly, for spontaneous neutron emission

$$\frac{B_i}{B_{i-1}} > \frac{(Zo_{i-1})^2 + Ao_{i-1} - Ao_i + \delta_{i-1} - \delta_i + n}{(Zo_i)^2},$$

for non-proton emission

$$\frac{B_i}{B_{i-1}} < \frac{(Zo_{i-1})^2 + Ao_{i-1} - Ao_i + \delta_{i-1} - \delta_i + p}{(1 - Zo_i)^2},$$

and for spontaneous proton emission

$$\frac{B_i}{B_{i-1}} > \frac{(A_i - 1 - Zo_{i-1})^2 + Ao_{i-1} - Ao_i + \delta_{i-1} - \delta_i + p}{(A_i - Zo_i)^2}.$$

Both, nuclides containing only two protons and nuclides having only two neutrons may not undergo spontaneous alpha emission. For these,

$$\frac{B_i}{B_{i-1}} < \frac{(A_i - 4 - Zo_{i-4})^2 + Ao_{i-4} - Ao_i + \delta_{i-4} - \delta_i + \alpha}{(A_{i-2} - Zo_i)^2},$$

and

$$\frac{B_i}{B_{i-1}} < \frac{(Zo_{i-4})^2 + Ao_{i-4} - Ao_i + \delta_{i-4} - \delta_i + \alpha}{(2 - Zo_i)^2}.$$

Thus, using predetermined values for Zo and Ao, the maxima and minima relative curvatures for all isobars may be calculated. Based on the non-spontaneous neutron emission data, it appears that there may not be as much scatter in the curvatures as there appears to be in Fig. 4. More significantly since the ratios $\dfrac{B_i}{B_{i-1}}$ for non-spontaneous proton-rich alpha-emission are almost all less than one, it appears, provided that the assumption of no alpha emission is correct, that, except for low A's, the function of B versus A must be monotonically decreasing.

The choice of a monotonically decreasing function for B versus A should probably be governed by considerations of best fit. In Fig. 4 the solid line is the curve for the relationship derivable from the Weizsäcker Mass Formula, $$B = a_3 A^{-1} + 3 e^2 A^{-1/3}/5 r_0,$$

with a_3 and r_0 as free parameters. These were determined from a least squares fit as 76.1 MeV and 1.32 Fermis respectively. An even better fit, especially in the $A = 40$ to 100 isobar region is obtainable by treating the function as a sum of two exponentials. This has no bearing on any theory that I am aware of but is tentatively useful.

Now that one parameter of the parabola, B, has a fixed form, and its best values have been determined, the values of Ao, Zo, and δ may be redetermined. Only two masses are required for each odd-A isobar and only three are required for each even-A isobar. Sufficient data is in the 1961 Mass Table to determine all of the parameters for 245 of the 255 parabolas. Only two of those missing are in the isobar region 13 through 246. The unknown parameters were determined by the interpolation procedure, and an extended mass table* was constructed. Nineteen calculated masses deviated from the known masses by more than one MeV. All but one were associated with closed-shell effects, for which no correction at all has been made. The largest deviation is 2.0 MeV and the standard deviation for 763 masses was 0.35 MeV. Certainly further refinements are possible.

Coryell's[3] work involved a more extensive analysis of the Zo-A relationship. Based on only a few known masses, he found that his procedure determines masses with an average deviation of 0.4 to 0.5 MeV. Seeger's[1] method yields masses with a standard deviation of about 0.7 MeV. Seeger's method offers the advantage of allowing an estimate of the predictability of unknown masses to be determined. It is unfortunate that the method reported in the present work does not allow the determination of an estimate of predictability. The reliability can only be determined in time.

References

[1] P. A. SEEGER, Semiempirical Atomic Mass Law, Nuclear Phys. 25, 1 (1961).

[2] N. BOHR and J. A. WHEELER, The Mechanism of Nuclear Fission. Phys. Rev. 56, 426 (1939).

[3] C. D. CORYELL, β-Decay Energetics. Ann. Rev. Nucl. Sci. 2, 305 (1953).

[4] H. B. LEVY, New Empirical Equation for Atomic Masses. Phys. Rev. 106, 1265 (1957).

[5] C. F. VON WEIZSÄCKER, Zur Theorie der Kernmassen. Z. Physik 96, 431 (1935).

[6] L. A. KÖNIG, J. H. E. MATTAUCH, amd A. H. WAPSTRA. 1961 Nuclidic Mass Table. Nuclear Phys. 31, 18 (1962).

* The mass table will be available as a Brookhaven National Laboratory report.

Status Report on Atomic Masses

By

Aaldert Hendrik Wapstra

Instituut voor Kernphysisch Onderzoek, Oosterringdijk 18, Amsterdam

and

Josef H. E. Mattauch

Max-Planck-Institut für Chemie (Otto-Hahn-Institut), Mainz

(Presented by A. WAPSTRA)

With 1 Figure

1. Improved Precision

Fig. 1 shows in the first place the precision with which the atomic mass of the main isotopes of each element was known in 1961[1]. It should be repeated here that these errors were mainly determined by mass spectroscopic measurements; (measurements of reaction energies, decay energies, or microwave mass ratios mainly determine mass differences of nuclides with nearby mass numbers) and also that it was found necessary to multiply by 2.65 the errors assigned to mass doublets on the basis of the spread in several measurements of a single doublet before they were acceptable in a general least squares adjustment.

New mass spectroscopic measurements have been made in Russia and used by KRAVTSOV[2] to improve the mass list in the mass number regions 85–104 and 111–125; his values are consistent with those discussed above in that he also multiplied the mass spectroscopic errors by 2.65. The presision assigned to his data has been indicated in Fig. 1 by an interrupted line.

Even with respect to these data, a very considerable improvement in the mass number region 70–130 is obtained in the mass spectroscopic measurements reported at this conference by RIES, DAMEROW, and JOHNSON, and by DUCKWORTH, BARBER et al. For the purpose of the present discussion, Fig. 1 also shows the precision of atomic masses following from their data if their errors are multiplied by 2.65.

Of the other improvements, the most important ones are those that promise a considerably better value for the mass of the neutron, or more

Table 1. Mass difference of neutron and H-atom (in keV)

^{14}C (β^-) — ^{14}C (p, n) = $\quad$156.07 ± 0.35 + $\quad$626.55 ± 0.39 = 782.61 ± 0.40
^{3}T (β^-) — ^{3}T (p, n) = $\quad\quad$18.13 ± 0.08 + $\quad$764.48 ± 0.40 = 782.61 ± 0.40
B (D) — (H$_2$ — D) = 2224.71 ± 0.40 — 1442.10 ± 0.10 = 782.61 ± 0.40

^{3}T (β^-) — ^{3}T (p, n) = $\quad\quad$18.61 ± 0.02 + $\quad$763.82 ± 0.08 = 782.43 ± 0.09
^{3}T (β^-) — ^{3}T (p, n) = $\quad\quad$18.61 ± 0.02 + $\quad$764.11 ± 0.15 = 782.72 ± 0.15
B (D) — (H$_2$ — D) = 2224.52 ± 0.20 — 1442.10 ± 0.10 = 782.42 ± 0.22

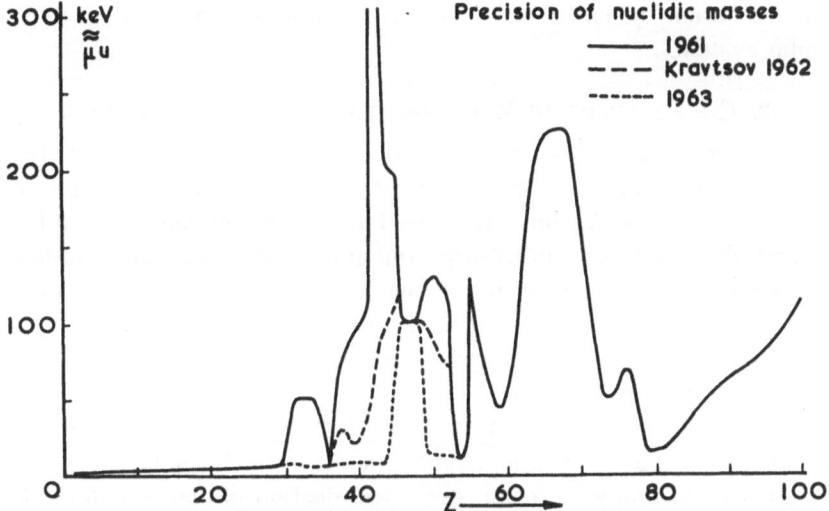

Fig. 1. Precision of atomic masses in our mass table of 1961, in Kravtsov's partial mass table of 1963 (interrupted line) and for a mass list to be made using the most recent experimental data (dotted line)

precisely, the mass difference between the neutron and the hydrogen atom. Table 1 gives the beta decay energies, the (p, n) reaction energies, the mass doublet and the deuteron binding energy that were the most important data influencing the $(^1n - ^1H)$ mass difference in our earlier mass adjustment. Below the dotted line, the new data have been given. The new value[3] for the tritium decay energy deviates considerably from the old value. The last one was mainly determined by one low result 18.0 ± 0.1 keV[4]; without it, the adjusted old value would have been about 18.40 ± 0.15, in agreement with the new result which has been obtained in very careful work. Two new values[5, 6] have been given for the $^3T(p, n)$ reaction energy, in approximate agreement with one another but somewhat lower than the earlier value, which, however, should probably be lowered somewhat as a result of the change in calibration that will be discussed on this conference by MARION. The new value of the deuteron binding energy[7] has been determined by measuring very accurately the energy of the gamma ray in the capture of neutrons by protons. It is seen that these new data promise to give the mass difference under consideration with a precision better than 0.1 keV.

What are the main points left to be desired? This is clear from Fig. 1. The first two humps will now occur for Pd-Ag and Cs-Ba-La. The heavier rare earth isotopes (Sm to Hf) still form the most poorly known region. A few more accurate measurements in the tungsten-to-platinum region would also be very welcome. And some measurements on the more long lived Th, U or Pu isotopes could perhaps be added in order to obtain a better-than-50-keV accuracy throughout the whole periodic system.

2. The Precision in Mass Spectroscopic Determinations

It has been the custom that the errors assigned to mass spectroscopic results were determined from the spread in repeated measurements of the same doublet. This custom can indeed be recommended; yet, it has to be expected[4] that in combining large amounts of such data the consistency will appear to be somewhat worse than would follow from these errors. The most simple remedy is to multiply the given errors by a consistency factor following from a least squares analysis. In our earlier work, this factor was found to be 2.65.

It is by no means evident that this factor would also apply to new data: these will have to be analysed carefully. Such analysis might lead to another consistency factor or, even, to rejection of the new data. Thus, the new doublets by Friedman et al.[8] deviate in the average about three times their errors from the values following from our earlier mass adjustment[1], but unfortunately the deviation is systematic: without exception, all their mass differences are larger. Since in addition the errors assigned to their doublets, certainly after multiplication with a consistency factor, are rather larger than those following from the earlier adjustment, we will be tempted to ignore these data in a future adjustment.

In studying the new data of Duckworth et al. and Ries et al., it is clear that the first group did not follow the above custom: an uniform error has been assigned to most of their doublets. In a sense, this is a pity since checking their errors by consistency considerations is now somewhat more difficult. Most of their doublets are of the type $^{A}Z\ ^{37}Cl - ^{A+2}Z\ ^{35}Cl$ and, therefore, yield the sum of the mass excess difference $(^{A}Z - ^{A+2}Z)$ and that of $(^{35}Cl - ^{37}Cl)$. Thus they can perhaps best be compared with mass differences from reaction energy measurements. Such comparisons seem to indicate a reasonable consistency.

Yet, comparisons can also be made with the recent results of the Minnesota group. Evidently, most sensitive will be comparisons of sums of Hamilton mass differences with the difference in mass of initial and final nuclide as found by Ries et al. The differences between these data are given in the column Δ in Table 2, adopting the value $(^{35}Cl - {}^{37}Cl) =$ $= 2959 \pm 3\,\mu u$ from our 1960 adjustment[1]. The errors given in these data do not include the errors assigned to the Hamilton results. This is done primarily since Table 2 shows clearly a systematic deviation which could most easily be explained by assuming that the above chlorine mass excess difference is large by about twice the error; indeed, the column Δ' obtained using a value $(^{35}Cl - {}^{37}Cl) = 2952\,\mu u$ shows a decidedly better agreement.

Table 2. Differences in Mass excess differences

Nuclides	n	Δ	Δ'
$^{47}Ti-^{49}Ti$	1	$+ 14 \pm 8$	$+ 7$
$^{46}Ti-^{48}Ti$	1	0 ± 6	$- 7$
$^{66}Zn-^{70}Zn$	2	$+ 30 \pm 9$	$+ 16$
$^{90}Zr-^{96}Zr$	3	$- 2 \pm 8$	-23
$^{92}Mo-^{100}Mo$	4	$+ 16 \pm 6$	-12
$^{116}Sn-^{124}Sn$	4	$+ 26 \pm 12$	$- 2$
$^{117}Sn-^{123}Sb$	3	$+ 57 \pm 16$	$+ 36$

Checking the 1960 mass adjustment, we find that a shift of about $7\,\mu u$ in the ^{37}Cl mass should be entirely credible: this mass is there obtained from two rather inconsistent doublets which had to be corrected by $+ 8 \pm 3.2$ and $- 9 \pm 3.5\,\mu u$ respectively. A remeasurement of these masses with a precision of $1\,\mu u$ or better would be a very valuable supplement to the Hamilton work.

Comparison of the value Δ' with the errors suggests that again the real errors are somewhat larger than the assigned ones. The question whether this is so for the Hamilton work, or the Minnesota data, or both is not easily solved. DAMEROW, RIES and JOHNSON tried to do so by measuring overdetermined sets of data and analyzing them statistically. Yet, for most elements, they could do so only by combining normal mass doublets with mass differences over about one or two mass units having considerably larger errors; their analysis demonstrates that the otherwise amazing precision assigned to these differences is nearly correct, but does not say too much about the more important doublets. The only two exceptions are Ge and Sn; in these cases they find $R_e/R_i = 0.71$ and 0.94.

It is not clear from this analysis whether full consistency between the Minnesota measurements and the Hamilton ones should be reached by multiplying the errors in the last ones by a factor of about 4, or those in the first ones by about 2, or both by even slightly less. This situation does not change if we consider the results of the "absolute" doublets for ^{117}Sn and ^{121}Sb, as shown in Table 3. In fact, the agreement is here made better if the mass of ^{35}Cl is taken $4\,\mu u$ smaller and that of ^{37}Cl

3 μu larger, in order to bring the mass excess difference of these two isotopes in agreement with the suggestion above (second line in Table 3). In the comparison it should be remembered that for DUCKWORTH et al. only the errors in their doublets are given; an error of almost 10 μu due to the uncertainly in the Cl-masses should be added. We also could not derive a more positive conclusion by a more detailed comparison between the results of the Minnesota and Hamilton groups.

Table 3. Masses from new mass doublets

	^{117}Sn	^{120}Sn	^{121}Sb
DUCKWORTH...	— 97032 ± 2	— 97799 ± 3	— 96192 ± 3
DUCKWORTH...	— 97044	— 97804	— 96190
DAMEROW.....	— 97059 ± 13	— 97806 ± 11	— 96183 ± 4
KRAVTSOV	— 97060 ± 100	— 97930 ± 90	— 96356 ± 70
1961	— 96940 ± 190	— 97870 ± 140	— 96250 ± 140

Also, comparison with the Russian results does not decide this question; but it does give an insight in the reliability of the errors assigned to the results of DEMIRKHANOV et al. This can already be seen in Table 3: Kravtsov's masses[2] are derived mainly from these results. He already multiplied errors in mass spectroscopical results by 2.65, and yet the differences with the masses as determined by the two groups discussed above are somewhat larger. We have the impression that in general the consistency factor that should be applied to these measurements is between 3 and 4; since the reported errors are already considerably larger than those of the two above groups, this will probably mean that these Russian measurements will carry almost no weight in a future mass adjustment, notwithstanding the undoubtedly very impressive accuracy reached by these fine measurements. Regretfully, one is obliged to say the same of the measurements of OGATA and those of MORELAND and BAINBRIDGE reported at this conference*. On the other hand, these represent the first new measurements of these two groups, and one may hopefully expect that, in the near future, they will turn out very helpful and significant results.

Thus, in mass spectroscopy, there has not been a significant increase in measuring accuracy since our last report, but the new more accurate methods have now been applied to a large number of elements.

3. Reaction and Decay Energies

Some significant advances have been made in determining (p, n) thresholds, as was already clear in the discussion of Table 1. They will

---

* The last remark refered to a preprint; the final errors assigned by MORELAND and BAINBRIDGE make their results significant.

be discussed later in this conference. A very useful technique was also applied to the $^{27}Al(p, \alpha)$ ^{24}Mg reaction, where the resonance energy for protons on aluminium were compared with those of alpha particles on magnesium leading to the same states of the compound nucleus. Unfortunately, this technique has only limited applicability. For the rest, the accuracy in measuring nuclear reaction energies has not been essentially improved, though again a considerable number of important new reaction energies has been measured.

Yet, there is one improvement that has to be discussed here: namely the work on the calibration standards for nuclear reactions. Details about this work will be discussed later in this conference; here, we will just report that the new best value 5304.8 ± 0.5 keV for the ^{217}Po α-particle energy is somewhat higher than the value 5302.8 ± 0.5 derived in our earlier adjustment where this energy had been treated as variable*. The possibility of such a deviation had been anticipated[8]. Systematic deviations may easily have the effect of making reaction energies systematically large; and then an adjustment together with mass doublets in which the calibration energy is treated as a variable would result in a low value for this variable. But exactly in view of this possible explanation the question arises what should be done now. A new experimental investigation would, of course, be in order; yet, we fear that its result will not be available soon enough for the coming mass adjustment. Probably, this systematic error will not seriously disturb the mass adjustment, and this will be tried first. In fact, Dr. RYTZ will present here a table of ground state reaction energies carefully corrected with the new calibration data that we hope to use. But in the unfortunate case that this would generate difficulties we will probably have to ignore the new calibration in treating the reaction energies. Of course, this does not apply to alpha decay energies: these will have to be recalibrated in any case.

4. The Coming Mass Adjustment

We plan to undertake a new mass adjustment in the near future. The above considerations show that newer and more accurate data have been obtained that warrant such an undertaking. We propose to do this based on the following considerations. The final result of the calculation should be firstly a table giving mass excesses (both in keV and in μu), total binding energies and beta decay energies. Secondly, a table will be calculated of mass differences. In our 1960-calculation, the last table was a quite extensive list of reaction energies published by the National Bureau of Standards, which we ourselves found indeed quite convenient;

---

* The changes in the atomic constants reported by COHEN and DuMOND in this conference changes the above values to 5304.6 and 5303.0 keV respectively.

yet, the essential part of these tables were only the errors in the mass differences between any nucleus (Z, A) and the following other nuclei:

$$(A\text{-}4) - (Z\text{-}2); \ (A\text{-}3) - (Z\text{-}2 \text{ or } Z\text{-}1); \ (A\text{-}2) - (Z\text{-}2, Z\text{-}1 \text{ or } Z);$$
$$(A\text{-}1) - (Z\text{-}2 \text{ or } Z\text{-}1 \text{ or } Z) \text{ and } A - (Z\text{-}2 \text{ or } Z\text{-}1)$$

(this is true except for a few nuclides where the precision of the masses is comparable with those of the conventional projectiles). Thus, we consider giving a table with only one reaction energy representing each of the above mass differences (11 items instead of the 63 in 1960), and, of course, a table of the constants to be added to these values in order to derive the other reaction energies.

A third point in setting up the calculation could be the ease of making changes if new data become available. Making such changes is not easy for the socalled "Primary masses" which are involved in an essential way in a least squares adjustment: it would entail changes in other nuclides in a rather complicated way. Secondary masses, however, that are connected to other masses by essentially one link only (be it a reaction or decay energy, a mass doublet or a microwave mass ratio) can easily be changed (though it should still be checked whether no other secondary masses are derived from them, which can be done easily by the diagrams of connections that will accompany the tables). We might, therefore, firstly indicate clearly every secondary nuclide (preferably by a sign indicating to which other nuclide this one is linked). But also we might analyze the system of links very carefully in order to keep the number of primary masses as low as possible without sacrificing accuracy. Thus, we plan to discard links that do not appreciably improve the accuracy of atomic masses or of the derived quantities discussed above. There is one more case in which we will not disregard data. In the low mass region, mass spectroscopic accuracy is so high that with the criteria outlined above many reaction energies will be discarded; this, however, would imply that the necessary comparison between mass doublets and reaction energies would be based on fewer data. Thus, we will not discard such cases. As will be discussed by Mr. Thiele, we will try to program the computer making the adjustment in such a way that it will decide itself which data should be discarded; however, the discarded data will be checked to be sure that no important information gets lost.

References

[1] L. A. König, J. H. E. Mattauch, and A. H. Wapstra, Nuclear Phys. 31, 18 (1962).

[2] V. A. Kravtsov, Nuclear Phys. 41, 330 (1963).

[3] F. T. Porter, Phys. Rev. 115, 450 (1959).

⁴ F. EVERLING, L. A. KÖNIG, J. H. E. MATTAUCH, and A. H. WAPSTRA, Nuclear Phys. 25, 177 (1961).

⁵ P. M. ENDT and P. J. M. SMULDERS, Physica 28, 1093 (1963). — A. RYTZ, H. H. STAUB, H. WINKLER, and F. ZAMBONI, Nuclear Phys. 43, 229 (1963).

⁶ SALGE, private communication.

⁷ J. W. KNOWLES, Can. J. Research 40, 257 (1962).

⁸ L. FRIEDMAN, W. HENKES, and D. CHRISTMAN, Phys. Rev. 115, 166 (1959).

⁹ A. H. WAPSTRA, Nuclear Phys. 18, 587 (1960).

Discussion

Discussion of this paper was postponed until after the next paper.

Considerations about Programming Mass Adjustments

By
W. Thiele
Max-Planck-Institut für Chemie (Otto-Hahn-Institut), Mainz, Germany

With 2 Figures

Summary

Following this conference we will make a new adjustment of atomic masses. Because of the large amount of input data, we worked out now a new program for a IBM 7090 computer, that will handle automatically the division of the whole problem into subadjustments and that will also automatically prepare the data for these parts and then carry out these subadjustments themselves. The program will handle subadjustments with up to 200 masses, whereof up to 140 may be primary.

Introduction

In making mass adjustments, one may divide this work into four parts. The first one of these parts is the measurement of mass spectroscopic doublets and nuclear reaction energies and also of a few fundamental constants. The second one is the collection and ordering of these data. The third one is the arrangement of these data as initial data of the adjustment including the division into subadjustments and the division of the masses into primary and secondary ones. The last part of the work is the properly mathematical part containing the least squares method and some statistical procedures.

The first two parts of this work lie outside of the scope of this report, with the only exception that the result of the data collection must be present in a form that allows data processing. This, in our case, is a file of punched cards. Dr. L. A. König reported about the mathematical division of this work three years ago at the Hamilton Conference[1], and we did not make principal changes to this part of our work. Therefore, it is mainly the data processing part which is reported here.

In our previous adjustments we had to attach to each mass the number of its assigned row in the normal matrix of the respective subadjustment.

Since the inverted matrix should be extended to take into account the secondary masses, a second number has to be attached to each primary mass. Adding or deleting of data may change secondary into primary masses or vice versa, causing much work and the possibility of errors in the input data of the adjustments. Consequently, we had the idea of letting an electronic computer do also the third part of the mass adjustment procedure. We are pleased, therefore, that we now have at our disposal the IBM 7090 computer in the Deutsches Rechenzentrum in Darmstadt.

Our program for the IBM 7090, and a IBM 1401 computer, as peripheral equipment for the IBM 7090 also installed in the Deutsches Rechenzentrum, Darmstadt, is divided in three phases:

1. Loading, decoding and diagnosing the data.

2. Arrangement of the subadjustments (combined with a possibility to include consistency factors).

3. Carrying out the subadjustments and preparing the result ready for publication.

Data Arrangement

As a result of the previous collection, data are held in a file of punched cards, because punched cards are a medium, which allows adding, altering and deleting of data in a very simple way. Every mass spectroscopic doublet or nuclear reaction measurement and every initial mass approximation is given on a separate card. These cards contain only information about one piece of data itself. That means, no information about subadjustment arrangement or consistency factors; so it is possible to use these cards for different adjustments. Each card is divided into several fields, having the same format for all kinds of data. There are seven different kinds of data as described in Table 1.

Table 1. Classification of different kinds of data

1. Atomic Masses.
2. Isobaric Reactions.
3. Nonisobaric Reactions.
4. Doublets with one unknown mass ($^{16}O = 16$).
5. Doublets with one unknown mass ($^{12}C = 12$).
6. Other Doublets ($^{16}O = 16$).
7. Other Doublets ($^{12}C = 12$).

The distinction of the nuclear reactions as isobaric and non-isobaric ones is necessary only for listing properties, while the doublets are divided into two groups, for convenience in writing the computer program. Since there are still measurements published in the ^{16}O-scale, we have to subdivide these two groups into two subgroups each.

The first field of the (punched) data card contains a label of six digits. The first digit thereof is the number in Table 1; the other five digits are

a numerical description of the most important nucleus contained in the
data. $(100\,A + Z)$ is a very useful and significant expression for our
purposes. The second and third fields contain the mass doublet or Q-value
and its error, or the initial mass approximation, while the fourth and
fifth fields contain keys for the laboratory and reference as in our previous
publications[2-5], so that we can give complete lists as output from our
adjustments. The symbol of the atom, reaction or doublet will be punched
into the last field of the card in clear text, however, within the restrictions
by the standard punched card code. Fig. 1 shows an example of such a
data card for the reaction $^{41}K(p, n)^{41}Ca$.

204 119	—1219.	20.	WIS	59-6-89	$^{41}K(p, n)\ ^{41}Ca$
Label	Q	q	Lab.	Reference	Reaction

Fig. 1. Typical data card

Preparing the Input Data for the Adjustment

This paragraph will describe the first phase of the program. Since
it is not economic to load data by punched cards into such a fast machine
as an IBM 7090, we will first transfer the contents of the punched cards
onto a magnetic tape with the IBM 1401 computer. During this loading
process a special routine will decode the text in the last field of those cards
into numerical expressions using our special code $(100\,A + Z)$ for the
atoms. For instance, the doublet text $^{12}C_4 - {}^{32}S^{16}O$ will be converted
to the numerical chain $+4\ 1206\ -1\ 3216\ -1\ 1608$. The IBM 7090-
program will then sort the input data to ascending labels and diagnose
the data for rough errors. Thus with a β-decay the program expects that
the difference between the $(100\,A + Z)$-values of both particles is one.
This diagnosis prevents long computer runs with erroneous data. Of course,
not all possible errors can be detected by this program. Output from
this phase will be, of course, input data for the next phase and a list of
all input data with assignment of a running number to each piece of data,
that enables special modifications in the second program phase.

Building-up the Adjustments

We have to introduce consistency factors as explained in our earlier
papers[2]. There are three possibilities of weighting:

1. Weighting whole data classes, as shown in Table 1.

2. Weighting all the data of one laboratory, in accordance with the
laboratory field in the punched cards.

3. Weighting single data designed by their running number listed
in the previous phase of the program.

Other input to this phase is provided in order to influence the automatic set-up of the subadjustments, if needed. The program will now set up a mass table from the initial values of the atomic masses. It is possible to join subadjustments, or to force nuclides to be primary, by duty cards which are input to the program. This feature is added to the program especially for application in the subadjustment for the lightest nuclei; for it is necessary to make this subadjustment large enough, in order to take care of all important correlations. E. g., ^{1}H, ^{12}C and ^{16}O should not be calculated by different subadjustments. After arrangement of the mass table the program scans the data to detect the connections between the masses. Nuclear reactions will connect target and product nuclei.

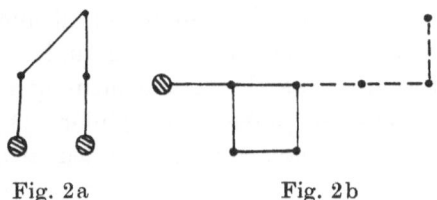

Fig. 2 a   Fig. 2 b

Connections between primary nuclides

◍ Nuclei determined by mass spectroscopic doublets. ● Nuclei not directly deter-mined by doublet measurements. ——— Connecting nuclear (primary) reactions. – – – Connecting nuclear (secondary) reactions

The projectiles are not used for this purpose, though, of course, they will be used in forming the normal equations, if they are variables in this subadjustment. Doublets with one unknown mass will only set a flag to the mass determined. This flag will be used later to distinguish between primary and secondary masses. Other doublets work similar to nuclear reactions. When, in this way, the network of nuclides and data is arranged, it will be split into the subadjustments, if it is not already split by the absence of correlations in some regions. Additional separations have to be made at connections, that are determined with relatively small precision. Therefore, a weight $p = 1/q^2$ is assigned to each connection, formed by one or more nuclear reactions or doublets. These weights will be summed for each nuclide. If one of the connections is too small compared with this sum, it will be called weak. If the connection is weak with respect to both nuclides it connects, it will be deleted, if it will not be found to be a secondary one. If wanted, such deletions can be suppressed by duty cards. Now the program can assign each nuclide and each input data to at most one subadjustment. The next step is distinguishing nuclides and data as primary and secondary ones. Therefore, the program will step through the network, starting from masses, which are determined by a doublet (flags) or now known as primaries (duty cards). There are two conditions consisting in a "logical—or fashion" for a connection in a chain of connections to be primary (Figs. 2a and 2b).

Firstly, if the program stepping through the network finds another nucleus which is determined by a doublet or primary itself, all nuclides from the last primary or doublet-determined mass up to and including this point are primary nuclides. On the other hand, if the program finds a blind alley in the network, all nuclides following the last branching point up to the end point are secondary ones. The second condition for masses to be primary is that they are arranged in a cycle. If the program finds that such a cycle has been closed, all tabulated nuclides of this cycle are primary ones. All nuclides found to be primary are indicated by a primary flag.

This phase can now be closed by omitting all nuclides whose masses are not determined and then preparing the input for the next phase and the output of a list that shows the arrangement of the subadjustments. If this arrangement does not agree with opinions of the physicist who supervises this work, this phase will be repeated after changing only a few duty cards. Therefore, we break up the program at this point.

The Adjustments themselves

The third phase of the program carries out the adjustments themselves and is thus the main part of the program. As said earlier, we used in general the same procedure as in our previous adjustments[1]. Only the organization of the program and the solution of the system of normal equations are changed. In the latter case, we added the well-known method of using pivotal elements[6]. This became necessary because of the extension of the program to matrices of degrees up to 140. (Unfortunately, we could not yet test, whether our program will handle such large matrices successfully; a problem, that is also influenced by the condition of the matrices to be inverted. But our previous experiences with the condition of normal matrices for mass adjustments were very good.) Since handling such matrices occupies a large part of the core storage, it is impossible to hold the whole program of this phase and all data vectors in core storage simultaneously. Therefore, additional magnetic tape operations become necessary. We could minimize these operations by constructing the program as a program chain containing 6 chain links. Each link will be brought into core storage only one time and will then handle all subadjustments, one following another. The functions of these various chain links are the following:

1. Preparing data for the construction of the system of normal equations as reducing mass spectroscopic doublets to the ^{12}C scale or treating those nuclides with masses that are taken as constants with respect to the specific subadjustment.

2. Setting up the system of normal equations.
3. Solving the normal equations.
4. Preparing primary results and their combinations.
5. Extending the error matrices by secondary measurements.
6. Preparing output of the result.

As additional output a magnetic tape will be produced, containing the new masses and all of the extended matrices (5) of all subadjustments as input for another program, not yet working, that will compute tables of nuclear reactions as WAPSTRA[4] published for masses with $A \geq 200$. The output from point 6 of the table above will be handled with a special routine on the peripheral IBM 1401 computer to produce lists suitable for photographic reproduction. This will avoid errors and loss of time by transcribing these lists.

Acknowledgments

The methods for carrying out the subadjustments were developed by Dr. L. A. KÖNIG[1]. We are very grateful to the Deutsches Rechenzentrum, that enabled us to make use of their IBM 7090 and IBM 1401 computers, and to Mr. G. MIEDEL who prepared the system tapes for the IBM 7090 for our purpose.

References

[1] L. A. KÖNIG, Mathematical Details of the Mass Computation. Proc. Int. Conf. Nuclidic Mases, p. 39. Toronto: University of Toronto Press. 1960.

[2] F. EVERLING, L. A. KÖNIG, J. H. E. MATTAUCH, and A. H. WAPSTRA, Adjustment of Relative Nuclidic Masses (I) $A \leq 70$. Nuclear Phys. 25, 177 (1961).

[3] L. A. KÖNIG, J. H. E. MATTAUCH, and A. H. WAPSTRA, Adjustment of Relative Nuclidic Masses (II) $70 < A < 200$. Nuclear Phys. 28, 1 (1961).

[4] A. H. WAPSTRA, Adjustment of Relative Nuclidic Masses (III) $A \geq 200$. Nuclear Phys. 28, 29 (1961).

[5] L. A. KÖNIG, J. H. E. MATTAUCH, and A. H. WAPSTRA, New Relative Nuclidic Masses. Nuclear Phys. 31, 1 (1962).

[6] A. RALSTON and H. S. WILF, Mathematical Methods for Digital Computers. New York-London: J. Wiley & Sons, 1960.

Discussion

K. T. BAINBRIDGE: I have two remarks regarding Dr. Wapstra's paper. First, I believe that his statement about the errors of mass spectroscopic results was too general. His statement was that the error of a doublet measurement was the standard error of the average of all individual measurements for the doublet. This is correct for the results of SMITH, and FRIEDMAN, and SMITH. If I have read correctly, the Minnesota group takes 20 mesurements for one datum. Then many such data are obtained for one doublet. Then the statistical error, used as part of the final published error, is the standard error of the mean of these data which corresponds to Birge's R_{ext}. While Smith's statistical error corresponds to Birge's R_{int}. There is a difference and one statement could not apply to both.

A. H. Wapstra: The point made was that errors based on consistency in repeating measurements of one doublet will inevitably turn out to be smaller than those based on consistency of a set related doublets. This is even true if the measurements to determine the first kind of errors are extended over a longer time etc., though it may be expected of course that the differences between the two kinds of errors then become less pronounced.

K. T. Bainbridge: Some criticism was made of the results of Moreland and Bainbridge and of Ogata. Moreland and I submitted results on the doublets H^3—He^3, HD—He^3, HD—H^3 and H_2—D. If a statement is made that these results don't agree with the 1961 Mass Table, that is fine, but I believe further statements as were made, touching on quality, are not warrented at this time.

A. H. Wapstra: The results of Moreland and Bainbridge were not criticized at all, certainly not because of disagreement with earlier results. The agreement with the 1961 Mass Table is quite reasonable. A remark was made about their reported precision, based on an earlier preprint. This precision was less than that reported at this conference.

H. Staub: The neutron-proton mass difference is substantially lowered by the new values of the T (p, n) threshold in conjunction with a weighted average of the tritium β-decay energy. This latter value is low compared to the recently determined value by Porter mainly on account of Popov's value (18.00 keV) which however, has a small quoted error. Is it correct to discard this value, which admittedly disagrees with all others, but is stated to be quite accurate and moreover obtained by a different method?

A. H. Wapstra: This question is discussed in our paper but was by accident omitted from the talk. In addition, I hope soon to make a new measurement in our Amsterdam Laboratory. Also, Moreland and Bainbridge's doublet give a value 18.47 ± 0.26 keV agreeing better with Porter's value than with the low one.

N. Zeldes: I would like to comment on your multiplying all mass doublets errors by 2.65. It seems to me unjustified for the following reason: How did you obtain this factor? You made a least squares adjustment of Minnesota's doublets for the light nuclei, given with quoted experimental errors of a few keV, and you found a consistency factor of 2.65. This means that besides their statistical errors these doublets have systematic errors of an order of magnitude of about twice or three times the statistical quoted errors. By multiplying the quoted errors by this factor you somehow randomize these errors and are able to combine the mass doublets with other data in one big mass adjustment. But then you take values of Demirkhanov and co-workers, or Minnesota values measured with their old instruments, with quoted statistical errors of about 100 keV, and multiply them by the same factor. This means that you suspect here systematic errors of about 100–200 keV, because in the light elements and with the new instrument there were systematic errors of a few keV. I don't think it's fair. It seems to me more justified to determine separately consistency factor for each group of doublets measured under the same conditions, and use it for that group. If you have a datum which you suspect of a systematic error it is better to throw it away altogether, rather than use it with a reduced weight.

A. H. Wapstra: Indeed I would prefer to determine separately the "consistency factor" for each group of data. Unfortunately, this cannot be done in a very convincing way since in most of these groups the number of interrelations is small. To see whether a similar factor should be applied to less precise measurements, several sets of such data with as many interrelation as possible were studied in Mainz. Consistency factors between 2 and 3 were always found here also. Thirdly, I am of the opinion that a consistency factor of this order of magnitude sounds very plausible, considering the method of assigning the original errors. Lastly, I consider it unjustifiable to throw out single data solely on the basis of bad consistency. If in a group of data, determined in the same way, a few disagree badly, this throws a doubt on the whole group. In many cases where better measurements later became available, use of this consistency factor has almost always been found to be fully justified.

N. ZELDES: I would like to ask Professor MATTAUCH whether the experiments made in Mainz with consistency factors of mass doublets just mentioned by Professor WAPSTRA, were made with light nuclei only, or also with heavy nuclei, whose quoted statistical errors were much larger.

J. MATTAUCH: The doublets measured by HEBEDA and by EL KHOLI in Mainz all lay in the region $A = 3$ and $A = 40$. It seemed most important to us to make the over-determination as large as possible. Always the consistency factor turned out to be between 2 and 3. Of course this says something only about our specific instrument.

K. WAY: When will values from the present mass adjustment be ready and what provision is being made to keep it up to date?

A. H. WAPSTRA: We hope that the new values will be available in December 1963. In order to allow corrections when new data become available, "secondary" data (those which can be corrected easily) will be clearly indicated. We hope to repeat the adjustment every two years or so.

J. W. DEWDNEY: Will it be possible to provide, in the new mass table, mass differences for nuclides for which absolute masses are not possible because of a missing link?

A. H. WAPSTRA: There are a few of such mass differences, some isolated β-decay energies and a set of α-decay energies mainly in the light rare earth region (around magic number $N = 82$). If desired, they could be included.

Since Prof. BREITENBERGER is going to make comments on the procedures used in setting up atomic mass formulae, we may state the purpose of our formula and our philosophy in adjusting it to the data. Our main purpose is to have a means for extrapolating mass differences (mainly neutron and proton binding energies and alpha and beta decay energies) to as yet unknown cases. Obtaining accurate extrapolated masses is an entirely secondary purpose.

In order to obtain these extrapolated values with as high a degree of confidence as possible, the formula to be used has to have something to do with reality. To be more specific, we want to have it obey three requirements:

a) The Coulomb term should agree with measurements (e. g. those made in Stanford) on the charge distribution in nuclei.

b) The deformation energy should agree with deformations as obtained from quadrupole moments, excitation energies of collective excited nuclear levels, etc.

c) The remaining terms should be (nearly) symmetric in the numbers of neutrons and protons in the nucleus, in agreement with the observed charge symmetry of the nuclear forces.

Then, we have to realize that in the present state of knowledge no atomic mass formula is exact. We can at best require it to represent the atomic mass data with a certain precision, σ. But this means that before making a least squares adjustment of any formula to the experimental data, one should add (quadratically) σ to these initial data. We think that in principle σ should be determined in such a way that the value χ^2 obtained in the adjustment equals the number of degrees of freedom. This is somewhat difficult to program for a computer; therefore we start by adopting a value of 100 keV for σ with the intention to increase this in our final adjustment if found necessary on the basis of the above criterion. Then since we want to adjust mainly to mass differences and since we realized that the atomic masses are correlated, we planned to use as input data for the adjustments not the nuclidic masses from any least squares adjustments but the reaction and decay energies themselves and atomic masses derived from mass doublets. We still consider doing this in our final adjustment. For the many preliminary adjustments needed to find a reasonable mass formula, however, we used as input values only adjusted values for masses, reasoning that after adding 100 keV or more to their errors their correlations are probably negligible.

Some Statistical Problems in the Computation of Nuclidic Mass Formulae

By

Ernst Breitenberger*

University of South Carolina

With 1 Figure

Introduction

The volume of the literature on nuclidic mass formulae bears witness to much vexation. There is no agreement on the terms to be used. Simple drop-model terms account for the major portion of each mass (v. WEIZSÄCKER 1935) and thus are a natural zeroth approximation. But how may we improve them into a first or second approximation while we are not quite certain about the structure of the nuclear surface ? There are also strong shell model terms (discovered experimentally by DUCKWORTH and collaborators, 1950/51), but again we do not yet know a universally acceptable procedure to refine these in higher orders. Anyway, how do we reconcile the collective and the individual particle effects while basic nuclear theory is still unsettled ?

Beside these theoretical worries there are technical ones, too. Whatever the terms one is interested in, they must be extracted from a great mass of data and be proven to exist beyond reasonable doubt. This is the first rather than the last problem in computing a mass formula. It has unique ramifications; pedants would speak of puzzles of "scientific method". Much of what has been said or written about it was not based upon rational analysis and can be discarded. Most authors have mechanically employed procedures given in texts on "least squares" dating from the Twenties or earlier, and thus obtained inadequate results. Indeed, nearly all the figures published so far can be legitimately called into question.

In this paper I shall outline the more important points to be watched. My emphasis will not be on decimal places but on logical consistency of

---

* Now at Ohio University, Athens, Ohio. This paper was sponsored by the National Science Foundation.

procedure. In this problem, as elsewhere, a good deal of mental discipline is needed in addition to a large computer.

The Data

Roughly one thousand nuclidic masses are known. They are obtained from a great number of experimental data in an adjustment by means of least sqares. The adjusted values conform to t-distributions, but the number of degrees of freedom is so large that we can regard the distributions as truly normal, or gaussian.

The standard errors attached to the masses are a bit uncertain, mainly because the assignment of precise weights to the input data is no simple affair. The errors given in the mass tables of MATTAUCH and collaborators may therefore be low, perhaps by 10% in some regions. This is a small price to be paid for the success of a vast undertaking and need not be bewailed, but it should not be forgotten altogether.

The adjusted masses, in contrast to the input data, are not independent. The input data (nuclear reaction and disintegration energies, and mass spectrometric doublets) consist essentially of mass *differences*; hence the masses themselves are correlated, and essentially in the positive sense. The full covariance matrix of one older adjustment of 70 masses has been published[1]. The covariances are not in general negligible even over large distances in the N, Z-plane, and over small distances they are of the same order as the variances of the masses.

Data generated by the successive addition of independent differences which are subject to error are often called "cumulative data" in statistical literature. Their behaviour shows unbelievable features which were not fully appreciated until recent times, are nowhere mentioned in the older literature, and are still not familiar to the run-of-the-mill physicist. One among many is the "zero-crossing problem" of operations research and the theory of stochastic processes; for a simple instance which is not totally irrelevant to the present problem, see FELLER[2].

Some years ago, many mass values were afflicted with severe bias, either singly or in batches. The mass surface was quite disfigured by warts and wrinkles. The underlying systematic errors of measurement have gradually been identified and eliminated. Simultaneously, many new workers entered the field with new techniques and instruments. Whereas they also brought new sources of bias with them (in the form of "departmental constants") the chances are that the new systematic errors are now much smaller and in the final adjustment cancel each other to a considerable extent because of their larger number; thus they are likely to contribute only a random, and therefore harmless, component to the standard errors of the adjusted masses.

Owing to the recent affluence of data the mass table of MATTAUCH *et coll.* had to be updated by adjustments in separate batches. So the mass values are in a state of flux at this moment, and no complete covariance matrix exists. However, the continued hunt for bias and the desire for overall consistency both call for a fresh, uniform adjustment of all masses; I take it for granted that this will be made in the near future.

The Aims

An hour or so of graphical analysis shows that the mass values lie close to a somewhat corrugated surface over the N, Z-plane, with fairly irregular deviations averaging perhaps 300 keV (a little less at high Z, and more at low Z). This figure may be deceptive, just as the impression of randomness, but it is a useful yardstick.

We want to find some analytical expression for the mass surface. There is no precedent anywhere for such a "regression analysis" on correlated data with a given covariance matrix. Our problem is truly unique. It is all the more important to study the required methods *before* embarking on the computations which are unavoidably extensive and call for rational planning. The purpose of the mass formula should be decided upon first of all, and adhered to mercilessly afterwards.

One may wish to use the formula to predict unknown masses. Regression formulae are usually good for interpolation but always bad for extrapolation. Anti-aircraft fire control is a familiar example. The standard error associated with an extrapolated value is very large indeed; this fact is poorly discussed in most books (where prime emphasis is usually given to the rather different problem of assigning confidence limits to the regression curve itself) but at least one modern work[3] contains tables and graphs. Interpolation, on the other hand, can be done quickly by graphical and arithmetical smoothing methods (e. g. [3-5]). On the whole, prediction alone appears not to be a worthwhile purpose for a mass formula.

Next, one may wish to use a mass surface to investigate the apparently random straggling of the mass values and to spot suspiciously large deviations. Owing to the large number of data, graphical or other simplified methods can hardly be advocated for this purpose; proper regression analysis seems indispensable. The type and number of terms to be included in the formula is not important in principle. Decisive is that one should achieve a high "goodness of fit" in the *special* sense that the residual variance of the data about the regression surface should correspond closely to the suspected average deviation of 300 keV or so. Please note that this criterion remains unaltered even if all masses are known with errors of only a few keV at some future date. We are *not* here asking for goodness of fit in the *usual* sense, i. e. for a close approximation to within the errors of measurement; we are concerned with the *natural* variance of the masses.

Thirdly, we may want to verify a theoretical formula, or to test different theories against each other, or to improve a theoretical formula by finding further terms in an empirical fashion. There is no end to the possibilities. The criterion for acceptance or rejection of terms, singly or in groups, should be a high level of significance, i. e. a small probability that the terms under test are simulated by pure chance (in the form of natural and experimental variance of the mass values).

Goodness of fit and significance cannot be altogether divorced from each other, but it is clear that one must be emphasized over the other according to the chosen purpose of the mass formula. How many terms are required in any case is quite another matter. Several hundred terms have been employed by some authors, but the formula of Kümmel *et al.* (as described at this conference and other meetings) demonstrates that a close representation of all masses above $A = 40$ is possible with no more than three or four dozen terms.

Maximum Likelihood

The parameters in a chosen formula will be determined such that the joint probability of the residuals v_j (experimental value minus predicted value of mass no. j) is at a maximum. If the mass values were independent the joint distribution of the v_j would be a gaussian with $-1/2$ times

$$\sum v_j^2/\sigma_j^2$$

in the exponent. It would then be sufficient to find the minimum of this sum ("least squares"). However, the joint distribution is the general multi-dimensional gaussian for *correlated* variables which contains instead of the simple sum of squares the quadratic form

$$\frac{1}{\Lambda} \sum \sum \Lambda_{jk} v_j v_k,$$

where $\Lambda = \det(\lambda_{jk})$, the Λ_{jk} are the minors of Λ, and the λ_{jk} are the elements of the covariance matrix of the masses; see any advanced text (e. g. [6]). It is this form which must be minimized.

When the resulting "normal equations" for the most probable parameter values are written out in matrix notation it is found that they have the weight matrix

$$W = \text{matr}\,(\Lambda_{jk}/\Lambda),$$

i. e. the inverse of the covariance matrix (e. g. [3, 7]). In the uncorrelated case W becomes diagonal (elements $\Lambda_{jj}/\Lambda = 1/\sigma_j^2$, others $= 0$). This occurrence of off-diagonal weights is the only difference from the familiar "least-squares" normal equations. It is quite important, of course, because it greatly complicates the algebra. Even so, two basic results of ordinary least-squares theory obviously remain true when the data are correlated: the values of the regression parameters have t-distributions, and the minimum value of the above quadratic form has a χ^2-distribution (of the appropriate number of degrees of freedom, or d. f.).

Testing

The χ^2 is not very important in itself because we are hardly interested in the goodness of fit which it can serve to test. However, it can also be used to assess the significance of terms by means of the variance-ratio, or F-test. I must apologize for talking ponderously about most elementary matters, but I cannot help deploring the fact that this simple tool has not been utilized in any of the work on mass formulae.

Suppose we fit 800 masses by a formula containing 50 terms and obtain a χ^2 of 2000 (which then has 750 d. f.). Suppose that by adding five more terms to the formula the χ^2 is reduced to 1900 (and 745 d. f.). Now split $2000 = 1900 + 100$; from the definition of χ^2 it follows that the part 100 conforms to a χ^2-distribution of 5 d. f. Then compute the "variance ratio"

$$\frac{100/5}{1900/745};$$

this conforms to the F-distribution which has been amply tabulated so that the significance of the group of 5 terms can be directly read off some handbook.

The formation of a quotient has one appreciable advantage: an unknown factor affecting χ^2 drops out. Since the errors of the mass values are presumably a little too small, the weights are too large, and so is χ^2. This bias counts in the χ^2-test but not in the F-test. Still, there is no guarantee that a group of terms under test, like the five of the example, does not depend principally on a particular batch of mass values whose errors are subject to more than average uncertainty. Hence we must anticipate that the numerator in the variance ratio fluctuates more than the theory of the F-distribution accounts for. A mathematically inexact but practically adequate remedy is to use the standard F-tables at a higher level of significance; e. g. to set one's goal at 2% rather than the hallowed 5%.

Testing various terms against each other, singly and in groups, is a most laborious business which has to be faced with patience. Solving the normal equations at each step can be somewhat simplified by short cuts (e. g. [8]); similar methods are available for the expulsion of one or more suspicious mass values (e. g. [9]).

Testing is made even more cumbersome by the obvious (but frequently belittled) fact that terms of rather different functional form can easily mask or imitate each other. A recent instance occurred in the determination of the earth's gravitational potential from satellite observations which has now been carried to terms of the twelfth order in the zonal harmonics expansion[10]. When the tenth-order term was included in the regression, it came out highly significant but simultaneously the fourth-order term dropped by one-half its value. Thus here an $r^{-4} P_4 (\cos \theta)$ mimicks an $r^{-10} P_{10} (\cos \theta)$. Clearly such ghosts should be laid. Conversely, one must never omit from a regression any terms which are known to be there, even if they are thought to be insignificant; their suppression falsifies the significance of other terms. Thus it is wrong to disregard the atomic binding energy which goes roughly as $Z^{5/2}$ (e. g. [11]) and amounts to more than 800 keV at $Z = 100$.

An arithmetical complication arises from the nature of the errors of the mass values. They are generally small at the bottom of the mass valley but large on its side slopes. Consequently those terms which represent mainly the slopes have coefficients subject to much variance. The normal equations can then be satisfied approximately by a fairly wide range of values for these critical coefficients. In geometrical language, some of the hyperplanes represented by the normal equations intersect at very small angles. Of course the point of intersection is difficult to compute for it will be affected by every rounding-off error. There are many methods for dealing with such "ill-conditioned" equations (e. g. [12]); two less well-known ones[13] are perhaps readily adaptable to the present problem. Of course such extra tasks put an even higher burden on one's computer. Ultimately they must be faced. In preliminary studies they could be avoided by the simple expedient of adding 300 keV or so (quadratically) to all errors while leaving the covariances unaltered. This means that we promote the natural variance of the masses to the status of experimental error; the disparities amongst the errors then disappear, the normal equations are well-conditioned, and we can even apply the standard χ^2-test of goodness of fit if desired. For all I know, the natural variance may not be normal, and the yardstick of 300 keV may be false, but if that should be so it would readily (and profitably) become apparent. A few of the awkward questions of the analysis of heterogeneous variances which then arise have been tackled in the literature (e. g. [14]).

The Correlations

Few theoretical papers on the regression analysis of correlated data have been written. A couple of matters of basic principle have been settled conclusively (e. g. [7, 15, 16]). There is also some work on problems arising in biostatistics, but it seems so far removed from our problem in scope that it may be disregarded. We are well-nigh left to our own devices. In particular, there exists no precedent to guide us in the more plebeian details of the analysis.

Instead of trying to elucidate the effects of correlation by epsilontics, I propose to discuss a transparent model case which has been fully studied[16] and convincingly exemplified[16, 17]. Referring to Fig. 1, suppose we have two variables connected by a (theoretical, exact) relationship $y = c\,x$. Now let us measure y at equidistant abcissae x_j, but not in the usual fashion, by independent determinations of the y_j themselves; let us rather make independent measurements of the differences $y_j - y_{j-1}$ with a

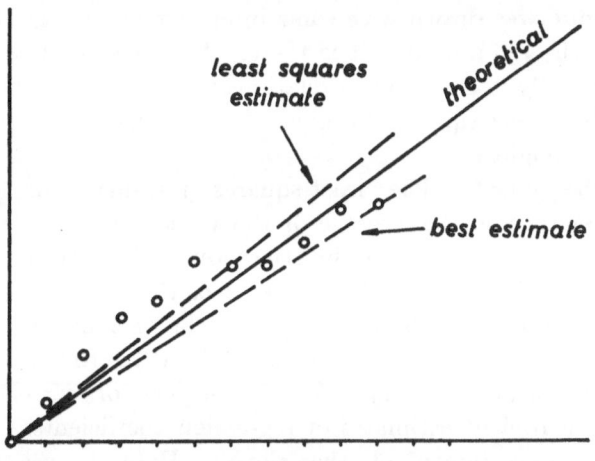

Fig. 1. Cumulative data

fixed standard error, and then build up the y_j by successive addition. The lowest point y_0 lies at the origin by definition. The first point y_1 follows from the first measured difference; let us assume that it comes to lie above the true straight line. The second point y_2 is obtained from y_1 by adding the new independent datum $y_2 - y_1$; it clearly has a better than even chance to lie again above the line, and an even chance to lie still farther from the line than y_1. If y_2 lies high, the next point y_3 may well lie still higher; and so on. After a while, chance will of course tend to correct its own excesses and bring the points back to the line, but then it may well produce an excess in the downward direction.

Thus the points do not scatter around the true line as independent data would; they queue up behind each other. The phenomenon is easily reproduced by means of a sheet of ruled paper and a pair of dice. I have no better advice to unbelievers than to try this; they will soon become convinced that the behaviour of cumulative data defies all intuition. See also [2] and ([16], Table II). We can generally state as *Rule 1:* positive correlations often simulate spurious trends.

How does one estimate the slope c of the underlying relationship if a sample of points as in Fig. 1 is given? The answer[16] is surprisingly simple: the best straight line passes through the last point of the sample, irrespective of the intermediate points. However, the formula for the estimation of the variance of this estimate of c involves all points, as one should expect. Needless to say, the last point can lie rather far off the true line because it may have been caught by one of the mentioned trends; hence the variance of the estimated slope is quite large.

When a physicist sees a set of points like that in Fig. 1 he immediately smells a rat. However, not every set of cumulative data looks so suspicious; indeed the figure was drawn with some intent to emphasize. So we might not recognize the rat although it is there. What will be the results if we overlook or wilfully neglect the correlations and treat our sample of points by the familiar least-squares formulae for independent, equally spaced data of equal weight?

Firstly the slope[16]. The least-squares formula still provides an unbiased estimate, i. e. the average of the values obtained from separate samples of points tends towards the true value. This estimate is however a little inefficient in the technical sense, i. e. the variance of the values from separate samples is a little larger than the variance of the estimates which would be furnished by the correct ("best") formula. Both results obviously have wider validity. *Rule 2:* neglect of correlations yields unbiased but inefficient estimates of regression coefficients.

Next the uncertainty of the slope[16]. Here the use of the least-squares formula for the standard error yields a result which is too small by a factor $2\sqrt{n}$, where n is the number of sample points. The figure is quite astonishing. For example, if we have a sample of 100 points we underestimate the error of the slope by a factor of 20 (repeat: twenty). Clearly such bias is intolerable. It arises, of course, because least-squares theory takes each point at its face value, instead of allowing for the possibility that it is harnessed into a spurious trend. Hence the bias is of general occurrence. *Rule 3:* neglect of positive correlations yields gross underestimates for the errors of regression coefficients.

Thirdly the χ^2. The exact behaviour of the residuals is complicated[16]. Still, since many samples of cumulative data contain spurious trends and therefore exhibit spuriously large residuals, the sum of the squares of the residuals must on the average be larger than the value of the more complicated quadratic form which behaves as a χ^2 in the correct theory. Thus the usual least-squares expression again fails to estimate the desired figure without bias. *Rule 4:* neglect of positive correlations yields an overestimate for χ^2.

What has this model to do with nuclidic masses? They present a two-dimensional problem. This does not seem to make much difference;

instead of trends we may find spurious bulges and dimples, that is all. However, the data are not simply obtained from measurements of mass differences between nearest neighbours. The doublets of mass spectrometry effectively establish differences over wide distances in the N, Z-plane. Clearly the partners in a doublet can no more be displaced relative to each other by some spurious pile-up of errors in intervening nearest-neighbour mass differences; a mass linked by doublet to the primary standard ^{12}C, in particular, is thereby firmly anchored to the true mass surface. Thus the doublets are so many fences for the bulges which could otherwise develop freely. In this regard the model gives us distinctly unfavourable impressions. The mass values cannot behave all that badly.

The effects of the correlations cannot be neglected in principle, though. Where are they most important? They are obviously diluted so as to become almost negligible in regions where the experimental variance of the mass values is much smaller than the natural variance. Everywhere else they count. In particular, the areas comprising the deformed nuclei and the heavy α-emitters are critical; there the experimental errors are rather large.

All mass formulae published so far were computed neglecting the correlations. The verdict is clear. The parameter values are somewhat inaccurate because of inefficiency. The errors are much too small and therefore useless. The smaller terms representing deformation energies and shell model effects in regions with relatively large mass errors are doubtful, to say the least, because they may represent spurious correlation effects. On the whole, none of these formulae embodies clear-cut, safe results such as would be required to help the nuclear theorist.

Owing to the complexity of both the data and the possible mass formulae, it is very hard to assess just how much bias the neglect of the correlations creates. I have no more than a guess which I formed after a study of the 1961 mass table (as completed by some plausible assumptions about the covariances). I believe that for the errors of the coefficients biassing factors of 2 are plausible, of 3 or more not impossible; a factor 2 is plausible for χ^2 too. This is not meant as a statement *ex cathedra*, and even if now it should be right it may be wrong tomorrow in the face of improved data.

The guess is nonetheless helpful. After all, we possess at this time no complete covariance matrix and thus are compelled to work with a faulty method. Someone may also wish at a future date to carry out preliminary analyses with a restricted aim, neglecting the correlations for simplicity. What should one do under such circumstances? The above figures suggest that errors for the coefficients need not be computed at all; they are too nearly meaningless. It is also clear that one should not go in search of very small terms; they could all too easily be spurious.

100   E. Breitenberger:

Significance is once more the leading issue. It can be tested by means of the variance ratio despite the bias in χ^2, as shown above. However, that bias is so severe that the numerator in the variance ratio now seems highly suspect; indeed the terms under test might merely account for a spurious bulge in the mass values and therefore embezzle a strong portion for the χ^2 in the numerator. Thus one should use the F-tables at unusually high levels of significance. I would be inclined to operate at better than 1 %. Tables for these levels are not common in textbooks but can be found in the literature[18].

Ad personam

I conclude by emphasizing that I am not a statistician. Some of these arguments may prove to be too amateurish; they should then be supplanted by competent professional advice. A fleet quarry calls for accurate guns.

References

[1] J. Mattauch, L. Waldmann, R. Bieri, and F. Everling, Die Massen der leichten Nuklide. Z. Naturf. 11 a, 525–548 (1956). The Masses of Light Nuclides. Ann. Rev. Nucl. Sci. 6, 179–214 (1956). In both references the factor 10^1 on top of the columns from ^{17}F to ^{21}Ne should read 10^{-1}.

[2] W. Feller, An Introduction to Probability Theory and its Applications, 2nd. ed., especially chapter 3. New York: Wiley. 1957.

[3] P. G. Guest, Numerical Methods of Curve Fitting. Cambridge University Press. 1961.

[4] M. H. Quenouille, Associated Measurements. London: Butterworth. 1952.

[5] T. N. E. Greville, On Smoothing a Finite Table: A Matrix Approach. J. Soc. Industr. Appl. Math. 5, 137–154 (1957).

[6] H. Cramér, Mathematical Methods of Statistics, chapters 22 and 24. Princeton University Press. 1946.

[7] A. C. Aitken, On Fitting Polynomials to Weighted Data by Least Squares. Proc. Roy. Soc. Edinb. 54, 1–11 (1933). On Fitting Polynomials to Data with Weighted and Correlated Errors. Ibid. 54, 12–16 (1933).

[8] W. G. Cochran, The Omission or Addition of an Independent Variate in Multiple Linear Regression. Suppl. J. Roy. Stat. Soc. 5, 171–176 (1938).

[9] R. L. Plackett, Some Theorems in Least Squares. Biometrika 37, 149–157 (1950).

[10] D. G. King-Hele, G. E. Cook, and Janice M. Rees, Earth's Gravitational Potential: Even Zonal Harmonics from the 2nd to the 12th. Nature 197, 785 (1963).

[11] L. L. Foldy, A Note on Atomic Binding Energies. Phys. Rev. 83, 397–399 (1951).

[12] A. D. Booth, Numerical Methods. London: Butterworth. 1955.

[13] P. B. Madić, A Method of Solving Ill-Conditioned Systems of Linear Simultaneous Algebraic Equations. Bull. Inst. Nucl. Sci. "Boris Kidrič" 6, 75–86 (1956). Transformation of Ill-Conditioned Systems into Well-Conditioned. Ibid. 6, 87–91 (1956).

[14] A. S. C. Ehrenberg, The Unbiased Estimation of Heterogeneous Error Variances. Biometrika 37, 347–357 (1950). — E. J. Williams, Applications

of Component Analysis to the Study of Properties of Timber. Austral. J. Appl. Sci. 3, 101–118 (1952). — J. P. SUTCLIFFE, Error of Measurement and the Sensitivity of a Test of Significance. Psychometrika 23, 9–17 (1958).

[15] G. S. WATSON and E. J. HANNAN, Serial Correlation in Regression Analysis: II. Biometrika 43, 436–448 (1956). And further references given there.

[16] J. MANDEL, Fitting a Straight Line to Certain Types of Cumulative Data. J. Amer. Stat. Assoc. 52, 552–566 (1957).

[17] F. S. ACTON, Analysis of Straight-Line Data, chapter 10. New York: Wiley. 1959.

[18] M. MERRINGTON and C. M. THOMPSON, Tables of Percentage Points of the Inverted Beta (F) Distribution. Biometrika 33, 73–88 (1943).

Discussion

J. MATTAUCH: I think the heavy criticism which we have just heard calls for at least some remark or excuse. Let me make clear what I mean by a little comparison. At the beginning of the "Fifties" there was a mass adjustment made by a group in the Kellogg Radiation Lab, and we had made one already by the method of least squares. When I pointed out to TOMMY LAURITSEN that his mass adjustment was not made by the method of least squares as he had meant it to be, he admitted that it was made by the method of least work. Now we are in a similar position with the mass formula. I think we know how to make mass adjustments in a way which is not criticized by Dr. BREITENBERGER. When we began to work on this mass formula, we knew about correlations from our work on mass adjustments. However, we do not have a correlation matrix 1500×1500, 1500 is the number of nuclidic masses. We don't have any machine that could handle this matrix. So we tried first to use instead of the masses, the reaction energies and doublets themselves, as Professor WAPSTRA pointed out. I have to admit that later on we found it easier to work with the masses themselves and therefore, if you will, I admit that this is not a method of least squares but a method of least work. But at least, we were cautious enough to give in our list of parameters no errors. You crossed the errors out and said they have no meaning. If you will read our paper in the Proceedings, you will find the list of the values of the parameters which, of course, you say have no significance, but you at least won't find any errors.

H. G. KÜMMEL: I would like to comment on several things Prof. MATTAUCH has just said. We have omitted the errors since we have learned in discussion with you that they don't make any sense as long as we use too simple a statistical method. Yet, I do not believe that you are right in saying that the parameters themselves are meaningless. The parameters that we and others found yield the masses with a certain precision. Perhaps we cannot claim to have the best parameters or even the best choice of terms. That you are too pessimistic also follows from the observation we have made that it does *not* matter very much whether we put in masses or nuclear data plus mass doublets. We did both. Finally, you raised some doubts concerning the feasibility or validity of any extrapolation. This is a point where more human aspects enter. If we have a formula which employs the correct physics, the statistics simply becomes less important or even meaningless. We believe we have a good physical picture.

E. BREITENBERGER: You touch upon several topics; I wonder if I can keep them apart. Firstly, and emphatically: there is nothing more practical than a good theory. If we possessed a sound, unified nuclear theory we would not have to bother about vast computations nor about the large extrapolation error from regression formulae. Next, I do not maintain that the parameters in any published formula including yours are meaningless. I only insist that their meaning is not sufficiently well defined for any safe and detailed theoretical conclusions to be drawn. You used shell model refinements of your making, Zeldes and Levy use others, I would use still others, and so on. Who is most nearly right? Only formulae obtained by means of an objective significance criterion could furnish a decision and help us understand how to improve the shell model in the most satisfactory manner. Thirdly, I don't feel pessimistic. I am a pedestrian physicist who would like to know just how much can be gotten from his data, and who entertains the hope that what he gets is useful for nuclear theory.

H. G. Kümmel: I agree with you that one has to use an objective significance test, and I admit that we did not do it. But this concerns the last and smallest correction. As long as I see the influence of new terms is very large we can safely say that they are reliable, important, and significant.

E. Breitenberger: I wonder if you can, for you have left out a couple of things from your formula: for example, a correction term for the possibility of a diffuse nuclear surface, and also the atomic binding energy. Leaving out such terms overburdens the smaller terms and makes them more doubtful.

N. Zeldes: Many theories can explain equally well the same region of phenomena. If you want to test between them you must expand the region more and more. Then, ultimately you can have the hope to decide between them but not otherwise.

E. Breitenberger: I was insisting on the fact that for such a decision we need objective criteria, because otherwise we merely sit here, the three of us, and say "No, I am good, he is not as good".

N. Zeldes: I agree, but it can happen that three theories will be equally good for a given phenomenon even according to your tests.

E. Breitenberger: That would emerge only after applying the tests.

A. de Shalit: I think what Zeldes is referring to is the well-known phenomenon, namely that for any finite number of results you can propose a number of theories which on statistical tests will give you equal reliability but will differ probably with respect to prediction of additional results, whereupon you can make further experiments and test between them.

K. Bleuler: May I suggest that we use only expressions in which all the parameters have a well defined physical significance. This would have as a consequence that the same parameters, the same empirical values, would also play a role in the interpretation of quite different nuclear properties (excitation energies, transitation probabilities, moments, and so on). We really believe a formula only if it is related to or in accordance with a large number of experimental facts which are based on different types of experiments.

E. Breitenberger: My paraphrase of these remarks would be quite simply that bad statistics is worse than no statistics at all. Indeed, I'm perfectly convinced that by careful graphical or arithmetical analysis it is possible to obtain safer results than by an incorrectly led, extensive least squares procedure.

K. BLEULER: Let me amplify my last remark. What is important to check a theory is that with certain assumptions you explained quite different facts. If you are able to explain the properties of the mass surface and in addition, the excitation energies you would believe your theory more.

A. DE SHALIT: I think I could paraphrase Bleuler's remarks in a different way. Namely, if you have a certain picture behind a mass formula like the one that KÜMMEL or ZELDES described and then you derive parameters, it is not enough that these parameters will give you the masses with such and such an accuracy and with such and such a reliability as far as probability goes. It is much more important that the coefficients which you get from such a fit are reasonable within the framework of the picture from which you started. In other words, what I would have expected to see, as an outsider, was that after a certain picture was presented and its parameters derived with more or less reliability from the empirical data, that one would proceed to an analysis which would tend to derive these parameters from two-body nucleon-nucleon interaction, and if this comes out reasonably well. This I would call a success of the formula. If this test is off by three orders of magnitude, then even if the fit is excellent and the χ^2 test couldn't be better, I say that a picture which leads to such a mass formula whose parameters do not fit is missing something.

The Proton Gyromagnetic Ratio as a Nuclear Standard

By

Horst A. Capptuller

Physikalisch-Technische Bundesanstalt Braunschweig, West Germany

Theoretical Considerations

In addition to the well-known properties of mass m, charge e and intrinsic angular-momentum p or spin I, the atomic nucleus possesses in general a magnetic moment μ involved by circulating electric currents. A spinning spherical shell with charge and mass uniformly distributed over its surface has a magnetic moment $\mu = \dfrac{e}{2\,m}\,p$. The nucleus does not conform accurately to this model but the nuclear resonance phenomenon provides one of the experimental methods for learning the relation between the nuclear magnetic moment and the angular momentum. This relation is called the gyromagnetic ratio γ which differs only in magnitude from the absolute theoretical value due to a non-uniform distribution of charge. It is customary to write $\mu = g\,\dfrac{e}{2\,m}\,p$, where the Landé-factor g depends on the effective shape of the nucleus.

If such a nucleus is placed in a magnetic field B, a torque is exerted on its magnetic dipole; it then behaves like a gyroscope with a precession frequency $\omega = \dfrac{\mu}{p} \cdot B = \gamma \cdot B$. Therefore γ is related to the other terms such as

$$\gamma = \frac{\mu}{p} = \frac{2\,\pi\,\mu}{I\,h} = g\,\frac{e}{2\,m}.$$

If γ is well-known by precision experimental methods it will be easy to evaluate a series of other atomic constants or to get corrections for the results of other independent measurements.

The constant $\mu_0 = \dfrac{e}{2\,m}\,\dfrac{h}{2\,\pi}$, called the nuclear magneton, may be found.

The Bohr-magneton may be found as well by substituting the electron mass for the nuclear mass. By comparing the cyclotron-resonance-frequency of protons with their precession-frequency, we get an equation for the specific charge of the proton and in a similar way the equation for the specific charge of the electron. By dividing both results we get a numerical value for the mass-ratio proton-electron. By multiplying the specific charge of the proton with the proton's atomic weight, we get the important Faraday-constant multiplied by the Avogadro-constant.

In the years 1948–1953 Du Mond and Cohen[1] and Bearden and Watts[2] used the results of experimental comparisions as a base for correcting the numerical values of many nuclear constants.

Experimental Procedures and their Results

The phenomenon of nuclear magnetic resonance may, in principle, be detected with any nucleus. However, the proton has the largest magnetic moment and proton-resonance has been found most suitable on account of the high signal-to-noise ratios obtained. Therefore the proton has become a standard nucleus.

The determination of its gyromagnetic ratio requires two measurements: the precession frequency and the external magnetic field strength. The detection of the so called Larmor-precession is possible by the effect of induction of an electromotive force in a receiver-coil or by the absorption of energy in a transmitter-coil.

Both methods and their results shall be discussed independent of their historical succession.

Free nuclear induction

At the homogeneous center of a magnetic standard-field, produced in a solenoid or between a pair of Helmholtz-coils, is a spherical container with water, the hydrogen atoms of which are the protons to be used. A receiver-coil surrounding the container is for detecting the electromotive force induced by the precessing protons if their average magnetic moment has a component perpendicular to the standard field. In order to achieve this, a polarizing field about a hundred times stronger than the standard field turns a majority of the protons at right angles to their final direction. The polarizing force is applied for a few seconds and then switched off and now the protons perform their gyrations but they decrease exponentially during about 3 s.

The induced alternating voltage of frequency f in the receiver coil of about a millionth part of a volt is amplified and then applied to a frequency or period measuring device. The flux-density of the standard field may be evaluated from the geometrical dimensions C_B of the solenoid or Helmholtz-coils and by measuring the exciting current J.

Thus the gyromagnetic ratio results from these measurements

$$\gamma_p = \frac{\omega}{B} = \frac{2\pi f}{C_B \cdot J}. \tag{A}$$

This method of free nuclear induction has been used several times in weak magnetic fields in the order of 1 to $100 \cdot 10^{-4}$ Wb/m². BENDER and DRISCOLL[3] using a solenoid with a flux density of $1.2 \cdot 10^{-3}$ Wb/m² obtained as result

$$\gamma_p = (2.67513 \pm 0.00002) \cdot 10^8 \text{ m}^2 \text{ s}^{-1} \text{ Wb}^{-1}.$$

YANOVSKII, STUDENTSOV, TICHOMIROVA[4] published the numerical value

$$\gamma_p = (2.67520 \pm 0.00012),$$

recently VIGOUREUX[5], using a solenoid, which is also a part of his current-balance, got the result

$$\gamma_p = (2.675171 \pm 0.000013).$$

YANOVSKII and STUDENTSOV[6] determined the gyromagnetic ratio in low fields produced by several pairs of Helmholtz-coils to be

$$\gamma_p = (2.67506 \pm 0.00006).$$

WILHELMY[7], using a modified method with a superregenerative oscillator, got the numerical value

$$\gamma_p = (2.67549 \pm 0.00008).$$

This result however will not be discussed further during international conferences due to the possibilities of systematic errors exceeding the estimated accidental errors.

Nuclear resonance absorption

The detection of nuclear resonance-precession in strong magnetic fields in the order of 0.1 to 1 Wb/m² requires less electronic equipment and mainly the effect of energy absorption is used.

An oscillator coil with the proton sample inside is placed in the air-gap of an electromagnet, the coil axis perpendicular to the main field. The oscillator is set near the threshold of oscillation; its amplitude is then very dependent on damping due to resonance absorption in the tank coil. If the magnetic field and frequency are set in the correct relation $\omega = \gamma \cdot B$ then proton resonance will occur, the effect of energy absorption will lower the oscillation level. In order to display the resonance on a cathode-ray-oscilloscope the magnetic field is modulated by passing an alternating current of low frequency (for example 50 c/s) through auxiliary coils. The exact relation between precession frequency and main dc-field may then be determined by observing the symmetrical position of the resonance peaks. Frequency measurement is very simple and practically without any error. However, the absolute determination of the flux density of the main field with sufficient accuracy is only possible by measuring the interacting force F between a current-carrying conductor and the main field. This equipment is well-known as a magnetic balance or a

Cotton-balance[8]. The current carrying circuit consists of a long rectangular coil with vertical sides. The lower horizontal sides are placed in the homogeneous region of the air gap, the upper sides in a region where any stray field is compensated to zero. The electromagnetic interaction force can be evaluated from the geometrical dimensions of the rectangular coil, the distribution along the circuit of the relative values of the magnetic flux density $\dfrac{B}{B_0}$ and the current J in the circuit.

$$F = J \cdot B_0 \int_{\text{coil}} \frac{B(x, y)}{B_0} \, dx.$$

Knowing the effective width $\bar{x}$ of the coil, the flux density B_0 at the point of the proton sample is given by

$$B_0 = \frac{F}{J \cdot \bar{x}}$$

and the gyromagnetic ratio γ_p is then given by

$$\gamma_p = \omega \frac{\bar{x}}{F} J. \tag{B}$$

Rectangular coils supported on glass formers were used many times for measuring magnetic flux densities[8, 9] before they were employed for determining nuclear moments to high accuracy. THOMAS, DRISCOLL, and HIPPLE[10] made a determination of γ_p in the years 1949/50 by using a Cotton-balance for measuring the flux density of 0.45 Wb/m² in the air gap of an electromagnet with the result

$$\gamma_p = (2.67523 \pm 0.00006) \cdot 10^8 \ (\text{m}^2 \ \text{s}^{-1} \ \text{Wb}^{-1}).$$

Recently, YAGOLA, ZINGERMAN, and SEPETYI[11] using several rectangular coils with one and two turns got the numerical value

$$\gamma_p = 2.67505 \pm 0.00005.$$

However with these fixed rectangular coils described above it is necessary to measure with high precision the width of the coil along its total length. The possible errors in measuring and therefore in evaluating the effective width may be increased by using a special rectangular coil whose geometrical width can be varied with a high degree of accuracy by means of quartz-length-standards[12]. The difference of the results of two force measurements is directly proportional to the difference in width, therefore a precision measurement of the actual width of the total coil can be avoided. Even relatively large deviations from the ideal rectangular size effectuate only small corrections. The two force measurements are then related to the flux density B_0 of the main field by the equation

$$\frac{F_1}{C_{01}} - \frac{F_2}{C_{02}} = N \cdot J \cdot B_0 \cdot \Delta x \cdot C_{xy}.$$

The corrections arising with a coil of N turns due to the inhomogeneity around the center of the air gap (C_{01} and C_{02}) and the variations of geometrical width (C_{xy}) are to be evaluated by numerical integration with knowledge of the field distribution along the coil. CAPPTULLER[12] published a calculation of the efficiency of this device and showed that it is possible to reduce the errors of magnetic field measurements by paying attention to some constructive pecularities. In the years 1960/61 he used such a special coil for the magnetic balance as part of the equipment for determining the gyromagnetic ratio of the proton[13] with the result

$$\gamma_p = (2.67522 \pm 0.00010) \cdot 10^8 \ (\mathrm{m^2 \, s^{-1} \, Wb^{-1}}).$$

Absolute Accuracy of the Experimental Methods

The measurement of the precession frequency by comparision with standard-frequencies of quartz-clocks has such a high degree of precision that it may be practically excluded of all estimated errors. The evaluation of the electromagnetic constant of one layer solenoid from its measured geometrical dimensions has an upper limit of errors in the order of 3 parts in a million. Besides this the value of the effective magnetic field may be affected by the parasitic stray fields, the earth's magnetic field and magnetic moments of ferromagnetic bodies in the neighbourhood of the main coil. Their influences may be eliminated if they remain constant during the time of measurement by reversing the exciting current and averaging the results. Impurities of ferromagnetic or even strong paramagnetic materials in all parts of the resonance-field region are a source of systematic errors and have an influence on the band-width of the resonance-signal. By carefully selecting all materials of the equipment these systematic errors may be reducted to 2 parts in a million.

Similar considerations have to be made if measurements are carried out in a strong field of an electromagnet, although the influence is much smaller. Field distribution, effective width of the rectangular coil and its magnetic influence on the pole-pieces may produce a sum of possible errors in the order of 5 parts in a million. Systematic errors may arise in the uncertainty of adjusting the position of the coil with respect to the magnetic field vector. The calculations are based on the assumption that the coil plane is perpendicular to the magnetic induction vector in the lower part of the turns. An error in coil adjusting of only $0.25°$ will cause a systematic error of 10 parts in a million. The earth acceleration g due to the force of gravity in the place where the measuring equipment is located is another source of uncertainty but may be limited to a few parts in a million.

Since the method of free nuclear induction has less sources of error, it is obvious that it will be preferred. But both methods have a common uncertainty, that is, the deviation from the absolute current unit. All measurements mentioned above have been carried out with respect to the national current units. However, simultanous measurements with both methods may eliminate this systematic error. According to eq. (A) and (B) using a current through the solenoid and the rectangular coil in series and measuring both precession frequencies ω_1 and ω_2 the gyromagnetic ratio results in

$$\gamma = \sqrt{\gamma_1 \cdot \gamma_2} = \sqrt{\frac{\omega_1 \cdot \omega_2 \cdot \bar{x}}{C_B \cdot F}}$$

and is independent of the current.

In spite of all modern experimental techniques, the use of sensitive electronic control and measuring equipments and a large degree of carefulness, the absolute limit of error at this time will be in the order of 10 to 20 parts in a million. From this point of view a good accordance between all measurements of the last years is attained.

International Average

According to a proposition of the Comité Consultatif d'Electricité[14] in 1961 a working group has been established which was charged to prove critically all results of the last years and to find out a preliminary average in terms of the absolute electrical units of the "Bureau International des Poids et Mesures". This international working group, consisting of experts in these measurements, had its first meeting in spring 1963 before the 10th session of the CCE[15]. The preliminary average of the last 5 measurements[3, 5, 6, 11, 13] has been stated for protons in pure water to

$$\gamma_p = (2.67513 \pm 0.00006) \cdot 10^8 \text{ m}^2 \text{ Wb}^{-1} \text{ s}^{-1}.$$

It is to be expected that it will be possible to increase the accuracy in the next years, when a number of new measurements have been carried out.

Appendix

Table of the results of the determination of the proton's gyromagnetic ratio based on the unit of current defined by the absolute units of electromotive force and resistance conserved within the Bureau International des Poids et Mesures

May 1963

Reference	Method	Result $\text{m}^2 \text{ Wb}^{-1} \text{ s}^{-1}$	Deviation Δ from average
3		$2.67515 \cdot 10^8$	$+ 2 \cdot 10^3$
5	Free nuclear induction	$2.67515 \cdot 10^8$	$+ 2$
6		$2.67504 \cdot 10^8$	$- 9$
11	Strong field, Cotton-balance	$2.67508 \cdot 10^8$	$- 5$
13		$2.67521 \cdot 10^8$	$+ 8$

Average: $2.67513 \cdot 10^8$

$$\sqrt{\frac{\Sigma \Delta^2}{5}} = \pm 0.00006 \cdot 10^8$$

References

[1] J. W. M. DUMOND and E. R. COHEN, Rev. Modern Phys. 20, 82 (1948); 25, 691 (1953).

[2] J. A. BEARDEN and H. M. WATTS, Phys. Rev. 81, 73 (1951).

[3] P. L. BENDER and R. L. DRISCOLL[5], Free Precession Determination of the Proton Gyromagnetic Ratio. Trans. IRE Instr. (Dec. 1958), 176.

[4] B. M. YANOVSKII, N. V. STUDENTSOV, and T. N. TICHOMIROVA, Izmeritelnaja Technika 2, 39 (1959).

[5] P. VIGOUREUX, Proc. Roy. Soc. A, 270, 72–89 (1962).

[6] B. M. YANOVSKII and N. V. STUDENTSOV, Izmeritelnaja Technika 6, 28–31 (1962).

[7] W. WILHELMY, Eine Neubestimmung des gyromagnetischen Verhältnisses des Protons. Ann. Phys. 19, 329 (1957).

[8] COTTON, L'éclairage électrique 24, 257 (1900).

[9] G. H. BRIGGS and A. F. A. HARPER, J. Sci. Instr. 13, 119 (1936).

[10] H. A. THOMAS, R. L. DRISCOLL and J. A. HIPPLE,, Measurement of the Proton Moment in Absolute Units. Phys. Rev. 78, 787 (1950).

[11] G. K. YAGOLA, V. I. ZINGERMAN, and V. N. SEPETYI, Izmeritelnaja Technika 5, 24–29 (1962).

[12] H. CAPPTULLER, Ein Differenzverfahren zur absoluten Bestimmung der Kraftflußdichte im Luftspalt von Laboratoriumsmagneten. Z. Instr.-Kde. 69, 133–140 (1961).

[13] H. CAPPTULLER, Bestimmung des gyromagnetischen Verhältnisses des Protons. Z. Instr.-Kde. 69, 191–198 (1961).

[14] Comité Consultatif d'Electricité 9e Session Oct. 1961, S. 14 und S. 85—89. Paris: Gauthier-Villars & Cie.

[15] Comité Consultatif d'Electricité 10e Session May 1963 (to be published).

[16] N. F. RAMSEY, Magnetic Shielding of Nuclei in Molecules. Phys. Rev. 78, 699 (1950).

[17] H. A. THOMAS, The Diamagnetic Correction for Protons in Water and Mineral Oil. Phys. Rev. 80, 901 (1950).

Discussion

H. STAUB: Would Dr. CAPPTULLER and Dr. COHEN kindly comment on the fact that the limits of errors stated for the average of γ_p in their respective reports differ appreciably?

E. R. COHEN: Capptuller's quoted error represents the mean deviation of the data and gives the average uncertainty of the experiments based on equal weights for all experiments. In our adjustment we have selected the *best* experiments (BENDER and DRISCOLL, VIGOUREUX) which are in excellent agreement and of high precision. Furthermore, the COHEN and DUMOND least squares adjustment includes other experimental data which also improves the accuracy of the quoted result.

It should also be remarked that the table which I recently circulated quotes 3 σ limits of error whereas the preliminary 1961 table quotes σ.

E. BREITENBERGER: Isn't it correct that the adjusted values are obtained on a large number of degrees of freedom whereas CAPPTULLER gives an average on four degrees of freedom so that the meaning of the quoted errors is rather different?

J. DuMOND: This is essentially what Dr. COHEN was saying. Our least squares adjusted values depend on more than just these results.

A. McNISH: The result and the error reported by Dr. CAPPTULLER are based on averaging all experimental determinations with equal weight. It was my privilege recently to examine the Soviet experiments in considerable detail. I do not regard them as being as accurate as the experiments of BENDER and DRISCOLL and of VIGOUREUX which were high precision experiments. Therefore, we in the United States have been inclined to regard the best experimental value of γ_p as that given by BENDER and DRISCOLL and VIGOUREUX which agree with each other to one part in 10^6 when corrections are applied for the differences in the ampere as maintained by these two national laboratories.

Precision γ-Ray Wavelength Measurement by Crystal Diffraction

By

J. W. Knowles

Atomic Energy of Canada Ltd. Nuclear Laboratories Chalk River, Ontario

With 9 Figures

The double flat crystal spectrometer was used as early as 1917 by A. H. COMPTON[1a] to study the reflectivity of X-rays by crystals. In the following ten years the spectrometer was used by many workers for comparing X-ray wavelengths[1]. In particular the careful analysis of the two crystal geometry by DAVIS and PURKS (1927)[2] and independently by EHRENBERG and MARK (1927)[3] showed that the spectrometer could be used for precision measurement. By 1930 MARK and VON SUSICH, and ALLISON and WILLIAMS[1b], used a two crystal spectrometer with an angular precision of a fraction of a second of arc to study the relative shapes of X-ray lines. With such precision, X-ray wavelengths could be compared with a standard error of a few parts in 10^5. In recent years, the double crystal spectrometer, supplemented by modern technical development, has been found useful for the precise comparison of nuclear radiations, which have wavelengths 10 to 200 times shorter than the standard (Mo $K\alpha$) X-ray[4].

A wavelength determination by the method of crystal diffraction is based on the Bragg relationship between wavelength λ, interplanar crystal spacing d_H, and diffraction angle θ_B inside the crystal:

$$\lambda = 2 d_H \sin \theta_B. \tag{1}$$

It can be shown, theoretically, that for diffraction by transmission through a crystalline lamina, with diffraction planes perpendicular to the incident surface, that eq. (1) also relates the wavelength outside the crystal and the external diffraction angle θ_B (the angle θ_B corresponds to the intensity in the peak of the diffracted line) to a precision of a few parts in 10^6. Most measurements involve a comparison of energies or wavelengths of two radiations. In such cases, it is not necessary to know the crystal spacing d_H but only the relative diffraction angles θ_{B_1} and θ_{B_2}. The relationship is

$$E_2/E_1 = \lambda_1/\lambda_2 = \sin \theta_{B_1}/\sin \theta_{B_2}, \tag{2}$$

where E_1 and E_2 are the energies corresponding to wavelengths λ_1 and λ_2 respectively.

It is important to measure the Bragg angle for γ-rays ($\leqq 3000$ keV) with a precision comparable to or greater than that previously attained

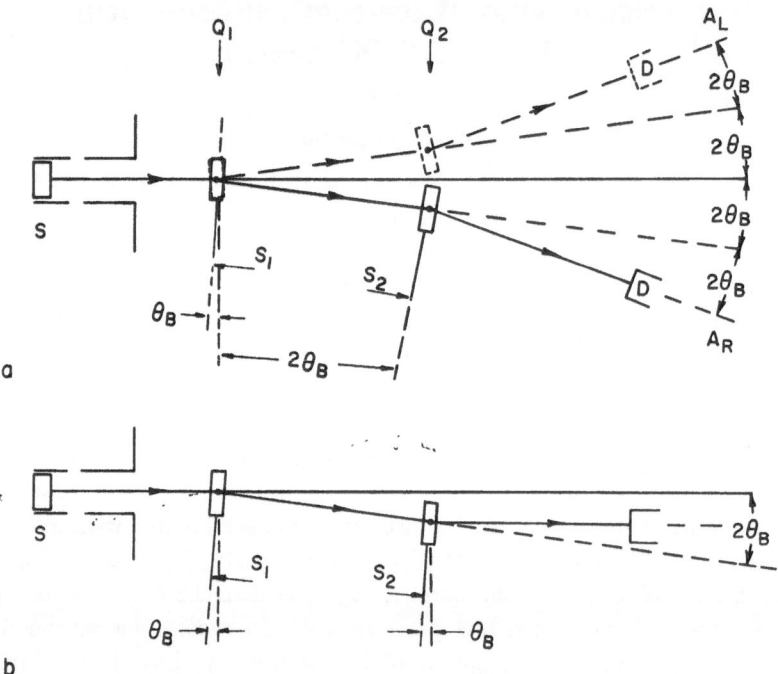

Fig. 1. (a) and (b) illustrate the diffraction of radiation by the double flat crystal spectrometer with crystals arranged, (a) in the anti-parallel position and (b) in the parallel position

for X-rays. A precise measurement becomes increasingly more difficult for the higher energy γ-rays as the diffraction angle becomes smaller. The characteristic (Mo $K\alpha_1$), 17.5 keV, X-ray diffracts from the d_{211}, 3.03 Å planes of calcite, at $\theta_B \approx 10°$ and the $H^1(n, \gamma)H^2$ (Deuterium) 2224 keV γ-ray diffracts from the same crystal planes at an angle $\theta_B \approx 180$ seconds of arc. For γ-rays of moderately high energy the resolution R is given in terms of the diffraction line width w at half maximum intensity and the Bragg diffraction angle θ_B:

$$R = \Delta E/E = \Delta \lambda/\lambda \approx w/\theta_B. \tag{3}$$

A smaller diffraction angle corresponds to a lower spectrometer resolution R.

The diffraction line width, w can be expressed as a function of a number of characteristic widths,

$$w = w(w_f, w_M, w_\lambda - -). \tag{4}$$

The most prominent of these are: w_f a width associated with the spectrometer geometry, w_M the crystal mosaic width associated with the angular

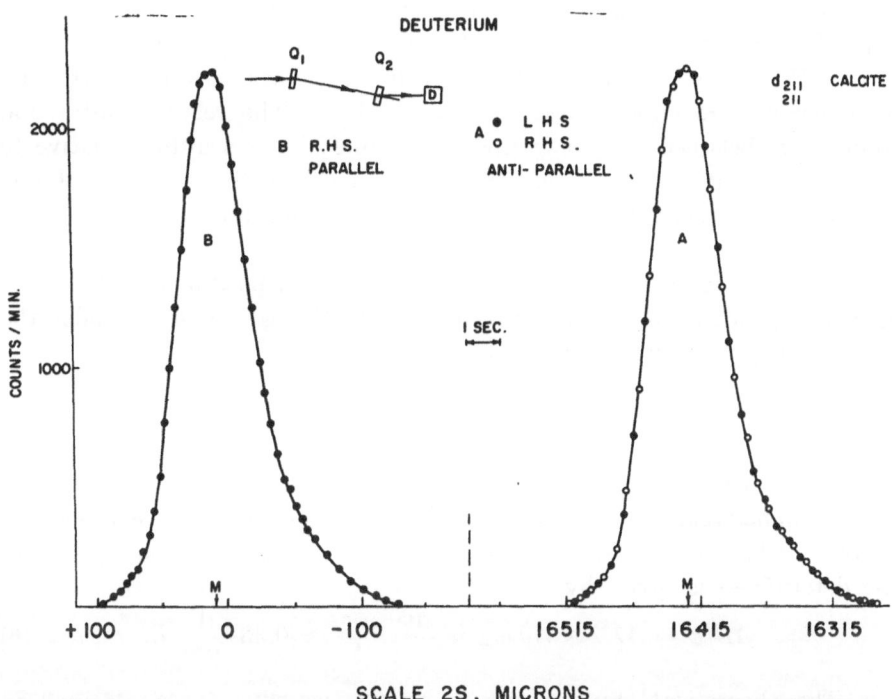

Fig. 2. It shows the deuterium γ-ray line shape following diffraction from the (211), (2̄1̄1̄) planes of two calcites. The measured line B, the two crystals in the parallel position, has the same shape as the line A, the two crystals in the antiparallel position. The right-hand curve O and left hand curve ●, plotted in standard scale microns are superimposed relative to their common median M. The pronounced asymmetry in the wing of the lines is due to crystal imperfection. For these early measurements the median is located to better than 2% of its width at half-maximum intensity

distribution of imperfections in the crystals and w_λ a width associated with the natural energy spread of the γ-ray. If these widths affect the line shape in a complicated manner it is difficult to define the angular position of the line to a small fraction of its width. However, the symmetry of the crystal diffraction geometry allows one to circumvent this particular difficulty. This is seen by referring to the drawing of the double flat crystal spectrometer Fig. 1. A γ-ray diffracted to the right A_R and to the left A_L of the main beam, Fig. 1a, has identical line shapes. The shapes are assymetric with respect to the direct beam if they result from crystal imperfection, then $w \approx w_M$. On the other hand, the line shapes are symmetric with respect to the main beam if they result from the natural energy spread of the radiation, then $w \approx w_\lambda$. If sufficient care is taken to adjust the spectrometer, w_f is much less than w_M and only widths w_M and w_λ contribute to the total width w of the doubly diffracted line. The assymetrical character of a diffracted line resulting from crystal mosaic spread is shown in Fig. 2, the diffraction of the 2.224 MeV (Deuterium) γ-ray from the (211, 211) planes of two calcite crystals. In the right hand graph, Fig. 2 the left and right hand diffracted lines shown by light and dark circles, are superimposed for comparison. The relative diffraction angle, $4\,\theta_B$ between these two lines is obtained by measuring, relative to a standard scale, the distance of separation of the line medians. Using this method the angle $4\,\theta_B$ is determined without interpretation of the line shape.

The standard error $\Delta w / w$ of the relative angular position of a diffracted line is dependent, in theory, only on the statistical count N under the line. It is given by the approximate equation

$$\Delta w / w \approx \pm\, 0.33 \, \frac{1}{\sqrt{N}}. \tag{5}$$

If $N \approx 10^4$ counts, the standard error with which the line position may be determined relative to its width w, is about $1/2$ per cent. The corresponding error to which the diffraction angle θ_B, wavelength λ and energy E can be determined is given by

$$\Delta E / E = \Delta \lambda / \lambda \approx \Delta w / \theta_B \approx \frac{\Delta w}{w} \cdot \frac{w}{\theta_B} \approx 0.33 \, \frac{1}{\sqrt{N}} R. \tag{6}$$

The theoretical standard error, eq. (6), usually is not attained in practice because of three technical limitations; (a) insufficient sensitivity for angle measurement, (b) insufficient mechanical stability for angle measurement and (c) a lack of knowledge of the relation between the scale used in the measurement and the diffraction angle, θ_B. The sum of these uncertainties may be expressed as an uncertainty in angle Δw_s. Eq. (6), modified to include this uncertainty, becomes

$$\Delta E^1 / E \approx 0.33 \, \frac{1}{\sqrt{N}} R + \frac{\Delta w_s}{\theta_B}. \tag{7}$$

Table 1 shows some of the most precise diffraction measurements made prior to 1959. The best angular precision, standard error 0.19 seconds, was attained by MULLER et al.[6] in the measurement of the 412 keV γ-ray of Au[198]. This precision has not been exceeded by other workers using the same focusing type of instrument.

Since that time some of the limitations on precision measurement of diffraction angles have been considerably reduced for parallel beam,

Table 1. Precision γ-ray measurements with diffraction spectrometers

Spectrometer description	Line width w secs.	d_H Å	Resolution %	Isotope	Best measurement keV*	Best fractional precision	Best angular precision sec. Δw_s
DuMond (1947), r. c.** 200 cm., quartz, d_{310}	22	1.178	1.04	Cu^{64}	510.979 ± 0.067	± 1.3 : 10⁴ ***	± 0.23
			0.84	Hg^{198}	411.801 ± 0.036	± 0.8 : 10⁴ ***	± 0.19
Beckman (1958), r. c.** 200 cm., quartz, d_{310}	15	1.178	0.57	Hg^{198}	411.770 ± 0.033	± 0.8 : 10⁴ †	± 0.19
	27		0.08	$La\,K_{\alpha_2}$	33.0340 ± 0.0004	± 1.2 : 10⁵ ††	± 0.3
	25		0.13	$Lu\,K_{\alpha_1}$	54.0698 ± 0.0009	± 1.7 : 10⁵ ††	± 0.3
Knowles (1959), double flat crystal calcite, d_{211}	2.7	3.028	0.5	Mg^{24}	1368 ± 1.	± 1 : 10³ †††	± 0.46
	1.6		0.6	Mg^{24}	2750 ± 3.	± 1 : 10³ †††	± 0.30

* A γ-ray energy is calculated from its corresponding wavelength λ_x in X-units using the (1955) Cohen et al.[7] conversion relation $\lambda_x E = 12372.44 \pm 0.16$ keV X.U. A wavelength is measured relative to the average $(W K\alpha_1)_x$ X-ray, 208573 ± 0.004 X-units, which in turn refers to the standard $(Mo K_{\alpha_1})_x$ X-ray.

** r. c. = radius of curvature.

*** These γ-ray energies, 1 : 10⁴ greater than those originally quoted by Muller et al.[6] in (1952) have been recalculated using the (1955) Cohen et al.[7] conversion relations.

† Bergvall (1959)[8a].

†† Bergvall (1960)[8b].

††† The flat crystal measurements, made prior to the development of the optical technique, have angular precision comparable to those made with bent crystal spectrometers[9].

flat crystal spectrometers. For example, the angular sensitivity for diffraction measurement has been improved for γ-ray measurements at Chalk River[5] and for X-ray measurements at Johns Hopkins University[10]. Both groups have developed optical systems for angle measurement with flat crystal spectrometers which are sensitive to angular changes of 0.02 seconds of arc. Recently at Chalk River this limit in sensitivity has

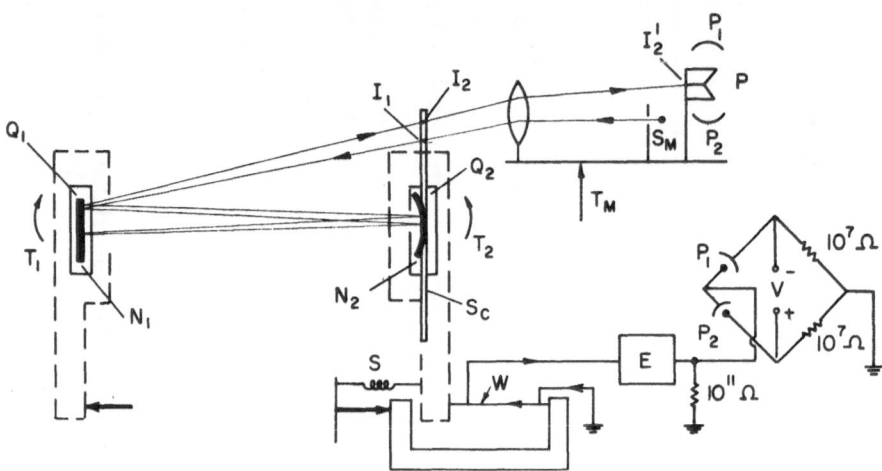

Fig. 3. A schematic drawing of the mechanical optical system used for measuring the relative angle between crystals Q_1 and Q_2. Mirror N_1 is an optical flat rigidly attached to crystal turntable T_1. Spherical mirror N_2 and standard scale S_c are similarly attached to turntable T_2. The focal length of mirror N_2 is twice the distance between the axes of rotation of T_1 and T_2. A travelling microscope T_M moves parallel to standard scale S_c and contains a line light source S_M. The light from this source focuses at $I_1 \propto I_2$ in the plane of the standard scale before and after reflection from the N_1, N_2 mirror system. The return image I_2 is again imaged at I_2' in the plane of the split prism P. The light balance is measured by photocells $P_1 \propto P_2$. The error signal controls a $1/4$ amp. electric current flowing in a nichrome wire W. A change in current causes a change in wire temperature followed by a corresponding change in wire length so that the spectrometer arm T_2 rotates to balance the light image I_2' on the split prism

been extended still further, to ≥ 0.001 seconds of arc using relatively simple equipment. A description of the 0.02 second equipment used at Chalk River has been reported[5]. It is similar to that shown in Fig. 3.

Improvement in angular sensitivity is of value only if it is combined with a corresponding improvement in the mechanical stability of the spectrometer. A high degree of stability has been achieved for the flat crystal spectrometer by the addition of feedback. The pertinent details are shown in Fig. 3. The angular position of the spectrometer arm T_2 is determined by the length of a 0.001″ diameter nichrome wire W kept under constant tension by a spring. The length of the wire depends on its temperature which is held at about 200° C by passing through it about 1 ampere of electric current. The strength of the current is proportional to a voltage produced in the photo-cell circuit by an unbalance of the light focused on the split prism P. The overall gain of the feed-back system is 10^4. The system effectively isolates the relative angular setting of turntables T_1 and T_2 from the external environment. By this means the angular stability of the spectrometer is better than 10^{-3} seconds of arc, a factor of 50 improvement over the same system without feed-back.

The result of these two improvements has been to reduce the mechanical limitation on angular measurement so that the precision of measurement is limited only by statistical counts, eq. (6). However, the actual measurement of a particular γ-ray energy may be in error because of systematic effects. The main source of systematic error is in the transfer from the measurement scale to the angle scale. Such a transfer is always necessary since Bragg diffraction angles are not compared directly. What are compared are lengths, i. e. distances on the arcs of divided circles or linear distances tangential to the angular rotation of the diffracting crystals. For example, with the double flat crystal spectrometer, the position of the median of a diffracted line, diffracted to the right and left of the direct beam, is related to the corresponding position of the light focus I_2 on the standard scale S_c (Fig. 3). The linear separation $2\,S$, on the standard scale, of the right and left light focus I_2 is related, by geometrical considerations, to the change in angle between the two diffraction positions. For diffraction at small angles from equivalent planes of two crystals the standard scale separation $2\,S$ and Bragg diffraction angle θ_B are related by

$$2\,S = 16\,R_T\,\theta_B\,(1 + 13.39\,\theta_B{}^2).   \tag{8}$$

$R_T = 1117.747 \pm 0.009\,\text{mm}$ at $28.0 \pm 0.2°\text{C}$ is the effective distance* between the axes of rotation of the two crystal turntables. Eq. (8) is believed accurate to a few parts in 10^5.

The comparison of two widely separated angles from the corresponding measurements of length has not been satisfactorily solved where precisions greater than $1 : 10^5$ are required. There may be systematic effects in the optical system which if they vary with angle could effect relative wavelength determinations. Much effort has been made to put an upper limit on the magnitude of such effects and to describe qualitatively their variation with diffraction angle. For the flat crystal spectrometer, the most important of these is the effect of irregularities on the flat mirror surface N_1, Fig. 3. This effect can not be ignored since the light is incident on a slightly different area of the flat mirror for each angle of rotation. The mirror is known to be flat to better than half a wavelength of light over its surface. However, irregularities in the surface, less than 2000 Å are too small for direct observation and yet may effect the angle measurement. The sharpness of the focussed image in the plane of the split prism allows

---

* The determination of R_T is considered in detail in the original paper[11].

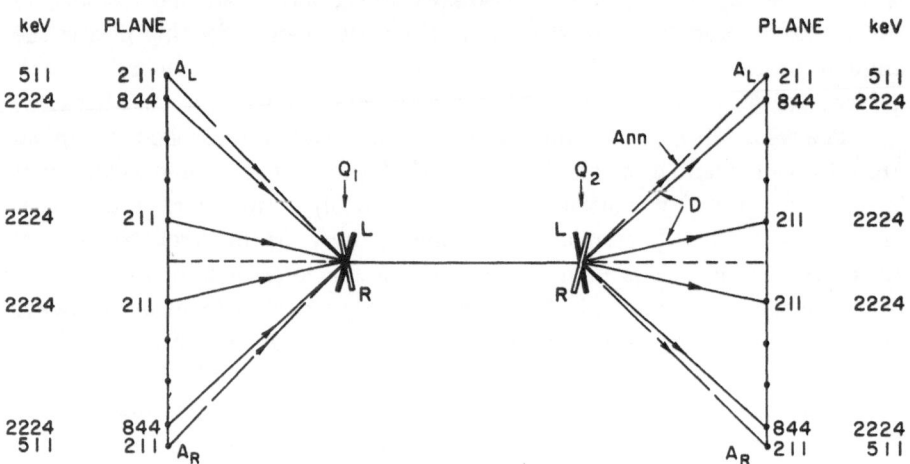

Fig. 4. A schematic drawing of the radiation paths for deuterium and annihilation γ-rays following diffraction from different crystal planes of calcite used in the double flat crystal spectrometer. A_L, A_L and A_R, A_R represent double diffraction from crystals Q_1 and Q_2 to the left and to the right of the main beam respectively and L and R are the corresponding angular settings of the crystals

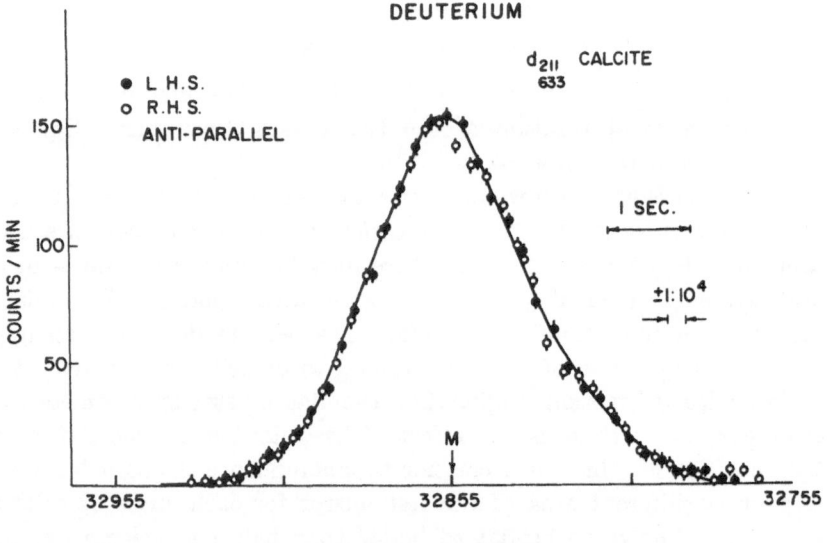

Fig. 5. The deuterium γ-ray diffracted from the (211), (633) crystal planes of two calcites, the crystals set in the antiparallel geometry. The left- and right-hand curves, ● and O respectively, are plotted in standard scale microns, relative to their common median M. The median is located to better than 3% of the width of the line

an upper limit to be put on the perfection of the mirrors used in the optical system. The limit obtained corresponds to an effect on the diffraction angle at 800 seconds of $3:10^5$ but the same limit would have negligible effect for a diffraction angle of 200 seconds. The effect is also negligible

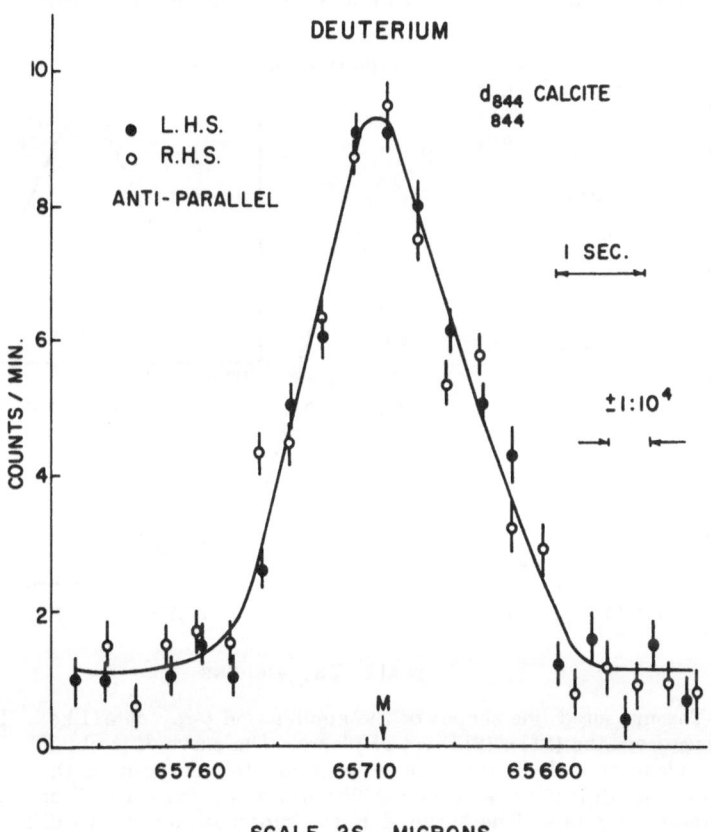

Fig. 6. The deuterium γ-ray diffracted from the (844), (844) crystal planes of calcite, the crystals set in the antiparallel geometry. The left- and right-hand curves, ● and O respectively, are plotted in standard scale microns relative to their common median M. The statistical count is not sufficient to define very precisely the shape of the lines, but it is sufficient to determine the median M, and therefore the superposition of the two diffracted lines to better than 5% of their width at half-maximum intensity

in the wavelength comparison of two γ-rays diffracted at large angles but with ray paths which differ by only a few hundred seconds. An example is the determination of the relative angles of the deuterium and annihilation γ-rays by diffracting the deuterium γ-ray from different crystal planes[11]. This is illustrated in Fig. 4 which shows different ray paths following diffraction from different crystal planes in the double crystal spectrometer. Double diffraction from the (211, 211) planes occurs at small angles while diffraction from the (844, 844) planes occurs at 4 times larger angles. This latter ray path is within 10 per cent of that of the annihilation radiation diffracted from the (211, 211) planes.

The deuterium diffraction lines[11] obtained by diffraction from different combinations of planes of the two crystals are shown in Figs. 2,

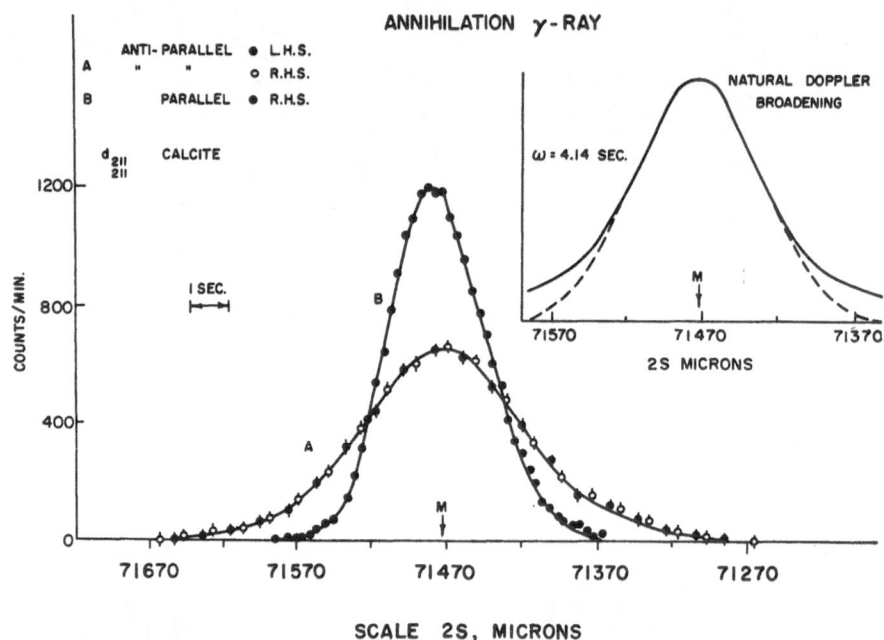

Fig. 7. The measured line shapes of the annihilation γ-ray at 511 keV. Diffraction takes place from the (211), (211) set of planes. The curve B is obtained by setting the two calcite crystals in the parallel position. In this position the spectrometer has zero energy dispersion, so that the line width is a measure of the perfection of the diffracting crystals. The curve A is the superposition of two diffraction lines measured to the right, the clear points, and left, the solid points, of the direct beam. The points are plotted in microns with reference to their common median M. Curves A and B have the same area. The doppler broadening of the annihilation γ-ray, apparent in curve A, is "unfolded" by making use of curve B, and the natural annihilation γ-ray shape obtained thereby is plotted in the inset. Its shape deviates from a Gaussian, the dotted line in the wings, and has a width at half-maximum intensity of 4.14 seconds of arc

5, and 6. In these early measurements, which precede the use of feedback in the angle measuring system, the precision is limited by mechanical instability. This is particularly true for diffraction from the (211, 211) crystal planes Fig. 2. The counting rate decreases with increasing order of diffraction and is particularly low for diffraction from the (844, 844) set of planes, Fig. 6. For diffraction from the (844, 844) planes the number of statistical counts determines the precision. Fig. 7 shows the measurement of the annihilation γ-ray, curve A, with which the deuterium γ-ray was compared. Determinations of the energy of the deuterium γ-ray are plotted in Fig. 8 as a function of standard scale distance $2\,S$. It shows that the energy comparison at different angles is constant within the limits of error. The deuterium binding energy is estimated by adding to the average energy of the deuterium γ-ray, the energy of recoil, 1.34 keV,

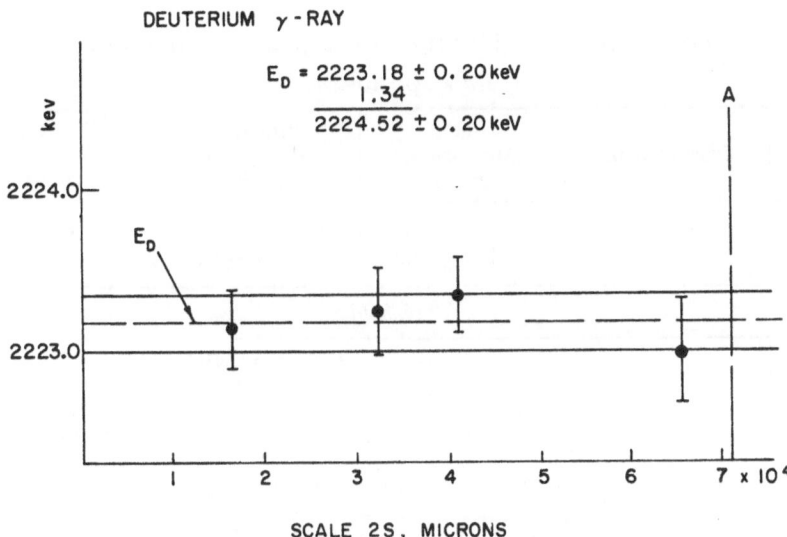

Fig. 8. The deuterium de-excitation γ-ray energy is plotted as a function of the standard scale separation $2\,S$ in microns. The angular position of the annihilation γ-ray is given by the vertical broken line A. The vertical bars on the points are the probable errors of the measurement at each angle. The horizontal broken line is the arithmetic mean energy. The horizontal solid lines indicate the probable limits of error for this average measurement. However the limit of error on the numerical value of the deuterium binding energy is the standard error

Table 2. Recent measurements of the deuterium binding energy*

Reference	Instrument	Reference energy (keV)	Deuterium binding energy (keV)
Motz et al. (1959)[12]	Compton spectrometer	Na[24] 2753.5 ± 0.3	2224.3 ± 1.0
Monahan et al. (1961)[13]	NaI scintillator	Na[24], Y[88], Bi[207] precision 1 : 10³	$2219\begin{smallmatrix}+3\\-2\end{smallmatrix}$
Jewell and John (1961)[14]	2-meter curved crystal	(annih.) 510.976 ± 0.007 Hg[198] 411.770 ± 0.036	2224.6 ± 1.5
Kazi et al. (1961)[15]	6-meter curved crystal	(annih.) 510.976 ± 0.007	2225.5 ± 1.5
Knowles (1961)[11]	Double flat crystal	510.976 ± 0.007	2224.52 ± 0.20

* The limits of precision are standard errors.

of the deuterium nucleus following emission of the γ-ray. The average precision of the deuterium energy measurement, expressed as a standard error, considering all measurements, is $1:10^4$.

Table 2, a summary of the deuterium γ-ray energy measurements made prior to 1962, shows the relative consistency of deuterium measurements made by different workers[12, 13, 14, 15].

Table 3. Estimated values of n-p mass difference*

Mass spectrometer

No.	Measurements	Nier** (Minnesota) (keV)	Smith** (Princeton) (keV)	Everling selfconsistent (keV)
1	H_2-D	1441.2 ± 0.8	1442.6 ± 0.2	1442.10 ± 0.09
2	n-p***	783.3 ± 0.8	781.9 ± 0.3	

Threshold

		$H^3\ (p, n)\ He^3$† $H^3\ (\beta^+)$	$C^{14}\ (p, n)\ N^{14}$†† $C^{14}\ (\beta^+)$	
3	n-p	782.3 ± 0.7	784.6 ± 0.8	782.6 ± 0.4

  * The limits of precision are standard errors.
 ** Wapstra (1960)[16].
*** n-p estimated from (2224.52 ± 0.20)-$(H_2$-D) keV.

 † Van Patter and Whaling (1954); Porter (1959)[17, 18].
 †† Sanders (1956); Pohm et al. (1955)[19, 20]; Everling et al. (1960); Wapstra (1960)[21a, b].

The direct measurement of the deuterium binding energy combined with that of the H_2-D mass difference is sufficient for a determination of the n-p mass difference. The determination is given in Table 3, row (2), using the H_2-D measurements of Nier, and of Smith, row (1)[16]. The n-p mass difference is obtained with somewhat less precision from threshold measurements plus β-decay end-point determinations Table 3, row (3), columns (2) and (3). In row (3) column (4) is given the "self-consistent" estimate of Everling et al. (1960)[21a, b] which depends on the "adjusted" value of the H_2-D mass difference, row (1), column (4)*.

A second example of γ-ray measurement is the determination of the wavelength $(\lambda_c)_x$ of the annihilation γ-ray in terms of the standard wavelength $(\lambda\ Mo\ K\alpha_1)_x = 707.801$ X-units. Such a measurement determines the ratio of the fine structure constant, α and the X-unit to milli-Ångstrom conversion constant Λ through the relation

$$\alpha^2/\Lambda = 2\,R_\infty\,(\lambda_c)_x \tag{8}$$

---

  * These determinations are discussed in some detail in the original paper[11].

where R_∞ is the Rydberg constant. The measurement is important if made with sufficient precision since it checks the consistency of other measurements relating to the atomic constants[4]. $(\lambda_c)_x$ was measured with a standard error of $3.5:10^5$ by the author in collaboration with Dr. BEARDEN. The absolute diffraction angle θ_B of the annihilation radiation

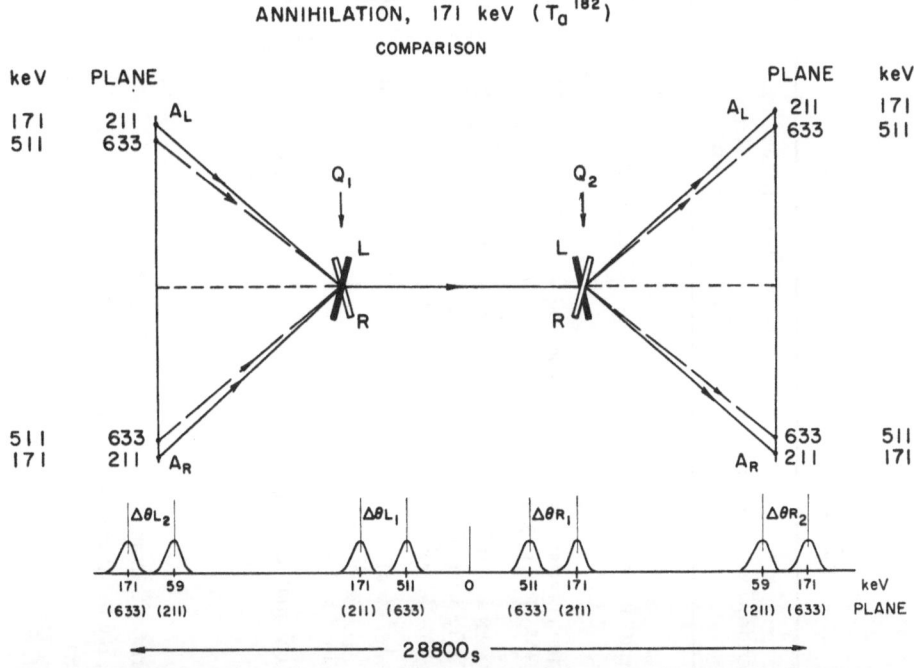

Fig. 9. A schematic drawing of radiation paths, to the right A_R, A_R and left A_L, A_L respectively, following diffraction from different crystal planes of the double flat crystal spectrometer. For these measurements the diffracting crystals, Q_1 and Q_2 assume positions R, R, and L, L, respectively. The 511 keV γ-ray is diffracted from the (633), (633) set of planes of calcite and the 171 keV γ-rays from the (211), (211) planes. The diffracted γ-rays, within a few percent, follow the same path. The angular differences $\Delta\theta_{L_1}$ and $\Delta\theta_{R_1}$ are measured in the comparison of the wavelengths of the 171 and 511 keV γ-rays. Equivalent measurements $\Delta\theta_{L_2}$ and $\Delta\theta_{R_2}$ are made in the comparison of the 171 keV γ-ray and the $W K\alpha_1$ X-ray

was determined at Chalk River by diffraction from the (211, 211) planes of two calcite crystals cleaved from the same crystal. The angle θ_B was calculated using eq. (8), and measurements; R_T, the distance between the rotation axes of the optical system and $2 S$, the optical measure of the linear separation of the medians of the right and left diffracted lines. The diffracted lines are shown superimposed in Fig. 7 A. The natural shape of the annihilation diffracted line obtained from diffraction measurements in the parallel and anti-parallel positions A and B respectively is shown in the inset. The finite width of this line, doppler

Table 4

Measurement*	Additional information	Annihilation wavelenght $(\lambda_C)_x$ (X-units)	α^2/Λ**
(1) Absolute diffraction angle precision $3:10^5$ $\theta_C = 7992.41 \pm 0.30$ micro-radians at 18.0° C	(a) Spacing of calcite in X-units	$24212.99 \pm .75 \times 10^{-3}$	$531{,}413.6 \pm 15 \times 10^{-10}$
	(b) Bearden comparison of calcite spacing with $(Mo\ K\alpha_1)_x$ X-ray	$24212.46 \pm .75$	$531{,}402.0 \pm 15$
(2) Measure of relative angles of annihilation γ-ray and $W\ K\alpha_1$ X-ray precision $1.5:10^5$ $9(\lambda_C)_x/(\lambda\ W\ K\gamma_1)_\gamma$	(a) BECKMAN et al.[22] comparison of $(W\ K\alpha_1)_x$ with $(Mo\ K\alpha_1)_x$ precision $2:10^5$	$24213.20 \pm .62$	$531{,}418.2 \pm 12$
	(b) WATSON et al.[23] comparison of $(W\ K\alpha_1)_x$ with $(Mo\ K\alpha_1)_x$ precision $4:10^5$	$24213.56 \pm .96$	$531{,}426.1 \pm 19$
The annihilation wavelength in X-units and α^2/Λ calculated using the adjusted values of Λ and α^2		$24212.14 \pm .26$	$531{,}395.0 \pm 5.8$

* all limits of error are standard errors.

** $\alpha^2/\Lambda = 2\ R_\infty\ (\lambda_C)_x$.

$\Lambda = 1.002063 \pm 0.000006$. conversion constant from X-units to milli-Angstroms.

10^8 x $R_\infty = 109737.31$ cm^{-1}, precision $0.1 : 10^6$. Rydberg constant for infinite mass.

$\alpha = 7.29720 \pm 0.00003 \times 10^{-3}$, fine structure constant.

broadened because of the thermal motion of the center of mass of the electron-positron pair, is 4.14 seconds or 3.2 eV*. Following this measurement, Dr. BEARDEN at Johns Hopkins University using a representative sample of the Chalk River calcite crystals measured the absolute diffraction angle of (Mo $K\alpha_1)_x$ radiation. From the two angle measurements the annihilation wavelength $(\lambda_c)_x$ was determined using eq. (2). The measurement is given in Table 4, row (1).

Subsequently, it was felt that the precision of the measurement of $(\lambda_c)_x$ could be improved and the possibility of systematic error reduced considerably by making use of diffraction from high order crystal planes of calcite. For the new measurement, illustrated in Fig. 8, the 511 keV γ-ray diffraction angle was compared with a neutron capture γ-ray of Ta[182] at 171 keV and in a separate experiment the 171 keV γ-ray was compared with the internal conversion (W $K\alpha_1$) X-ray of W[182]. W[182] is produced following β-decay of Ta[182]. These three radiations have energies in the ratio 9 : 3 : 1 to within a few per cent. The 171 keV γ-ray diffracted from the (211, 211) planes of two calcite crystals is at nearly the same angle as that of the 511 keV γ-ray diffracted from the (633, 633) planes. Likewise the 171 keV γ-ray diffracted from the (633, 633) planes is at very nearly the same angle as the (W $K\alpha_1)_x$ radiation diffracted from the (211, 211) planes. The difference in diffraction angle was measured for all pairs of lines on both sides of the direct beam in order to eliminate effects of line shape. The ratio $(9 \lambda_c)_x/(\lambda \text{ W } K\alpha_1)_x$ obtained from these measurements has a standard error of $1.5 : 10^5$ and is given in Table 4 rows 3 and 4. Measurements of the ratio of the (W $K\alpha_1)_x$ and (Mo $K\alpha_1)_x$ X-rays had been made previously by BECKMAN et al.[22] with standard error $2 : 10^5$ and by WATSON et al.[23] with standard error $4 : 10^5$. The value of $(\lambda_c)_x$ determined from these measurements is given in column (3). The measurements of $(\lambda_c)_x$ from the two experiments agree within their standard error. Because of the relatively large error in the measurements of $(\lambda_c)_x$ the value of α^2/Λ determined from the annihilation measurements is not as good a check of the (1962) atomic constants as one might wish. Both measurements, the diffraction angle of annihilation radiation relative to (W $K\alpha_1)_x$ and (W $K\alpha_1)_x$ relative to (Mo $K\alpha_1)_x$ can be improved. More accurate measurements are now being attempted by Dr. BEARDEN and the author. The expectation is that the fractional standard error of the measurement of α^2/Λ can be reduced still further possibly to $\leq 1 : 10^5$.

References

[1] (a) A. H. COMPTON, The Reflection Coefficient of Monochromatic X-rays from Rock-salt and Calcite. Phys. Rev. 10, 95 (1917).

---

* To the annihilation energy determination following the diffraction angle measurement must be subtracted 1.6 eV half the mean kinetic energy of the center of mass of the electron-positron pair.

[1] (b) COMPTON and ALLISON, X-rays in Theory and Experiment, 2nd. Ed., p. 710–750. New York: D. van Nostrand Co. (1943).

[2] B. DAVIS and H. PURKS, Measurement of the Molybdenum K Doublet Distances by means of the Double X-ray Spectrometer. Proc. Nat. Acad. Sc. 13, 419 (1927).

[3] W. EHRENBERG and H. MARK, Über die natürliche Breite der Röntgen-Emissionslinien. Z. Physik 42, 807, 823 (1927).

[4] E. R. COHEN, Most Probable Values of the Physical Constants. Paper presented at the Symposium on Standards and Physical Constants at the American Physical Society Meeting, Washington (1962). The paper is a summary of an analysis of the available data by DuMOND, McNISH and COHEN.

[5] J. W. KNOWLES, Measurement of γ-ray Diffraction Angles to ± 0.02 Second of Arc with a Double Flat Crystal Spectrometer. Canad. Journ. Physics 40, 237 (1961).

[6] MULLER, HOYT, KLEIN, and DuMOND, Precision Measurements of Nuclear γ-ray Wavelengths of Ir^{192}, Ta^{182}, RaTh, Rn, W^{187}, Cs^{137}, Au^{198}, and Annihilation Radiation. Phys. Rev. 88, 775 (1952).

[7] COHEN, DuMOND, LAYTON, and ROLLETT, Analysis of Variance of the (1952) Data on the Atomic Constants and a New Adjustment (1955). Revs. Modern Phys. 27, 363 (1955).

[8] P. BERGVALL, (a) Precision Measurement of $K \alpha$ X-ray Lines from Rare Earth Elements. Arkiv for Fysik 16, 57 (1959); (b) Precision Measurements of Gamma Energies and Intensities by Crystal Diffraction. Arkiv for Fysik 17, 125 (1960).

[9] J. W. KNOWLES, A High Resolution Flat Crystal Spectrometer for Neutron Capture γ-ray Studies. Canad. Journ. Physics 37, 203 (1959).

[10] J. G. MARZOLF, Interference Calibration of X-ray Spectrometer. Bull. Amer. Phys. Soc. II, 7, 339 (1962).

[11] J. W. KNOWLES, Diffraction Angle Measurements Bearing on the Deuterium Binding Energy and on the X-unit-to-milli-Ångstrom Conversion Constant. Canad. Journ. Physics 40, 257 (1961).

[12] MOTZ, CARTER, and FISHER, Gamma-Ray Energy Measurements with a Compton Spectrometer. Bull. Amer. Phys. Soc. II, 4, 477 (1959).

[13] MONAHAN, RABOY, and TRAIL, Measurement of the Energy of the Gamma Radiation from Neutron Capture by Hydrogen. Nuclear Phys. 24, 400 (1961).

[14] R. W. JEWELL and W. JOHN, Binding Energy of the Deuteron Measured with a Bent Crystal Spectrometer, UCRL-6095, University of California (1960).

[15] KAZI, RASMUSSEN, and MARK, Measurement of the Deuteron Binding Energy Using a Bent-Crystal Spectrometer. Phys. Rev. 123, 1310 (1961).

[16] A. H. WAPSTRA, Proceedings of the International Conference on Nuclidic Masses, McMaster University, Hamilton, p. 535. Edited by H. E. DUCKWORTH, University of Toronto Press (1960).

[17] D. M. van PATTER and W. WHALING, Nuclear Disintegration Energies. Revs. Modern Phys. 26, 402 (1954).

[18] F. T. PORTER, Beta Decay Energy of Tritium. Phys. Rev. 115, 450 (1959).

[19] R. M. SANDERS, Study of the $C^{14} (p, n) N^{14}$ and $C^{14} (\alpha, n) O^{17}$ Reactions. Phys. Rev. 104, 1434 (1956).

[20] POHM, WADDELL, POWERS, and JENSEN, Beta Spectrum of C^{14}. Phys. Rev. 97, 432 (1955).

[21] EVERLING, KONIG, MATTAUCH, and WAPSTRA, (a) Relative Nuclidic Masses. Nuclear Phys. 18, 529 (1960); (b) Nuclear Data Tables, Part I (Nuclear Data Project, National Academy of Sciences, National Research Council, Washington D. C.).

[22] BECKMAN, BERGVALL, and AXELSSON, A Precision Curved Crystal X-ray and Gamma-Ray Spectrometer. Arkiv for Fysik 14, 419 (1958).

[23] WATSON, WEST, LIND, and DUMOND, A Precision Study of the Tungsten K-Spectrum using the 2-Meter Focusing Curved Crystal Spectrometer. Phys. Rev. 75, 505 (1949).

Discussion

A. H. WAPSTRA: Have you already tried to check your precision by measuring gamma rays involved in a sum relation (Ritz' principle)?

J. W. KNOWLES: No, at least not to the precision of the annihilation measurement.

H. E. DUCKWORTH: Some of us are interested in having a very accurate energy value for some neutron-capture gamma ray leading to the ground state, say the one arising from neutron capture in Cd^{113}. Such a value would be of great use as standard for (d, p) reactions and for much mass spectroscopic work.

J. W. KNOWLES: The neutron capture γ-ray in Cd^{113} is a possibility, but other favorable possibilities are those in carbon and I think in boron. In these latter cases, there are well known decays from the capturing states to states above the ground state at 3 and 4 MeV respectively. These states subsequently decay to the ground state. The two cascade γ-rays in each of these isotopes are sufficiently low in energy to be readily measurable by crystal diffraction. The sum of the two γ-ray energies determines the binding energy.

Precision Measurements
on the Muon, Muonium, Positronium, and Helium
as Related to the Fundamental Atomic Constants

By

J. M. Bailey and V. W. Hughes

Gibbs Physics Laboratory, Yale University, New Haven, Connecticut, USA

(Presented by J. M. BAILEY)

With 2 Figures

1. Introduction

Quantum electrodynamics is our best understood and most precisely verified theory. Both the electron and the muon appear to be Dirac particles with conventional electrodynamic coupling. Electrons and muons partake in no strong interactions but they do have weak interactions. Precise measurements on these particles, both when free and when in the bound hydrogen-like systems of muonium ($\mu^+ e^-$) and positronium ($e^+ e^-$) have been vital to the establishment and verification of the theory of quantum electrodynamics. Such measurements are also useful for obtaining information about certain of the fundamental atomic constants. For the interpretation of these measurements only the electromagnetic interactions need be considered because the weak interactions are small by comparison. The effects of strongly interacting particles on higher order radiative corrections become important[1] only to an accuracy well beyond present experimental precision. In contrast, the effect of strong interactions on the hydrogen atom, which contains a proton, is considerably greater than the uncertainties of present-day experiments.

Many experiments, in particular measurements of the electron g-value g_e[2], the Lamb-shift[3], and positronium hyperfine structure[4], show that electrons are "Dirac-particles" (i. e., are completely described by Quantum Electrodynamics). An accurate measurement of the muon g-value[5], a comparison of high-energy electron-proton[6] and muon-proton[7] scattering, and muon-pair photoproduction[8] show that the muon is also a Dirac particle.

2. Properties of the Muon

The electromagnetic properties of the electron include its spin, charge, g-value, magnetic moment, and mass (which we denote respectively by J,

e, g, μ, and m), and they are regarded as fundamental atomic constants. Similarly the corresponding electromagnetic properties of the muon should be regarded as fundamental atomic constants. In addition, the lifetime, τ_μ, of the muon is an important constant in the theory of weak interactions. The best values of these constants are:

$$
\left.
\begin{aligned}
J_\mu &= {}^1\!/_2{}^9, \\
e_\mu/e_e &= 1^{10}, \\
g_\mu/2 &= 1.001162 \pm 0.000005 \ (5 \ \text{ppm})^5, \\
\mu_\mu/\mu_p &= 3.18338 \pm 0.00004 \ (13 \ \text{ppm})^{11}, \\
m_\mu/m_e &= 206.765 \pm 0.003 \ (13 \ \text{ppm})^{11}, \\
\tau_\mu &= 2.198 \pm 0.002 \ \mu \ \text{sec}^{12},
\end{aligned}
\right\}
\tag{1}
$$

where quoted errors in parentheses are in "parts per million" (ppm).

Since the measurement of the muon magnetic moment is of especial importance in the following pages, we consider it briefly. The precession frequency, in a known magnetic field, of muons stopped in any one of a number of hydrogen compounds is measured. A major problem is to decide how much the muon is influenced by its environment, and it is assumed (with some supporting evidence) that the muon replaces a proton. The quoted error does not question this assumption, but arises chiefly from uncertainties in the (rather indirect) calibration of the apparatus.

3. The Fine Structure Constant

The fine structure constant α is one of the most important and least accurately known of the fundamental constants. At present the best value of α is obtained from the measurement of the fine structure interval in the deuterium $2^2 P$ state[13]. Measurements on muonium, positronium and helium may soon improve our knowledge of α.

The quantity most suitable for precise measurement in the hydrogen-like (but short-lived!) atoms muonium and positronium is the hyperfine structure interval, $\Delta\nu$, of the ground state. We first consider muonium.

The theoretical value of $\Delta\nu$ for muonium is[14, 18]:

$$
(\Delta\nu)_M = \left(\frac{16}{3} \alpha^2 c R_\infty \frac{\mu_\mu}{\mu_e} \right) \left(1 + \frac{m_e}{m_\mu} \right)^{-3} (1 + a_e)^2 (1 - \epsilon_1 - \epsilon_2)(1 - \delta_\mu), \tag{2}
$$

where

$$
a_e = \frac{\alpha}{2\pi} - 0.328 \frac{\alpha^2}{\pi^2}; \quad \epsilon_1 = \left[\left(\frac{5}{2} - \ln 2 \right) - \frac{3}{2} \right] \alpha^2 = (1 - \ln 2)\, \alpha^2,
$$

$$
\delta_\mu = \frac{3}{\pi} \alpha \frac{m_e}{m_\mu} \ln \frac{m_\mu}{m_e}; \quad \epsilon_2 = \frac{8\alpha^3}{3\pi} \ln \alpha \left(\ln \alpha + \frac{37}{96} + \frac{1}{5} - 2 \ln 2 \right).
$$

9*

Inserting the value of m_μ given in (1), together with[15, 16]

$$\mu_e/\mu_p = 658.2106 \ (0.3 \ \text{ppm});$$

$$R_\infty = 109{,}737.31 \ \text{cm}^{-1} \ (0.1 \ \text{ppm}); \tag{3}$$

$$c = 2.997925 \times 10^{10} \ \text{cm/s} \ (1.3 \ \text{ppm})$$

and

$$\alpha = 7.29719 \times 10^{-3} \ (9 \ \text{ppm}), \tag{4}$$

we find

$$(\Delta\nu)_M = 2.632936 \times 10^7 \, \alpha^2 \, (\mu_\mu/\mu_p) \ \text{Mc/s} \ (\pm 1.6 \ \text{ppm}). \tag{5}$$

If we insert the value (1) of the muon moment in (5) we get:

$$(\Delta\nu)_M = 8.381636 \times 10^7 \, \alpha^2 \ \text{Mc/s} \ (\pm 13 \ \text{ppm}), \tag{6}$$

$$= 4463.14 \pm 0.09 \ \text{Mc/s} \ (\pm 22 \ \text{ppm}). \tag{7}$$

We[17] have recently carried out a measurement of $(\Delta\nu)_M$ with the result

$$(\Delta\nu)_M \ (\text{expt}) = 4463.33 \pm 0.19 \ \text{Mc/s} \ (\pm 43 \ \text{ppm}). \tag{8}$$

Using (6) this gives

$$\alpha_M = 7.29735 \times 10^{-3} \ (\pm 22 \ \text{ppm}). \tag{9}$$

We hope to repeat the experiment soon with increased accuracy.

The salient features (which one must know to understand the sources of error) of the muonium experiment are as follows[18]. Muonium is formed by stopping polarized positive muons in argon:

$$\mu^+ + A \rightarrow (\mu^+ e^-) + A^+.$$

By observing, for different magnetic field values, the effect of microwaves on the angular distribution of decay positrons, we obtain a signal with a resonance lineshape as shown in Fig. 1.

The width of the line is determined by the muon lifetime—this happens to be quite long (compared to the lifetime of other unstable elementary particles and atomic states), and the line center is determined with an error much less than the linewidth, by investigating the detailed lineshape and by running for long enough to get good counting statistics—the data for Fig. 1 took about two days to collect.

To stop a reasonable number of the 40 MeV muons from the Nevis synchrocyclotron, we need high density argon gas; consequently each muonium atom makes frequent collisions with argon atoms, and although argon is very inert (which is why we use it), it perturbs the muonium wavefunction during the collision. The result is a shift in $(\Delta\nu)_M$ which is proportional to the density of the argon buffer gas for low densities, as shown in experiments[19] on hydrogen in argon. Some of our experimental results are shown in Fig. 2, where each point represents the result of a run like that shown in Fig. 1. The presence of a buffer gas density-shift is undeniable, but its linearity is questionable, although we are encouraged by the good agreement between the slope of our "best fit" line and the corresponding measurement[19] on hydrogen. ($\Delta \nu$ differs widely for the three hydrogen isotopes, but the three *fractional* density-shifts $d(\ln \Delta \nu)/dp$ agree very closely with each other and with our value.) The extrapolation to zero pressure can be made less dubious by making more measurements at lower pressures; unfortunately this becomes difficult (lower signal-

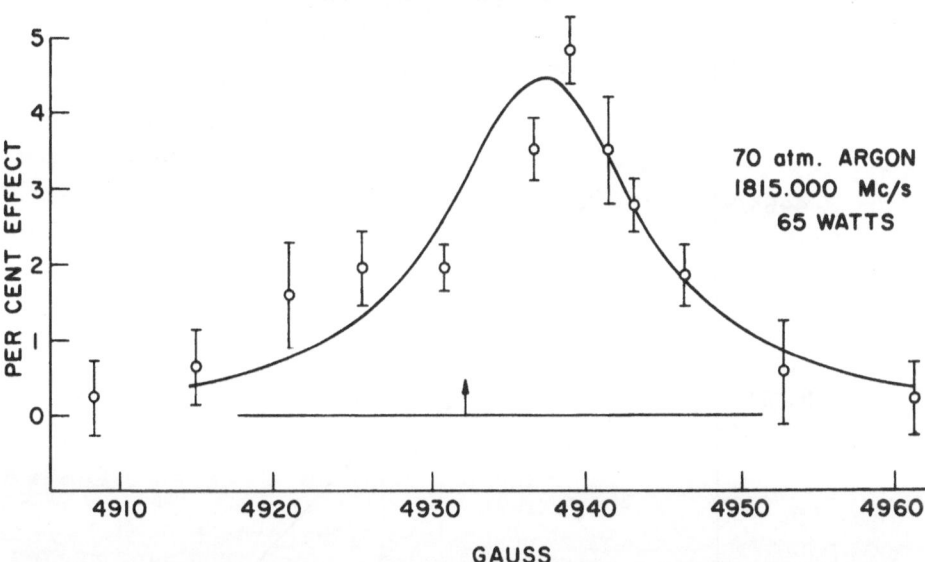

Fig. 1.   Resonance curve for the transition $(m_J, m_\mu) = (^1/_2, {}^1/_2) \rightarrow (^1/_2, -{}^1/_2)$ in strong magnetic field. Arrow indicates predicted center in absence of density-shift

to-noise) because of reduced stopping power. The error in our result comes almost entirely from statistics, magnified by the extrapolation procedure. (If we fit to our data a line with slope given by the hydrogen measurements, we obtain nearly the same value for $\Delta \nu$, but with an error one-half of the error quoted, which is based on fitting a line of non-fixed slope; this procedure is plausible, but we prefer the conservative error limit, which is one standard deviation.) Evidently Fig. 2 summarizes both the chief advantage of the muonium experiment (its great accuracy) and its chief obstacle (the extrapolation procedure). The theoretical value of the muonium measurements is summarized by eq. (5) which relates the three numbers $(\Delta \nu)_M$, (μ_μ/μ_p), and α^2; if α becomes better known the more fundamental form (2) must be used.

Positronium is a unique and remarkable atom from the theoretical standpoint. To begin with, its reduced mass of $m_e/2$ makes its gross structure quite different from other hydrogen-like systems. The calculation of its hyperfine structure must consider Feynman diagrams (involving virtual pair creation and annihilation) which do not occur in other hydrogen-like systems. To date $(\Delta\nu)_p$ has been calculated only to order α^3, with the result[20]

$$(\Delta\nu)_p = (\alpha^2 c \, R_\infty) \left[\frac{7}{6} - \left(\frac{16}{9} + \ln 2 \right) \frac{\alpha}{\pi} \right], \tag{10}$$

$$= 2033.7 \times 10^2 \text{ Mc/s}. \tag{11}$$

We may hope for a calculation of the higher terms of this important number in the near future by use of computer programs.

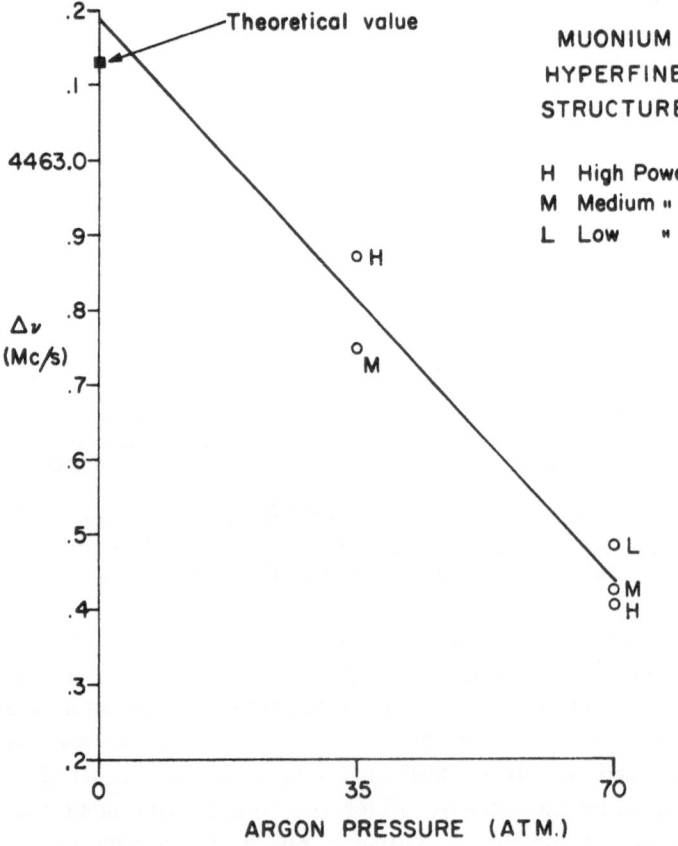

Fig. 2. Values of $\Delta\nu$ for muonium vs. argon gas pressure. Preliminary data for forward telescope only

From the experimental viewpoint positronium has several slight advantages over muonium. It is formed by stopping positrons in an inert gas; positron-emitting sources (e. g., Na^{22}) are more convenient than synchrocyclotrons, and provide a higher flux of particles, so that it is easier to get good statistics; moreover 500 keV positrons are easily stopped in low-density gas, so the buffer gas density-shift effect is less troublesome. But the natural linewidth is much higher than for muonium, and this completely offsets the advantages.

Experiment[4] gives

$$(\Delta \nu)_p = (2033.5 \pm 0.3) \times 10^2 \text{ Mc/s.} \tag{12}$$

A re-measurement of $(\Delta \nu)_p$ to much higher precision is now being carried out[21].

Another experiment which may give information of high accuracy on the fundamental atomic constants is the measurement of the fine structure of the $2^3 P$ levels of helium. The fine structure intervals $\Delta E_{J J'}$ have been calculated[22] only to terms of order α^3;

$$\Delta E_{J J'} = (\alpha^2 c R_\infty) f_{J J'}, \tag{13}$$

$$\Delta E_{01} = 29616.65 \text{ Mc/s}; \quad \Delta E_{12} = 2286.55 \text{ Mc/s.} \tag{14}$$

Accurate theoretical values for $\Delta E_{J J'}$ require calculations of Schroedinger atomic wavefunctions and of quantum-electrodynamic radiative corrections; both of these calculations are well understood in principle but the computations are laborious. Although the atomic structure problem is quite old, the most powerful methods of attacking it have appeared only recently[22, 23]. Better calculations of ΔE by these methods are in progress. The most accurate measurements of these intervals have been made by "level-crossing"[24] and atomic beams[25] methods, giving

$$\Delta E_{01} = 29619.1 \pm 1.9; \quad \Delta E_{12} = 2291.19 \pm 0.05 \text{ Mc/s} \tag{15}$$

and very much more accurate results are being obtained by both methods (unpublished).

The fine structure of the $2^3 P$ states of helium, like that of the 2 P states of hydrogen, depends on nuclear details only to an extent much smaller than present experimental accuracy. The helium experiment is capable of great precision; its only disadvantage is that its interpretation requires heavy calculations.

4. Hyperfine Structure of Hydrogen

The theoretical value of $\Delta \nu$ for hydrogen is[14, 26]:

$$(\Delta \nu)_H = \left(\frac{16}{3} \alpha^2 c R_\infty \frac{\mu_p}{\mu_e} \right) \left(1 + \frac{m_e}{m_p} \right)^{-3} (1 + a_e)^2 (1 - \epsilon_1 - \epsilon_2)(1 - \delta_p), \tag{16}$$

where δ_p is a relativistic proton recoil and structure term. This is the same expression as for muonium with the subscript μ replaced by p, apart from the term δ_p. The calculated value[26] of δ_p is a lengthy expression involving proton form factors and various questionable assumptions, whereas δ_μ, the corresponding term for muonium, is unambiguous. Inserting the values given above, and[15]

$$m_p/m_e = 1836.12 \ (10 \ \text{ppm}), \tag{17}$$

$$(\Delta\nu)_H = 2.667443 \times 10^7 \ \alpha^2 \ (1 - \delta_p) \ \text{Mc/s} \ (\pm 1.5 \ \text{ppm}). \tag{18}$$

If we use the calculated value[14] $\delta_p = 35 \times 10^{-6}$ we get:

$$(\Delta\nu)_H = 2.667351 \times 10^7 \ \alpha^2 \ \text{Mc/s} \ (\pm 1.5 \ \text{ppm}), \tag{19}$$

$$= 1420.338 \ \text{Mc/s} \ (\pm 18 \ \text{ppm}). \tag{20}$$

Now $(\Delta\nu)_H$ has been measured by several methods; the two most accurate results being:

Optical Pumping[19] $1420\,405\,749 \pm 6$ c/s,

Hydrogen Maser[27] $1420\,405\,751.800 \pm 0.028$ c/s. $\tag{21}$

One way to remove the discrepancy between (20) and (21) is to suppose that the value (4) of α (which depends chiefly on measurements[13] of deuterium fine structure), used in obtaining (20) from (19), is in error. But this is all subject to our fundamental uncertainty of the phenomena represented by δ_p, which may be seriously in error. This uncertainty is so far-reaching that we feel that numbers based on measurements of $(\Delta\nu)_H$ are of no value for determining α, but rather give information about proton structure.

5. Conclusion

At the present moment the best value of α is obtained by combining the value obtained from the fine structure of deuterium with the value obtained from the hfs of muonium to give

$$\alpha = 7.29721 \times 10^{-3} \ (7 \ \text{ppm}),$$

$$\alpha^{-1} = 137.0387 \pm 0.0010. \tag{22}$$

It can be hoped that a more precise value of α will soon be available from the methods for determining α discussed in Section 3.

References

We have attempted to give an up-to-date, but not necessarily complete, list of references; further references will be found in the papers listed.

[1] L. Durand, Phys. Rev. 128, 441 (1962).

[2] D. T. Wilkinson and H. R. Crane, Phys. Rev. 130, 852 (1963).

[3] W. E. Lamb, Jr. and R. C. Retherford, Phys. Rev. 86, 1014 (1952). — S. Triebwasser, E. S. Dayhoff, and W. E. Lamb, Jr., Phys. Rev. 89, 98 (1953). — For latest theoretical work see A. J. Layzer, J. Math. Phys. 2, 308 (1961).

[4] R. Weinstein, M. Deutsch, and S. C. Brown, Phys. Rev. 94, 758 (1954). — V. W. Hughes, S. Marder, and C. S. Wu, Phys. Rev. 106, 934 (1957).

[5] G. Charpak, F. J. M. Farley, R. L. Garwin, T. Muller, J. C. Sens, and A. Zichichi, Phys. Letters 1, 16 (1962).

[6] K. W. CHEN, A. A. CONE, J. R. DUNNING, Jr., S. G. F. FRANK, N. F. RAMSEY, J. K. WALKER, and RICHARD WILSON, Phys. Rev. Letters 11, 561 (1963).

[7] G. E. MASEK, T. E. EWART, J. P. TOUTONGHI, and R. W. WILLIAMS, Phys. Rev. Letters 10, 35 (1963).

[8] A. ALBERIGI-QUARANTA, M. DE PRETIS, G. MARINI, A. ODIAN, G. STOPPIN and L. TAU, Phys. Rev. Letters 9, 226 (1962).

[9] V. W. HUGHES, D. W. McCOLM, K. O. H. ZIOCK, and R. PREPOST, Phys. Rev. Letters 5, 63 (1960). — P. K. KABIR, Nuov. Cim. 22, 429 (1961).

[10] This is one of two possible viewpoints; for discussion (with obsolete numerical values) see G. SHAPIRO and L. M. LEDERMAN, Phys. Rev. 125, 1022 (1962).

[11] G. McD. BINGHAM, Nuov. Cim. 27, 1352 (1963). — D. P. HUTCHINSON, J. MENES, G. SHAPIRO, and A. M. PATLACH, Phys. Rev. 131, 1351, 1362 (1963).

[12] F. J. M. FARLEY, T. MASSAM, T. MULLER, and A. ZICHICHI, 1962 International Conference on High-Energy Physics at CERN, p. 415. — S. L. MEYER, E. W. ANDERSON, E. BLESER, L. M. LEDERMAN, J. L. ROSEN, J. ROTHBERG, and I.-T. WANG, Phys. Rev. 132, 2693 (1963).

[13] E. S. DAYHOFF, S. TRIEBWASSER, and W. E. LAMB, Jr., Phys. Rev. 89, 106 (1953).

[14] D. W. ZWANZIGER, Bull. Amer. Phys. Soc. 6, 514 (1961). — A. J. LAYZER, Bull. Amer. Phys. Soc. 6, 514 (1961). Private communication from D. W. ZWANZIGER.

[15] J. W. M. DU MOND, Ann. Phys. 7, 365 (1959).

[16] The value given for μ_e/μ_p is an average of the three best *published* results: S. H. KOENIG, A. G. PRODELL, and P. KUSCH, Phys. Rev. 88, 191 (1952). — E. R. BERINGER and M. A. HEALD, Phys. Rev. 95, 1474 (1954). — J. S. GEIGER, V. W. HUGHES, and H. E. RADFORD, Phys. Rev. 105, 183 (1957).

[17] R. PREPOST, J. M. BAILEY, W. E. CLELAND, M. ECKHAUSE, and V. W. HUGHES, Bull. Amer. Phys. Soc. 9, 81 (1964).

[18] K. O. H. ZIOCK, V. W. HUGHES, R. PREPOST, J. M. BAILEY, and W. E. CLELAND, Phys. Rev. Letters 8, 103 (1962).

[19] F. M. PIPKIN and R. H. LAMBERT, Phys. Rev. 127, 787 (1962). — For theoretical work on density-shifts, see G. A. CLARKE, J. Chem. Phys. 36, 2211 (1962).

[20] R. KARPLUS and A. KLEIN, Phys. Rev. 87, 848 (1952).

[21] E. D. THERIOT, H. G. ROBINSON, K. O. H. ZIOCK, and V. W. HUGHES, private communication.

[22] C. L. PEKERIS, B. SCHIFF, and H. LIPSON, Phys. Rev. 126, 1057 (1962).

[23] H. M. SCHWARTZ, Phys. Rev. 130, 1029 (1963).

[24] J. LIFSITZ and R. H. SANDS, Bull. Amer. Phys. Soc. 6, 424 (1961).

[25] F. M. J. PICHANICK, R. D. SWIFT, and V. W. HUGHES, Bull. Amer. Phys. Soc. 9, 90 (1964).

[26] C. K. IDDINGS and P. M. PLATZMAN, Phys. Rev. 113, 192 (1959); 115, 919 (1959).

[27] S. B. CRAMPTON, D. KLEPPNER, and N. F. RAMSEY, Phys. Rev. Letters 11, 338 (1963).

Discussion

J. W. KNOWLES: Could you elaborate on why the muon hyperfine structure splitting is not influenced by nuclear P term corresponding to that in hydrogen?

J. M. BAILEY: There are two pieces of experimental evidence that the muon does not have charge structure ("form factors") like the proton. One is the very precise measurement of the muon g-value by the Cern group; the agreement between theory and experiment is excellent. The other comes from a comparison of muon-proton scattering experiments (by MASEK et al.) with electron-proton experiments with the same (large) momentum transfer; agreement is good. A more accurate muon-proton experiment is now being done at Brookhaven.

J. DuMOND: Do you have any comments about this, Dr. COHEN?

E. R. COHEN: If one believes the theory of electrons and muons, then muonium cannot be afflicted with the corrections which in hydrogen are a result of virtual pion production in the strong interaction proton field.

J. DuMOND: Could you comment on the possibility of finding better methods of decelerating the muons than the one you have described using high pressure gases?

J. M. BAILEY: Most simple methods that immediately suggest themselves unfortunately result in a smaller flux of muons. Elaborate extraction schemes ("muon channels") are better, and give beams with energies nearer 1 MeV than 40 MeV. But I think a different experimental method for detecting the fast decay positrons is the answer. Making our gas target function as a gas scintillation chamber or as an ionization chamber, for example, would allow us to use a long target filled with low pressure gas (this is not feasible with the present system of plastic scintillators outside the target).

The Most Probable Up-to-Date Value
of the Normal Molecular Volume V_0 of an Ideal Gas

By

T. Batuecas

University of Santiago, Santiago de Compostela, Spain

The evaluation with all possible accuracy of the normal molecular volume, V_0 of gases in the ideal state is of great interest not only in order to know that fundamental constant but also because it is indispensable in order to establish rigorously the values of the fundamental constants R_0 (gas constant per mole) and k (Boltzmann's constant). Since in order to fix V_0 with desirable precision one must employ the method of limiting gas density, we believe it will be useful before going into the main subject to recall briefly the ideas on which the method is based.

I

More than a century ago, (1842), REGNAULT established the notion of *limiting density*. His experiments on the coefficients of expansion of several gases have shown that these coefficients all lie close to the same value, approximately 1/273, when the pressure diminishes. Therefore, at constant temperature, gaseous densities tend to limiting values which in each case depend only on the nature of the gas studied.

One half century later, Lord RAYLEIGH[1] called attention to the notion of limiting density in the course of his magnificent researches on gases, but to D. BERTHELOT[2] goes the merit of having demonstrated the great importance of that concept. If real gases obey strictly the laws of BOYLE, of GAY-LUSSAC and of AVOGADRO, the determination of 'molecular masses by means of gaseous substances could be accomplished very simply by use of the formula:

$$M = V_0 \cdot L_1, \tag{1}$$

where M and L_1 designate the molecular mass and the normal density of the gas respectively, while V_0 is the normal molecular volume, that is to say the volume which at 0° C, and under normal atmosphere pressure would be occupied by 1 g molecule of an ideal gas.

Now, since the experiments of Regnault have shown that the laws of Gay-Lussac and Avogadro are only limiting laws, holding only at very low pressures and for extreme rarefaction, the molecular volumes of all gases are equal. It is logical to assume with D. Berthelot[2]: "that the physical scale of molecular weights, based on the densities of gases, approaches the closer to the chemical scale as the pressure of the gas is lowered and that it will become identical with the latter if one substitutes for the normal densities (that is to say the densities measured at $0°$ C and under 1 atm. pressure) the limiting densities taken for an infinitely low pressure". It follows, therefore, that in order to rigorously calculate the molecular masses, formula (1) must be replaced by the following:

$$M = V_0 \cdot L_0, \tag{2}$$

L_0 designating the limiting density of the gas studied.

Let us note in passing that formula (2) only defines a scale of molecular masses to within an undetermined constant to the extent that the ratio $\dfrac{M}{L_0}$ always remains equal to V_0 for any gas whatever. Since the limiting density cannot be measured experimentally, D. Berthelot has shown that it is easy to deduce it from the mass of the liter, L_p, under a finite pressure, if we assume that the compressibility of real gases can be expressed by the formula:

$$1 - \frac{p \cdot v}{p_0 v_0} = A_{p_0}{}^p \cdot (p - p_0). \tag{3}$$

$A_{p_0}{}^p$ designating the mean departure from the law of Boyle in the pressure interval, $p - p_0$. Now, if the pressure p_0 becomes infinitely small, it is easy to see that according to formula (3), the molecular volumes of the different gases will take on, under pressure p, values proportional to: $1 - A_0{}^p \cdot p$, $1 - A_0'{}^p \cdot p$, $1 - A_0''{}^p \cdot p \ldots$, the constant of proportionality being evidently V_0. Finally, since the molecular mass M of any gas is equal to the product of the molecular volume by the mass of the liter at pressure "p" (expressed in atmospheres) one obtains

$$M = V_0 \cdot L_p (1 - A_0{}^p \cdot p), \tag{4}$$

which is the formula established by D. Berthelot[2], on the basis of his method of limiting densities.

This formula (4) similar to (2) furnishes for M only a system of proportional values, that is to say, defined to within a constant multiplier. However, being valid for any pressure whatever in the interval 1–0 atmospheres, formula (4) is of more general applicability than formula (2). Furthermore, comparison of formulae (4) and (2) leads to the following relation:

$$L_0 = L_p \cdot (1 - A_0{}^p), \tag{5}$$

which permits the determination of the limiting density, L_0, without the need for extrapolation. Because of the fact that for each gas the quantities L_0 and $A_0{}^p$ remain constant (so long as the temperature does not change) we see from relation (5) that $1/L_p$ rather than L_p ought to be a linear function of the pressure. The mass of a liter, defined by the expression $L_p = L_0/(1 - A \cdot p)$ can be expressed as a power series

$$L_p = L_0 [1 + A_0{}^p \cdot p + (A_0{}^p)^2 \cdot p^2 + (A_0{}^p)^3 \cdot p^3 + \ldots]$$

in which one can neglect the higher order terms above the second degree inasmuch as $A_0{}^p$ scarcely ever exceeds 10^{-2} in any gas. Furthermore, it is possible to write without appreciable error,

$$L_p = L_0 [1 + A_0{}^p \cdot P + (A_0{}^p)^2 \cdot P^2]$$

a quadratic formula which for permanent gases (He, Ne, H_2, N_2, O_2 and Ar, for which the value $A_0{}^p$ is less than 10^{-3}), reduces to the following, $L_p = L_0 + L_0 \cdot A_0{}^p \cdot P$. Only for the easily liquifiable gases such as NH_3, SO_2. $ClCH_3 \ldots$ for which the value $A_0{}^p$ exceeds 10^{-2} will there be slight departures from the linear formula for pressures in the neighborhood of 1 atm. This allows us, therefore, to employ a linear extrapolation even though it may not always be strictly rigorous.

II

Following these general considerations, let us look in a little greater detail at the question of the normal molecular volume, V_0, which at present may be considered most probable. Before attacking the subject, it seems to us instructive to review briefly the progress achieved since 1898 in determining a precise value of V_0. A fair number of experimental researches on gases since 1900 had for their object the determination of molecular and atomic masses with all possible precision, by using various physico-chemical techniques applicable to real gas, notably that of limiting densities. Now according to formula (4), with the exception of the measurements necessary for the evaluation L_p and $A_0{}^p$, the calculation of M requires a suitable knowledge of V_0. This explains the numerous efforts of eminent scientists to determine a value of this fundamental constant. It would seem that A. LEDUC[3] was the first who, with his method of molecular volume and his own measurements concerning O_2, succeeded in deducing an approximately correct value for the normal molecular volume, $V_0 = 22.405$ liters/mole. Later LEDUC[4] has indicated that the value will be in the neighborhood, $V_0 = 22.410$ liters/mole.

On the other hand, according to the then current most reliable data for L_p and $A_0{}^p$ regarding the 7 gases H_2, CO, O_2, CO_2, C_2H_2, HCl and SO_2, D. BERTHELOT[5] proposed (1904) the value, $V_0 = 22.412$ liters/mole, representing not the mean but the best value according to him of the

7 applicable results, among which the extreme values were 22.420_8 (H_2) and 22.398_3 (HCl) and the mean deviation $\pm 0.006_5$. Later D. BERTHELOT pointed out very properly[6] that since the value V_0 is not given a priori, once having chosen the convention that $O_2 = 32$, one is forced to adopt for V_0 the value derived from the ratio $32/L_p (1 - A_0^1)$ relative to oxygen.

The researches on gases which were carried on over a period of 20 years at Geneva under the initiative and direction of PH. A. GUYE, with the assistance of numerous collaborators, constituted a great advance because these researches, in which the greatest care was taken both as to the measurements and as to the purity of the substances studied, furnished numerous results concerning L_p and A_0^1 measured directly at $0°$ C, a very important point. Notably as regards the normal molecular volume PH. A. GUYE, after having used for a very long while a value $V_0 = 22.410$ liters/mole, finally adopted[7] the following value: $V_0 = = 22.414$ liters/mole, deduced on the basis of the measurements at $0°$ C on the compressibility (due to R. W. GRAY, on the one hand, and A. JAQUEROT and O. SCHEUER on the other) and those of A. GERMAN on L_1 for oxygen.

Later E. MOLES[8], making use of data at $0°$ C (those most worthy of confidence) on oxygen arrived at the value, $V_0 = (22,4148 \pm 0.0007)$ liters/ mole as the mean of the three values shown in Table 1. In this table, adapted from E. MOLES, the values of the departure from Avogadro's law $(1 + \lambda)$ have been replaced by their reciprocals, $1 - A_0^1 = 1/(1 + \lambda)$. Furthermore, the symbol used here to designate the normal liter and the mass of the normal molecular volume are different from those used by MOLES[8]. It should be noted that the preceding mean and also the value adopted by PH. A. GUYE had been deduced taking as base $O_2 = 32$ and the value $g_{45} = 980.616$ gal.

Table 1. The normal molecular volume data used by E. MOLES

L_1 g/liter	Author	$1 - A_0^1$	Author	V_0 liters/ mole
1.42891	GRAY	0.00097	GRAY and BURT	22.4162
1.42890	MOLES and BATUECAS	0.00085	GUYE and BATUECAS	22.4139
1.42892	BAXTER and STARKWEATHER	0.00091	BAXTER and STARKWEATHER	22.4143
			MEAN	22.4148

One may, however, ask whether in order to achieve calculation to the highest precision it is entirely justified to use values of L_p and A_0^1 which were not obtained by a single experiment. The only researches among all those performed on oxygen up to the present which can satisfy such a condition would be as far as we know the following, innumerated in chronological order: G. P. BAXTER and H. W. STARKWEATHER (1924 to 1926, Harvard, USA), E. MOLES et al. (1934–1937, Madrid), T. BATUECAS, F. L. CASADO, G. G. MALDE (1939–1941–1949, Santiago de Compostela, Spain). We shall use the data concerning L_p and A_0^1 coming from these researches to calculate in each case the normal molecular volume. Let us add that these calculations have been made on the one hand adopting for the molecular mass the value of the new scale, $O_2 = 31.9988$, and on the other hand referring all the values of L_p to a standard gravity $g_0 = 980.665$ gal.

Let us look in the first place at the work of G. P. BAXTER and H. W. STARKWEATHER[9], which because of the extreme care given to the purification of the oxygen and to the large number measurements, must be considered as monumental.

Table 2. The values of L_p from measurements of BAXTER and STARKWEATHER for various pressures

P (atm.)	L_p (g_{45}) g/liter	L_p (g_0)
1.00000	1.42896	1.42903
0.75000	1.42865	1.42872
0.50000	1.42830	1.42837
0.25000	1.42796	1.42803

Table 2 shows in column 2 the mean value found by BAXTER and STARKWEATHER for L_p at pressures of 1, $^3/_4$, $^1/_2$, $^1/_4$ atm., all values referred to gravity, g_{45} and after correction for adsorption (see E. MOLES, loc. cit., p. 45); in column 3 appear the values of L_p (g_0) referred to standard gravity. Using this value of L_p (g_0) and formula (5), $L_0 + L_p \cdot A_0^1 \cdot P - L_p = 0$, one calculates by the method of least squares the following values of limiting density and the mean departure from Boyle's law, $L_0 = 1.42769 \pm 0.00001$ g/liter and $A_0^1 = 0.00095 \pm 0.00001$ rounded off to the 5th decimal. Finally, by means of the value of L_0 and formula (2), $M = V_0 \cdot L_0$, the measurements of BAXTER and STARKWEATHER permit one to deduce for the normal molecular volume

$$V_0 = \frac{31.9988 \pm 0.0002}{1.42769 \pm 0.00001} = 22.4130 \pm 0.0002 \text{ liters/mole}.$$

The studies performed on oxygen between 1934 and 1937 at Madrid by E. MOLES and several collaborators[10] constitute another example of magnificent precision. Great efforts were made to obtain extremely pure gas and also in the measurement of pressure. MOLES maintained the entire mercury column of the manometer in crushed ice, thus rendering unnecessary any correction for temperature.

In Table 3 are shown, in addition to the mean values (after all corrections) of L_p (g_{45}) at pressures of 1, $^2/_3$, $^1/_2$, $^1/_3$ atm., those of L_p (g_0), that is to say referred to standard gravity. By means of the preceding values of L_p (g_0) and by a calculation entirely analogous to the one sketched above, one obtains by least squares:

$$L_0 = 1.42768 \pm 0.00001 \text{ g/liter and } A_0' = 0.00093 \pm 0.00001$$

rounded off to the 5th decimal place, and finally for the normal molecular volume

$$V_0 = \frac{31.9988 \pm 0.0002}{1.42768 \pm 0.00001} = 22.4131 \pm 0.0002 \text{ liters/mole.}$$

This value requires a small subtractive fractional correction of 1.3×10^{-5}, whose origin, as we shall see below, is due to the fact that the gravity used ($g = 979.953$ gal) suffered from a small systematic error of that fractional amount. As a result, the experiments of E. MOLES and collaborators yields for V_0 the value,

$$V_0 = 22.4128 \pm 0.0002 \text{ liters/mole.}$$

Finally, we now turn to our research on oxygen performed at the University of Santiago, Spain, in collaboration with F. L. CASADO and G. GARCIA MALDE[12]. As in the case of our predecessors, the concern as to purity of gas and precision of measurement has been our constant goal during this research. Thus, in order to better guarantee the calibration of the compressibility apparatus and the measurements of pressure, we have determined the density at $0°$ C of the highly pure mercury which we used. Also, a rigorous knowledge of the acceleration of gravity at the point where the experiments were made was indispensible in order to assure high precision. A determination of g at the laboratory of Physical Chemistry at the University, performed at our request in 1942 by the "Spanish Gravimetric Service" furnished us with a relative value (with respect to the Potsdam system), $g = 980.417$ gal., which has been utilized in all our calculations. Furthermore, our researches are the only ones we know of in which L_p and $A_0{}^1$ have been measured on identical samples of very pure oxygen. This is important to point out because it then becomes possible to calculate V_0 by two procedures; the one utilizing formulas (5) and (2), the other formula (4).

Table 3. Measurements of E. MOLES and Collaborators

P (atm.)	L_p (g_{45}) g/mole	L_p (g_0) g/mole
1.00000	1.42894	1.42901
0.66667	1.42849	1.42856
0.50000	1.42829	1.42836
0.33333	1.42804	1.42811

Table 4. The Measurements of F. L. CASADO and T. BATUECAS

P (atm.)	L_1 ($g = 980.629$) g/mole	L_p (g_0)	g/mole
1.00039	1.42881	1.42886	mean of 20 separate measurements
0.66621	1.42837	1.42842	mean of 24 separate measurements
0.33309	1.42804	1.42809	mean of 25 separate measurements

Let us examine first the result which one arrives at by applying the first of these procedures to the measurements on oxygen made by F. L. CASADO and the author at pressures of approximately 1, $^2/_3$ and $^1/_5$ atm. In Table 4 appears in addition to the mean values with respect to gravity $g_{45} = 980.629$ gal., those referred also to standard gravity (g_0).

Using formula (5) and the value $L_p(g_0)$ given in the table, we calculate by least squares the following results for the limiting density and for the mean departure from Boyle's law:

$$L_0 = 1.42768 \pm 0.00002 \text{ g/mole and } A_0{}^1 = 0.00082 \pm 0.00003.$$

With the help of this result for L_0 and formula (2) one obtains for the normal molecular volume:

$$V_0 = -\frac{31.9988 \pm 0.0002}{1.42768 \pm 0.00002} = 22.4131 \pm 0.0003 \text{ liters/mole.}$$

The other procedure for calculating V_0 which utilizes formula (4), evidently requires a suitable determination of the quantities L_p and $A_0{}^1$. As regards the *mass of the liter*, the 69 individual values found by F. L. CASADO and T. BATUECAS are amply sufficient for our needs. As for $A_0{}^1$ the two series of independent measurements, made by F. L. CASADO and T. BATUECAS (10 measurements), and by T. BATUECAS and G. GARCIA MALDE (11 measurements), furnish the following general mean: $A_0{}^1 = 0.00087 \pm 0.00002$. Now, because of the fact that according to formula (4) if M and $A_0{}^1$ are known, each value of L_p permits us to deduce another value for V_0, it has been possible to obtain 69 distinct values for the normal molecular volume. It goes without saying, that these values have been calculated using: $O_2 = 31.9988$, together with the mean value indicated above for $A_0{}^1$. Proceeding in this way, one finds for the general mean of the 69 distinct values, $V_0 = 22.4139 \pm 0.0009$ liters/mole, in good accord with that deduced by the other procedure for calculation.

However, these two means are still subject to a small substractive fractional correction of 1.3×10^{-5} to which we have referred in commenting on the measurements of E. MOLES of Madrid.

This correction has the following origin. As we have said before, the determination of g at the Laboratory of Physical Chemistry (University of Santiago) has furnished us with the value 980.417 gal. with respect to that at Potsdam (981.274 gal. established in 1906). Since it has been recently ascertained that this latter value was slightly high by 0.013 gal., it follows that all the values of g referred to the Potsdam system, and in particular those determined at Santiago and Madrid, must be corrected by the same fractional amount[13]. Let us point out in this connection that the small subtractive correction affects only the values of p; if these values diminish by the fractional amount 1.3×10^{-5}, those of L_p will increase by the same ratio and this will produce finally a decrease of 1.3×10^{-5} in the value of V_0.

Consequently, the two values which one obtains from the measurements at Santiago, (after all corrections):

$V_0 = 22.4128 \pm 0.0003$ liters/mole, and $V_0 = 22.4136 \pm 0.0009$ liters/mole,

lead to a general mean, $V_0 = 22.4132 \pm 0.0006$ liters/mole. The latter value represents the most probable result of our measurements on oxygen.

Table 5. Resume of normal molecular volume measurements

V_0 liters/mole	Author
22.4130 $\pm$ 0.0002	Baxter and Starkweather
22.4128 $\pm$ 0.0002	Moles and Collaborators
22.4132 $\pm$ 0.0006	Batuecas, Casado and Garcia Malde

In Table 5 we give a resumé of the results obtained for the normal molecular volume according to the high precision measurements on oxygen most worthy of confidence. In view of the good agreement of these results, it is permissible to believe that their mean, $V_0 = 22.4130 \pm 0.0003$ liters/mole, may be considered to be the most probably value at the present time of the normal molecular volume of an ideal gas.

It is worthwhile to emphasize that one arrives at the *same result* if one converts the value given in 1957 by E. R. Cohen, K. M. Crowe and J. W. DuMond[14] to the scale of ^{12}C.

III

In spite of the high precision of the result concerning V_0 indicated above, it would be desirable to have new measurements to evaluate that fundamental constant with still higher precision. The following brief commentary is intended to give a few useful suggestions in this direction.

As long as oxygen was the reference element for atomic masses, the rigorous evaluation of the normal molecular volume had to be made directly with respect to this gas. Now, however, after the adoption of the new scale based on the mass of the nuclide $^{12}C = 12$ the situation is changed and oxygen no longer plays a privileged role but has become an element like any other. This being the case, the evaluation of V_0 could be made, at least in principle, using any gas whatever provided that certain suitable conditions be fulfilled. The three principle conditions that a gas must satisfy in order to be utilized in such a precise evaluation of V_0 would be in our opinion:

Firstly — The isotopic composition of the element or elements constituting the gaseous molecule must be invariable or at least must have only a very small natural variation. This is in order to guarantee with sufficient precision that one knows the molecular mass of the gas to be studied.

Secondly — The magnitudes L_p and $A_0{}^1$ must be known in each case very exactly. This makes it necessary to exercise the greatest care in the purity of the materials and in the measurements themselves. Let us add once again that a knowledge of g, determined at the place where the experiments are performed and also the density at $0°$ C of the mercury used in the measuring apparatus seems indispensible.

Thirdly — The correction for adsorption of different gases on the glass walls of the vessel must be reduced to a minimum. This is because any experimental method directed at the correct evaluation of adsorption must furnish data referring to the particular surfaces of the receptacle used to measure the gaseous densities. Since the determination of the quantity of gas adsorbed by the walls of the vessel is by no means easy, workers have tried to avoid the difficulty by using an experimental technique[10] which is only capable of furnishing results to a mediocre approximation.

Consequently, only by the use of permanent gases for which adsorption is very small or negligible can one guarantee that the uncertainty due to an error of adsorption has been eliminated. But among the permanent gases, nitrogen is the most suitable for an evaluation of V_0, provided that its isotopic composition remains constant.

Along with nitrogen, other gases H_2, CO, He, Ar and Ne could also be studied for the rigorous evaluation of the normal molecular volume. If such a program of research could be accomplished, without doubt our knowledge of the fundamental constant V_0 could be made still much more exact than at the present day.

References

[1] Lord RAYLEIGH, Proc. Roy. Soc. **50**, 448 (1892).

[2] D. BERTHELOT, Compt. rend. **125**, 54 (1898); J. Phys., 7th series **15**, 263 (1899).

[3] A. LEDUC, Ann. Chim. Phys. (1898).

[4] A. LEDUC, Ann. Chim. Phys. 8, Ser. 19, 441—475 (1904).

[5] D. BERTHELOT, Z. Elektrochem. **10**, 621 (1904).

[6] D. BERTHELOT, Compt. rend. **145**, 180 (1907).

[7] PH. A. GUYE, J. Ch. Phys. **14**, 380 (1916).

[8] E. MOLES, Z. anorg. allg. Chem. **167**, 46 (1927).

[9] G. P. BAXTER and H. W. STARKWEATHER, Proc. Nat. Acad. Sci. **10**, 479 (1924); **12**, 699 (1926).

[10] „Les Determination physico-chimiques des Poids Moléculaires et Atomiques des Gaz", Collection Scientifique, Paris, 1939; Instit. Internat. Cooper, Intellect., pp. 1—75.

[11] F. L. Casado and T. Batuecas, An. Fis. Quim. 48, 4 (1952).

[12] T. Batuecas and G. Garcia Malde, An. Fis. Quim. 46, 517 (1950).

[13] I am grateful to Professor J. W. DuMond for having called my attention to this small systematic error affecting all of the measurements of gas densities made at Santiago.

[14] Cohen, Crowe, and DuMond, Fundamental Constants of Physics, p. 266 (1957).

The International Union of Pure and Applied Chemistry and its Interest in the Values of the Fundamental Constants

By

Frederick D. Rossini

(Representing the International Union of Pure and Applied Chemistry, through the Division of Physical Chemistry for the Commission on Physicochemical Data and Standards and the Commission on Thermodynamics and Thermochemistry)

I thank you, Mr. CHAIRMAN, for the opportunity to say a few words before this Conference on the subject of the fundamental constants.

I am present at the Conference as an Observer and Official Representative of the International Union of Pure and Applied Chemistry, through its Division of Physical Chemistry, for the Commission on Physicochemical Data and Standards and the Commission on Thermodynamics and Thermochemistry, in connection with the problem of values of the fundamental constants. I want to summarize briefly the position of the International Union of Pure and Applied Chemistry on the problem.

The progress of science throughout the world requires that our scientific community have an adequate system of communication. The sciences of physics and chemistry involve a tremendous amount of information which must be communicated. Much of the information to be communicated is quantitative in character.

An experimental investigator in his laboratory makes measurements of certain physical phenomena with instruments and apparatus calibrated in terms of the fundamental units of measurement—length, mass, and time. Most frequently, the quantity reported in the literature by an investigator is not just precisely alone the quantities he has measured, but these combined with certain fundamental constants to obtain quantities for comparison with related observations of other investigators. For proper and convenient communication, it is desirable that the different investigators use the same values of the fundamental constants in reducing their respective sets of data, otherwise the reported values will be different even though the quantities originally measured may actually have been in complete accord. It is important, therefore, for all of science that there be available a currently acceptable self-consistent set of "best" values of the fundamental constants.

The need for fundamental constants was recognized in chemistry many years ago when there was established the chemists' international scale of atomic weights, so that chemists throughout the world might communicate their observations in a quantitative language that would be readily understood by other scientists. In this connection, I would like to say that the science of chemistry owes a great debt to Professors MATTAUCH and NIER and their collaborators on the Commission on Nuclidic Masses under the International Union of Pure and Applied Physics for our first truly international scale of atomic masses, now that the sciences of physics and chemistry have agreed upon the carbon scale with ^{12}C as the reference.

In addition to a self-consistent set of "best" values for the atomic weights, the science of chemistry needs a self-consistent set of "best" values of the fundamental constants. This need is particularly strong in the field of physical chemistry where we have to communicate much quantitative information relative to values of the physical, thermodynamic, and spectral properties of the chemical substances.

In chemistry, we find it convenient to classify the fundamental constants somewhat arbitrarily in three categories: – the basic constants, the values of which are obtained from experimental measurements; the defined constants, the values of which are fixed by definition; and derived constants, the values of which are obtained from the foregoing and appropriate physical relations.

With the value of the absolute temperature of the triple-point of water having been fixed by definition several years ago by the International Committee on Weights and Measures, we need in chemistry the values only of five basic constants: the velocity of light; the Avogadro number; the Faraday constant; the Planck constant; and the pressure-volume product for one mole of a gas at standard pressure and 0° C. In this group of five basic constants one may use the charge on the electron in place of either the Avogadro number, or the Faraday constant, depending upon the relative uncertainties. The actual evaluation of these constants of course involves other measurements. With values for these five basic constants, one may proceed to evaluate the derived constants to obtain the complete set of fundamental constants.

The science of chemistry needs a self-consistent set of "best" values of the fundamental constants. For the large majority of the work in chemistry the need for high precision is about one magnitude less than that needed for physics. For chemistry, we need a set of values of the fundamental constants which are not only the "best" available for the given period of time but also which are self-consistent.

Under the International Union of Pure and Applied Chemistry, the Commission on Physical Data and Standards and the Commission on Thermodynamics and Thermochemistry, in the Division of Physical Chemistry, are most anxious to reach accord with the Commission on Nuclidic Masses and Related Constants of the International Union of Pure and Applied Physics, in the matter of values of the fundamental constants.

It is important to have a self-consistent set of "best" values of the fundamental constants which are available for use by the large majority of chemists and physicists who need to communicate quantitative scientific information with one another. It is to be emphasized that a given agreed-upon set of "best" values of the fundamental constants will be subject to revision at appropriate intervals as new data become available. Such revision would be done by an appropriate international body.

Appropriate groups under the International Union of Pure and Applied Chemistry have reviewed the set of fundamental constants which are being considered at this Conference and find them acceptable. IUPAC is prepared to reach international accord with IUPAP on this self-consistent set of "best" values of the fundamental constants.

Present Status of our Knowledge of the Numerical Values of the Fundamental Physical Constants

By

E. Richard Cohen

North American Aviation Science Center, 8437 Fallbrook Avenue, Canoga Park, California

and

J. W. M. DuMond

California Institute of Technology, 1201 East California Street, Pasadena, California

(Presented by E. R. COHEN)

With 2 Figures

Introduction

In 1955 we published a least squares adjustment of the fundamental atomic constants[1]. Since that time several corrections and revisions[2-5] have been applied to the 1955 data, but no complete reanalysis had been made. We have now completed a new analysis of all of the old data and of the host of new data which has become available since 1955. The present adjustment includes several new measurements of physical constants such as the FARADAY, the magnetic moment of the proton, and the proton gyromagnetic ratio. In addition, new measurements of the fine structure constant, α, indicate new experimental relationships among the physical constants which have been determined with sufficient accuracy that they warrant inclusion in any future adjustment. The present adjustment is based on data which was available in 1962. Some results which have come to our attention since then have been used only as confirmatory evidence to support our conclusions, but have not been analyzed in detail, in some cases because of lack of complete information concerning calibration standards, in other cases merely because we received the information too late for inclusion.

The changes which have occurred in the available experimental data since 1955 have been extensive. They can be classified roughly into three categories:

1. Definition of units.

2. Redetermination of auxiliary data which affect the input values to a least squares adjustment but do not significantly influence the accuracy of the experiment.

3. New experimental data of increased accuracy which replaces previous experimental values.

Changes in Definition of Units

The Unified Scale of Atomic Weights

After several years of consideration by the IUPAP and IUPAC, the new unified scale of atomic weights was adopted in 1960. This scale is based on the arbitrary assignment of the mass of exactly 12 units to the isotope ^{12}C. As such, this definition replaces both the physical scale of atomic weights based on the assignment of mass 16 to the isotope ^{16}O and the chemical scale of atomic weights which assigns the mass 16 to the "natural" isotopic mixture of oxygen isotopes. The transition to this scale has been greatly facilitated by the excellent and exhaustive computations of EVERLING, KÖNIG, MATTAUCH, and WAPSTRA, which were first reported at the First International Conference on Nuclidic Masses held at McMaster University in 1960[6], and to which revisions and additional data have been added since then[7].

In the unified scale of nuclidic masses the mass of ^{16}O is given[7] as $15.99491494 \pm 0.00000028$, so that we have the conversion factor from the unified scale to the old physical scale ($^{16}O = 16$)

$$\frac{\text{Mass on physical scale } (^{16}O = 16)}{\text{Mass on unified scale } (^{12}C = 12)} = 1.000317917 \pm 17. \tag{1}$$

The definition of the old chemical scale ($O = 16$) is, of course, confused by the uncertainty of the definition of "natural isotopic abundances" for oxygen in the face of known variations in these abundances, depending on the origin of the oxygen sample. These differences are possibly the result of biological processes since the $^{18}O/^{16}O$ ratio for oxygen from inorganic sources (FeO or water) is as much as 4% less than that from air or limestone. Based on measured abundances of the isotopes we find 1.000275 ± 0.000005 for the ratio of masses on the old physical scale to masses on the old chemical scale, although the factor has generally been used as an *exact* value with no assigned error to convert physically determined nuclidic masses to chemical atomic weights.

The Redefinition of the Thermodynamic Temperature Scale

The thermodynamic temperature scale in the past, as exemplified by the centigrade scale, was based on the arbitrary assignment of $0°$ and $100°$ to the melting point and boiling point, respectively, of water under a pressure of one atmosphere. This definition leads experimentally to a thermodynamic temperature scale in which the absolute zero temperature point is $-273.16 \pm 0.01°$ C. This scale has now been abandoned, and in its place the two fixed points are taken to be $-273.15°$ C for absolute zero and $0.01°$ C for the triple point of water. (In the official recommendation natural water is specified for the triple point determination, recognizing that the isotopic purity of the water will in principle affect the precise value of this equilibrium point. However, the present limits of measurement are such that the natural isotopic variation in water is of no importance. At such time as experimental techniques make the more precise definition necessary, a specific isotopic composition will have to be specified.) On this thermodynamic scale the ice point of water is $0.0000 \pm 0.0001°$ C, and the steam point is $99.9964 \pm 0.0036°$ C. Thus we can no longer say that we have a centigrade scale in the sense that there exists a $100°$ temperature difference between two fixed points, and the name "centigrade" should be abandoned, although the designation °C is to be retained, with the scale referred to as the Celsius scale. The Kelvin scale of temperatures is defined at the same time by adding 273.15 to the Celsius scale as here defined.

Revisions of Auxiliary Data

The Acceleration of Gravity

The relative values of gravity at different points on the earth's surface can be measured with great accuracy and reproducibility, and gravity differences between two stations can often be determined to accuracies of a tenth of a part per million[8].

The experimental difficulties of defining the absolute base for gravity measurements are considerable. For many years the absolute determination made with the reversible pendulum method in 1906 by KUHNEN and FURTWANGLER at Potsdam (981.2740 cm sec^{-2}) was regarded as the International standard. Since then, however, evidence has been accumulating that this Potsdam value is too high by approximately 15 parts per million. The U. S. Bureau of Standards has for twenty years or more, used a gravity standard for g_0 at Washington 980.082 cm sec^{-2}, which is 17 parts per million lower than the value on the Potsdam network[9]. This is the so-called "Dryden reduction".

The pendulum method of measuring g was devised to circumvent the difficulties of accurate small interval time measurement. With the recent improvements in time standards and the ability to measure nanosecond intervals, increasing attention has been given to gravity measurements utilizing the straightforward timing of a body in free fall. It now seems quite clear that the traditional "Potsdam value" of gravity should be corrected downward by 13.0 ± 0.4 milligals. This correction is important in our adjustments because of its effect on the absolute determination of the ampere. The change of 4 ppm from the Dryden reduction implies a change of 2 ppm in the value of the ampere as maintained by the Bureau of Standards during the past decade.

The Velocity of Light

In the past 15 years there have been a dozen or more measurements of the velocity of light, all of which have yielded measured values significantly larger than the pre-war measurements of MICHELSON and of ANDERSON. These determinations are summarized in Table 1.

Table 1. Velocity of Light

Method	Author	Value
Microwave Interferometer.	FROOME	299792.50 ± 0.1 km/sec
Geodimeter	BERGSTRAND, McKENZIE	299792.85 ± 0.2 ,,
Cavity Resonance	ESSEN	299792.50 ± 0.3 ,,
Shoran	ASLAKSON	$299793.4 \ \ \pm 2.0$,,
Adopted Value..........	(ISRU, IUGG)	299792.50 km/sec 186282.42 mi/sec

The most recent reliable measurements are those of BERGSTRAND[10] and of MACKENZIE[11] using visible light, and of ESSEN[12] and FROOME[13] using microwaves. Based primarily on the results of these microwave measurements, the International Scientific Radio Union has recommended the adoption of the value $c = 299792.5$ km per second, and this value has also been adopted by the International Union of Geodesy and Geophysics. This value is almost certain to be correct to better than one part per million, and as such, it may be considered to be sufficiently accurate that it does not enter into least squares adjustment as a variable.

The Revision of the Theoretical Formula for the Electron Magnetic Moment Anomaly

The anomalous magnetic moment of the electron was first calculated to second order in α by KARPLUS and KROLL[14]. The calculation has been shown to be in error by the more recent work of KROLL, and by SOMMERFIELD[15], PETERMANN[16], and SMRZ and ULEHLA[17], all of whom are now in agreement with regard to the value of the coefficient of the second order term.

$$\mu_e/\mu_0 = 1 + \frac{\alpha}{2\pi} - 0.328 \left(\frac{\alpha}{\pi} \right)^2 + \ldots \tag{2}$$

New Experimental Data

There are several different types of experiments which measure different combinations of fundamental constants. These different relationships are connected in such a way that no single experiment can be said, in general, to define any one of the fundamental constants α, e, N, Λ, uniquely. It is for this reason that statistical analyses, such as the method of least squares, must be used not only to extract the maximum information from the host of experimental data, but also to give an indication of those experiments which, because of their inconsistency with the totality of the remaining experiments, are likely to be afflicted with various systematic errors. We shall first briefly list and describe the experiments and their numerical results and then discuss more fully the relationships which exist among them.

Avogadro Constant

The use of crystal data to compute the Avogadro number has been explained many times in the literature. Because of the progressive increase, with the passage of time, in accuracy and precision in all the physical measurements bearing on the fundamental constants, the attitude toward this problem and hence also the approach to it have undergone notable changes over the past 30 years.

Basically, the idea is simple; crystals may be idealized as 3-dimensional periodic lattice structures built up out of unit cells whose geometry can be precisely determined by the methods of X-ray crystallography, so that the *volume* of the unit cell can be known very accurately *on the relative scale of X-ray wavelengths, i. e., in cubic X-units* (though unfortunately much less accurately in terms of our macroscopic cgs units, e. g., angstroms or centimeters, because of the imprecision in our knowledge of the conversion factor, Λ, between the two scales. The X-unit was originally intended to be 1 milliangstrom or 10^{-11} cm; however, through ruled grating diffraction measurements it has been shown to be larger than the milliangstrom by about 0.2%.)

Each unit cell of a crystal may consist of a whole number of molecules of its chemical formula or a fraction of one molecule, depending on the crystal symmetry system. In any case, this number, which we shall call f, is exactly known (e. g., f may equal $1/2$ or 8, or some other specifiable value.) Thus the product of f and the molecular mass, M, gives the molecular mass of the unit cell, $M f$. The product of the density, ϱ, of the crystal by the absolute volume, v, of its unit cell should ideally give the absolute mass, ϱv, of the unit cell, and the quotient of $M f$ by ϱv should thus give the Avogadro number, N. However since we do not, in fact, measure the absolute volume of the unit cell in cm^3, but do measure the density in grams cm^{-3}, it is necessary to recognize that

XRCD (X-ray crystal density) data alone do not, in fact, yield a value of N but rather a value of the product $N \Lambda^3$.

We define the conversion factor, Λ, as the ratio of the milliangstrom unit to the X-unit. The determination of Λ depends on difficult precision measurements of X-ray emission lines reflected in grazing incidence on artificial ruled diffraction gratings, and its true value is doubtful to at least several tens of parts per million. Different precise determinations have fluctuated over a range much in excess of 100 ppm.

The question immediately arises as to why we treat the XRCD determination of the conceptually awkward composite quantity, $N \Lambda^3$, and the determination of the conversion constant, Λ, as two separate data rather than combining them so as to determine the conceptually far more interesting Avogadro number directly. The reasons are twofold: (1) because, as we shall see, the present XRCD data yield a mean value of $N \Lambda^3$ with an uncertainty of order substantially less than $\pm$ 20 ppm, whereas the value of Λ is only established at present with much poorer accuracy than this, and (2) because the present state of our knowledge is such that, in any general least squares adjustment of fundamental atomic constants, Λ must enter explicitly or implicity, as one of the unknowns subject to adjustment, and will do so in more than one of the fundamental observational equations.

There are at least 17 high precision XRCD measurements which are worthy of mention here. Since some of these measurements were made as much as 30 or more years ago, often with a purpose other than the primary one, for our purposes, of the high precision determination of the atomic constants, it has been necessary in all cases to go back to the original observational data in order to extract the maximum amount of information from the measurements. In all cases the molecular weights of the crystals used were recomputed on the basis of current best values of nuclidic masses and measured isotopic abundances. Except for crystals of anisotopic composition, such as Al, the limiting factor in the molecular weight determination is the abundance of the isotopes rather than the nuclidic masses. In all cases those physical molecular weights so computed are more precise than the adopted chemical weights, but generally not inconsistent with the latter. The only exception to this statement is in the case of germanium, for which the computed physical atomic weight is 72.630 ± 0.006, whereas the chemical atomic weight, as determined by gravimetric methods, is 72.60*. It has also been necessary, in many cases,

---

* It is to be emphasized that these numbers are *both* on the unified (^{12}C) scale, and that by "physical" and "chemical" we imply only the basis of the methods of determination, i. e., isotopic abundances and nuclidic masses in the one case and gravimetric determination of chemical equivalent weights in the other.

to recompute lattice spacings from the measured X-ray diffraction angles, including, where appropriate, a correction for the crystal index of refraction.

Table 2. Comparison of Wavelength Ratios of λ(Mo $K\alpha_1$) and λ(Cu $K\alpha_1$)

Taken From Tables	Bearden's Laboratory
Mo $K\alpha_1$ 707.831 x. u.	707.845 x. u.
Cu $K\alpha_1$ 1537.396 x. u.	1537.400 x. u.
Mo $K\alpha_1$/Cu $K\alpha_1$ = 0.460409	Mo $K\alpha_1$/Cu $K\alpha_1$ = 0.460417

Ratio of ratios, 0.460417/0.460409 = 1.000017

Table 3. Data on $N\Lambda^3$ (1962)

Crystal	Value of $N\Lambda^3$	Weight w_i	Author
Using Mo $K\alpha_1$; tabular value 707.831 X-units			
KCl	6058.229	4	YUCHING TU[a]
Calcite	6059.600	20	G. BROGREN[b]
Diamond	6059.595	21	YUCHING TU[a]
Calcite	6059.601	20	J. A. BEARDEN[c]
Diamond	6059.810	23	YUCHING TU[a]
Calcite	6059.982	35	YUCHING TU[a]
Quartz	6060.064	3	G. BROGREN[b]
Rocksalt	6060.115	5	YUCHING TU[a]

Weighted Average Value $N\Lambda^3 = 6059.725 \begin{array}{l} \pm\ 0.081\ \text{internal} \\ \pm\ 0.109\ \text{external} \end{array}$

Using Cu $K\alpha_1$; tabular value 1537.396 X-units			
Silicon	6059.896	23	SMAKULA et al.[d]
CaF$_2$	6060.055	10	SMAKULA et al.[d]
CsI	6060.080	2	SMAKULA et al.[d]
TlCl	6060.145	5	SMAKULA et al.[d]
Aluminum	6060.168	28	SMAKULA et al.[d]
TlBr	6060.282	5	SMAKULA et al.[d]
Germanium	6060.418	3	SMAKULA et al.[d]
LiF	6060.812	12	STRAUMANIS[e]

Weighted Average Value $N\Lambda^3 = 6060.184 \begin{array}{l} \pm\ 0.106\ \text{internal} \\ \pm\ 0.107\ \text{external} \end{array}$

[a] Y. TU, Phys. Rev. 40, 662 (1932).
[b] G. BROGREN, Arkiv for Fysik 7, No. 4, 47 (1953).
[c] J. A. BEARDEN, Phys. Rev. 38, 2089 (1931).
[d] A. SMAKULA and J. KALNAJS, Il Nuovo Cimento, Suppl. Ser. X, 6, 214 (1957); Phys. Rev. 99, 1737 (1955). — A. SMAKULA and V. SILS, Phys. Rev. 99, 1744 (1955). — A. SMAKULA, J. KALNAJS, and V. SILS, Phys. Rev. 99, 1747 (1955).
[e] M. STRAUMANIS, A. IEVINS, and K. KARLSONS, Z. physik. Chem., B 42, 143 (1939).

A vital source of uncertainty in this data is the wavelength of the characteristic X-ray line used to measure lattice spacing. X-ray measurements can determine only angles, and hence not only does the measurement give us dimensions in X-units, it does so only relative to an adopted wavelength for the line used. We have used the values 707.831 X-units for the peak of the Mo $K\alpha_1$ line and 1537.396 X-units for the peak of the Cu $K\alpha_1$ line. These values are those listed in the most reliable wavelength tables, such as those of CAUCHOIS and HULUBEI[18] or the more recent tabulation of SANDSTROM[19]. J. A. BEARDEN has, on the other hand, recently measured the Mo $K\alpha_1$ line relative to Cu $K\alpha_1$ on the basis of which, if we retain our quoted value for the Mo $K\alpha_1$ wavelength, we would infer the value 1537.370 for Cu $K\alpha_1$ (see Table 2). Because of this disagreement in calibration it is best to maintain the separation between data measured on the one hand with Mo radiation and on the other with Cu radiation. Table 3 thus lists 16 independent determinations of $N \Lambda^3$ grouped into two sets of 8 measurements each. HENINS[20] has recently reported a value for $N \Lambda^3$ measured in Bearden's laboratory. Corrected to a Cu $K\alpha$ wavelength of 1537.400 x. u., this gives $N \Lambda^3 = (6059.76 \pm 0.24) \times 10^{20}$, which is in excellent agreement with the Mo data of Table 3, but possibly in disagreement with the Cu data.

If, however, we use Bearden's measurements for the wavelengths (from Table 2), the average of the Mo data of Table 3 is reduced to $N \Lambda^3 = 6059.36$, since an increase of 1 ppm in the assignment of the wavelength decreases $N \Lambda^3$ by 3 ppm. Similarly, the Cu data would be reduced to 6060.15 and Henin's measurement reduced to 6059.72.

Until such time as more complete data is presented on Bearden's measurement of the ratio of Mo $K\alpha_1$/Cu $K\alpha_1$ and on Henin's measurements of $N \Lambda^3$ (these have been reported only in abstract form so far), we shall retain only the data as given in Table 3.

Conversion Factor from Siegbahn X-units to Milliangstroms

The wavelengths of X-ray spectral lines measured relative to each other by the high precision methods of crystal defraction are known with a precision which often exceeds one part in 100,000. Unfortunately, the accuracy with which these wavelengths can be expressed in centimeters or angstrom units is poorer by approximately a factor of 10. We thus have two scales of X-ray wavelengths—a relative scale of X-units by which wavelengths can be compared accurately, and a conversion factor, Λ, which is known perhaps to only 30 parts per million. In 1945, R. T. BIRGE[21] obtained the weighted average value of the converion factor to be 1.002030 ± 0.000020. In 1947, BRAGG, SIEGBAHN, WARREN, and LIPSON[22] recommended for general adoption the somewhat smaller value 1.002020 ± 0.000030. The chief argument for this downward revision was based on the measurements of F. Tyren[23]. In this work Tyren used a concave grating vacuum spectrometer to compare the wavelengths of X-ray lines with the wavelengths of hydrogenic Lyman series spark lines from highly ionized atoms. The wavelengths of these calibration lines were calculated from the Sommerfeld-Dirac theory. In 1947, however, the discovery of the Lamb Shift[24] invalidated this entire scheme. The corrections to the Lyman series lines resulting from this are shown in Fig. 1. Tyren's computed wavelengths required corrections varying from 25 to 100 parts per million. Unfortunately, it is difficult to make an after-the-fact correction of Tyren's data. The X-ray lines were measured in several orders, and one cannot tell whether the calibration of an 8 angstrom X-ray line was based on a second order comparison with O^{VIII}, third order with N^{VII}, fourth order with C^{VI}, fifth order or sixth order with B^V, or eighth order with Be^{IV}. Tyren's work must, therefore, be rejected.

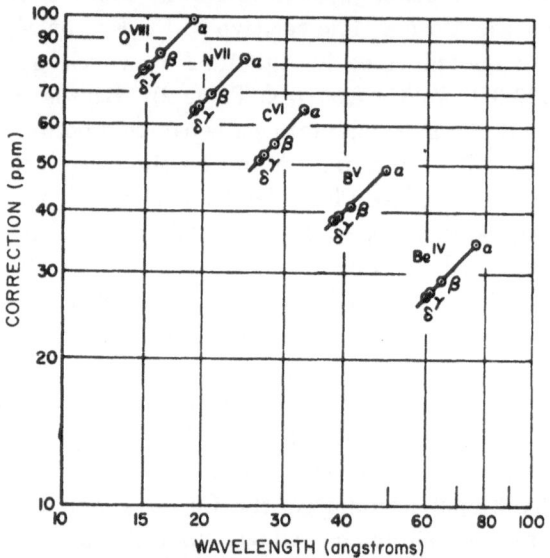

Fig. 1. Lamb Shift Correction to Lyman Series Lines for Hydrogenic Spark Spectra

The remaining useful data is that of Bearden[25] and Bäcklin[26] who used plane gratings whose grating constant was determined by direct comparator measurement, so that the wavelengths of the X-ray lines were determined absolutely. The results of these measurements are shown in Table 4.

Table 4. Measurements of Λ

(Conversion factor from X-units to milliangstroms)

J. A. Bearden ...	1931	1.002020 ± 0.000035
J. A. Bearden ...	1935	1.002110 ± 0.000075
Erik Bäcklin	1935	1.002011 ± 0.000033

Because of these uncertainties, and the great importance of this conversion constant, an attempt to redetermine it has been underway by Kirkpatrick and DuMond for the past several years. They are in essence repeating Tyren's measurements using a concave vacuum grating spectrograph, but only preliminary results have so far been obtained.

Ratio of the Proton Magnetic Moment to the Nuclear Magneton

This ratio has been independently determined with high precision by four different groups. In chronological order, these measurements are:

a. SOMMER, THOMAS, HIPPLE[27]	1951	2.792757	$\pm$ 0.000025
b. BLOCH, JEFFRIES, TRIGGER[28]	1950–1956	2.792750	$\pm$ 0.000100
c. BOYNE, FRANKEN[29]	1961	2.792906	$\pm$ 0.000056
d. SANDERS, DELLIS, TURBERFIELD[30]	1962	2.792770	$\pm$ 0.000070

The method consists of determining the proton spin resonance frequency and the proton cyclotron frequency in the same magnetic field. SOMMER, THOMAS, and HIPPLE used the Omegatron—an apparatus which is essentially a small cyclotron in which protons are accelerated to a maximum orbital radius of about one centimeter. The condition of cyclotron resonance was determined by detecting the ions as they spiraled out to the full radius of the cavity. It was necessary to correct for space charge forces and for the radial components of the trapping field, which was present in order to prevent axial drift. These effects altered the forces acting on the electron from the value it would have in an ideal cyclotron. However, by measuring the observed resonance frequency for ions of different masses, it was possible to extrapolate these perturbations to zero. BOYNE and FRANKEN used a similar arrangement, but the condition of cyclotron resonance was detected by the change in impedance of the oscillator supplying the high frequency electric field, thus detecting the increase in power supplied to accelerate the protons. TURBERFIELD, SANDERS, and DELLIS used a modification of the inverse cyclotron of BLOCH and JEFFRIES in which the protons are injected at high energy and are decelerated by the electric field. Protons were injected tangentially near the outside of the electrodes and were decelerated inward until they approached a radius at which the time taken to cross the central conductor was equal to one cycle of the alternating voltage, so that the proton reached a final orbit of constant radius. The original experiments in 1955 have been augmented by an extensive set of measurements using H^+ and H_2^+ ions, and only the results based on these final measurements are given here.

Gyromagnetic Ratio of the Proton

(i) *General Considerations.* In a magnetic field of intensity, H, a proton has two quantum states: parallel and antiparallel to the field; the states are separated in energy by $2\mu_p H$, where μ_p is the magnetic moment of the proton. The protons are usually those contained in a sample of hydrogenous liquid such as water or mineral oil at room temperature. The sample is placed in a steady magnetic field, and a coil about the sample supplies a small alternating field of radiofrequency, ν, at right angles to the steady field. When ν is close or equal to the value

$$\nu_n = (2\,\mu_p/h)\,H, \tag{3}$$

it will cause transitions between the two energy states. In thermal equilibrium the population of protons in the lower state slightly exceeds that in the higher state, the population ratio of the two being $\exp[2\,\mu_p\,H/(k\,T)]$, where $2\,\mu_p\,H/(k\,T)$ is of order 10^{-8} to 10^{-4}, and the observation of a net energy absorption relies on this small population difference. The resonance may also be detected by signals induced in a coil at right angles to the exciting coil (so as to be decoupled from the latter). In the method of free precession used by Bender and Driscoll for their case of weak magnetic fields, the proton sample is strongly polarized initially in an intense field and then quickly transferred to the weak field to be measured. The larger Boltzman factor resulting from the strong field insures a sufficiently large population ratio so that the sample retains its polarization several seconds while the transfer to the weak field is being effected. This is done by shooting the sample pneumatically down a tube several meters long from one field region to the other. During transfer the polarized protons keep themselves aligned in a continuous way with whatever instantaneous field they experience from point to point. Once in the weak field to be measured, the exciting signal is applied to the sample very briefly, after which the protons are left to themselves to precess freely about the applied weak field, gradually losing energy to the liquid. While doing so they induce a signal in the pick-up coil whose frequency, essentially the free precession frequency for that field strength, can thus be measured.

The constant of proportionality, $\gamma_p = \nu_n/H$, between the resonant frequency for protons and the field, H, at which that frequency occurs is called the gyromagnetic ratio. The phenomenon clearly affords an extremely accurate and reproducible way of measuring magnetic field intensities.

(ii) *Available Data on* γ_p, *the Gyromagnetic Ratio of the Proton.* There are at present six independent determinations of the gyromagnetic ratio of the proton. These measurements are not all of the same accuracy, and at least one is afflicted with doubts as to the possible existence of systematic error. The results are listed in Table 5, in which the data have been corrected to the gyromagnetic ratio of a bare proton. There is some uncertainty in this data, furthermore, with regard to the calibration of the standard ampere as maintained in the various national laboratories in terms of the absolute definition of the ampere. Vigoureux's measurement, performed at the National Physical Laboratory in Teddington, England, was, however, based on electrical standards directly intercompared with the standard ampere maintained at the U. S. National Bureau of Standards. It is, therefore, reassuring that these two measurements are in such excellent agreement.

Table 5. Gyromagnetic Ratio of the Proton

$$\gamma_p = \omega_p/H = 2\,\mu_p/\hbar = \left(\frac{\mu_p}{\mu_e}\right)\left(\frac{\mu_e}{\mu_0}\right)\frac{e}{m\,c}$$

	Field (gauss)	$\gamma_p{}'$ (sec^{-1} gauss^{-1})	$\gamma_p{}'$ Corrected to BIPM Units[g]	γ_p With Diamagnetic Correction cgs units
THOMAS, DRISCOLL, and HIPPLE[a] ...	4800	26752.7 $\pm$ 0.6		26753.4 $\pm$ 0.6
H. CAPPTULLER[b] ..	2800	26752.2 $\pm$ 1.0	26752.1	26752.5 $\pm$ 1.0
YAGOLA, ZINGERMAN, and SEPETYI[c] ..	2400–4700	26750.5 $\pm$ 0.2	26750.8	26751.2 $\pm$ 0.2
BENDER and DRIS-COLL[d]	12	26751.19 $\pm$ 0.08	26751.5	26751.92 $\pm$ 0.08
VIGOUREUX[e]	10–20	26751.15 $\pm$ 0.08	26751.5	26751.88 $\pm$ 0.08
YANOVSKII and STUDENTSOV[f] ...	0.6–1.2	26750.6 $\pm$ 0.3	26750.4	26750.8 $\pm$ 0.3

1 NBS amp = 1.000012 amps.

1 BIPM amp = 1.000027 amps.

[a] H. A. THOMAS, R. L. DRISCOLL, and J. A. HIPPLE, J. Res. Nat. Bur. Stand. 44, 569 (1950); Phys. Rev. 78, 787 (1950).

[b] H. CAPPTULLER, Z. Instr.-Kde 69, 191 (1960).

[c] G. K. YAGOLA, V. I. ZINGERMAN, and V. N. SEPETYI, Izmeritelnaya Tekhnika 1962, No. 5, 24.

[d] P. L. BENDER and R. L. DRISCOLL, Trans. IRE Instr. I-7, 176 (1958).

[e] P. VIGOUREUX, Proc. Roy. Soc., A 270, 72 (1962).

[f] B. M. YANOVSKII and N. V. STUDENTSOV, Izmeritelnaya Tekhnika 1962, No. 6, 28. — B. M. YANOVSKII, N. V. STUDENTSOV, and T. N. TIKHOMIROVA, ibid. 1959, No. 2, 39.

[g] Consultative Committee on Electricity, International Committee of Weights and Measures, Paris, May 1963.

THOMAS, DRISCOLL, and HIPPLE measure the proton resonance at a frequency of 20 megacycles in a field of approximately 4800 gauss. The field was set up between the pole pieces of an electromagnet 12.5″ in diameter with a 2″ gap. Although care was taken to insure the pole faces were accurately parallel and vertical, no precautions were taken to insure that the two cyclindrical pole pieces were coaxial. Thus, the extent to which a vertical displacement of one of the pole faces relative to the other produced a tilting of the magnetic flux lines in the gap was not determined. The field intensity near the center of the gap was measured by weighing the force exerted on a rectangular current carrying coil wound on a rectangular glass coil form which hung from one arm of an analytical balance. The field could be measured accurately in terms of the absolute ampere, the acceleration of gravity, and the dimensions of the coil.

A small proton resonance probe explored the field distribution in the gap across the pole face diameter to correct for small variations from uniformity and for the difference in magnetic field between the position of the coil and the position of the proton resonance sample. Now, the proton resonance frequency measures the total field intensity, but the weighing procedure measures only the horizontal component of the field. An obliquity of the magnetic field of forty minutes of arc would be sufficient to account for the discrepancy of fifty parts per million between this experiment and the measurement by Bender and Driscoll. Since it is undetermined whether this obliquity actually existed or how large it might be, one is forced to reject this measurement from consideration.

Sommerfeld's Fine Structure Constant

Triebwasser, Dayhoff, and Lamb[31] have measured the frequency separation of the $2 P_{1/2}$ and $2 P_{3/2}$ levels in deuterium. This famous measurement is still the best source of information available for determining the Sommerfeld fine-structure constant, alpha. The most recent theoretical formula for this frequency shift has been given by Layzer[32, 33] accurate to terms of order $\alpha^7 m c^2$ in the energy.

$$E = \frac{\alpha^2}{16} R_\infty\, c \left(\frac{M}{M+m}\right)^3 \left[2\frac{\mu_e}{\mu_0}\left(\frac{M+m}{m}\right) - 1 + \frac{5}{8}\,\alpha^2 - 2\,\frac{\alpha^3}{\pi}\ln 137 + \dots\right]$$

(4)

where M is the mass of the nucleus and m the mass of an electron. Taking the frequency separation as 10971.59 ± 0.10 megacycle and the Rydberg constant[34] as $109737.31\ \mathrm{cm}^{-1}$, we obtain for $1/\alpha$ the value 137.0388 ± 0.0006.

In order to achieve the full accuracy available in this experiment it was necessary to develop a complete theory of line shape of the transition so that the energy difference between the two levels could be determined in the presence of natural width and doppler broadening corrections. The measurements were made on deuterium rather than hydrogen in order to reduce doppler broadening by using the heavier nucleus. Because of this it was possible to measure the transition to an error which corresponds to less than $1/_{100}$ of the line width. Some questions have, however, been raised as to the possibility of systematic errors which might have been present in the experiment at this level of precision. These doubts, although not fully justified, have been inspired by the disagreement between the value of the fine structure constant determined from these measurements and the values of the fine structure constant as deduced from the hyperfine structure splitting in hydrogen. The hfs splitting is expressed by the formula[33]

$$\Delta \nu = \frac{16}{3} \, \alpha^2 \, R_\infty \, c \left(\frac{\mu_p}{\mu_0} \right) \left(\frac{\mu_e}{\mu_0} \right) \left(\frac{M}{M+m} \right)^3 \left[1 + \frac{3}{2} \, \alpha^2 - \left(\frac{5}{2} - \ln 2 \right) \alpha^2 - X \, \alpha \, \frac{m}{M} \right]$$

(5)

where X represents a correction factor for the finite extension of the electromagnetic field "inside" the proton. The experimental measurements of the hyperfine splitting are perhaps the most precise physical measurements ever made[35]:

$\Delta \nu$

		$\Delta \nu$		
a.	NAFE, NELSON	1420410	$\pm$ 6 kc.	(1947)
b.	PRODELL, KUSCH	1420405.73	$\pm$ 0.05	(1955)
c.	WITTKE, DICKE	1420405.72	$\pm$ 0.04	(1956)
d.	KLEPPNER, GOLDENBERG, RAMSEY	1420405.762	$\pm$ 0.004	(1962)
e.	PIPKIN, LAMBERT	1420405.7491	$\pm$ 0.0060	(1962)

In spite of accuracies approaching 1 in 10^9 the applicability of the data is confused by the uncertainty in the theoretical formula. A calculation of the coefficient, X, due to IDDINGS and PLATZMANN[36], based on a rather literal interpretation of the Hofstadter form factor for the proton structure, plus additional correction terms representing the effects of virtual photon production calculated by ZWANZIGER and by LAYZER[37], yields a value of α some 17 ppm higher than that obtained from the measurements of the fine structure separation. No way is known of assigning a numerical uncertainty to the value $\alpha^{-1} = 137.0352$, which is computed from the hfs data in this way. JULIAN SCHWINGER has stated that in his opinion the correction for nuclear structure of IDDINGS and PLATZMANN could even be incorrect in algebraic sign!

In order to avoid the uncertain problems of nucleon structure, the hyperfine splitting has been measured in muonium by V. W. HUGHES[38]. The muonium "atom", contains only a positive μ-meson and an electron, and, therefore, is unconfused by any difficulties with a virtual pion field*. The theoretical expression for the hyperfine structure splitting of muonium should, therefore, be quite accurate and is given by[39]

$$\Delta \nu = \frac{16}{3} \, \alpha^2 \, R_\infty \, c \left(\frac{\mu_\mu}{\mu_0} \right) \left(\frac{\mu_e}{\mu_0} \right) \left(\frac{m_\mu}{m_\mu + m} \right)^3 \cdot$$

$$\cdot \left[1 + \frac{3}{2} \, \alpha^2 - \left(\frac{5}{2} - \ln 2 \right) \alpha^2 - \frac{3}{\pi} \frac{\alpha}{m_\mu} \frac{m}{m_\mu} \ln \frac{m_\mu}{m} \right]$$

(6)

with

$$\mu_\mu / \mu_0 = \frac{m}{m_\mu} \left[1 + \frac{\alpha}{2\pi} + 0.745 \left(\frac{\alpha}{\pi} \right)^2 + \ldots \right].$$

(7)

---

* The finite lifetime of muonium leads to an uncertainty in the energy states of this atom, and hence the hfs transition is not as sharp as and cannot be measured with the accuracy of the hfs transition in hydrogen.

The magnetic moment of the μ-meson has been measured relative to the magnetic moment of the proton by Hutchinson et al.[40], who find $\mu_\mu/\mu_p = 3.18334 \pm 0.00005$. When this is combined with Lambe's measurement[41] of $\mu_e/\mu_p = 658.2105$, we find $m_\mu/m = 206.768 \pm 0.003$. From the measured value of $\varDelta\nu = 4463.10 \pm 0.10$ Mc, we can then compute $1/\alpha = 137.0392 \pm 0.0015$.

Note Added in Proof. Hughes has recently reported more complete measurements of the muonium hyperfine structure splitting. This work includes a careful analysis of the effects of pressure of the argon buffer gas [J. Bailey, W. Cleland, M. Eckhause, V. W. Hughes, R. Prepost, Bull. Amer. Phys. Soc. II-9, 81 (1964)].

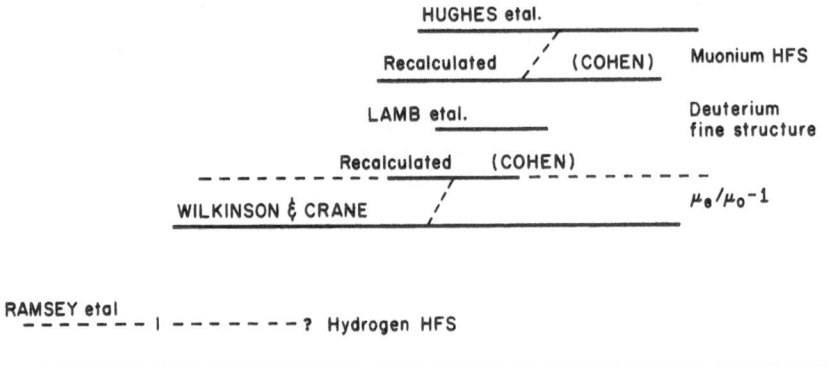

Fig. 2. Measurements of the Fine Structure Constant
Heavy lines represent experimental uncertainties. Dashed lines are intended to indicate the existence of theoretical uncertainties to which no real limits can be set. The recalculated values represent independent evaluations, by the authors of the present report, of the published experimental data

Five resonance curves with argon pressures of 35 atmos. and 70 atmos. extrapolated to zero pressure yielded $\varDelta\nu = 4463.33 \pm 0.19$ Mc/sec. The value obtained for $1/\alpha$ then becomes 137.0360 ± 0.0030. This revised value is in agreement within its statistical error with the value obtained by Lamb from duterium fine structure and is also consistent with the value calculated from hydrogen hyperfine structure.

Wilkinson and Crane[42] have measured the g-factor of the free electron by measuring the frequency difference between the cyclotron and the spin precession frequency. Because they are able to measure the difference directly, they determine the ratio of the two frequencies with an accuracy of better than 3×10^{-8}. They find $\mu_e/\mu_0 = 1.001159622 \pm 0.000000027$, and hence the measurement is of sufficient accuracy to give directly either a verification of the accuracy of the theoretical formula or, if we believe the accuracy of the formula, an independent determination of the fine structure constant, α. If we assume that the theoretical formula is correct (although to this accuracy eq. (2) should be calculated to terms in $(\alpha/\pi)^3$ in order to be sure that these are indeed small), and that the experimental result is not further confused by the existence of an electron dipole moment, Wilkinson and Crane's measurement gives a value for $1/\alpha = 137.0381 \pm 0.0010$, where the quoted error has been reduced to represent only the statistical accuracy of the data with no provision for systematic error or other uncertainties. The data on the fine structure constant is summarized in Fig. 2, which compares the various calculations of $1/\alpha$. The figure gives two results (one as quoted by the authors and the other as recomputed here) for the muonium hfs and for the magnetic moment anomaly. For muonium hfs the recalculation involved using a newer value of the muon mass. For the WILKINSON and CRANE anomaly measurement we have re-evaluated the statistical accuracy of the experiment and separated it from the theoretical uncertainty of calculated theoretical formula.

Annihilation Radiation

The wavelength associated with the energy released in the annihilation of an electron is just $\lambda = h/mc$, the Compton wavelength. This may also be written

$$\lambda = \alpha^2/2\,R. \tag{8}$$

We know the Rydberg constant accurately in reciprocal centimeters, but if we measure the annihilation wavelength with crystal diffraction methods we measure λ in X-units, so that we have a means of measuring only $\alpha^2/\Lambda = 2\,R\,\lambda_{A,x}$.

The original measurement of the annihilation wavelength by DuMond et al.[43], using a bent quartz crystal spectrometer, has been recently surpassed in accuracy by KNOWLES[44] at Chalk River, using the large positron flux from the NRU reactor and a double flat crystal spectrometer. The measured Bragg angle for the annihilation radiation was found to be 3996.205 microradians at $18°$ C using calcite crystals reflecting from the (211) planes. Using the value 707.831 x. u. for Mo $K\alpha_1$ as the definition of the X-unit implies 3029.440 X-units for the calcite grating constant. We thus find a value for α^2/Λ, and if we use a value for α (which is much more accurately known than α^2/Λ) we find a value of $\Lambda = 1.002049 \pm \\ \pm 0.000031$, which is in good agreement with the data of Table 4.

Remeasurement of the Faraday Constant

For more than 30 years it has been realized that the work on the FARADAY done early in this century using the silver coulometer might be in error. Im 1929, R. T. BIRGE[45] had arrived at significantly different values of e/m depending on whether he computed that constant from electron beam deflection experiments or by so-called "spectroscopic" methods. The early work on the FARADAY was for obvious reasons directed more towards obtaining reproducibility than towards values of absolute significance. The objective, of course, was primarily one of establishing the arbitrary standard of electrical current. As emphasis is placed on absolute significance, however, the question of the exact mechanism of transfer becomes important, and we must be able to answer the following questions, among others:

a. What is the mechanism of charge transfer? Is all the charge carried by simple silver ions, or is there a possibility of complexes?

b. Is there a fractionation by electrolysis of the two silver isotopes?

c. Have we measured only the weight of the silver deposited on the cathode, or does the deposit also contain inclusions of other matter from the electrolyte?

d. Have we measured all of the deposit, or has some of the silver dissolved in the electrolyte or otherwise become detached before weighing?

Recent measurements at the National Bureau of Standards by CRAIG et al.[46] appear to have met with all of these questions. Mass spectroscopic abundance determinations of electrolytically deposited silver have shown that the fractionation effect is negligible. The electrochemical equivalent was determined by weighing the silver removed from the anode rather than measuring the silver plated onto the cathode, thereby removing any possibility of inclusions in the determination. The electrochemical equivalent of silver was determined to be

$$1.117972 \pm 0.000007 \text{ mg/coul.}$$

where the error is the statistical error only; no systematic error greater than one part per million is expected. The Faraday follows from this number if we know the atomic weight of silver. The table of nuclidic masses by EVERLING, KÖNIG, MATTAUCH, and WAPSTRA gives us accurate mass values for the individual isotopes. We need, therefore, only a measure of the isotopic abundance. The measurements by SHIELDS, CRAIG, and DIBELER[47] and by SHIELDS, GARNER, and DIBELER[48] are in excellent agreement and yield a value of 1.07589 ± 0.00052 for the ratio of ^{107}Ag to ^{109}Ag. The atomic weight of silver is thus found to be $107.86828 \pm \pm 2.7$ ppm. This result is significantly more accurate than the adopted chemical value based primarily on gravimetric chemical methods. The value of the FARADAY becomes

$$96485.66 \pm 0.66 \text{ NBS coul/mole.}$$

To this must be applied a conversion factor to express the result in absolute coulombs; 1 NBS coulomb = 1.000012 absolute coulombs, so that the final value of the FARADAY on the unified scale of atomic masses is

$$96486.82 \pm 0.66 \text{ coul/mole.}$$

Proton Moment in Bohr Magnetons

The proton moment has been measured in Bohr magnetons to sufficient accuracy that it enters our analysis as an auxiliary constant rather than as a variable of the least squares adjustment. It is possible to measure the proton moment directly in Bohr magnetons by measuring the ratio of the cyclotron frequency of the free electron to the proton resonance frequency in the same magnetic field, as in Items 1 to 4 of Table 6, or to measure the ration of the g-factors for the electron and proton (Items 5 to 8), which must then be corrected by the ratio μ_e/μ_0.

The earlier measurement of μ_0/μ_p by GARDNER and PURCELL[49] was repeated with increased accuracy by HARDY and PURCELL[50]. This measurement, which claims an accuracy (standard deviation) of 0.0005, disagrees with the measurement by LIEBES and FRANKEN[51] by 0.0071 $\pm$ $\pm$ 0.0030. On the other hand, the recent measurement by SANDERS, TITTEL and WARD[52] is in excellent agreement with the Liebes and Franken result. The most accurate determination of μ_0/μ_p, however, comes from the measurement of the ratio of the electron and proton g-factors. The

Table 6. Proton Moment in Bohr Magnetons
(Corrected to the Bare Proton)

(Diamagnetic Corrections: — 29.7 ppm, mineral oil, spherical sample, — 26.0 ppm, H_2O spherical sample.)

	ω_e/ω_p	μ_0/μ_p
1. GARDNER and PURCELL[a]	657.475 $\pm$ 0.008 (oil, H_2O)	657.4565
2. LIEBES and FRANKEN[b]	.462 $\pm$ 0.004 (oil)	.442
3. HARDY and PURCELL[c]	.4676 $\pm$ 0.0005 (gas)	.4501
4. SANDERS, TITTEL, and WARD[d] .	.4620 $\pm$ 0.0024 (oil)	.4436

	g_J/g_p'	g_s/g_p
5. KOENIG, PRODELL, and KUSCH[e] .	658.2171 $\pm$ 0.0004 (oil)	658.2096
6. BERINGER and HEALD[f]	.2181 $\pm$ 0.0003 (oil)	.2106
7. GEIGER, HUGHES and RADFORD[g]	.2169 $\pm$ 0.0004 (oil)	.2094
8. LAMBE and DICKE[h]	.21591 $\pm$ 0.00002 (H_2O)	.2105

$$g_J/g_s - 1 = {}^1/_3\, \alpha^2 = 17.75 \text{ ppm}$$

$$\mu_0/\mu_p = (g_s/g_p)_{\text{Lambe}}/(\mu_e/\mu_0) = 657.4481$$

[a] Phys. Rev. 76, 1262 (1949); 83, 996 (1951).
[b] Phys. Rev. 104, 1197 (1956); 116, 633 (1959).
[c] Bull. Amer. Phys. Soc, 114, 37 (1959).
[d] Proc. Roy. Soc., A 272, 103 (1963).
[e] Phys. Rev. 88, 191 (1952).
[f] Phys. Rev. 95, 1474 (1954).
[g] Phys. Rev. 105, 183 (1957).
[h] Thesis, Princeton 1959 (unpublished).

measurement by E. B. D. Lambe[41] gives $g_J/g_p' = 658.21591 \pm 0.00002$ with the proton g-factor uncorrected for diamagnetism of the spherical H_2O proton sample. This measurement represents an accuracy of one part in 30 million. The bound electron correction is $+ 17.75$ ppm, and the proton diamagnetic correction is $- 26.0$ ppm. Although this latter correction may be inaccurate by several parts in 10 million, if we use it as an adopted correction factor, Lambe's measurement yields $g_s/g_p = 658.2105$. Using the theoretical value for μ_e/μ_0, we then calculate

$$\mu_0/\mu_p = 657.4481.$$

This value is smaller than the Hardy and Purcell measurement by 0.0030 and larger than the Sanders, Tittel, Ward measurement by 0.0045. We shall, therefore, use it as our best estimate as an auxiliary constant in the reduction of the experimental data to be subjected to the least squares analysis.

Selection and Rejection of Data for Least Squares Analysis

Uncertainties Afflicting Precision X-Ray Spectroscopic Data

X-ray emission line spectroscopy, following the discoveries of Laue, Bragg, and Moseley, was studied intensively for some two or three decades before it was abandoned by the physicist because of the greater interest and challenge in nuclear physics. The wavelengths of some 3000 or more X-ray emission lines have been measured and tabulated using crystal diffraction methods.

Early X-ray spectroscopy, however, failed to recognize the fact that the lattice spacings, d, of different samples of the same crystal species can vary by many parts per million from sample to sample. In the case of calcite, Straumanis[53], Bearden[54], and others have shown that these variations are correlated with chemical purity of the crystal, but purity is by no means an unambiguous guarantee as to the value of the lattice constant of a crystal sample, as this may also depend on its history of mechanical treatment and the density of dislocations and other imperfections in its structure. Lonsdale[55] has shown by her divergent beam method that apparently perfect samples of diamond may exhibit small but definite differences in grating spacing.

The definition of the X-unit accepted by most X-ray specialists is given by stating that at a specified temperature the "standard grating constant of purest calcite" is so-and-so many X-units. In the operational sense this is unsatisfactory since it prescribes no test for recognizing whether a given calcite sample has the "standard grating constant of purest calcite". On the level of precision attainable in many modern measurements in other fields of physics, the imprecision of this definition as a basis for the entire scale of X-ray emission line wavelengths represents a serious limitation to the accuracy of X-ray data. As long as "purest calcite" is no more than a vague abstraction there is no clearly defined, unambiguous, and reproducible unit in terms of which to measure X-ray wavelengths*.

An additional source of confusion is worthy of mention, namely the lack of any agreement among X-ray spectroscopists themselves as to just what feature of an X-ray line profile is to be taken as the fiducial point for which the wavelength is quoted. Most often it is the position of the peak of the line which has been taken as the fiducial references point for reporting the wavelength. The recommended method for locating the peak of a line profile, frequently attributed to Lord RAYLEIGH, is to extrapolate up to the peak the locus of mid-points of chords parallel to any background upon which the profile may be superposed. Many, though far from all, X-ray emission lines exhibit symmetrical profiles, and this locus then turns out to be quite accurately straight and vertical. In such cases it may not be an exaggeration to say that the peak position may be reproducibly located, by means of careful repeated work, to within a few thousandths of the line's breadth at half maximum height. The Cu $K\alpha_1$ line, however, exhibits marked asymmetry, the locus of chordal mid-points is distinctly curved, and it is doubtful whether 'this high a degree of reproducibility would be possible for locating its peak position.

There has also been a recent tendency to use the *centroid* of the *observed* line profile and attribute to it the *tabulated* wavelength value. Since the observed line profiles in such cases are broadened by the folded-in effects of slit widths and other instrumental sources of line-breadth, the centroid of the observed profile can be related to the center of the natural line shape only to the accuracy with which the centroid of the instrumental "window" or resolution function can be measured. Furthermore, one is still confronted with the difficulty that, for a Lorentzian natural line shape, the centroid does not exist in a strict sense, and hence the observed centroid is dependent upon the choice of initial and final points over which to extend the integration to determine it, as well as upon erratic fluctuations, both statistical and real, in the extreme wings of the profile.

Many of the tabulated emission line wavelengths have been determined photographically with single crystal spectrometers. In such cases microphotometer traces may or may not have been made of the line profile. Frequently, measurements were made by merely setting a microscope cross hair on the "center" of the streak of blackened silver grains. It is difficult to determine whether such measurements correspond to the peak, the centroid, or the median of the instrumental line profile! Which of these

---

* One could, of course, define the X-unit in terms of the grating constant, at some fixed temperature, of the *pure, perfect calcite lattice,* but although such a definition is exact, it can not be achieved by measurement, since the substance does not exist.

three it is may depend largely on the degree of exposure of the emulsion. The result has certainly little relation to any one of these three features of the *natural* line profile.

In summary then, it is easy to understand why, for the purposes of high precision, X-ray spectroscopic data tend to be unsatisfactory and unreliable. A complete revision of emission line data is needed with much more careful attention to, and universal adoption of, conventions which define measured quantities in an operational manner before the present state of affairs can be improved.

X-Ray Data in the Present Least-Squares Adjustments

A preliminary adjustment of the available data reported last year[56] included X-ray data of three kinds: (1) Direct determinations of the conversion constant, Λ (Bearden, Bäcklin); (2) numerous determinations of $N \Lambda^3$ from crystal data (densities, molecular weights, and X-ray lattice spacing measurements) in two groups—those made by using the Mo $K\alpha_1$ emission line and those made using the Cu $K\alpha_1$ emission line; (3) the two-crystal measurement at Chalk River by J. W. Knowles of the wavelength of the annihilation radiation using calcite crystals in transmission.

We have already stated that a dichotomy was manifested among the data on $N \Lambda^3$ if one used accepted tabular wavelength values, depending on which emission line, Mo $K\alpha_1$ or Cu $K\alpha_1$, had been employed in measuring the d of the crystals. That is to say, the wavelength ratio, $\lambda(\text{Mo } K\alpha_1)/\lambda(\text{Cu } K\alpha_1)$, calculated from the tabular values of these wavelengths, disagrees with the weighted mean value of this ratio implied by the totality of the 16 crystal diffraction measurements used for determining $N \Lambda^3$.

The discrepancy gap is widened still more if one uses, instead of the generally accepted tabular wavelengths, the most recent results of J. A. Bearden (Table 2) with the two-crystal spectrometer directed expressly at comparing with high precision the Mo $K\alpha_1$ and the Cu $K\alpha_1$ wavelengths.

This discrepancy of 17 ppm in wavelength ratio between Bearden's results and the ratio of the tabular values corresponds to three times as large a discrepancy in $N \Lambda^3$, or 51 ppm. It is in such a direction as to worsen the already existing discrepancy of 75 ppm between the tabular values and the two groups of $N \Lambda^3$ data to 126 ppm. So large a discrepancy seems, indeed, quite intolerable. To date, however, we do not know what accuracy to assign to Bearden's measurements.

The $N \Lambda^3$ discrepancy we are discussing need not necessarily, of course, be due to an error in the wavelength ratio of the Mo $K\alpha_1$ and Cu $K\alpha_1$ lines based on the above tabular values. Alternatively, it could be due to a common systematic error afflicting all of the $N \Lambda^3$ data of at least one class. Some suspicion of this latter explanation arises from the fact that all but one of the eight Cu $K\alpha_1$ data on $N\Lambda^3$ come from one group of workers and one X-ray crystallographic technique (A. SMAKULA). In our correspondence with Prof. SMAKULA on these points, however, we learn that his crystal spacing determinations have, in the case of Ge, been corroborated very closely by BOND and COOPER at Bell Telephone Laboratories. If Smakula's and Cooper's germanium crystals were closely similar (a not unlikely supposition), the suspicion of a common source of systematic error from Smakula's goniometric measurements is largely removed. At our request, Dr. BOND and Mrs. COOPER have consented to make a comparison of the Mo $K\alpha_1$-to-Cu-$K\alpha_1$ wavelength ratio using a germanium crystal under conditions similar to those under which the $N\Lambda^3$ data of SMAKULA were observed.

Dr. L. PARRATT (Cornell University) called our attention to a case in which the peak value of the Cu $L\alpha$ line was reported[57] to be shifted by 5×10^{-4} (approximately $^1/_4$ of the line width) as the exciting voltage was raised from 1.1 kV (threshold) to 6.6 kV. This effect may be due to satellite lines resulting from the production of multiple ionization at higher voltages. Evidence of prominent satellite-like structure on one side of the line profile in the high voltage case suggests this. If the X-ray emission line spectrum is eventually to furnish a really reliable set of fixed points in the electromagnetic spectrum, it is, therefore, of great importance to study these possible effects of shifts in peak position under different conditions of excitation. In the case of the Cu $K\alpha_1$ line, no such marked structure is evident, but the line profile *is* asymmetric, and the fractional wavelength shift in the peak position required to explain the present discrepancies is far smaller than that observed by VAN DEN BERG. It should be pointed out that in the case of the $N\Lambda^3$ determinations, the internal discordances we have been discussing correspond to a very small wavelength shift of the Cu $K\alpha_1$ line peak relative to its half-width. A shift of 100 ppm in $N\Lambda^3$, for instance, would require for its explanation a shift of only 0.02 half-widths of the Cu $K\alpha_1$ line.

Because of these uncertainties we have felt it necessary to delete all of the experimental data which involve X-ray wavelengths. We thus delete from consideration as a variable in our least squares adjustment the value of the conversion factor, Λ. Fortunately, the nuclear magnetic resonance data and the Faraday constant are sufficient to determine the Avogadro constant so that we do not lose too much by rejecting the X-ray crystal density measurements of $N\Lambda^3$. As a subsidiary calculation we will be able to compute a value of Λ from the X-ray crystal density measurements combined with our least squares adjusted value of the Avogadro constant, and, in fact as we shall see, this is in principle the most accurate method for determining the value of the conversion factor, Λ, if we have a clear unambiguous definition of the X-unit.

Least Squares Analysis of the Available 1962 Data

Motivation and General Approach for This Analysis

It is apparent that the present situation regarding the total budget of available data on the constants does not permit their evaluation by completely objective methods without bringing into play personal judgments which must unavoidably involve some subjective elements. Under such circumstances it seems to us that the best we can do is to analyze as thoroughly as the circumstances warrant the results of utilizing all or various judiciously chosen parts of the available data in order to evaluate the effect of the individual data on the overall adjustment and to determine if possible whether discrepancies exist in the data, and if so to attempt to identify their origins.

Such a program, in which a considerable list of more or less discrepant data is broken down into subgroups for least squares analyses is often referred to as an "analysis of variance". A somewhat similar analysis was made on the data available in 1955[1].

Consider n independent observational data leading to n equations in m unknowns $(m < n)$. From these data one adjustment with n equations in m unknowns can be made. Then n different adjustments can be formed consisting of subsets of $n - 1$ equations in each of which a different one of the original n equations has been omitted from the set. Next, we can omit from the original set pairs of equations and thus form $n(n - 1)/2$ subsets, each consisting of $n - 2$ equations. The process may be continued down to the point where each subset comprises only one more equation than the number of unknowns. Beyond this point, overdetermined least-squares solutions certainly cannot be formed.

The equations with which we must deal (see Table 7 for the input items) are far from being of the most general (non-degenerate) form. The nine equations developed from these data yield only four different functional relationships connecting the three unknowns α, e, N. The redundancy results from the fact that most of these kinds of equations have three representatives corresponding to measurements of the same physical quantity by different people, different methods, or both. They also exhibit a further degeneracy in that the proton moment determinations, the gyromagnetic ratio determinations, and the Faraday form a subset in which any pair taken together determines the same function of the unknowns as the third member. There are still, however, several hundred over-determined sets which could be formed if we wished to explore all possible combinations. We have instead used an approach in which we successively eliminate those input data which exhibit the largest departures from fit with the general consensus, and after each such rejection reexamine all of the remaining data for goodness of fit. In this way we have carried out over 100 least-squares analyses of various possible subsets of the input data. This cannot, therefore, be said to constitute a *complete* analysis of variance, but we believe it to be probably complete enough to furnish a fair picture of the present state of knowledge of the constants.

Measures of Incompatibility of an Input Datum

The analogy between a least-squares adjustment and an over-determined mechanical structure consisting of elastic members is a valuable one to consider; the more rigid elastic members correspond to the determinations of high accuracy to which greater statistical weight must be attached. Because the individual input data suffer from errors, the over-determined set of equations is, however, more or less incompatible, and the analogous situation in the case of the mechanical structure results in different amounts of elastic energy stored in the various members. The squared normalized residual, r_i^2, then may be though of as the contribution to the total elastic energy of strain, χ^2, contributed by the ith datum when the system has found equilibrium at its minimum energy state, i. e., condition of least squares. Large contributions, those for which $r_i^2 \gg 1$, lead us to suspect that a systematic error was present in that datum or (what is the same thing) that too small an a-priori or "internal" error (from information "internal" to that particular datum) had been assigned to it. This test is valuable as a rough indication; another criterion is $r_i'^2$, which is the analogue of the elastic energy required to stretch the rejected i^{th} datum back into forced accord with the value of that datum implied by the adjustment from which the ith datum was dropped.

A third useful measure is the difference between the experimental value and the least-squares-adjusted value relative to the standard error *of this difference*. Such a quantity is easily calculated from the results available from two least squares analyses, one which includes the item in question, and the other which deletes it. Let the value of the function resulting from the second least squares adjustment be y_2, and let its variance be σ_2^2. Since this is the value of an experimental quantity calculated from a set of data from which that specific experiment has been deleted, we can call this the "indirect" value, although the appropriateness of the term can be questioned if the set contains other independent determinations of the same quantity. If the direct experimental value is y_0 (with variance σ_0^2), it is clear that the complete least squares analysis must give a result, y_1 and σ_1^2, which can be computed by simple statistical rules since y_0 and y_2 are independent. We can write

$$\frac{1}{\sigma_1^2} = \frac{1}{\sigma_0^2} + \frac{1}{\sigma_2^2}; \qquad \sigma_1^2 = \frac{\sigma_0^2 \sigma_2^2}{\sigma_0^2 + \sigma_2^2} \tag{9}$$

and

$$y_1 = \sigma_1^2 \left(\frac{y_0}{\sigma_0^2} + \frac{y_2}{\sigma_2^2} \right). \tag{10}$$

The statistics r and r' of the preceeding paragraph are given by

$$r = (y_0 - y_1)/\sigma_0, \tag{11}$$

$$r' = (y_0 - y_2)/\sigma_0 \tag{12}$$

and it is, therefore, easy to show that

and

$$r'/r = (\sigma_0^2 + \sigma_2^2)/\sigma_0^2 = \sigma_2^2/\sigma_1^2 \tag{13}$$

$$\sigma_1^2/\sigma_0^2 = (r' - r)/r', \tag{14}$$

$$\sigma_2^2/\sigma_0^2 = (r' - r)/r. \tag{15}$$

We must remember that y_0 and y_2 are statistically independent, while y_0 and y_1 are not, since the least squares solution, y_1, is computed from data which includes y_0. The variance of the difference $y_0 - y_2$ can be written immediately as $\sigma_0^2 + \sigma_2^2 = (r'/r) \sigma_0^2$, but the variance of the difference $y_0 - y_1$ must be computed by first expressing $y_0 - y_1$ in terms of $y_0 - y_2$. When this is done we then find that this variance is $(r/r')^2 (\sigma_0^2 + \sigma_2^2) = (r/r') \sigma_0^2 = \sigma_0^2 - \sigma_1^2$. A result which may be at first somewhat surprising, but which is actually obviously necessary, is that the difference $y_0 - y_1$ divided by its standard deviation is equal to $y_0 - y_2$ divided by *its* standard deviation. Hence, either expression can be used as a measure of the consistency of the experimental measurement, y_0, with the remainder of the data. The square of this quantity is $r r'$. It is clear that $r r'$ is the squared normalized residual associated with the addition of one experimental datum to the least squares adjustment, while $\chi^{2'}$ is the sum of the squared normalized residuals of the least squares adjustment without this datum; hence we have

$$\chi^2 = \chi^{2'} + r r'. \tag{16}$$

In keeping with our analogy of an elastic structure, we can make a direct physical interpretation of $r r'$. We recognize first of all that the normalization of the residuals renders them equally well an analogue of the stress (force) or the strain (elongation) of a member of the truss. (The normalization is equivalent to using units such that the elastic stiffness constant of the member is unity.) Then r is a measure of the initial stress in the member before it is removed from the structure, and r' is a measure of the change in length of the member upon removal. Thus $r r'$ is twice the elastic energy of strain stored in the member. (The factor 2 corresponds to the identification of χ^2 with twice the total elastic energy of the system.) We shall base our arguments on consistency of the data on a discussion of the magnitudes of r, r', and $r r'$.

Discussion of the 1962 Analysis

We are now in a position to describe the results of our least squares adjustments. The 9 input items, which together form the basis for our analysis, involve three unknowns, which we can take to be α, e, N. The actual choice of variables is immaterial since these variables merely form the *coordinate system* for our description of the function space. The *geometry* of this space will be independent of the coordinate system which describes it, and we have, for some of our calculations, used the variables $1/\alpha$, α^3/e, F.

Table 7. Data for 1962 Least Squares Analysis

A. *Fixed Auxiliary Constants*

1. Rydberg Constant, R .... 109737.31 cm^{-1}
2. Velocity of Light, c .... 299792.5 km sec^{-1}
3. 1 US-NBS Coulomb .... 0.1000012 ± 0.0000004 absolute emu
4. Electron Moment, μ_e/μ_0 .... 1.001159615
5. Proton Moment, μ_p/μ_0 .... $1/657.4481 = 0.0015210325$
6. Atomic Masses ($^{12}C = 12$)

$$H \dots\dots\dots\dots\dots 1.00782522$$
$$H/M_p \dots\dots\dots\dots 1.000544607$$
$$D \dots\dots\dots\dots\dots 2.01410219$$
$$D/M_d \dots\dots\dots\dots 1.000272448$$

B. *Experimental Data Subject to Adjustment*

1. Hyperfine structure separation in hydrogen (RAMSEY et al.; PLATZMANN and IDDINGS) .... 1420406 ± 35 kc/sec*
2. Fine structure separation in deuterium (LAMB et al.)..................... 10971.59 ± 0.10 Mc/sec
3. Magnetic moment of the proton in nuclear magnetons (BOYNE and FRANKEN)....................... 2.792906 ± 0.000056
4. Magnetic moment of the proton in nuclear magnetons (SOMMER, THOMAS, and HIPPLE) .... 2.792757 ± 0.000025
5. Magnetic moment of the proton in nuclear magnetons (SANDERS and TURBERFIELD) .... 2.792770 ± 0.000070
6. Faraday Constant – Silver (NBS) ... 9648.682 ± 0.066 coul/mole
7. Gyromagnetic ratio of the proton (BENDER and DRISCOLL) .... 26751.92 ± 0.08 sec^{-1} gauss^{-1}
8. Gyromagnetic ratio of the proton (VIGOUREUX) .... 26751.88 ± 0.08 sec^{-1} gauss^{-1}
9. Gyromagnetic ratio of the proton (THOMAS, DRISCOLL, and HIPPLE) .. 26752.80 ± 0.25 sec^{-1} gauss^{-1}

* The error quoted here is not the error of measurement but the much more poorly defined uncertainty in the theoretical interpretation of the relationship between the measured number and the fundamental constants.

The actual data we have used is given in Table 7. The numerical values listed here differ slightly from those used in the 1961 analysis[56].

The elimination of all of the X-ray data allows the system of 9 equations to be decomposed into two independent subsets—two equations which determine α and seven equations which determine Ne and α^3/e. These seven equations can give us no information regarding a value of the fine structure constant and cannot help us in choosing between a value for this constant derived from the hyperfine measurements or the fine structure measurements. However, the muonium hyperfine structure measurements as well as the theoretically more uncertain determination of α from the electron magnetic moment anomaly both lend support to the correctness of the Lamb measurement and indicate the existence of an error in the calculation of the structure factor in the interpretation of the hyperfine structure result.

It is also clear that the proton magnetic moment measurement of BOYNE and FRANKEN is inconsistent with the other measurements of this quantity. The measurement of BLOCH and JEFFERIES, as corrected by TRIGGER, although consistent with the other two, is considerably less accurate and may still be suspected of suffering from systematic error.

The various least squares analyses which we have carried out by successively deleting one equation at a time from the complete set and evaluating the χ^2 associated with each deletion, indicate clearly that the discrepancy between the gyromagnetic ratio measurement of THOMAS, DRISCOLL, and HIPPLE, on the one hand, and the measurement of the same quantity by BENDER and DRISCOLL, and VIGOUREUX is to be resolved in favor of the latter, and strongly indicate that both the Thomas, Driscoll and Hipple measurement of the gyromagnetic ratio and the Boyne and Franken measurement of the magnetic moment of the proton in nuclear magnetons may contain systematic errors and should be deleted from our analysis.

The conclusion that systematic errors exist in these measurements is, however, not based completely on statistical arguments, but arises also from an independent evaluation of the experiments themselves. We, therefore, base our least squares analysis on Items 2, 4, 5, 6, 7, and 8 of Table 7. The adjustment of these equations leads to the following recommended best values of the physical constants:

$$1/\alpha = 137.0388 \pm 0.0006,$$
$$e = (4.80298 \pm 0.00006) \times 10^{-10}\,\text{esu},$$
$$h = (6.62559 \pm 0.00015) \times 10^{-27}\,\text{erg sec},$$
$$m = (9.10908 \pm 0.00013) \times 10^{-28}\,\text{g},$$
$$N = (6022.52 \pm 0.09) \times 10^{20}\,\text{mole}^{-1}.$$

For completeness we should, of course, calculate a best value of the conversion factor, Λ. We may use for this the weighted average of the direct measurements of the conversion factor, $\Lambda = 1.002024 \pm 0.000023$. However, we can also calculate Λ from the formula

$$\Lambda = [N \, \Lambda^3/N]^{1/3}.$$

We then have a choice as to what value of $N \, \Lambda^3$ to use. The weighted average of the molybdenum data of Table 3 gives $N \, \Lambda^3 = 6059.725 \pm 0.109$, and using $N = 6022.52 \pm 0.09$ we obtain

$$\Lambda_{Mo} = 1.002055 \pm 0.000010$$

whereas, if we use the weighted average of the copper data, $N \, \Lambda^3 = 6060.184 \pm 0.107$, we find

$$\Lambda_{Cu} = 1.002081 \pm 0.000010.$$

If we believe that there is an inconsistency between the quoted wavelengths for Cu $K\alpha_1$ and Mo $K\alpha_1$, then these two conversion factors represent distinct numbers. One refers to the X-unit defined so that the peak wavelength of Mo $K\alpha_1$ is 707.831 X-units, whereas the other refers to an X-unit defined so that the peak wavelength of Cu $K\alpha_1$ is 1537.396 X-units. We cannot expect to get the same numerical value of Λ in the two cases because the two groups of $N \, \Lambda^3$ data (8 measurements with the Mo line and 8 with the Cu line) do not imply the same wavelength ratio for these two lines as the ratio the tabular values (Mo $K\alpha_1 = 707.831$; Cu $K\alpha_1 = 1537.396$), used in computing the data in the two groups. If there were no uncertainties concerning impurities, vacancies, dislocations and other imperfections in crystals, Λ_{Mo} and Λ_{Cu} would differ numerically simply because they do not refer to "X-units" of exactly equal size. This would not, of course, invalidate either one of the two values, Λ_{Mo} or Λ_{Cu}, provided the corresponding wavelength value is taken *as the definition* of one "X-unit".

Table 8 presents a partial list of recommended physical constants. These numbers have been computed from the least squares adjusted values given above and the auxiliary constants listed in Table 7. The errors listed in this table are not standard deviations, but limits of error arbitrarily taken to be three times the computed standard deviation. The standard deviations of quantities computed from the least squares adjusted variables must be computed in terms of the variance matrix (Table 9) since the errors of the least squares adjusted values are not independent of each other, but are, in fact, strongly correlated. The errors in these constants may be either greater or less than those computed by simple propagation of errors, depending on whether the correlation is positive or negative. Thus the relative uncertainty in F, the FARADAY, is smaller than the uncertainty in either N or e. The uncertainty in the computed value of a constant not given in the table may be obtained from the following

12*

Table 8. Recommended Values of the Physical Constants — 1963

(Errors are 3 σ-limits)

	Symbol or Formula	Value	Error Limits	Units
Velocity of Light	c	299792.5	0.3	10^3 m sec^{-1}
Elementary Charge	e	4.80298	0.00020	10^{-10} esu
	$e' = e/c$	1.60210	0.00007	10^{-20} emu
Avogadro Number	N	6022.52	0.28	10^{20} mole^{-1}
Electron Rest Mass	m	9.1091	0.0004	10^{-28} g
Planck's Constant	h	6.6256	0.0005	10^{-27} erg sec
	$\hbar$	1.05450	0.00007	10^{-27} erg sec
Faraday Constant	$F = N e$	2.89261	0.00005	10^{14} esu mole^{-1}
		9648.70	0.16	emu mole^{-1}
Atomic Mass of Electron	$N m$	5.48597	0.00009	10^{-4}
Fine Structure Constant	α	0.729720	0.000010	10^{-2}
	$1/\alpha$	137.0388	0.0019	
	$\alpha/2\pi$	1.161385	0.000016	10^{-3}
	α^2	53.2492	0.0014	10^{-6}
	$1 - (1 - \alpha^2)^{1/2}$	26.6242	0.0007	10^{-6}
Charge to Mass Ratio Electron......	$e/m c$	1.758796	0.000019	10^7 emu g^{-1}
	e/m	5.27274	0.00006	10^{17} esu g^{-1}
	h/e	1.37947	0.00004	10^{-17} erg sec esu^{-1}
	$h c/e$	4.13556	0.00012	10^{-7} erg sec emu^{-1}
Compton Wavelength of Electron	$h/m c$	2.42621	0.00006	10^{-10} cm
	$\hbar/m c$	3.86144	0.00009	10^{-11} cm

Compton Wavelength of Proton	$N h / M_p c$	1.32140	0.00004	10^{-13} cm
	$\lambda_{c\,p}/2\pi$	2.10307	0.00006	10^{-14} cm
Rydberg Constant	$R_\infty = \dfrac{\alpha^2 m c}{2 h}$	109737.31	0.03	cm^{-1}
First Bohr Radius	$a_0 = \hbar/(m c \alpha) = \alpha/(4\pi R_\infty)$	5.29167	0.00007	10^{-9} cm
Classical Electron Radius	$r_0 = e^2/m c^2$	2.81777	0.00011	10^{-13} cm
	r_0^2	7.9398	0.0006	10^{-26} cm^2
Thomson Cross Section	$\dfrac{8}{3}\pi r_0^2$	6.6516	0.0005	10^{-25} cm^2
Bohr Magneton	$\hbar e/2 m c$	9.2732	0.0006	10^{-21} erg gauss^{-1}
Nuclear Magneton	$\hbar e N/2 M_p c^2$	5.0505	0.0004	10^{-24} erg gauss^{-1}
Proton Moment	μ_p	1.41049	0.00013	10^{-23} erg gauss^{-1}
Energy equivalent of				
1 Electron Mass		511006	5	ev
1 Atomic Mass Unit		931.478	0.015	MeV
1 Proton Mass		938.256	0.015	MeV
1 Neutron Mass		939.550	0.015	MeV
Gas Constant	R_0	8.3143	0.0012	joules mole^{-1} deg^{-1}
		82.056	0.011	cm^3 atmos mole^{-1} deg^{-1}
Standard Volume of perfect gas	V_0	22413.6	3.0	cm^3 mole^{-1}
Boltzmann Constant	$k = R_0/N$	1.38054	0.00018	10^{-16} erg deg^{-1}
	R_0/F	8.6170	0.0012	10^{-5} ev deg^{-1}
	$1/k = F/R_0$	11604.9	1.5	deg ev^{-1}
First Radiation Constant	$c_1 = 8\pi h c$	4.9921	0.0003	10^{-15} erg cm
	$8\pi h c^2$	1.49660	0.00011	10^{-11} watts cm^2
	$2 h c^2$	1.19096	0.00009	10^{-12} watts cm^2 ster^{-1}

Table 8, continued

	Symbol or Formula	Value	Error Limits	Units
Second Radiation Constant	$c_2 = h\,c/k$	1.43879	0.00019	cm deg
Atomic Specific Heat Constant	$c_2/c = h/k$	4.7993	0.0006	10^{-11} sec deg
Wien Displacement Constant	$c_2/4.965114$	0.28978	0.00004	cm deg
Stephan-Boltzmann Constant	$\sigma = \left(\dfrac{\pi^2}{60}\right)\left(\dfrac{k^4}{\hbar^3 c^2}\right)$	5.6697	0.0029	10^{-5} erg cm^{-2} sec^{-1} deg^{-4}
	$a = \dfrac{4\sigma}{c} = \left(\dfrac{\pi^2}{15}\right)\left(\dfrac{k^4}{\hbar^3 c^3}\right)$	7.565	0.004	10^{-15} erg cm^{-3} deg^{-4}
Zeeman Splitting per Gauss	$e/4\pi m c^2$	4.66858	0.00004	10^{-5} cm^{-1} gauss
Anomalous Electron Moment	$\mu_e/\mu_0 - 1 = \dfrac{\alpha}{2\pi} - 0.328\,\dfrac{\alpha^2}{\pi^2}$	1.159615	0.000015	10^{-3}
Proton Moment — Bare	μ_p/μ_n	2.79276	0.00007	
Proton Moment — Spherical H$_2$O Sample	μ_p'/μ_n	2.79268	0.00007	
Proton Gyromagnetic Ratio	γ	26751.9	0.2	sec^{-1} gauss^{-1}
		4257.70	0.03	cycles sec^{-1} gauss^{-1}

Table 9. Variance Matrix

Variances are given, in units of (ppm)^2, on and below the diagonal. Correlation coefficients are given in parentheses above the diagonal. Therefore, this table actually presents the two symmetric matrices: The variance matrix and the correlation coefficient matrix. Since there are only three statistically independent variables, these matrices are degenerate and are only of rank 3.

	α	e	N	h	m	F
α ...	20.8	(0.96)	(— 0.88)	(0.95)	(0.88)	(0)
e ...	62.3	199.4	(— 0.93)	(0.99)	(0.97)	(— 0.06)
N ...	— 62.3	— 204.4	239.5	(— 0.93)	(— 0.93)	(0.41)
h ...	103.8	336.5	— 346.4	569.2	(0.98)	(— 0.08)
m ...	62.3	211.9	— 221.8	361.5	236.9	(— 0.12)
F ...	0	— 5.0	35.1	— 9.9	— 9.9	30.2

formula if the constant is computed from the elemental values appearing in the error matrix, using the following formulas:

$$f = f(x_1, x_2, \ldots x_n),$$

$$\sigma_f{}^2 = \sum_{i=1}^{n} \sum_{j=1}^{n} \left(\frac{x_i}{f} \frac{\partial f}{\partial x_i} \right) V_{ij} \left(\frac{x_j}{f} \frac{\partial f}{\partial x_j} \right) \tag{17}$$

where $\dfrac{x_i}{f} \dfrac{\partial f}{\partial x_i}$ is the "modulus of elasticity" of the function, f, with respect to the variable x_i. Since the matrix V_{ij} is given in units of (ppm)^2, the the computed value of σ_f is then also expressed in parts per million[58].

Acknowledgments

We wish to acknowledge the contributions of A. G. McNish of the U. S. National Bureau of Standards, whose aid in assembling the input data, and whose comments and criticism of problems of metrology have been invaluable in the preparation of this analysis.

We also wish to thank J. S. Thomsen and J. A. Bearden for careful and considered comments on a preliminary version of this adjustment.

References

[1] E. R. Cohen, J. W. M. DuMond, T. W. Layton, and J. S. Rollett, Revs. Modern Phys. 27, 363 (1955).

[2] E. R. Cohen and J. W. M. DuMond, Phys. Rev. Letters 1, 291 (1958).

[3] J. W. M. DuMond, IRE Trans. 1–7, 136 (1958).

[4] J. W. M. DuMond, Ann. Phys. 7, 365 (1959).

[5] J. W. M. DuMond, Proc. Nat. Acad. Sci. 45, 1052 (1959).

[6] F. Everling, L. A. König, J. H. E. Mattauch, and A. H. Wapstra, Nuclear Phys. 15, 342 (1960); 18, 529 (1960).

[7] L. A. König, J. H. E. Mattauch, and A. Wapstra, Nuclear Phys. 31, 1 (1962); 31, 18 (1962).

⁸ C. MORELLI, Absolute and First Order World Gravity Net, Special Study Group No. 5, Report to IUGG, Aug. 1959.

⁹ P. R. HEYL and G. S. COOK, J. Res. Nat. Bur. Stand. 17, 805 (1936). — H. JEFFREYS, M. N. Roy. Astr. Soc. (Geophysical Suppl.) 5, 219 (1948).

¹⁰ E. BERGSTRAND, Ann. Français Chronometrie 11, 97 (1957).

¹¹ I. C. C. MACKENZIE, Ordnance Survey Professional Papers, No. 19, HMSO (1954).

¹² L. ESSEN, Proc. Roy. Soc., A 204, 260 (1950); Proc. Roy. Soc., B 66, 189 (1953). — L. ESSEN and K. D. FROOME, Proc. Roy. Soc., B 64, 862 (1951).

¹³ K. D. FROOME, Proc. Roy. Soc., A 247, 109 (1958).

¹⁴ R. KARPLUS and N. M. KROLL, Phys. Rev. 81, 73 (1951).

¹⁵ C. M. SOMMERFIELD, Phys. Rev. 107, 328 (1957).

¹⁶ A. PETERMANN, Nuclear Phys. 5, 677 (1958).

¹⁷ P. SMRZ and I. ULEHLA, Czech. Journ. Phys. 10, 966 (1960).

¹⁸ Y. CAUCHOIS and H. HULUBEI, Longueurs d'Onde des Emissions X et des Discontinuities d'Absorption X. Paris: Herman et Cie Ed. 1947.

¹⁹ A. E. SANDSTROM, Handbuch der Physik, XXX, p. 164. S. Flügge, Ed. Berlin: Springer-Verlag. 1957.

²⁰ I. HENINS, Bull. Amer. Phys. Soc. II, 7, 339 (1962) (Abstract U 8).

²¹ R. T. BIRGE, Amer. J. Physics 13, 69 (1945).

²² W. L. BRAGG, J. Sci. Instr. 24, 27 (1947); Phys. Rev. 72, 437 (1947).

²³ F. TYREN, Z. Physik 109, 722 (1938); Thesis, unpublished. Uppsala (1940).

²⁴ W. E. LAMB Jr. and R. C. RETHERFORD, Phys. Rev. 72, 241 (1947).

²⁵ J. A. BEARDEN, Phys. Rev. 37, 1210 (1931); 48, 385 (1935).

²⁶ E. BÄCKLIN, Z. Physik. 93, 450 (1935).

²⁷ H. SOMMER, H. A. THOMAS, and J. A. HIPPLE, Phys. Rev. 82, 697 (1951).

²⁸ E. BLOCH and C. D. JEFFRIES, Phys. Rev. 80, 305 (1950). — C. D. JEFFRIES, Phys. Rev. 81, 1040 (1951). — K. R. TRIGGER, Bull. Amer. Phys. Soc. II, 1, 220 (1956).

²⁹ H. S. BOYNE and P. A. FRANKEN, Phys. Rev. 123, 242 (1961).

³⁰ D. J. COLLINGTON, A. N. DELLIS, J. H. SANDERS, and K. C. TURBERFIELD, Phys. Rev. 99, 1622 (1955). — J. H. SANDERS and K. C. TURBERFIELD, Proc. Roy. Soc., A 272, 79 (1962).

³¹ S. TRIEBWASSER, E. S. DAYHOFF, and W. E. LAMB Jr., Phys. Rev. 89, 98 (1953).

³² A. J. LAYZER, Phys. Rev. Letters 4, 580 (1960).

³³ R. P. FEYNMAN, "The Present Situation in Quantum Electrodynamics". Solvay Conference Jubilee, 1961 (privately circulated).

³⁴ W. C. MARTIN, Phys. Rev. 116, 654 (1959). — J. W. M. DuMOND and E. R. COHEN, Revs. Modern. Phys. 25, 691 (1953).

³⁵ J. E. NAFE and E. B. NELSON, Phys. Rev. 73, 718 (1948). — A. G. PRODELL and P. KUSCH, Phys. Rev. 79, 1009 (1950); 88, 184 (1952); 100, 1188 (1955). — J. P. WITTKE and R. H. DICKE, Phys. Rev. 96, 530 (1954); 103, 620 (1956). — D. KLEPPNER, H. M. GOLDENBERG, and N. F. RAMSEY, Applied Optics 1, 55 (1962). — F. M. PIPKIN and R. H. LAMBERT, Phys. Rev. 127, 787 (1962).

³⁶ C. K. IDDINGS and P. M. PLATZMANN, Phys. Rev. 113, 192 (1959).

³⁷ D. E. ZWANZIGER, Bull. Amer. Phys. Soc. II, 6, 514 (1961), Abstract K 1. — A. J. LAYZER, Bull. Amer. Phys. Soc. II, 6, 514, Abstract K 2.

³⁸ V. W. HUGHES, Bull. Amer. Phys. Soc. 118, 33 (1963). — K. ZIOCK, V. W. HUGHES, R. PREPOST, J. BAILEY, and W. CLELAND, Phys. Rev. Letters 8, 103 (1962).

[39] R. KARPLUS and A. KLEIN, Phys. Rev. 85, 972 (1952). — N. M. KROLL and F. POLLACK, Phys. Rev. 86, 876 (1952). — R. ARNOWITT, Phys. Rev. 92, 1002 (1953).

[40] D. P. HUTCHINSON, J. MENES, G. SHAPIRO, A. M. PATLACH and S. PENMAN, Phys. Rev. Letters 7, 129 (1961).

[41] E. B. D. LAMBE, Thesis, Princeton University, 1959 (unpublished).

[42] D. T. WILKINSON and H. R. CRANE, Phys. Rev. 130, 852 (1963).

[43] J. W. M. DuMOND, D. A. LIND, and B. B. WATSON, Phys. Rev. 75, 1226 (1949). — D. E. MULLER, H. C. HOYT, D. J. KLEIN, and J. W. M. DuMOND, Phys. Rev. 88, 790 (1952).

[44] J. W. KNOWLES, Canadian Journ. Physics 40, 257 (1962).

[45] R. T. BIRGE, Revs. Modern Phys. 1, 1 (1929).

[46] D. N. CRAIG, J. I. HOFFMAN, C. A. LAW, and W. J. HAMER, J. Res. Nat. Bur. Stand 64 A, 381 (1960).

[47] W. R. SHIELDS, D. N. CRAIG, and V. H. DIBELER, Proc. Int'l. Conf. on Nuclidic Masses, H. E. Duckworth, Ed., p. 519. Hamilton, Ontario: University of Toronto. 1960; J. Amer. Chem. Soc. 82, 5033 (1960).

[48] W. R. SHIELDS, E. L. GARNER, and V. H. DIBELER, J. Res. Nat. Bur. Stand. 66 A, 1 (1962).

[49] J. H. GARDNER and E. M. PURCELL, Phys. Rev. 76, 1262 (1949) 83, 996 (1951).

[50] W. A. HARDY and E. M. PURCELL, Bull. Amer. Phys. Soc. 114, 37 (1959).

[51] S. LIEBES and P. FRANKEN, Phys. Rev. 104, 1197 (1956); 116, 633 (1959).

[52] J. H. SANDERS, K. F. TITTEL, and J. F. WARD, Proc. Roy. Soc., A 272, 103 (1963).

[53] M. E. STRAUMANIS, Acta Cryst. 2, 82 (1949); Phys. Rev. 92, 1155 (1953).

[54] J. A. BEARDEN, Bull. Amer. Phys. Soc. II, 7, 339 (1962).

[55] K. LONSDALE, Trans. Roy. Soc., A 240, 219 (1947).

[56] E. R. COHEN, Bull. Amer. Phys. Soc. II, 7, 305 (1962).

[57] C. B. VAN DEN BERG, Thesis, University of Groningen, 1957, p. 62.

[58] E. R. COHEN, K. M. CROWE, and J. W. M. DuMOND, Fundamental Constants of Physics, Chapter 7. New York: Interscience Publishers, Inc. 1957.

Discussion

A. McNISH: One of the slides that was shown indicated that the variables subject to adjustment were α, e, N and Λ. I am afraid this may lead to some confusion.

E. R. COHEN: Yes, that was an older slide. The Λ was calculated essentially on a secondary stage from the adjustment. In fact, it is probably true that this is now the best way to measure the conversion factor from X-units to Angstroms. We can calculate Avogadro's number quite accurately from data that does not involve X-ray wave lengths and then if we measure the lattice spacing of a crystal and weigh it to determine how much mass is in it, if we know Avogadro's number we know how many atoms are in it and, therefore, we can calculate the lattice spacing and this, then, can give us the conversion factor from X-units to centimeters to perhaps ten parts per million if we can get a clear definition of what the X-unit is in the first place.

J. W. KNOWLES: If you reject the annihilation measurement of KNOWLES and BEARDEN on the basis that an X-ray line cannot be measured to one hundredth of its width, with certainty, then it seems without purpose to define a conversion constant for conversion from X-unit to milliangstroms since part of this definition requires the definition of an X-ray line shape center to one hundredth of a line width or better.

E. R. Cohen: Knowles most recent measurement, reported here, has not been previously available. His older data is uncertain because of difficulties in measuring the lattice constant of the planes he actually used. Furthermore, if Knowles' measurement of α^2/Λ is the *only* measurement retained which involves Λ, it may be dropped from the least squares adjustment and introduced only as a subsidiary calculation of Λ from the adjusted value of α. Hence it is not rejected, but need not be included in the primary analysis.

J. W. Knowles: If old data is thrown out, based on the fact that it is not well defined, it should be noted, that recent data is much better defined. One must, at some point, define a standard of acceptance of more recent data.

H. Staub: Does one know the physical reasons for the discrepancy of Franken and Boyne's values of μ_p and are there any plans to repeat these measurements? I would consider this as highly desirable.

E. R. Cohen: Boyne and Franken used an inhomogeneous electric field; field divergence and space charge were evaluated by extrapolating the resonance frequency as a function of $\dfrac{1}{H^2}$ to infinite magnetic field (zero orbit size). The extrapolation was dependent on operating conditions in a way not fully understood. Measurements at higher fields are probably required to resolve this. The experiment was indicated as "preliminary" by Boyne and Franken. Boyne, now at USNBS, may repeat this measurement at some time in the future using higher magnetic fields and heavier ions, such as H_2^+, D^+.

A. McNish: There are plans to repeat this experiment but Boyne is very much interested in another experiment just now so there will be some delay. I also want to make a statement about rejection of data. I should like to point out that data were not rejected merely because they were discrepant. Each experiment was carefully examined and a decision was reached as to the credence which could be placed in the result. If, by such internal criticism there was cause to doubt the reliability of the experiment, and if, in addition, the result was discrepant with more reliable experiments, then the result of that experiment was rejected.

Systematics of Single Particle Levels[1]

Department of Physics, University of Pittsburgh, Pittsburgh 13, Pa.

With 1 Figure

The basic assumption of nuclear shell model is that from the stand-point of any one nucleon, the forces between it and all the other nucleons can be represented by a potential well. Given the parameters of this well, the locations of all shell model states should be readily calculable. However, after many years of such calculations it has become clear that there is no simple set of well parameters that can accurately fit even a small fraction of the data. It is thus important to accumulate data on the location of these states in order to find what effects are important in determining their location. These data are also very useful since they are inputs in a great many calculations in nuclear structure theory.

A very useful method for experimentally locating shell model states is provided by stripping reactions such as (d, p). Since this reaction is essentially equivalent to the insertion of a neutron into a nucleus, and since an inserted neutron must go into a single particle (i. e. shell model) state, this reaction excites levels in proportion to how much they contain of the configuration "ground state of target nucleus plus a neutron". Moreover, the angular distribution of the emitted proton may be analysed to determine the orbital angular momentum with which the stripped neutron entered the nucleus. Except for the ambiguity over whether $j = l + 1/2$ or $j = l - 1/2$, this determines which shell model state is contained in each nuclear level. The location of the shell model state is then taken as the "center of gravity" of the nuclear states containing parts of it, weighting each in proportion to how strongly it is excited.

In general, this location is determined not only by the shell model potential, but also by the residual interactions. However, if we deal with nuclei having one particle outside of a closed shell, the residual interactions vanish and we obtain direct determinations of the shell model states. It is therefore expedient to use only such nuclei. This has the corollary advantage that the $j = (l + 1/2)$ and $j = (l - 1/2)$ members of the spin-orbit doublet are very widely split in these nuclei, so that a decision between these two possible assignments is usually straightforward.

All that has been said of the (d, p) stripping reaction applies equally well to the inverse reaction, (p, d) except that here we study single hole levels rather than single particle levels. The information obtained is thus complementary. It has been found that the (d, t) reaction serves as an adequate substitute for (p, d) if high energy protons are not available.

Before going on to the results, a few more detailed remarks on the method may be interesting:

The analysis of the angular distributions and cross sections is now done almost exclusively with distorted wave Born approximation (DWBA) calculations. In terms of these, the measured cross section on an even-even target nucleus, $d\sigma/d\Omega$, is

$$\sigma(d, p) = (2I + 1)\,\sigma_T\,S,$$
$$\sigma(d, t) = \sigma_T\,S, \tag{1}$$

where I is the spin of the final state, σ_T is the theoretical cross-section obtained from DWBA calculations, and S is the spectroscopic factor. When a closed shell nucleus is bombarded, the sum of the S values for all levels corresponding to a given shell model state is unity. This is often useful in resolving ambiguities between $j = l + 1/2$ and $j = l - 1/2$ assignments. It is also a valuable guide in ascertaining whether all appreciable components of a shell model state have been located. This is necessary in order to find the center of gravity which we interpret as the location.

Additional information on this problem may be obtained from the theoretical prediction of the energy interval over which a state may be spread. This is given by giant resonance theory as $2W$, where W is the depth of the imaginary optical model potential. An extrapolation of existing data on this gives

$$W \simeq 1/3\,E^*,$$

where E^* is the excitation energy. Thus a level at 3 MeV excitation may be expected to have almost all its components between 2 and 4 MeV.

In all of this work, it is impossible to overemphasize the need for good energy resolution, especially when working with the higher lying levels and in heavy nuclei. Unresolved doublets give meaningless angular distributions or worse still, distributions that may be wrongly interpreted. These lead to misassignments and hence to break down of the sum rules and thus to great confusion.

Measurements have been made on all closed neutron shell nuclei, and the available data is shown in Fig. 1. First let us look at the data for Ca and Zr where the results are plotted directly. The trend within these groups is toward less binding with increasing A. This is clearly counter to the general trend with A. It represents a symmetry energy effect; that is, the depth of the shell model potential decreases with increasing neutron excess. This effect is well known for the optical model potential[2] and the magnitudes of the effects found by the two methods are in good agreement.

In order to remove the complications of the symmetry energy, it is convenient to correct all data for symmetry energy to the mass of maximum

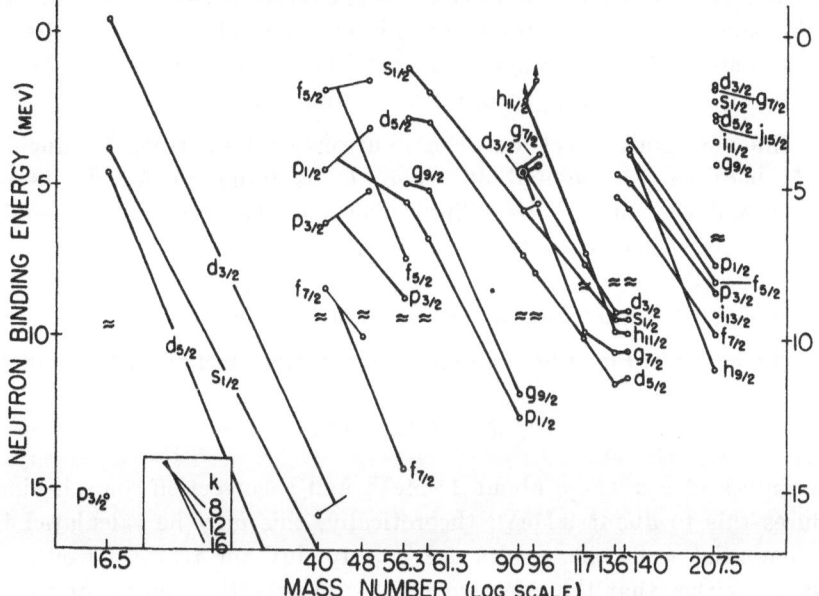

Fig. 1. Locations of single particle levels for various masses. Data are corrected to mass of maximum beta stability. $\approx$ is top of Fermi Sea

beta stability for that element. For Ca and Zr, this is shown in Fig. 1. All other data in Fig. 1 has been corrected for this before plotting.

One interesting observation from Fig. 1 is the locations predicted for the neutron giant resonances. These should occur where the $S_{1/2}$ states go through zero binding energy. From Fig. 1, they should be at $A \approx 50$ and at $A \approx 150$, in agreement with experiment.

Some interesting effects which are important in determining the character of Fig. 1 may be noted as follows:

Self-binding.

It may be noted that in the Ca and Zr isotopes, the states that are filling—$f_{7/2}$ and $d_{5/2}$ respectively—move down with increasing A, in contrast to the behavior of the states that are not filling. This is due to an extra binding from the good overlap in their wave functions; this effect has been discussed by others[3] and is referred to as self-binding energy. Its magnitude may easily be estimated from Fig. 1.

n-p interaction between nucleons of same l.

It may be noted from Fig. 1 that the $g_{7/2}$ state moves down more rapidly than the others between $A = 90$ and 136. This is believed to be due to the extra attraction between $g_{7/2}$ neutrons and $g_{9/2}$ protons which are filling in this region, due to the strong overlap between wave functions of the same l. Other examples of this are the behavior of $h_{9/2}$ between $A = 140$ and 207, where $h_{11/2}$ protons are filling, and of $f_{5/2}$ and $f_{7/2}$ between $A = 40$ and 61, where $f_{7/2}$ protons are filling.

The analogous effect between neutrons and neutrons is much less for $f_{5/2}$ between $A = 40$ and 48 where $f_{7/2}$ is filling or for $d_{3/2}$ between $A = 90$ and 96 where $d_{5/2}$ is filling. This may be due to the "repulsion between half-shells" expected from tensor forces.

dE/dA, Change of binding energy with mass number.

The points in Fig. 1 lie approximately on straight lines, which indicates that

$$dE/dA \approx -k/A.$$

The values of k average about 11 MeV, and a correction for self-binding reduces this to about 9 MeV; theoretically, this may be calculated from the potential well; the results are $k \simeq 16$ MeV for realistic wells. This indicates either that the effective mass is greater than unity, or that the well depth decreases with increasing A, or both.

Spin-orbit Splittings.

From Fig. 1, the energy difference between states with $j = l + 1/2$ and $j = l - 1/2$ is readily obtained; the results are listed in Table 1 (Col. 3). If this splitting arises from a term in the potential proportional to $(l \cdot S)$, the splitting for a given mass should be proportional to $(2l + 1)$. This is tested in Col. 4, where it is seen that there are large discrepancies for cases (marked by asterisks) where one member of the doublet is empty and the other full. This may be explained by the self-binding energy; quantitative estimates from Table 1 agree with estimates discussed above. The final column of Table 1 tests the assumption that the spin-orbit splitting is proportional to $A^{1/3}$, as expected for a volume interaction[4]. The results suggest that an $A^{2/3}$ dependence is operating; this is expected from a surface interaction.

Table 1. Spin-orbit Splittings

Mass		ΔE (MeV)	$\Delta E/(2l+1)$	$[\Delta E/(2l+1)] A^{1/3}$
208	1	0.90	0.30	1.77
	2	0.96	0.19	1.14
	3	1.78	0.26	1.51
	4	2.47	0.28	1.62
	6	5.71	0.44*	2.6
140	1	1.37	0.46	2.4
	2	2.2	0.44	2.3
	3	1.88	0.27	1.40
	5	6.05	0.55*	2.9
138	2	2.4	0.48	2.5
96	2	3.32	0.66	3.0
90	2	2.70	0.54	2.4
	4	6.74	0.75*	3.4
54	1	~ 1.8	~ 0.6	4.0
	3	7.0	1.0*	3.8
48	1	2.03	0.68	2.5
	3	8.17	1.17*	4.3
40	1	1.8	0.60	2.0
	2	~ 7.0	1.4	4.1
	3	6.5	0.93	3.2
16	1	6.16	2.05	5.2
	2	5.08	1.02	2.6

Spacings between major shells.

Spacings between major shells may be readily obtained from Fig. 1; these may be readily calculated for a given potential well. The experimental spacings are found to be about 25% smaller; this can best be explained as due to an effective mass greater than unity.

Dependence on l.

It is commonly believed that in a given major shell, higher l states lie lower[4]. This effect was studied from Fig. 1; it was found that there is little evidence for this among states that are empty, but strong evidence for this among states that are full. This is interpreted as due to a dependence of self-binding energy on l, as expected from the fact that higher l states contain more particles.

References

[1] The work described here has been discussed in more detail in: B. L. COHEN, P. MUKHERJEE, R. H. FULMER, and A. L. McCARTHY, Rev. Modern Phys. 35, 332 (1963). — B. L. COHEN, Phys. Rev. 130, 227 (1963).

[2] F. G. PEREY, Proceedings of Padua Conference on Nuclear Reactions Mechanism (to be published).

[3] M. BARANGER, Phys. Rev. **120**, 957 (1960).

[4] S. G. NILSSON, Kgl. Danske Videnskab Selskeb, Mat.-Fys. Medd. **29**, No. 16 (1960).

Discussion

K. BLEULER: FEYNMAN has introduced an effective mass (larger than the free particle mass) in the case of liquid helium and his theory might also apply in the case of nuclear structure. We once thought that such an effect (with a much larger effective mass) could play an important role in the case of the single particle ground states. We then realized that in this case the BCS-Theory gives already a large correction which makes up to a large extent for the characteristic differences between theoretical and experimental values. On the other hand, it can be easily seen that these corrections from pairing-theory become extremely small for higher excited single particle states. It is, therefore, very interesting that Dr. COHEN finds for this case an effective mass larger than the free mass in accordance with Feynman's theory.

F. MALIK: You said that the central part of the optical model potential and the shell model potential are the same. In view of the recent work of BUCK and PEREY, it seems to me that the optical model potential could be quite different from the shell model one. Their work is certainly very interesting because they could find energy independent and mass number independent parameter. Their depth of the real part of the potential is about 71 MeV. This is quite different from the usual 50 MeV one. Are we sure at all about the depth of the central potential then? Particularly, there is ample reason to believe from the unified model of the nuclear reactions that the nuclear potential has a non-local part. It will be very interesting to see a non-local shell model potential in the line of PEREY and BUCK.

B. L. COHEN: This is carrying the problem one step beyond where we have carried it. The effective mass is, of course, equivalent to a non-local potential, but you want to go beyond this. I think it is first important to understand things as well as possible on a simple level before going to more conplicated things. But your point may very well be correct.

Separation Energies of Nucleon Pairs

By

H. T. Tu and K. Way

Nuclear Data Project, National Academy of Sciences – National Research Council, 2101 Constitution Avenue, N. W., Washington 25, D. C.

(Presented by K. WAY)

With 9 Figures

In order to study the implications for nuclear structure of the data on nuclear masses, the separation energies of neutron and proton pairs have been plotted in various ways. S_{2n} is used to designate the separation energy of a neutron pair and is given here only for even-N nuclei. S_{2p} designates the separation energy of a proton pair and is given only for even-Z nuclei. Thus the pair separation energies presented always pertain to two nucleons in the same orbit.

BEINER and BLEULER[1] have already shown pair separation energies as a surface above the Z, $N - Z$ plane. (In their paper the last particle of a pair is either odd or even.) It seemed possible that features additional to those discernable in their plots might be emphasized by plotting sets of traces along the separation energy surface. It has turned out that each such set which we have plotted brings out points not revealed by the others. Undoubtedly still other facets important for nuclear structure will be brought to light by plots other than those presented here.

First, neutron-pair separation energies, S_{2n}, were plotted as functions of N while Z was kept constant. Such plots show the change in the binding energy of a pair as successive pairs are added to a given nucleus. It was hoped that such plots would give indications of shell effects in addition to showing the large discontinuities, now so familiar, at $N = 20$, 50, 82, and 126. Fig. 1 shows the results in the region between $N = 20$ and $N = 50$. Marked changes in slope are noticeable after the neutron numbers 20, 28, and 50 but nothing is noticeable at $N = 32$, 34, 38, or 40. The line for Ge ($Z = 32$) is peculiar but the errors are large. Other irregularities do not seem to be associable with any particular N-numbers.

Thus there seems to be a definite shell or subshell at $N = 28$ but not at other possible neutron numbers in this region.

Additional plots for higher neutron numbers reveal no subshell effects at 56, 58, 64, or indeed elsewhere. An anomaly just the opposite from a subshell effect appears at $N = 90$. One concludes that, except for the $f_{7/2}$ subshell which closes at 28, either subshells are so close together in energy value that the order of their filling is not detectable through mass measurements, or else the amount of configuration mixing is so great that the order has no real meaning.

A similar conclusion can be drawn about proton subshells. Figs. 2 and 3 show plots of proton-pair separation energies, S_{2p}, as functions of Z, for values of Z between 20 and 50. For each line N is kept constant. The changes at $Z = 20$ and 28 are less pronounced than for neutrons but still definite. Again 32, 34, and 40 do not stand out. In higher regions no consistent changes are noted at 56, 58, or 64. One concludes, as in the neutron case, that there are no subshells except at 28. The marked anomaly found for $N = 90$ does not show up for $Z = 90$. Thus in the second rotational region the deformed shape does not seem to have a marked energy advantage over the spherical shape. This should result in an overlapping of the vibrational and rotational levels.

Next, a search was made for traces of constant pair-separation energy. On the neutron plot lines were drawn connecting the values for nuclei differing by two neutrons and one proton (instead of two neutrons as in Fig. 1). Fig. 4 shows such a plot in the region from $N = 50$ to $N = 82$. Here the constancy of S_{2n} is very striking. There is not more than 1 MeV difference in the neutron pair values as you proceed in steps from a given nucleus to one which differs from it by a triton, that is as you go along a path of $2Z - N = $ const (which is often parallel to the bottom of the mass valley). In other words, the separation energy of a neutron pair decreases steadily as one adds neutron pairs to a given nucleus but remains fairly constant if, each time a neutron pair is added, a proton is also put in. The decrease in the binding of a neutron pair due to increasing the number of neutron pairs is almost exactly compensated by the addition

---

In all figures:

● Value from measurements with error about size of dot.

⬤ Value from measurements with error known.

⬦ Value from measurements with error unknown.

○ Value found from one extrapolation of β-systematics and one well-known measured value.

⬡ Value found from one extrapolation of β-systematics and one approximately known measured value.

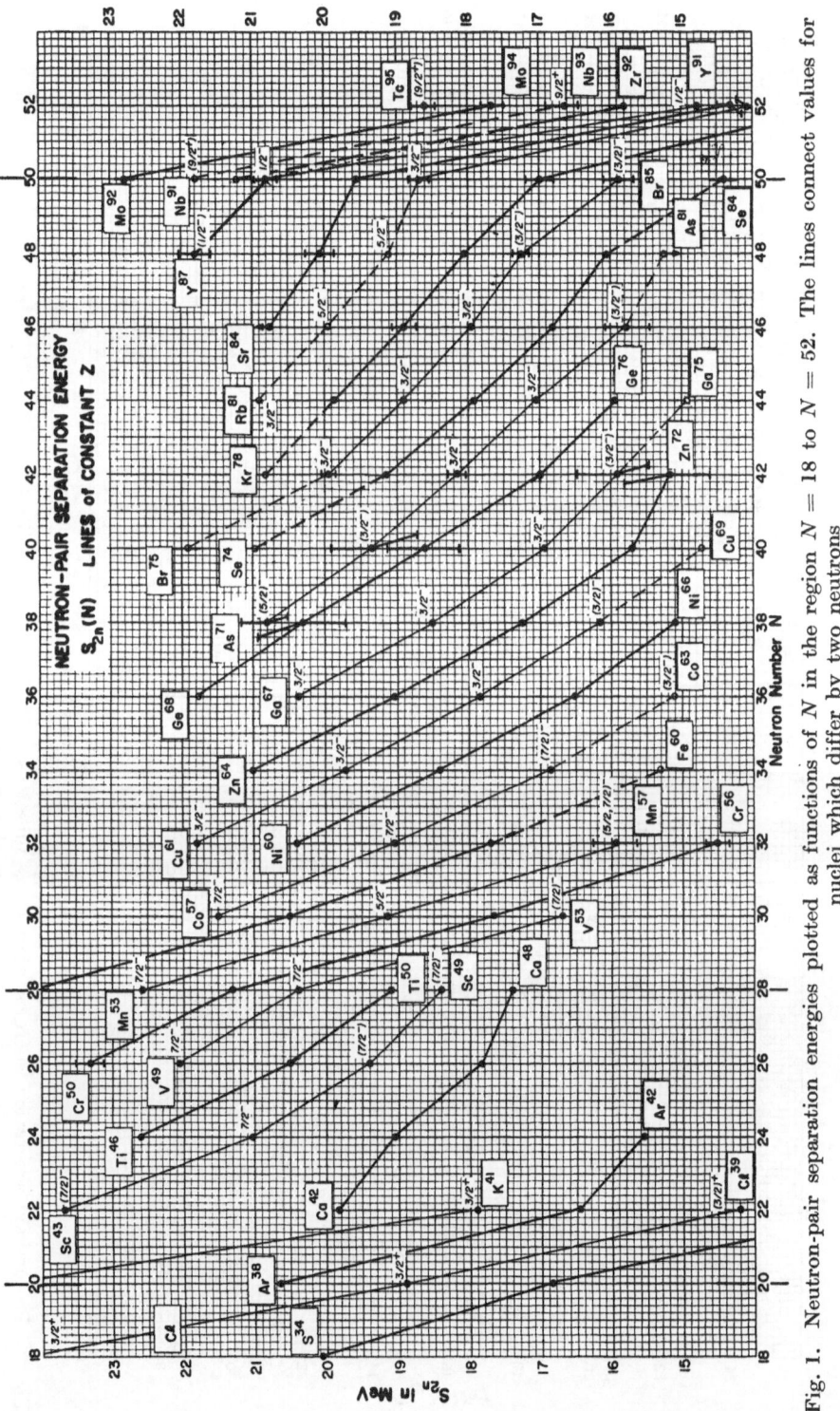

Fig. 1. Neutron-pair separation energies plotted as functions of N in the region $N = 18$ to $N = 52$. The lines connect values for nuclei which differ by two neutrons

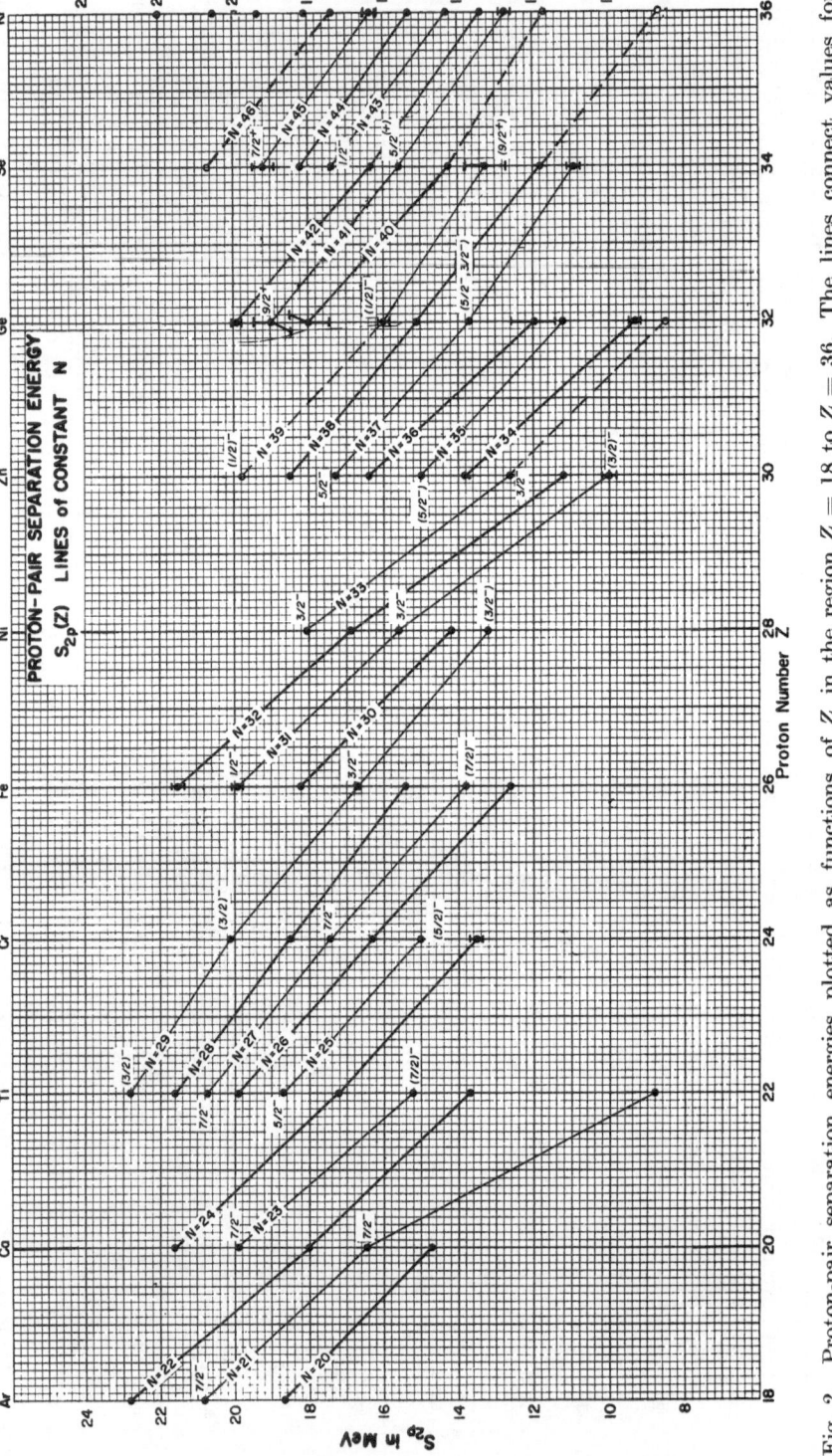

Fig. 2. Proton-pair separation energies plotted as functions of Z in the region $Z = 18$ to $Z = 36$. The lines connect values for nuclei which differ by two protons

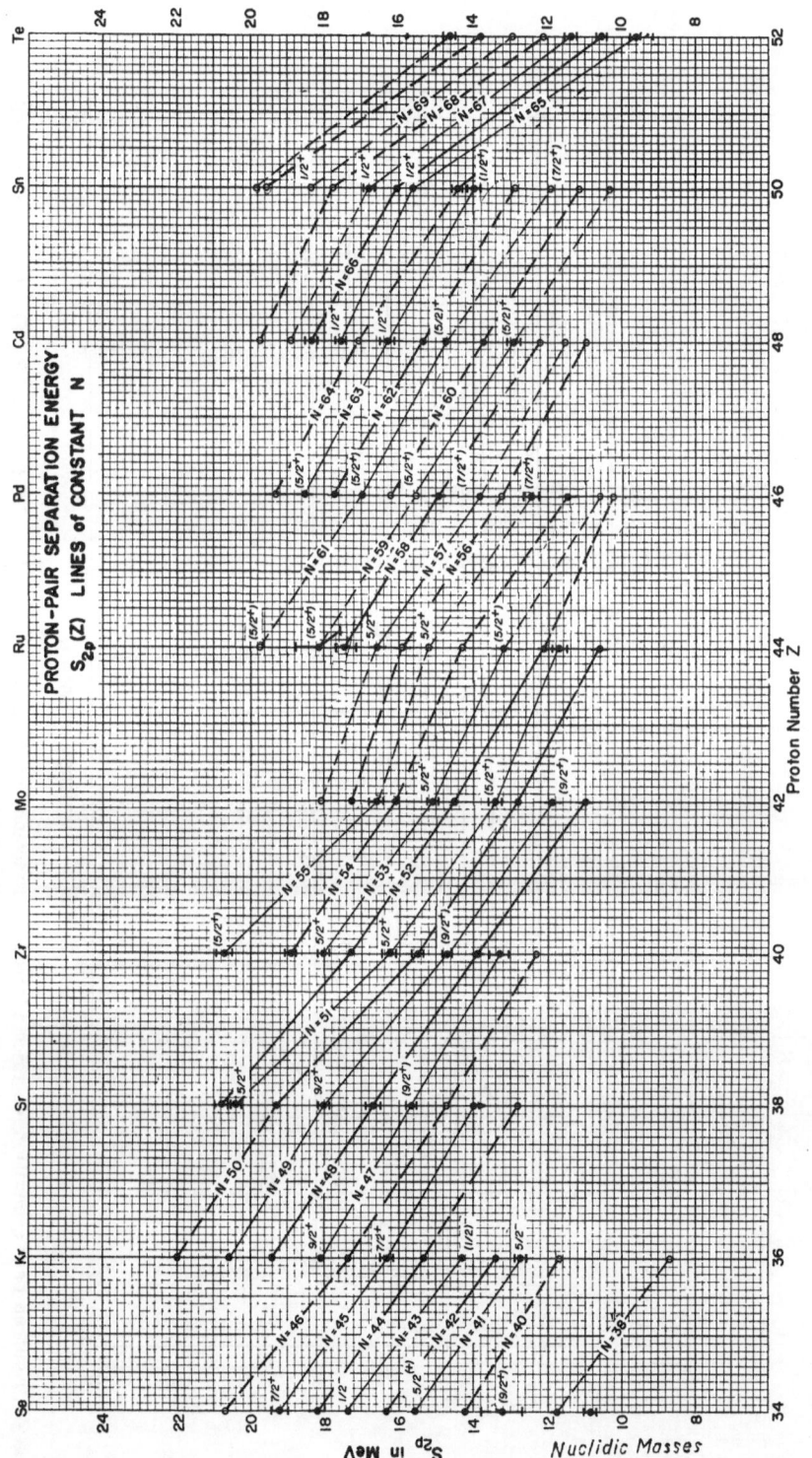

Fig. 3. Proton-pair separation energies plotted as functions of Z in the region $Z = 34$ to $Z = 52$. The lines connect values for nuclei which differ by two protons

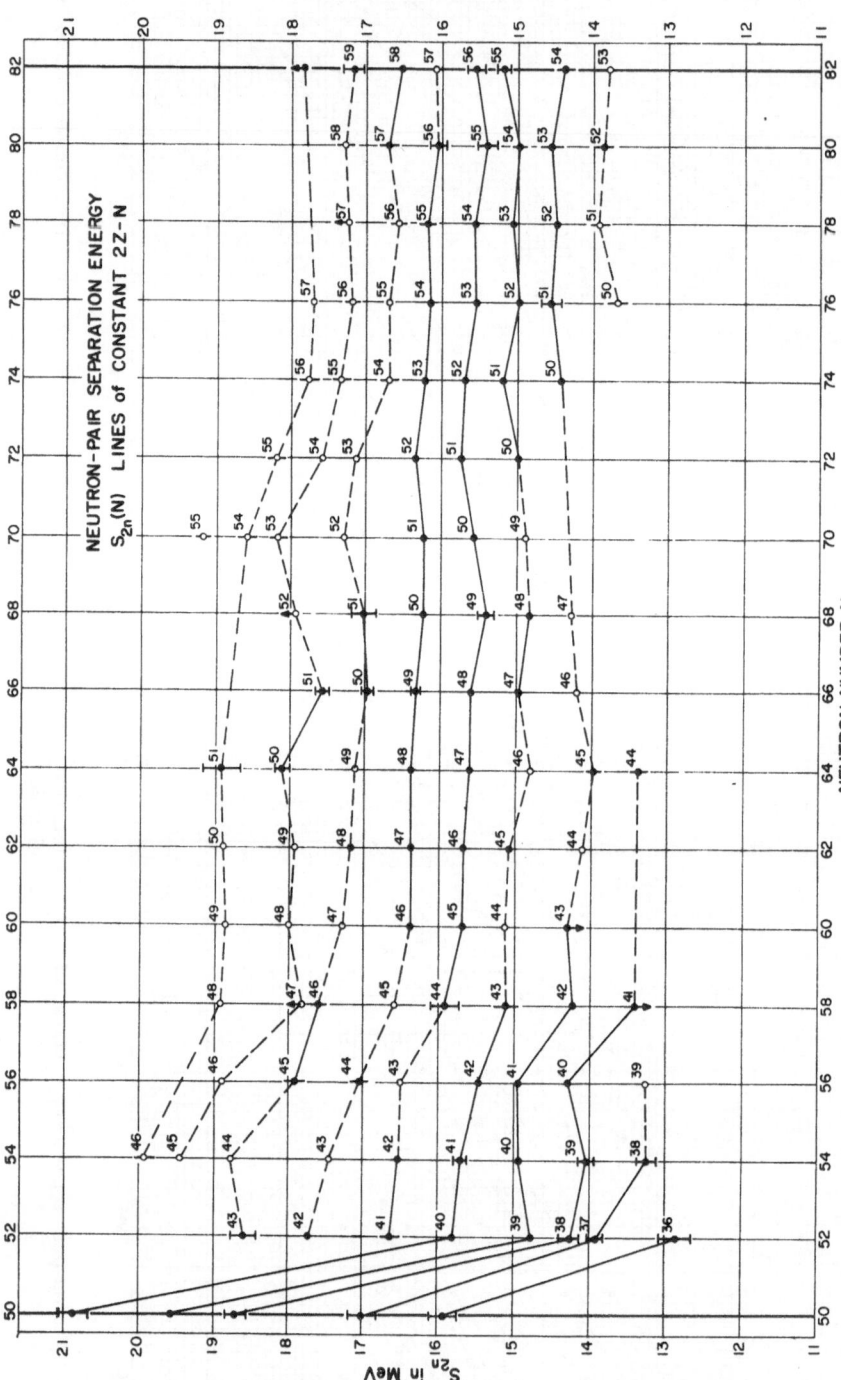

Fig. 4. Neutron-pair separation energies plotted as functions of N in the region $N = 50$ to $N = 82$. The lines connect values for nuclei which differ by two neutrons and one proton

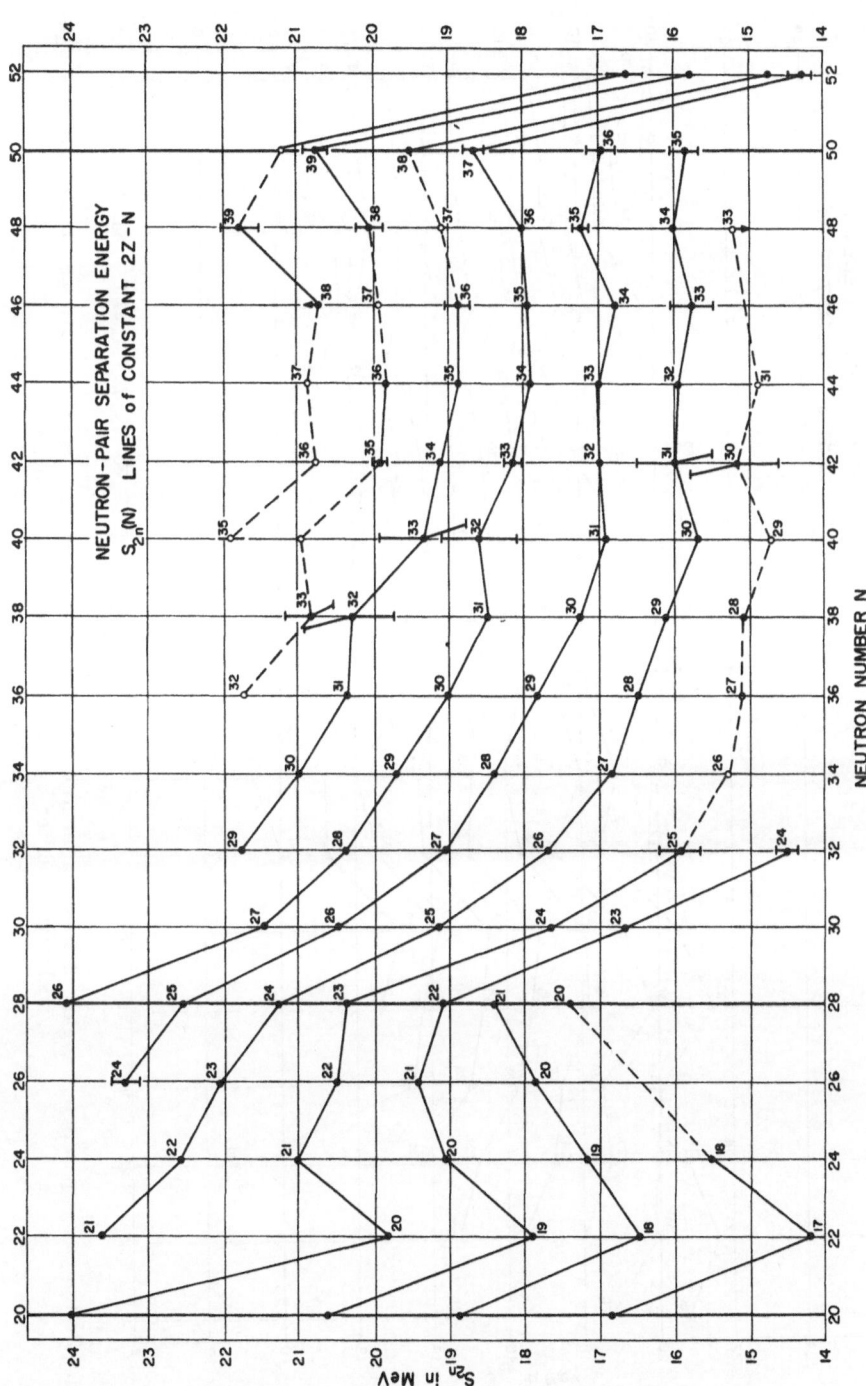

Fig. 5. Neutron-pair separation energies plotted as functions of N in the region $N = 20$ to $N = 52$. The lines connect values for nuclei which differ by two neutrons and one proton

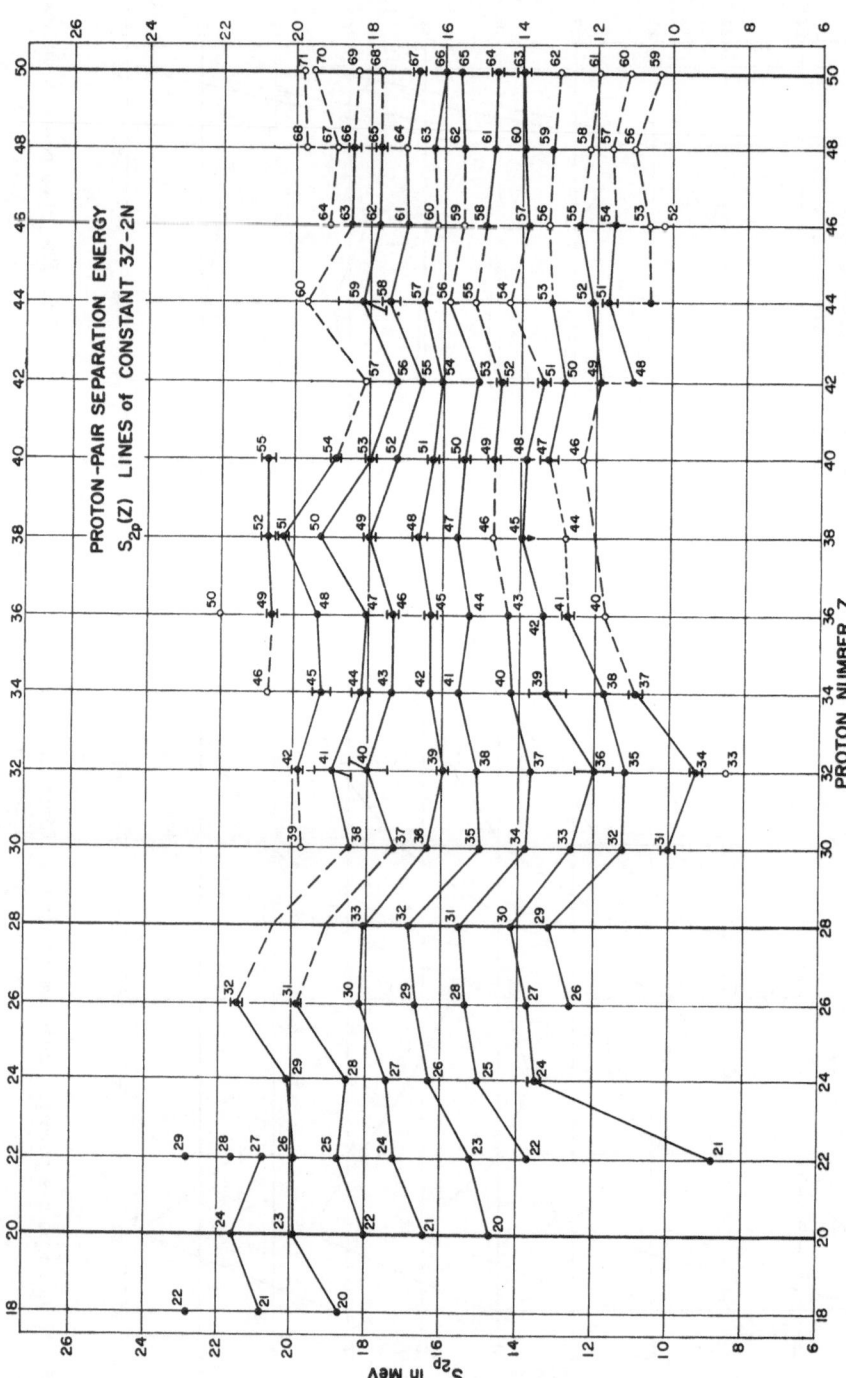

Fig. 6. Proton-pair separation energies plotted as functions of Z in the region $Z = 18$ to $Z = 50$. The lines connect values for nuclei which differ by three neutrons and two protons

of a single proton. The simplest form of the Weizsächer semi-empirical mass formula[2] predicts a constant decrease in S_{2n} as two neutrons and a proton are added to a given nucleus. From the formula, Dr. N. B. GOVE[3] has derived the following expression which predicts that the decrease diminishes with increasing A.

$$\Delta S_{2n}/\Delta(2n+p) = 270\,(N-Z-1)/A^2 + 25/A^{4/3} - 80/A \text{ MeV}. \qquad (1)$$

Fig. 5 shows the same type of plot in the region of Fig. 1. Here a large decrease in S_{2n} is observed. However, a discontinuity at $N = 28$ is quite marked and a distinct change in pattern at $N = 40$ is noticeable. In the region $N = 82$ to $N = 126$, (not shown) the S_{2n} lines have a hump in the region $N = 90$ to $N = 110$. In the region above $N = 126$ an upward tendency begins immediately after the magic number, showing again that the energy behaviour in the second rotational is different from that in the first.

In a search for lines of constant proton-pair separation energy it was found that there is a good deal of constancy along lines joining nuclei which differ by three neutrons and two protons. (These lines are along steps of He[5], that is along a path of $3Z - 2N =$ constant). This means that it takes three neutrons to compensate for the drop in proton-pair separation energy resulting from the addition of two protons. Fig. 6 shows the plot in the region $Z = 18$ to $Z = 50$. In most cases the separation energy of two protons varies by not more than 1 MeV as, for instance, along the line $_{30}\text{Zn}^{64}_{34}$ to $_{50}\text{Sn}^{114}_{64}$. It is also to be noted that discontinuities at $Z = 20$ and $Z = 28$ stand out more clearly here than in Fig. 2. However, nothing is marked at $Z = 40$. In the region from $Z = 50$ to $Z = 82$, the lines have a downward trend while above $Z = 82$ a hump appears. However, it does not start sharply at $Z = 88$ where the second rotational region seems to begin.

We next made a systematic attempt to discover how particles of the other kind affect the separation energy of a given nuclear pair by plotting $S_{2n}(Z)$ and $S_{2p}(N)$. Graphs of this type were first made for single nucleon separation energies by YAMADA and MATUMOTO[4]. These authors noted that the effect of adding particles of the other kind seemed to be quite regular, even when passing across one of their magic numbers. Our results confirm this observation. A few exceptions are to be noted, also observable in their plots, which we suggest may be due to large changes in angular momentum.

Fig. 7 shows S_{2n} for constant neutron number plotted as a function of Z. The slopes of the $S_{2n}(Z)$ curves for the neutron pairs (123, 124) and (125, 126) decrease markedly at $Z = 82$. Presumably the neutron in the pairs involved are in low angular-momentum, $p_{3/2}$ or $p_{1/2}$, states. The decrease occurs when the added protons change from low angular-

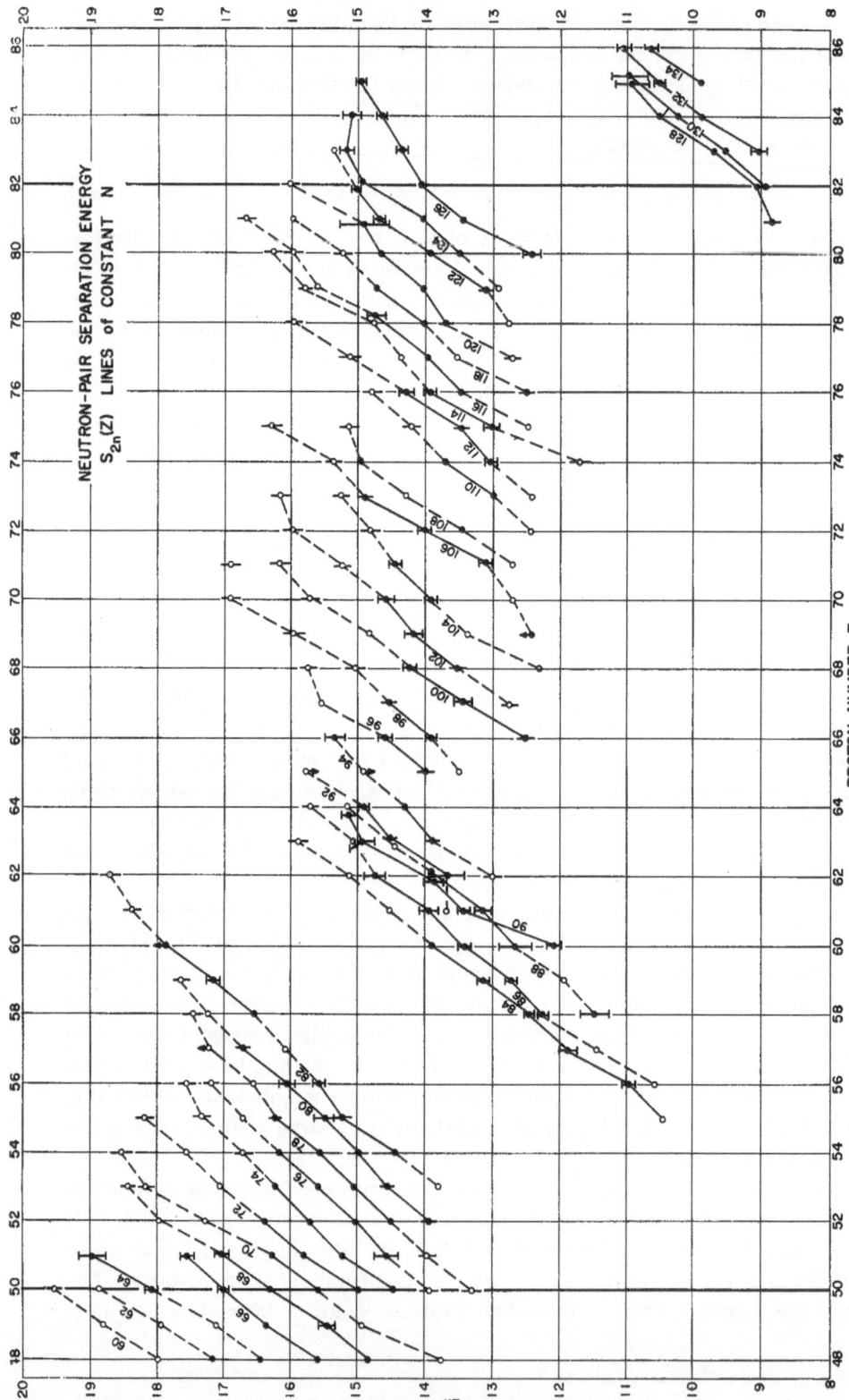

Fig. 7. Neutron-pair separation energies plotted as functions of Z in the region $Z = 48$ to $Z = 86$. The lines connect values for nuclei which differ by one proton

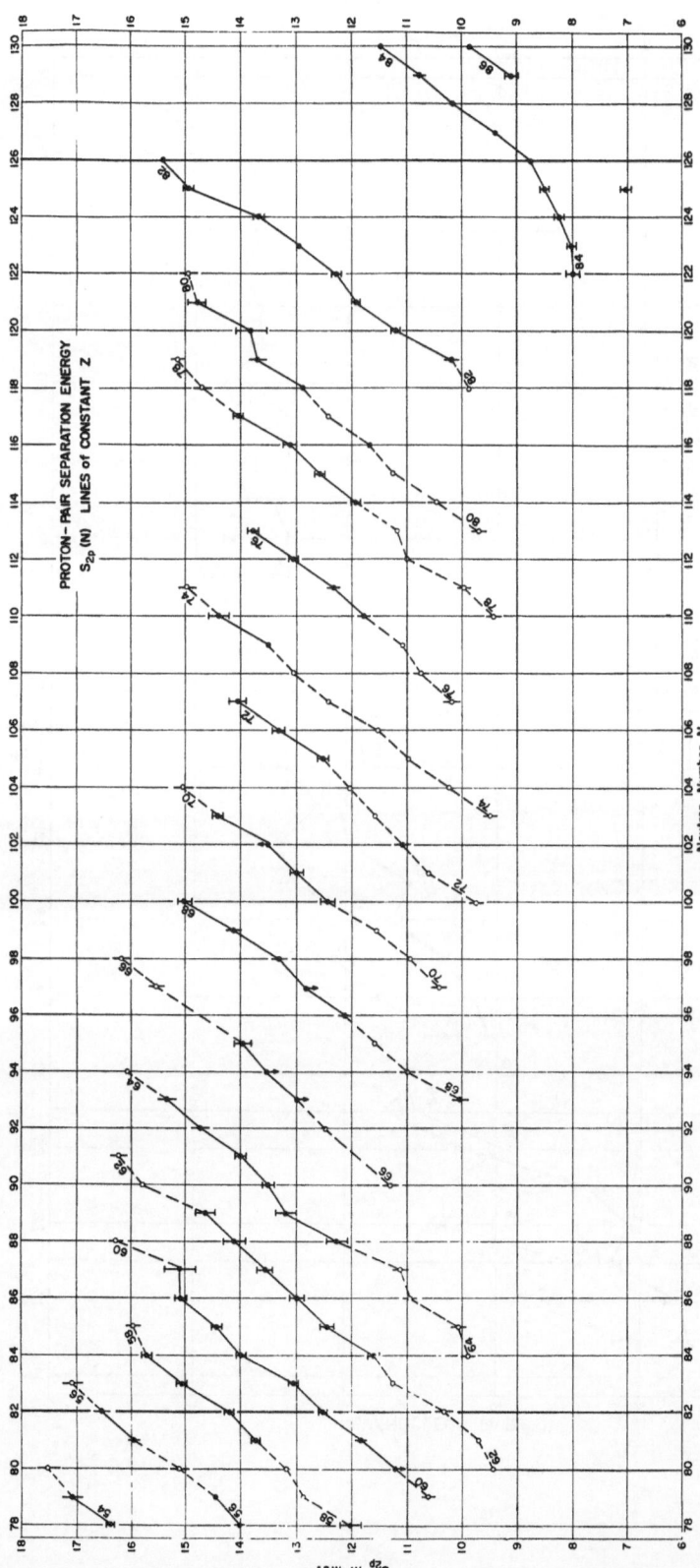

Fig. 8. Proton-pair separation energies plotted as functions of N in the region $N = 78$ to $N = 130$. The lines connect values for nuclei which differ by one neutron

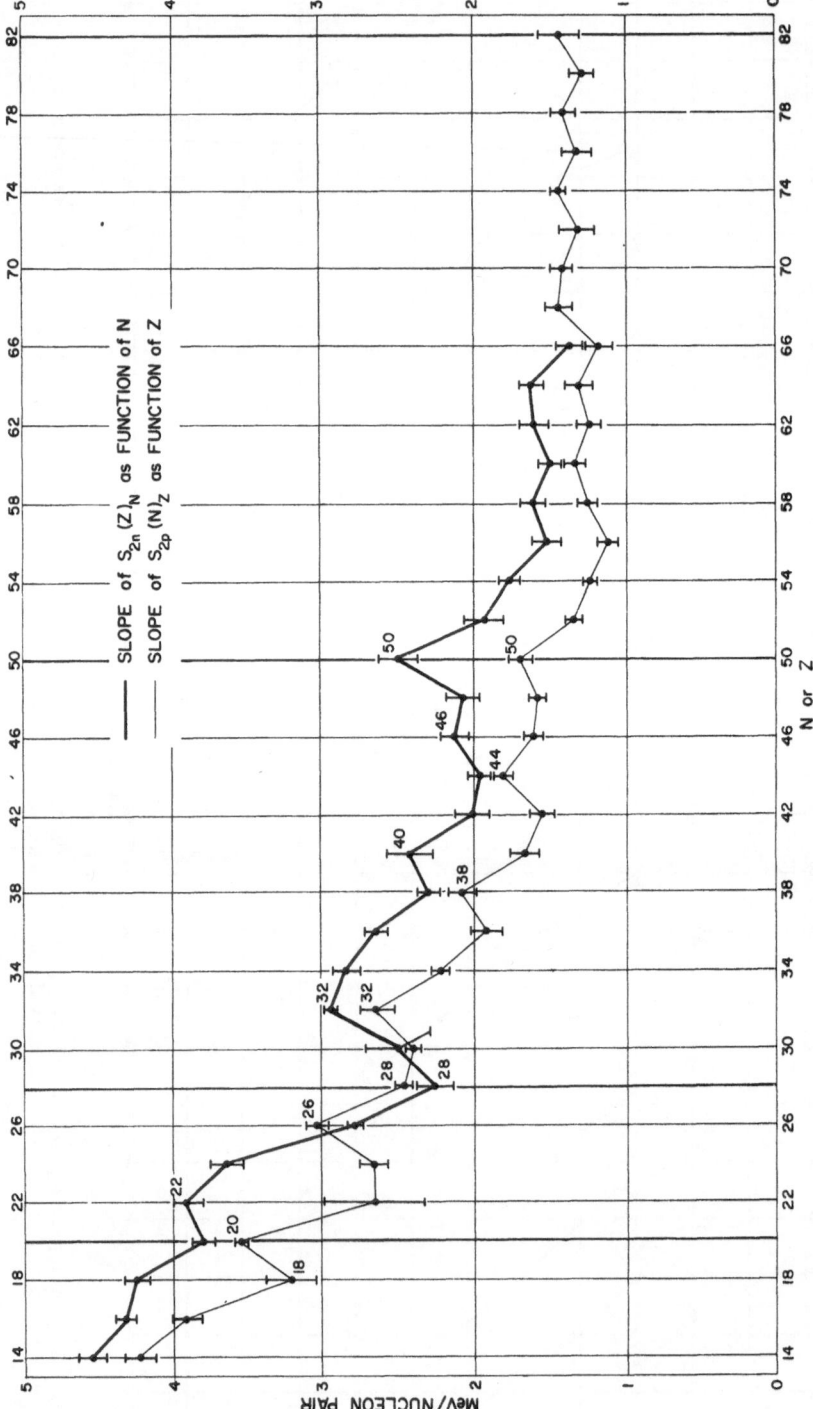

Fig. 9. Slopes of $S_{2n}(Z)$ and $S_{2p}(N)$ plotted as functions of N and Z respectively in the region 14 to 82

momentum, $s_{1/2}$ or $d_{3/2}$, states for $Z \leq 82$ to $h_{9/2}$ states for $Z > 82$. When $N = (127, 128)$, the neutrons of the last pair are probably in high angular-momentum $g_{9/2}$ states. In this case the slope increases after $Z = 82$ when the protons enter high angular-momentum states.

An analogous effect is to be seen on the $S_{2p}(N)$ curve for the proton pair (83, 84) on passing through the magic N-number 126. The plot is shown in Fig. 8.

Here again the binding of the high angular momentum $h_{9/2}$ proton increases when the added neutrons change from $p_{1/2}$ or $p_{3/2}$ for $N \leq 126$ to $g_{9/2}$ for $N > 126$.

The assumption has often been made that the attraction between neutrons and protons in similar angular momentum states is greater than when they are in dissimilar states. The $S_{2n}(Z)$ and $S_{2p}(N)$ curves support this assumption when the angular-momentum difference is large. However they offer no such support when the difference is small. For example, no change in the slope of $S_{2n}(Z)$ for n's in the range 32 to 36 ($p_{3/2}$ or $f_{5/2}$) is noticeable on passing through $Z = 28$ where the protons change from $f_{7/2}$ to $p_{3/2}$ states.

It is striking that the overall slopes of the $S_{2n}(Z)$ and $S_{2p}(N)$ lines do not vary smoothly. Some are noticeably greater or smaller than those of their neighbors. We have plotted these slopes, obtaining values for them by eye. This plot is shown in Fig. 9. A peak in the slope plots means that the separation energy of a given nucleon pair increases, as particles of the *other* kind are added, faster than do the separation energies of neighboring nucleon pairs. For example, the separation energy of the 31st and 32nd neutron pair increases, as protons are added to a nucleus in which this is the last neutron pair, faster than does the separation energy of the 29th and 30th or the 33rd and 34th neutron pair. The results may be summarized as follows:

Large slopes of $S_{2n}(Z)_N$ for $N = 32, 40?, 50,$

of $S_{2p}(N)_Z$ for $Z = 20?, 26, 32, 38, 44, 50.$

Small slopes of $S_{2n}(Z)_N$ for $N = 28, 38?, 42, 44,$

of $S_{2p}(N)_Z$ for $Z = 18, 28$ or $30, 36, 42, 56.$

There were not enough good data to draw conclusions for higher values of N and Z.

The magic number 50 is a place of large slope while 28 has small slope. For both N and Z, 32 has large slope and 42 small slope. The meaning of these maxima and minima in the slope curves is not yet clear. An attempt is being made to correlate them with other nuclear properties.

In summary, the conclusions we have been able to draw from a study of neutron- and proton-pair separation energies are the following:

1. There are no clear indications from nuclear masses of shell effects except at 20, 28, 50, 82, and 126. (The very light nuclei were not included in the study.) No clear mass discontinuities are discernable at 32, 34, 38, 40, 56, 58, or 64. There is, however, a slight change in the neutron-pair separation energy pattern at $N = 40$.

2. There is a marked increase in neutron-pair separation energies at $N = 90$. However, there is no sudden change in pattern in the region $N = 136$ where the second rotational region begins. Neither is there any marked effect for $Z = 88$ or 90. This implies that in this region the energies of the spherical and deformed shapes are nearly equal.

3. There is a striking constancy between major shells of S_{2n} along lines of constant $2Z - N$ (nuclei differing by two neutrons and one proton).

4. There is a similar constancy between major shells of S_{2p} along lines of constant $^3/_2 Z - N$ (nuclei differing by three neutrons and two protons).

5. Stronger binding between neutrons and protons when both have either high or low angular momentum is implied by curves for $S_{2n}(Z)_N$ for $N = (123, 124)$, $(125, 126)$, and $(127, 128)$ in the neighborhood of $Z = 82$ and for $S_{2p}(N)_Z$ for $Z = (83, 84)$ and N near 126.

6. Marked maxima are observed in the slope of $S_{2n}(Z)_N$ and $S_{2p}(N)_Z$ which are sometimes associated with magic numbers but oftentimes are not. In particular maxima occur for both N and $Z = 32$ and minima for N and $Z = 42$.

Appendix

Our results are based on data available to us to the end of June 1963. For $A \leq 70$ and $A \geq 212$ values from the 1961 Mainz tables were used with few exceptions. In the region $70 < A < 212$ the separation energies were generally found by the easily corrected "mass-link" system which has been used by the Nuclear Data Group for sometime. This system consists in finding the best mass difference between two consecutive A-chains from all available data *directly* connecting the two chains and computing other values from β-disintegration energies. Differences between non-consecutive A-chains can be taken into account but are not easily handled in this system.

References

[1] M. BEINER and K. BLEULER, Geometrical Representation of Nuclear Properties. Nuclear Phys. 22, 589 (1961).

[2] H. A. BETHE and R. F. BACHER, Nuclear Phys., Rev. Modern Phys. 8, 165 (1936).

[3] N. B. GOVE, private communication.

[4] M. YAMADA and Z. MATUMOTO, Nuclear Ground-State Energies. J. Phys. Soc. Japan 16, 1498 (1961).

Discussion

H. G. Kümmel: I would not be surprised at your findings. The θ_{ik} terms that we introduced do just what you observe. The computer chooses the θ_{ik} such that there is a crack in the neutron binding energy for a magic number in protons and no systematic trend in between. This is described by our θ_{ik} terms which are couplings between neutrons and protons.

K. Way: Does it depend on the angular momentum of the pair?

H. G. Kümmel: No, the angular momenta do not enter into the mass formula.

K. Way: But then you would see this at other shells, but you don't.

H. G. Kümmel: You see this at two shells.

K. Way: At $N = 82$, it is really hard to say that you see anything.

H. G. Kümmel: Sometimes the computer chooses large θ_{ik}, sometimes small ones so that it may happen that in some shells you have this effect and others not.

K. Way: But you are just stating data, you are not explaining it.

H. G. Kümmel: I won't say that we can explain it except that we have some arguments for why such terms occur. We cannot say, of course, why they occur sometimes and sometimes almost not at all.

K. Bleuler: This small, quasi-negligible influence of the proton magic number on the behavior of neutrons (and visa-versa) seems to me rather satisfactory. However, if a magic number is related to a large spin-change, a corresponding change in configuration mixing should occur which might influence the behavior of the energy-surface in accordance with the experimental facts.

K. Way: It seems to me that if Dr. Cohen's stabilization effect exists that it would show up more in plots of this type. Perhaps it is something that is so subtle that you can't yet see it in mass measurements but it shows up in level positions.

K. Bleuler: I am very much interested in the fact that no irregularities could be found in relation to the closure of the subshells corresponding to the different individual single particle states, whereas the irregularities related to the major shells of course remain. This startling effect is explained in a natural way by BCS pairing theory: There, it is seen that the characteristic pairing energy Δ, which actually gives a measure of the energy interval at which mixing between single particle states occurs, is *large* with respect to the energy intervals between states within a major shell whereas it is *small* with respect to the energy interval which separates two different major shells. This numerical relation appears to be a crucial property of shell structure.

K. Way: In Dr. Cohen's last paper, the separation between the subshells seems to be sufficient that you should notice it.

B. L. Cohen: The answer to that is a long story so I won't go into it here but there is, in line with this discussion, one new closed shell we did find. We found a closed shell at 56 neutrons for Zr, but not in Mo as the $g_{7/2}$ by this time was moved down far enough to mix. The evidence that Zr^{96} is a closed shell is very convincing: There is a break in the masses of about $1/2$ to $3/4$ MeV, the

2^+ collective state moves suddenly up from 0.9 MeV to 1.7 MeV, there are no low lying $5/_2^+$ states in Zr^{97}, etc. None of these effects occur for Mo^{98}.

Note Added in Proof. The results of RIES, DAMEROW, and JOHNSON on Atomic Masses from Gallium to Molybdenum, Phys. Rev. 132, 1662 (1963), which were not available at the time of the meeting, change considerably the slopes of the $S_{2n}(Z)$ and $S_{2p}(N)$ lines from the values shown in Fig. 9. The new data lead to much smoother curves in the region 28 to 42 for N or Z and almost eliminate the peaks at N or $Z = 32$.

Mass Systematics Involving Low-Lying Excited States*

By

F. Everling

Institute for Atomic Research and Department of Physics,
Iowa State University, Ames, Iowa, USA

With 3 Figures

Introduction

For the evaluation of the nuclear binding energy surface it is helpful to consider the excited states together with the ground states[1, 2]. The (positive) total nuclear binding energy for any state is $B = B_{g.s.} - E_x$ where "g. s." refers to the ground state and E_x is the excitation energy.

For simplicity the mass excess Δ can be used instead, since the total nuclear binding energy is

$$B(A, T_\zeta) = \tfrac{1}{2} A[\Delta(n^1) + \Delta(H^1)] + T_\zeta[\Delta(n^1) - \Delta(H^1)] - \Delta(A, T_\zeta).$$

Both brackets are constant; the first term depends linearly on the mass number A and the second is constant since only sequences of nuclides with constant ζ-component of the isobaric spin, T_ζ, are considered. Therefore a straight line in the mass excess $\Delta(A)$ for $T_\zeta = $ const. means a linear relation in the total binding energy $B(A)$.

The steep increase of Δ in the light mass region can be compensated for by plotting the sum of the mass excess with a straight line, $\Delta + p\,A$, where again linear relations are conserved.

Results

Fig. 1 presents such a graph for self-conjugate even nuclides with $p = 1.5$ MeV for convenient display.

The 2+ and 4+ levels of Ne20, Mg24, and Si28 lie on almost straight trends, but the ground states do not. Weak indication has been given by MORITA and TAKESHITA[3] for a level in Ne20 at 0.65 MeV — it happens to be just in the right position to fit a trend — which has not been found in

* Contribution No. 1356. Work was performed in the Ames Laboratory of the U. S. Atomic Energy Commission.

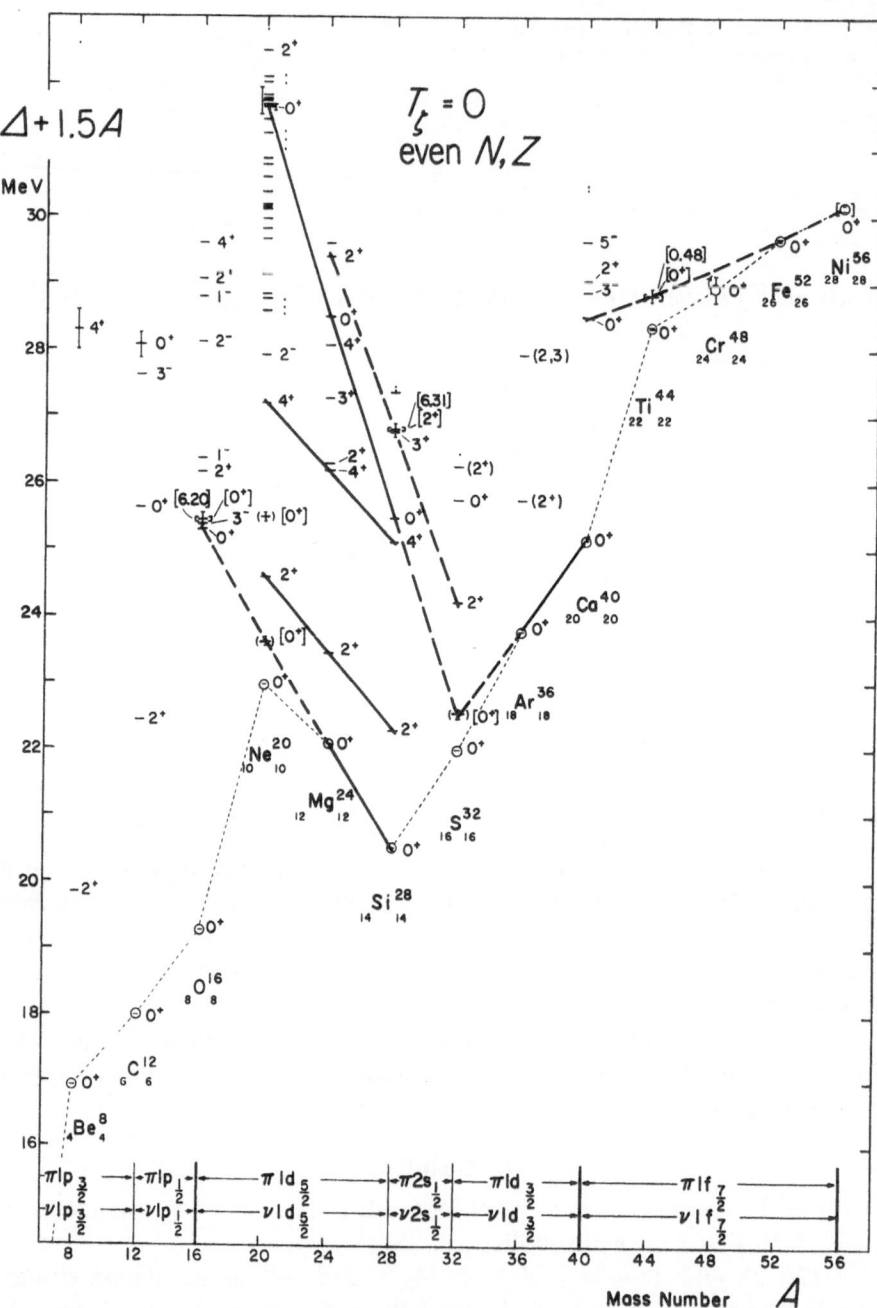

Fig. 1. Nuclidic mass excess Δ for ground states (circles) and excited states (dashes), with a linear function $1.5\,A$ added, versus the mass number A. *Parentheses* indicate doubtful experimental data, *brackets* indicate expectations if this systematics holds without exceptions. *Uncertainties* are given separately for ground state mass excesses and excitation energies (≤ 30 keV if not shown). *Heavy lines* show possible trends (dashed if doubtful)

other reactions. However, it is not yet ruled out by using the same reaction $F^{19}(d, n)Ne^{20}$. Therefore there is either a peculiar shift of the Ne^{20} ground state or an additional 0^+ level which would lead the trend to O^{16} 6.05 MeV, 0^+. In any case, a step occurs at $N, Z = 8$. Similarly,

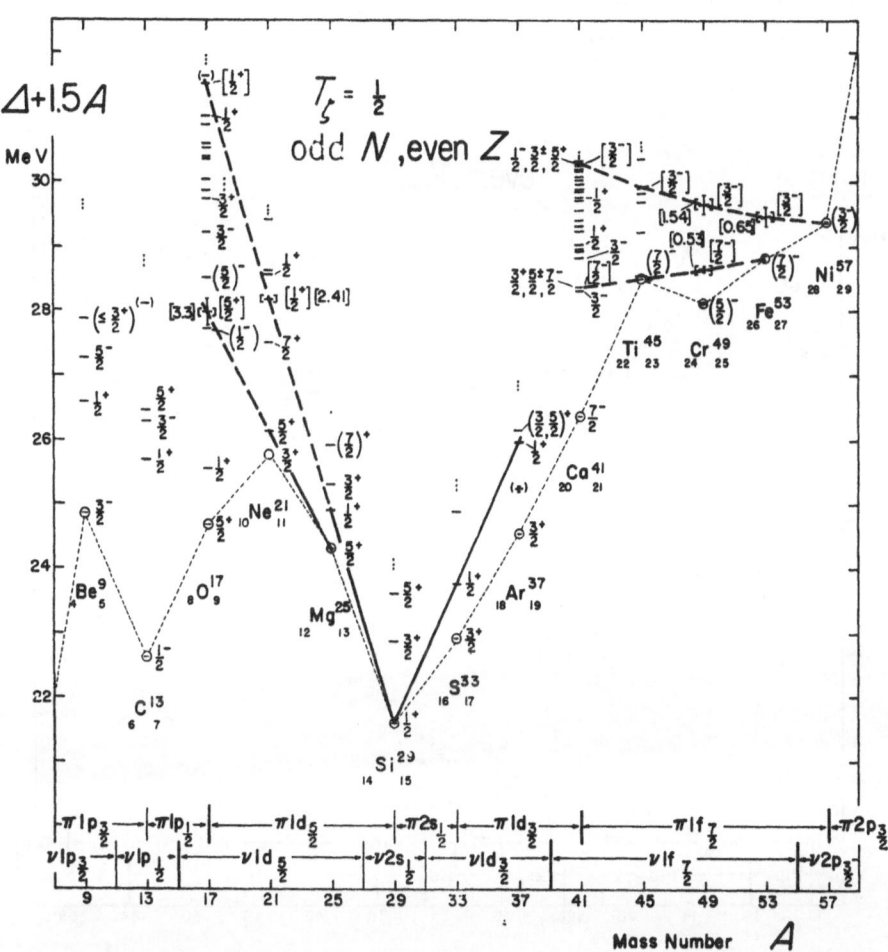

Fig. 2. Odd-neutron nuclei with $T_\zeta = {}^1/_2$. For conventions, see Fig. 1

there is a step at $N, Z = 20$ with Ti^{44} corresponding to Ne^{20}. The mass excess of Ni^{56} is our prediction from systematics of nuclides with $N = 28$.

The 0^+ trend through *excited* states of Si^{28} and Mg^{24} apparently represents configurations with a filled $2 s_{1/2}$ subshell (for protons and neutrons) while the $1 d_{5/2}$ subshell is being occupied. If the tentatively indicated level Si^{28} 6.31 MeV, 2^+ exists, the energy required for exciting these 2^+ states from the 0^+ states below them would be independent of the additional $2 s_{1/2}$ alpha-particle. In S^{32}, the ground state is shifted

14*

down by 0.41 MeV if there is no additional 0+ level. Earlier reports of a level at 0.5 and 0.43 MeV respectively have been mainly ruled out except for a slight asymmetry in the $P^{31}(d, n)S^{32}$ ground state group. From the position S^{32} 0.41 MeV a straight trend leads over Ar^{36} to Ca^{40}, representing the occupation of the $1 d_{3/2}$ subshell.

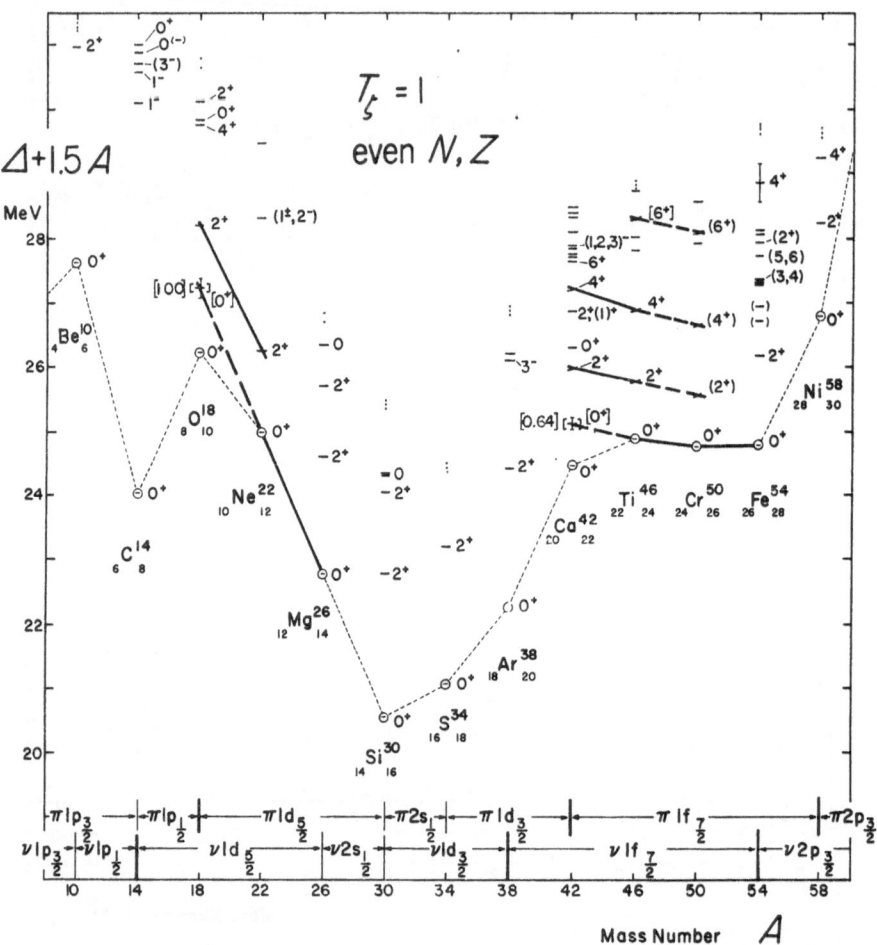

Fig. 3. Even nuclei with $T_\zeta = 1$. For conventions, see Fig. 1

Fig. 2 gives these nuclei with a neutron attached. Here the filling of the $1 d_{3/2}$ subshell (while a $2 s_{1/2}$ neutron remains single) is an established straight trend.

A trend through the lowest Si^{29} and Mg^{25} $1/_2^+$ states leads in O^{17} to a level 6.0 MeV above the first $1/_2^+$ level, apparently reproducing the step noted earlier in O^{16}. The $5/_2^+$ states lead to 3.3 MeV above ground state in O^{17} where no level is known. In the corresponding Ca^{41} there is a level at the right position for which $7/_2^-$ is possible.

Fig. 3 shows the nuclides with two neutrons more than the self-conjugate ones. The Ca^{42} ground state is shifted down by 0.64 MeV from a smooth trend which is suggested by the sequence of the ground states of Fe^{54}, Cr^{50}, and Ti^{46} as well as by the 2+ and 4+ trends indicated. An interesting open question is whether the known 6+ level in Ca^{42} belongs to the band on the known 1.84 MeV, 0+ level or destroys the pattern suggested. The 2+ and 4+ levels of Fe^{54} apparently do not belong to these trends because of the neutron shell closure $N = 28$.

Since the $1\,d_{5/2}$ subshell is shorter than the $1\,f_{7/2}$ shell, we cannot see anything there, except that the downward shift of the Ca^{42} ground state has its analogue in O^{18}. The level schemes are still too incomplete for locating the states with two neutrons occupying the $2\,s_{1/2}$ subshell.

Conclusions

1. There is a perfect analogy between magic numbers 8 and 20.

2. Smooth trends are likely to exist within subshells which may be destroyed, however, by configuration mixing or other unknown effects immediately after magic numbers 8 and 20.

The almost linear trends mean that about the same energy is liberated when the constituents of an alpha-particle are added repeatedly to a nucleus. The Coulomb repulsion apparently is compensated by the nuclear interactions with particles already in the same subshell.

3. A step exists at these magic numbers in the mass excess (and total nuclear binding energy) surface.

4. If additional levels should turn out to exist and complete these trends, the ground states of O^{17} and Ca^{41}, O^{18} and Ca^{42}, as well as Ne^{20} and Ti^{44} may have to be considered as cluster configurations of O^{16} and Ca^{40} in their ground states plus one neutron, two neutrons, and an alpha-particle respectively. The additional states may then contain O^{16} and Ca^{40} respectively in their first excited state. This may be the reason that they are hard to populate — if they exist at all.

References

[1] D. R. INGLIS, Revs. Modern Phys. 25, 390 (1953).
[2] F. EVERLING, Nuclear Phys. 40, 670 (1963).
[3] S. MORITA and K. TAKESHITA, J. Phys. Soc. Japan 13, 1241 (1958).

Discussion

K. BLEULER: In the case of the $7/2$ levels in light nuclei we have also realized that excited states exhibit a rather continuous behavior as a function of the mass number whereas the position of the corresponding ground states is rather irregular. I was very much interested that you found more examples of this kind.

F. Everling: We have tried to extend the curves as far as possible, but when going from the 1 $f_{7/2}$ shell to lighter nuclei, of course, it does not work so accurately because of the residual interaction. The fact that there are sometimes anomalous ground states is a difficulty in the work on mass formulas. One of these, the ground state of Cr^{49}, was shown in Fig. 2, and I predicted the unknown $7/2^-$ single particle level by interpolation. In the odd-proton nucleus Mn^{51}, I did the same and learned recently from Arnell and Sterner that they just found a level at this position for which the spin $7/2^-$ is not yet confirmed, however. This state could also be predicted from the parabola of nuclei with 26 neutrons which leads to the same value. For states immediately after magic numbers, such trends are probably destroyed by configuration mixing.

Nuclidic Masses and Structure of Nuclei

By

V. A. Kravtsov

USSR, Leningrad, Leningrad Polytechnical Institute of M. I. Kalinin

(Presented by A. I. ABRAMOV)

With 2 Figures

Nuclidic masses are a very important characteristic of nuclei. The binding energies of nuclei calculated from nuclidic masses enable us to form an idea of the structure of nuclei and they provide us with valuable information for verifying nuclear models. Thanks to the efforts of L. KÖNIG, F. MATTAUCH, and A. WAPSTRA[1] we have now excellent complete tables of nuclidic masses and of binding energies of nuclei based on the most recent data. We have used the most recent measurements of DEMIRKHANOV, DOROKHOV, and DZKOUYA[2, 3] in order to complete these tables in a region where these data were the least precise, namely for nuclides ranging from strontium to ruthenium and for nuclides of tin and antimony.

We consider it important for the present International Conference to outline the direction of future work in order to improve the precision of measurements of masses of various nuclides. What is, however, the situation with regard to the nuclidic masses today? The masses of the light nuclei for $A \leq 70$ are known with a high degree of precision and do not seem to require any further improvements. There are some minor discrepancies as compared with measurements by L. FRIEDMAN and al.[4], but they are not of a fundametnal character and they will not lead to significant changes when new computations are being completed. The results of new measurements by A. NIER and W. JOHNSON of the Minnesota group for $70 \leq A \leq 140$ are being completed and will be published soon. When computing masses in this region the question will immediately arise as to the methods of verification of mass-spectrometric data which usually seem to be not quite reliable. Here the number of measured energies of nuclear reactions is rather limited and therefore it seems to us that it should be of great importance to intensify the work of the groups which are engaged in measurements of energies of nuclear reactions. The Massachusset's Institute of Technology, the Ukranian Physical and Technical Institute in Kharkov and other Institutes should make themselves helpful in carrying out new measurements in this field. It seems that it would be appropriate to pass over to measuring of Q's for ground states of samples enriched by certain isotopes and of nuclei with a large Z.

In particular, with all my respect to Professor H. DUCKWORTH, I must express some doubts as to the new measurements of masses of tin nuclides by R. BARBER, H. DUCKWORTH, et al.[5]; it seems to me that these measurements must be verified by sufficiently new measurements of

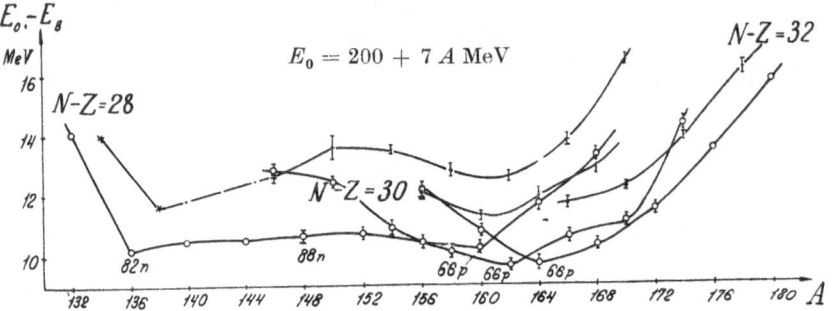

Fig. 1. The sections of the energy surfaces with a reduced slope ($E_0 = 200 + 7\,A$ MeV, E_B binding energy, A mass number) by planes $N - Z = 28$, 30 and 32. ⊙ points for even-even nuclei, + points for odd A nuclei, I errors

Q-values for (d, t) and (d, p) reactions for tin isotopes, i. e. more precise than those which have been carried out by B. COHEN and R. PRICE[10]. It seems that it would be advisable to carry out these measurements on samples enriched by various tin isotopes. In spite of an excellent methodology of measurements used by R. DEMIRKHANOV and V. DOROKHOV at SUKHUMI, I am somewhat skeptical about their measurements of ruthenium nuclides. They must also be verified by measurements of Q-values for (d, p) and (d, t) reactions on enriched samples of ruthenium isotopes.

At some future time the measurements of Q-values for nuclear reactions should be carried out for heavier nuclei with a precision comparable to that of mass-spectrometrical methods. The times have changed now so that also in the field of heavy nuclei the "reactionists" should catch-up, with regard to the precision of the "mass-spectrometrists".

It would be advisable to repeat, for the sake of verification, also the mass measurements of thallium nuclides; in particular it would be important to measure here Q for reactions on thallium isotopes, since the data now at our disposal, as has been shown by the author of the present report[6], are greatly dubious. This would be of a particular interest, since thallium nuclei are situated very close to the double magic ^{208}Pb and since some theoreticians, such as SLIV et al. have learned how to compute such nuclei.

Up-to-date knowledge of masses and of binding energies of nuclei enables us to draw some new conclusions with regard to the structure of

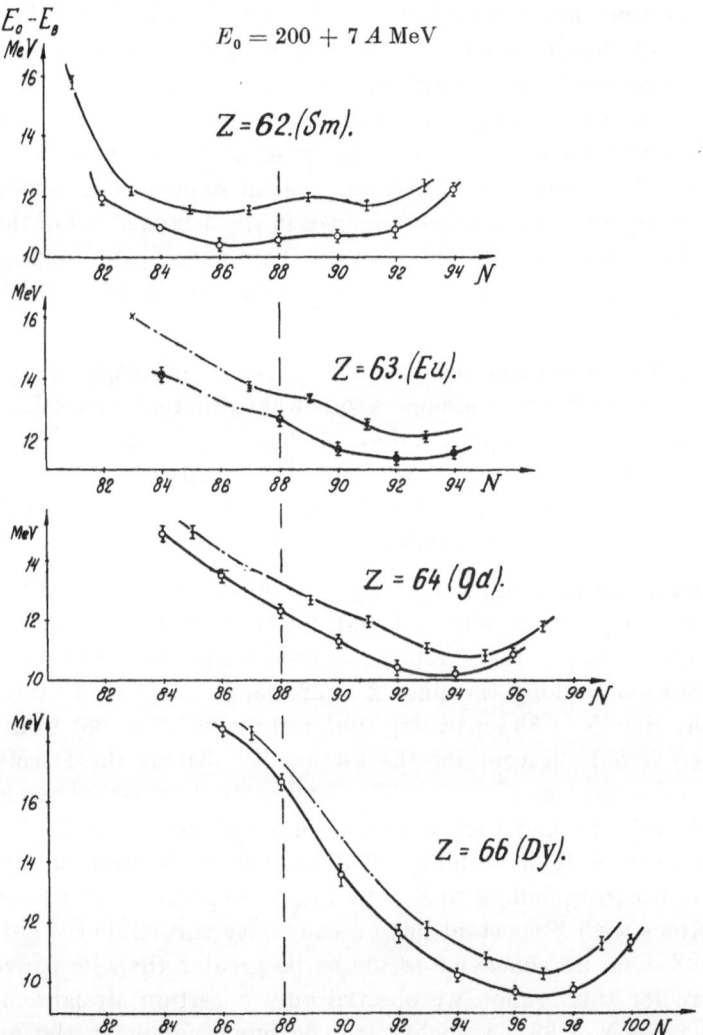

Fig. 2. The sections of the energy surfaces with a reduced slope ($E_0 = 200 + 7\,A$ MeV, E_B binding energy, A mass number) by planes $Z = 62$, 63, 64 and 66. ○ points for even-even nuclei on sections 62, 64 and 66, points for odd A nuclei on section 63. + points for odd A nuclei on sections 62, 64 and 66, points for odd-odd nuclei on section 63

rare-earth nuclides. These conclusions are deduced from the tables published by L. König, F. Mattauch, and A. Wapstra[1] in the region of rare-earth nuclei, which have been computed on the basis of measurements by V. Bhanot, W. Johnson jr., and A. Nier[7] (USA) and also on the basis of measurements by R. DEMIRKHANOV, T. GOUTKIN, and V. DOROKHOV[8] (USSR). Fig. 1 represents sections of energy surfaces with a reduced slope in this region for nuclei with an equal number of excess neutrons, and namely for $N - Z = 28$, 30 and 32. All these sections mark a very clearly apparent valley following the nuclei with 66 protons. It is known that on energy surfaces the binding energy of which is considered to have a negative value, the valley follows the nuclei having an enhanced stability; therefore the number 66 for protons is at least a semi-magic number. In order to give an explanation of this number it is necessary to introduce some changes in the arrangement of the protons' levels. According to P. KLINKENBERG[9] the protons, after closing the shell of 50 protons, fill consequently the levels $1\,g\,^7/_2$, $2\,d\,^5/_2$, $1\,h\,^{11}/_2$, $2\,d\,^3/_2$ and $3\,s\,^1/_2$.

In order to explain the existence of the semi-magic number 66 for protons it is necessary to assume a somewhat different succession of proton levels, for instance: $1\,g\,^7/_2$, $2\,d\,^5/_2$, $3\,s\,^1/_2$, $1\,h\,^{11}/_2$, and $3\,d\,^3/_2$. It is of great interest that this semi-magic number belongs to the sub-shell filled with protons in the region of deformed nuclei, in the interval between the two neutronic magic numbers.

Another interesting fact in this region of nuclei is the influence of the boundary between spherical and deformed nuclei on the structure of the energy surfaces. Fig. 2 represents sections of the energy surfaces with a reduced slope along the lines $Z = 62$, 63, 64 and 66. All these sections cross the line $N = 88$ separating spherical nuclei ($N < 88$) from deformed nuclei ($N \geq 88$). Except for the section $Z = 64$ all these sections show a clear convexity at $N = 88$. This means that spherical nuclei have a reduced stability and that a transition to deformed nuclei leads to an enhancement of their stability. The absence of a crest or inflexion for the section corresponding to $Z = 64$ can be explained by the fact that all the sections with Z-constant have a convexity turned downward. For the section $Z = 64$ its concavity seems to be greater than its convexity and therefore for this section we observe only a certain straightening of the section near $N = 88$. A slight crest or bump following the nuclei with 88 neutrons can also be observed for the sections $N - Z = $ const., see, for instance, the section $N - Z = 28$ on Fig. 1. This crest is more apparent on the upper curve for odd-odd nuclei than on the lower curve for even-even nuclei.

It is also possible to distinguish a boundary crest for the sections $N - Z = $ const.; this crest has been formed on the energy surface at the boundary of the region of deformed light nuclei. Along the line $N = 114$ at the upper boundary of the deformed nuclei of the rare-earth group one can also distinguish formation of a boundary crest.

All these crests are scarcely distinguishable, contrary to distinct and clear valleys which cross the energy surfaces along the lines with magic and semimagic numbers of nucleons. In order to provide definite evidence of the existence of these crests and in order to, determine their nature, it is necessary to measure with a higher degree of precision nuclidic masses in the neighborhoud of nuclei with $N = 88$ and $N = 114$.

One would expect to see a similar crest or bump in the region of heavy nuclei along the line $Z = 88$, that is to say for the radium nuclei; but in view of our imperfect knowledge of masses of the heavy nuclei such a crest is not observable. It is understandable since the sharp variations of binding energies of light nuclei are smoothed for middle-mass nuclei and they become hardly noticeable for heavy nuclei. For studying the structure of heavy nuclei and for bringing out the existance of a crest for $Z = 88$, as well as for revealing the existance of new semi-magic numbers, besides $N = 152$ already known, it is necessary to acquire a more precise knowledge of masses of heavy nuclei. There are some reasons to believe that a sub-shell is also formed by 92 protons. For a more precise knowledge of evolution of energy surfaces it is essential to carry out more precise measurements of the alpha-decay energies by means of good magnetic alpha-spectrometers. We have to put forward a suggestion to the alpha-spectroscopists in the whole world and, first of all to PERLMANN and GHIORSO et al. (USA) and to GOLDIN et al. (USSR) to do such measurements. Moreover, it would be desirable that our Conference should make a suggestion to all the nuclear spectroscopists who are studying decay of nuclei of all kinds, inviting them in their enthusiasm for studying the positions of excited levels, not to forget also the energy difference of the ground states. We know particularly little about the energy differences of the ground state for a positron decay and even less for the electron capture. We must ask them to kindly fill the gaps in these data which are so badly needed.

I hope that our requests will be favourably considered by the Conference which will include them together with other suggestions in the Conference decisions. We hope that this will contribute to a more coordinated and systematic work of physicists engaged in the studies of masses and binding energies of nuclei.

References

[1] L. A. KÖNIG, J. H. E. MATTAUCH, and A. H. WAPSTRA, Nuclear Phys. 28, 1 (1961).

[2] R. A. DEMIRKHANOV, V. V. DOROKHOV, and M. I. DZKUYA, J. Exptl. Theoret. Phys. (USSR) 40, 1572 (1961). — R. A. DEMIRKHANOV, T. I. GUTKIN, O. A. SAMADASHVILI, and I. K. KARPENKO, Izv. Acad. Nauk USSR, Phys. Series 25, 882 (1961).

[3] V. A. KRAVTSOV, Nuclear Phys. 41, 330 (1963).

[4] L. FRIEDMAN and L. E. SMITH, Phys. Rev. 109, 2214 (1958).

⁵ R. C. Barber, R. L. Bishop, L. A. Cambey, W. McLatchie, and H. E. Duckworth, Canad. Journ. Physics 40, 1496 (1962).

⁶ V. A. Kravtsov, Izv. Acad. Nauk USSR, Phys. Series 25, 130 (1961).

⁷ V. B. Bhanot, W. H. Johnson, and A. O. Nier, Phys. Rev. 120, 235 (1960).

⁸ R. A. Demirkhanov, T. I. Gutkin, and V. V. Dorokhov, Izv. Acad. Nauk USSR, Phys. Series 25, 124 (1961).

⁹ P. Klinkenberg, Revs. Modern Phys. 24, 63 (1952).

¹⁰ B. L. Cohen and R. E. Price, Phys. Rev. 121, 1441 (1961).

Energies of Natural Alpha Radiators

By

A. Rytz

Bureau International des Poids et Mesures, Sèvres (S.-et-O.), France

With 1 Figure

More than 60% of the Q-value measurements of charged primary particle reactions used in nuclidic mass adjustments have been calibrated with natural α-particle energy standards, predominantly ^{210}Po; in some other cases, ^{212}Po and ^{212}Bi have been used. ^{214}Po has been used in a few relative measurements of reaction energy standards. Additional application of absolute α-energies is found in α-spectroscopy, where even less abundant α-groups are useful, in order to cover the full range between 4 and 10 MeV.

Early absolute measurements have been made by RUTHERFORD[1], I. CURIE[2] and BRIGGS[3]. The first precision measurements were the ones by LEWIS and BOWDEN[4], BRIGGS[5], who claimed for a precision of 10^{-4} for ^{214}Po, and ROSENBLUM and DUPOUY[6].

A considerable increase in accuracy was accomplished by the application of magnetic field measurements with the aid of the gyromagnetic ratio of the proton. Since 1953, all absolute measurements used the 180° magnetic deflection method[7-11]. Some precise relative measurements[12, 13] may be mentioned as well, but will not be considered specially. Best values of all important natural α-energies have been calculated by BRIGGS[14] and by WAPSTRA[15].

This paper should give a brief account of the measurements made at Orsay[9] and Zürich[11] and develop some ideas relating to comparisons with other results. A few remarks on calibration standards will be included.

The Orsay measurements, in 1959/60, were the first experiments made with the new permanent magnet having pole pieces of 1.6×1 m and a gap up to 11 cm. Mechanical and geometric considerations restricted the maximum radius of curvature of the α-particle orbits to 50 cm. Although the field of 8540 Gauss, in a gap of 4.8 cm, was only 0.6% below saturation, it was found to be sufficiently stable and reproducible. It could be measured point by point along the different particle orbits, in the vacuum chamber, by a proton resonance probe. After correction by shims made of suitably shaped iron foils, the rms deviation from the mean field strength did not exceed $3 \cdot 10^{-5}$ and the maximum deviation was $7.5 \cdot 10^{-5}$. The effective field was calculated using Hartree's correction and had an overall standard error of $2.3 \cdot 10^{-5}$.

The spectrograph had on one end a geometrically well defined source holder with a simple pair of baffles, and a holder for the C 2 nuclear track plates on the other end. The distance between the source and a fiducial mark on the plate holder was controlled by a quartz rod in order to minimize thermal expansion. The fiducial mark consisted of a narrow slit fixed at a quartz rod, which could be projected on the photo plate by a separate α-source. Length measurements were made by comparison with a standard meter bar from the Bureau International de Poids et Mesures. Measurements on the plate could be made with a special microscope which was used for track counting as well. The overall standard error of length measurements was $5.1 \cdot 10^{-5}$ and the combined standard error in $B\varrho$ was ± 20 Gauss $\cdot$ cm.

Table 1

α-Emitter	Primary material	Source preparation method		Thickness derived from line shape (keV)	Line width measured / calculated
		treatment	intercepted on		
^{210}Po	^{210}Po	electrolysis	Ta	1.09	1.30
		1st evaporation	Ta		
		2nd evaporation	support		
^{212}Bi ^{212}Po	^{228}Th	Activation .... evaporation	Pt support	0.52 0.51	1.13 1.09
^{214}Po	^{222}Rn	activation .... evaporation	Pt support	3.70	1.74
^{211}Bi	^{227}Ac	activation	support	1.52	1.34
^{223}Ra ^{219}Rn ^{215}Po ^{211}Bi	^{227}Th	extraction from soln.. distillation.......... evaporation.........	quartz quartz support	17 21.5 25 30.8	5.3 5.7 6.2 7.7

Table 1 lists the nuclides investigated, the methods of source preparation and the thickness as derived from line shapes. The ideal source strength was found to be $100\,\mu$ Ci divided by the number of exposure hours. The source support was a piece of polished tantalum sheet with an activated surface of 10×0.1 mm. In fact, source preparation is the most delicate part of such measurements and has repeatedly been the origin of serious systematic errors. However, the small deviation from the line shapes calculated for zero thickness, in the case of the thorium active deposit, is strong evidence in favour of thin, clean sources. The ^{214}Po source, in spite of its similar preparation, was considerably thicker. A direct reason for this behaviour could not be found. There are several arguments, yet, which strongly suggest that generally the preparation procedure used was very efficient and gave very clean sources:

1. The source material was not visible by naked eyes with exception of ^{214}Po which had to be by far the most intense of all the sources used, owing to the short life of ^{214}Pb and ^{214}Bi.

2. In many instances, the lines were only slightly wider than calculated. In the case of ^{211}Bi, however, it can be seen that even a very broad line may define the correct energy value. The directly obtained line was 8 times narrower than the line observed on the plate exposed to ^{223}Ra (Table 1). This was due to the successive α-recoils. Yet, the difference at the high energy edge between these lines was less than one standard error. Further, the ^{214}Po line was 75% wider than calculated, but, since there does not exist much difference between the active deposits of Th and Ra and since the sources were prepared similarly, it seems impossible that the ^{214}Po source was seriously contaminated.

3. The ^{223}Ra source contained ^{219}Rn, and this substance emanated very abundantly from the source, so as to give a dense background of α-tracks. Since it is well known[16], that Ra sources covered by a thin layer of Al lose their radon only very slowly, we may conclude that this rather thick source was perfectly clean.

4. The ^{210}Po source which gave the best line shape was used 5 times during a whole week. No systematic line shift could be detected. The number of particles integrated over the whole line was found to decrease with a half life about 6% shorter than 138 days. Moreover, the spectrograph was thoroughly contaminated by these sources. These observations again indicate that the source was not covered by a considerable inactive layer.

The photo plates were hit by the particles perpendicularly, and the tracks would not have been easily discernible. However, a small distortion of the emulsion obtained by a suitable drying procedure gave a perfect visibility. A determination of the maximum distortion in the 25 μm thick emulsion was found to be well below the uncertainty in length measurements.

The shape of the high energy edge of each line was calculated and fitted to the observed shape assuming a homogeneous source of non-vanishing thickness. Further, it was found that scattering by the residual gas is always negligible.

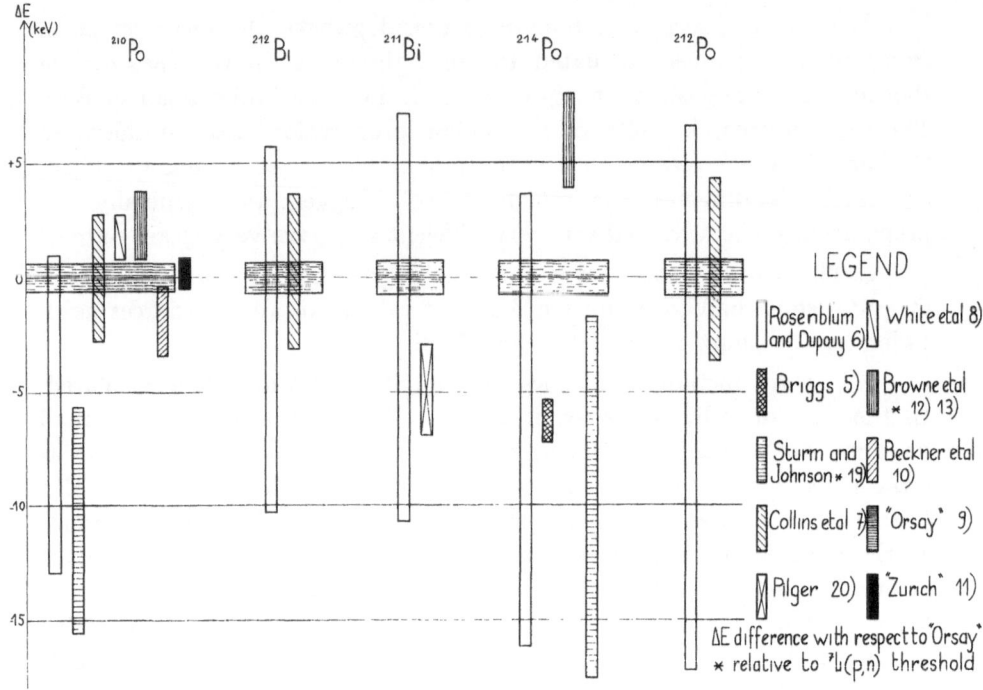

Fig. 1. Comparison of precision measurements of important natural α-radiators. The energy values have been adjusted to a common set of fundamental constants

Table 2. Results obtained at Orsay

α-Emitter		$B\varrho$ (Gauss · cm)	E_α (keV)
^{210}Po		331774 ± 20	5304.81 ± 0.62
^{212}Bi	α_0	354346	6050.60 ± 0.66
	α_1	355492	6089.77 ± 0.66
^{211}Bi	α_0	370720	6622.19 ± 0.69
	α_1	360936	6277.52 ± 0.68
^{214}Po		399442	7686.95 ± 0.75
^{212}Po		427060	8785.40 ± 0.80
^{223}Ra	α_4	345290 ± 28	5745.5 ± 1.0
	α_5	344350	5714.3
	α_6	341050	5605.3
	α_7	338960	5537.1
^{219}Rn	α_0	376160	$6817.6 \quad \pm 1.0$
	α_1	368720	6550.9
	α_2	365120	6423.9
^{215}Po		391490	$7384.1 \quad \pm 1.0$

Constants used:
$$\gamma_p = (26751.3 \pm 0.2) \text{ rad } s^{-1} \text{ Gauss}^{-1}$$
$$F = (9652.19 \pm 0.11) \text{ emu} \cdot \text{Mole}^{-1}$$

Table 2 gives the final results obtained at Orsay and Fig. 1 compares them to other measurements. We believe that the sources prepared in the manner described did not give an appreciable contribution to the error derived from $B\varrho$ measurements.

In a further experiment undertaken at Zürich[11] measuring the absolute energy of ^{210}Po α-particles, a somewhat different arrangement has been used. In a 210° electromagnet, specially designed for absolute precision measurements of charged particle energies[17], a ^{210}Po source was placed behind the entrance slit. In front of the exit slit, a solid state detector was installed. The slit distance was defined by a molybdenum rod which has been measured before and after the experiment by the SWISS FEDERAL BUREAU OF STANDARDS. The magnet field has been corrected, by shims made of magnetic recording tape, to a high degree of homogeneity. It was constant in time within less than ± 5 parts in 10^6. The count rate was measured as a function of proton resonance frequency and the line shape was calculated by a procedure analogous to the one applied to the permanent magnet spectrum. The calculated curve was fitted to the experimental points by least squares and the particle energy turned out to be $E = (5304.93 \pm 0.60)$ keV.

The ^{210}Po source of about $50\,\mu$Ci had been prepared by R. J. WALEN at Orsay in exactly the same manner as described before. The thickness was found to be as high as 12 keV, which was certainly due to impurities present in the primary material. The source age was 8 days. Although source age may cause an important line shift[18], we believe that the excellent agreement with the Orsay result strongly supports the observations mentioned above, that sources prepared by double evaporation do not alter the maximum energy of the particles emitted, during several days.

In comparing absolute α-energy measurements, it seems reasonable to consider only those magnetic deflection experiments, in which the field has been measured by nuclear magnetic resonance. Of course, a common set of fundamental constants should be used. However, it seems to me, that this has not always been done in comparing results from different laboratories.

It might be useful to call to our minds the formula for calculating α-particle energies. From the Lorentz force, we get

$$B\varrho = \frac{m\,v\,c}{2\,e},$$

where B expresses the magnet field in Gauss, $\varrho = \dfrac{D}{2}$ the radius of the particle orbit, $m \cdot v$ its momentum, e the elementary charge and c the speed of light, all in cgs units. The non-relativistic energy therefore is

$$E' = \frac{m}{2}\,v^2 = \frac{2\,e^2}{m\,c^2}\,(B\varrho)^2 \cdot \frac{c}{e} \cdot 10^{-11}\,\text{keV}.$$

Writing $\dfrac{N_A\,e}{c} = F$, the Faraday, in electromagnetic units, and $N_A\,m = M_\alpha$, the nuclidic mass of the α-particle in μu, we get

$$E' = \frac{2\,F}{M_\alpha}\,(B\varrho)^2 \cdot 10^{-11}\,\text{keV}.$$

B is measured by nmr:

$$B = \frac{2\,\pi\,\nu_0}{\gamma_p}.$$

Relativistically, we get:

$$E = E'(1 - \delta) = \frac{\pi^2}{\gamma_p{}^2} \frac{2\,F}{M_\alpha} D^2 \nu_0{}^2 (1 - \delta)\, 10^{-5}\,\text{keV},$$

where

$$\delta = \frac{1}{2} \frac{F}{M_\alpha c^2} \cdot 10^{17} \cdot E_{\text{keV}}.$$

The second order correction is always smaller than $3 \cdot 10^{-6}$, i. e. negligible. The Faraday and gyromagnetic ratio of the proton are not yet known with an accuracy high enough to be without influence on the results.

Table 3 lists the results of more recent ^{210}Po measurements and the adjustments due to a reduction to a common pair of values F and γ_p.

Table 3. ^{210}Po α-energy as measured by different authors and adjusted to the same fundamental constants

Author	Year	Published value (keV)	Standard error (keV)	Adjusted value (keV)
COLLINS et al.	53	5304.2	2.9	5304.56
WHITE et al.	58	5305.4	1.0	5306.40*
BROWNE et al.	60	5307.5	1.5	5306.94
BECKNER et al.	61	5302.5	1.5	5302.67
RYTZ (Orsay)	61	5304.81	0.62	5304.60
RYTZ et al. (Zürich)	61	5304.93	0.60	5304.73

Constants used: $\gamma_p = (26751.19 \pm 0.08)\ \text{rad} \cdot s^{-1} \cdot \text{Gauss}^{-1}$
$F = (9648.682 \pm 0.066)\ \text{emu} \cdot \text{Mole}^{-1}$

* As proposed by BROWNE et al.[12].

This table shows that the adjustments are small in most cases, but not negligible. Since the following paper is dealing with the Polonium problem, I shall not give further comments now. However, I should like to stress that ^{210}Po does not seem to be a calibration standard having all the properties desired. It is very probable that many Q-value measurements contain systematic errors due to ^{210}Po source condition which now are impossible to trace back. The logical consequence would be to abandon ^{210}Po as a calibration standard for precision work and to look for some other and more suitable standards. By far the best sources are obtained if ^{212}Pb is used, the daughter product of which give three intense lines of 8785, 6090 and 6051 keV in the intensity ratio $1 : 0.1 : 0.25$. The short half-life ($10.6\ h$) is outweighed by the absence of spectrometer contamination, which, for ^{210}Po, is often quite troublesome. Among the artificially produced α-emitters, some other suitable energy standards, such as ^{244}Cm or ^{241}Am, may be found. I do not have any experience in source preparation from these nuclides and, therefore, do not feel competent myself to make a well founded proposition.

Conclusion

Many reliable absolute measurements of the most important α-energy standards are now available. These standards, together with the new nuclear reaction energy standards, will provide a basis for a considerably improved nuclidic mass adjustment. Further improvement seems rather questionable unless ^{210}Po, as a calibration standard, is abandoned and extensive new measurements based on more reliable calibration standards are undertaken.

References

[1] E. RUTHERFORD and H. ROBINSON, The Mass and Velocities of the α Particles from Radioactive Substances. Philos. Mag. 28, 552 (1914).

[2] I. CURIE, Détermination de la vitesse des rayons α du polonium. C. R. Acad. Sci. (Paris) 175, 220 (1922).

[3] G. H. BRIGGS, A Redetermination of the Velocities of α Particles from Radium C, Thorium C and C'. Proc. Roy. Soc. (London) A 118, 549 (1928).

[4] W. B. LEWIS and B. V. BOWDEN, An Analysis of the Fine Structure of the α-particle Groups from Thorium C and of the Long Range Groups from Thorium C'. Proc. Roy. Soc. (London) A 145, 235 (1934).

[5] G. H. BRIGGS, A Determination of the Absolute Velocity of the Alpha-Particles from Radium C'. Proc. Roy. Soc. (London) A 157, 183 (1936).

[6] S. ROSENBLUM et G. DUPOUY, Mesures absolues des vitesses des principaux groupes de rayons α. C. R. Acad. Sci. (Paris) 194, 1919 (1932); J. de Phys. et Rad. 4, 262 (1933).

[7] E. R. COLLINS, C. D. MCKENZIE, and C. A. RAMM, A precise technique for the determination of some nuclear reaction energies. Proc. Roy. Soc. (London) A 216, 219 (1953).

[8] F. A. WHITE, F. M. ROURKE, J. C. SHEFFIELD, R. P. SCHUMAN, and J. R. HUIZENGA, Absolute Energy Measurements of Alpha Particles from ^{210}Po. Phys. Rev. 109, 437 (1958).

[9] A. RYTZ, Absolutmessung der Energie der wichtigsten natürlichen Alpha-Strahler. Helv. Phys. Acta 34, 240 (1961).

[10] E. H. BECKNER, R. L. BRAMBLETT, G. C. PHILLIPS, and T. A. EASTWOOD, Absolute Measurement of a Set of Energy Calibration Standards. Phys. Rev. 123, 2100 (1961).

[11] A. RYTZ, H. H. STAUB, and H. WINKLER, Absolute precision determination of several resonance and threshold energies and the α-particle energy of ^{210}Po. Part I. Helv. Phys. Acta 34, 960 (1961).

[12] C. P. BROWNE, J. A. GALEY, J. R. ERSKINE, and K. L. WARSH, Comparison of ^{210}Po Alpha-Particle Energy with the ^{7}Li$(p, n)^7$Be Reaction Threshold Energy. Phys. Rev. 120, 905 (1960).

[13] C. P. BROWNE, Comparison of Alpha-Particle Energies from ^{210}Po and ^{214}Po and the Energy of ^{210}Po Alpha Particles. Phys. Rev. 126, 1139 (1962).

[14] G. H. BRIGGS, The Energies of Natural Alpha Particles. Revs. Modern Phys. 26, 1 (1954).

[15] A. H. WAPSTRA, Energies of Alpha Particles. Nuclear Phys. 18, 587 (1960).

[16] G. BASTIN-SCOFFIER, Exposé de méthodes en spectrographie α. Application aux corps de la famille du Radium naturel. Thèse. Fac. des Sci. Univ. de Paris, 1961.

[17] H. WINKLER und W. ZYCH, 180°-Ablenkmagnet für die Absolut-bestimmung von Partikelenergien. Helv. Phys. Acta **34**, 449 (1961).

[18] C. P. BROWNE and T. A. EASTWOOD, Comparison of Alpha-Particle Energies from Various ^{210}Po Sources. Phys. Rev. **124**, 1494 (1961).

[19] W. J. STURM and V. JOHNSON, A comparison of Several Nuclear Absolute Voltage Determinations. Phys. Rev. **83**, 542 (1951).

[20] R. C. PILGER, Nuclear Decay Schemes in the Actinium Family. Thesis. Unif. of Calif., **1957**.

Discussion

B. KARLIK: On which materials were the ^{210}Po sources deposited?

A. RYTZ: All of the sources were on tantalum backing.

B. KARLIK: Did you investigate whether after a certain length of time, there was any diffusion into the tantalum because in many materials polonium diffuses very much and so the line shape and energy measurement is influenced.

A. RYTZ: We did not investigate that systematically. In the case of polonium, we could observe after taking off the polonium by aqua regia that it was still active.

B. KARLIK: Then there was a certain diffusion into the support. I suppose this might account for some of the discrepancies between measurements from various laboratories.

A. RYTZ: I think that is correct.

G. C. PHILLIPS: We have investigated the source preparation techniques of ^{210}Po on the energy determinations as has Dr. BROWNE. Dr. EASTWOOD of Chalk River helped both Dr. BROWNE and the Rice group in preparing the sources. We got consistent data in each of the laboratories and we do feel that ^{210}Po is a good, useful reproducible laboratory energy standard.

The Energy of Polonium Alpha Particles as Determined by Different Methods*

By

Cornelius P. Browne

University of Notre Dame, Notre Dame, Indiana

With 6 Figures

This paper discusses the best value for the particles emitted by ^{210}Po to be used as a standard for the measurement of nuclear reaction energies. A new value for this energy was proposed[1] at The First International Conference on Nuclidic Masses in 1960. Fig. 1 shows the situation at that

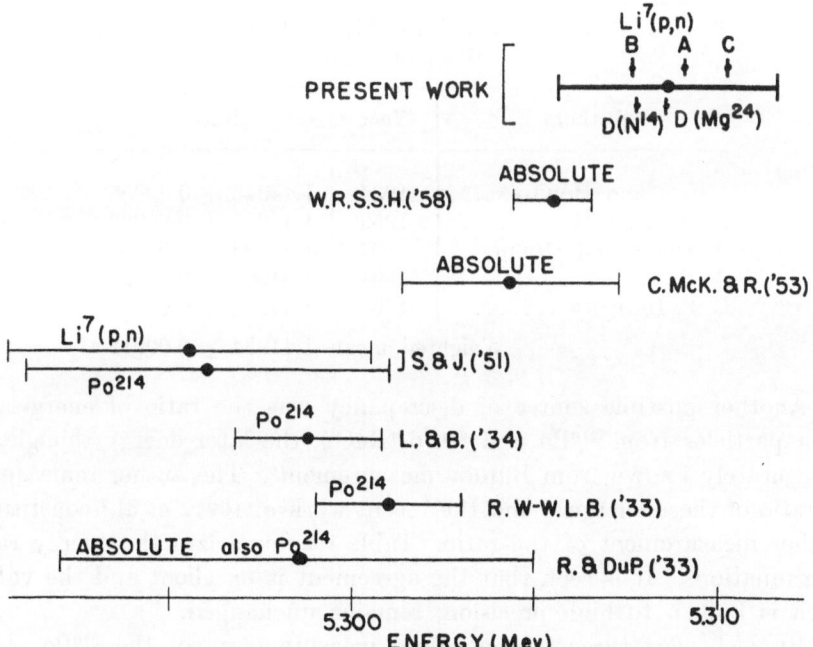

Fig. 1. Summary of determinations made up to 1960 of the energy of alpha particles from ^{210}Po. Reproduced from text reference 1. The original papers from which these values are taken are listed in reference 2

* Work supported in part by the Office of Naval Research.

time. The results of a number of determinations, made by several methods, are shown. Most values are labeled with the energy standard used, i. e., ^{214}Po, Li$^7(p, n)$Be^7 threshold, or absolute. The determination labeled "present work" was a comparison[2] with the Li$^7(p, n)$ threshold made by four different methods. The figure shows that the later value was considerably higher than any of the older determinations and was, in fact, 10 keV above the then accepted value. It will be shown below that the work in various laboratories since 1960 has greatly reduced these discrepancies.

One possible cause of the discrepancies is the method of source preparation. The work reported in reference 2 showed that polonium sources more than a few days old could give a lower alpha energy than fresh sources because of migration of polonium into the backing. T. A. Eastwood and the author, and Eastwood and the Rice group carried out comparisons of sources made by various methods[3, 4]. No effect was found. The consistency of data from different laboratories shown below also suggests that, as long as the source is freshly prepared and surface layers are avoided, the method of preparation is not critical. As discussed earlier[2] the measurements of Sturm and Johnson suffered from old polonium sources and are excluded from the averages in this paper.

Table 1. Energy Ratio of Alpha Particles from
^{214}Po and ^{210}Po

Authors	Year	Ratio
Lewis and Bowden ...	1934	1.44946 ± 0.00055
Collins et al.	1953	1.44927 ± 0.00082
Agapkin and Goldin ..	1957	1.44973 ± 0.00082
A. Rytz...............	1961	1.44905 ± 0.00027
C. P. Browne	1961	1.44954 ± 0.00040

Weighted mean 1.44934 ± 0.00012

Another possible source of discrepancy was the ratio of energies of alpha particles from ^{210}Po and ^{214}Po (RaC'), the later energy thought to be accurately known from Briggs measurement[5]. The author remeasured the ratio of these energies[6] and the recent work of Rytz et al.[7] constitutes another measurement of the ratio. Table 1 summarizes the energy ratio determinations. It is seen that the agreement is excellent and the value, which is known to high precision, remains unchanged.

By last year three new absolute measurements of the ^{210}Po alpha energy were reported[4, 7, 8] and one new absolute measurement of the ^{214}Po alpha energy[8]. The picture now changed radically. Six absolute measurements made over a period of 28 years agree very well and give an average value[6] of 5.3045 ± 0.0009 MeV. If one combines the above ratio for the ^{214}Po to ^{210}Po alpha energies with the recent absolute measurement[8] of the ^{214}Po energy a value of 5.3043 ± 0.0005 MeV is obtained, in perfect agreement with the absolute values. This suggests a systematic error in the BRIGGS value for the ^{214}Po alpha energy.

Until recently the remaining uncertainty seemed to be in the comparison of the ^{210}Po alpha energy with the $Li^7(p, n)$ threshold. The recent work of BECKNER et al.[4] constitutes a comparison of these two energies which disagree somewhat with the work of BROWNE et al.[2]. In

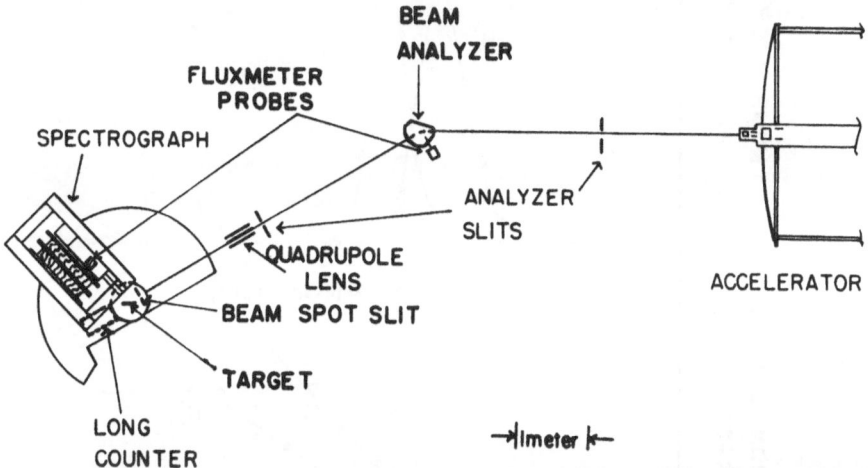

Fig. 2. Experimental arrangement for comparing threshold of the $Li^7(p, n)Be^7$ reaction induced by H^+ and H_2^+ beams

this later work, of the four methods used, the one of highest precision involved no change of magnetic fields but did require that the $Li^7(p, n)$ threshold be run with the diatomic beam. It is to be noted that this method gave the highest value for the energy ratio, that is, the one in most serious disagreement with the RICE work[4] and with the absolute determinations. Recent work of R. G. HERB and co-workers[9] on (p, γ) resonances measured with a diatomic beam suggested that an appreciable shift in the $Li^7(p, n)$ threshold might also be found with the diatomic beam. A series of measurements and a calculation have been carried out to test this hypothesis.

The experimental arrangement is the same as that used previously and is shown in Fig. 2. The spectrograph was used to measure the energy of protons scattered from targets bombarded by either a proton beam or an H_2^+ beam. A long counter was used to count neutrons from the $Li^7(p, n)Be^7$ reaction induced by either beam. It was assumed that a given output energy of protons, scattered from a monolayer of gold atoms, corresponded to a given input energy of protons whether from a monoatomic or diatomic beam. The procedure was to run the Li⁷(p, n) threshold with, say the proton beam, set the input energy at the observed threshold, slide a gold target in place of the lithium and record, on the spectrograph plate, protons scattered at 90°. This fixed a point on the plate corresponding

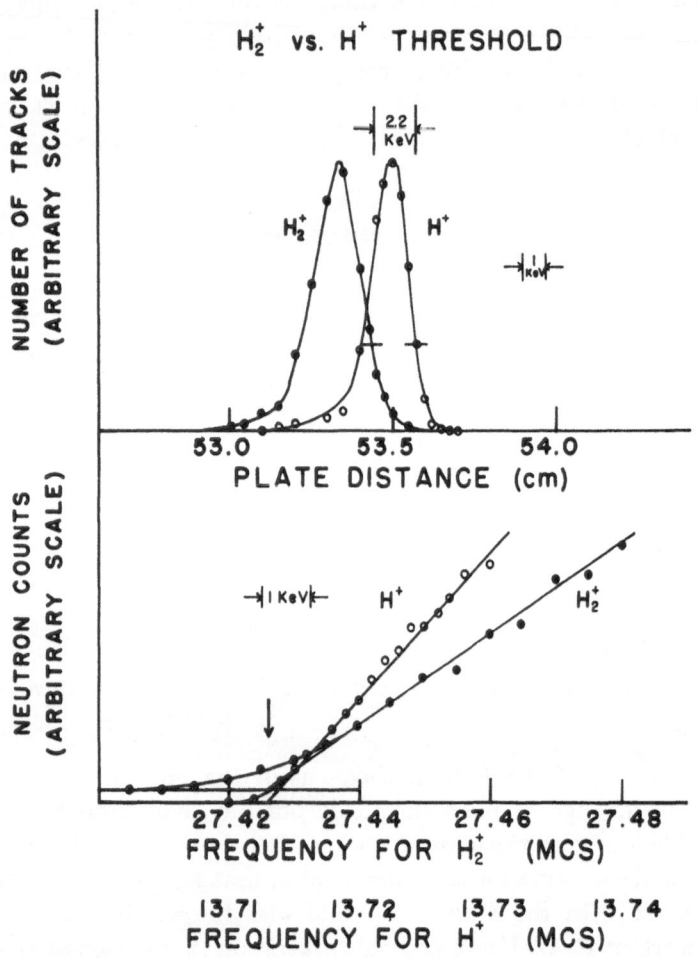

Fig. 3. Displacement of Li⁷(p, n)Be⁷ threshold induced by an H_2^+ beam from that induced by an H^+ beam. The frequency scales in the lower half of the figure are placed to make the apparent thresholds coincide. The groups of elastically scattered protons in the upper half of the figure are seen to be displaced from each other

to the threshold energy. The lithium target was now replaced and the threshold run with the diatomic beam. With the diatomic input energy set at the apparent threshold, the gold target was inserted and scattered protons again recorded in the spectrograph. The spectrograph field was kept fixed throughout the measurements. In the absence of any atomic effects in the threshold measurement or scattering, the scattered proton groups should, of course, lie at exactly the same place on the spectrograph plate.

Fig. 3 shows data from a typical run. In the lower half of the figure the number of neutrons is plotted against the frequency of the NMR fluxmeter used to measure the field of the beam analyzing magnet. The frequency scales for atomic and diatomic beams are placed to make the apparent thresholds coincide. The upper part of the figure shows the proton groups appearing on the spectrograph plate when the input frequency is set at threshold for each beam. Clearly the two groups do *not* coincide. Many such runs were taken with thresholds and scatterings done in different orders to check on possible surface layer build-up. In every case the protons scattered from the diatomic beam had a lower energy than those from the atomic beam.

An auxiliary experiment was done to test the effect of target thickness on energy of the scattered protons. One might suppose that with a thicker target atomic effects would appear with the diatomic beam. No appreciable difference was found in the shift measured using thick or thin gold or aluminum targets. Either the hypothesis of no effect with a monolayer target is correct or the effect is the same for different target thicknesses and different atomic numbers. Of course the ionization loss in the gold is much greater than in the lithium target and one might expect the Coulomb repulsion effect to be much less.

Table 2. Shift in Apparent Threshold for the $Li^7(p,n)Be^7$ Reaction Induced by H^+ or H_2^+ Beams

Run	E_H keV thick target	E_{H_2} keV thin target
1	2.2	2.6
2	1.5	1.3
3		1.6
4	1.7	1.7
5	1.5	1.2
6	2.5	2.5
7		1.8
Weighted mean	1.81	1.71

Final Average 1.75 ± 0.12 keV

Table 2 summarizes the data on the apparent shift in threshold. The data are listed separately for scattering from very thin and from thicker targets. In some runs one pair of thresholds was run whereas in others two pairs of thresholds were taken. The runs were appropriately weighted in obtaining the average apparent displacement of 1.75 ± 0.12 keV. Details of these measurements will be reported elsewhere. Results of the calculation based on the work of DAHL, HERB, and co-workers is given below. The calculated curve fits the experimental points very well near threshold.

It is seen that a very large effect is found with our experimental conditions. These conditions are the same as those used in the earlier comparison[2] of the ^{210}Po energy with the $Li^7(p,n)$ threshold. When the diatomic beam energy was set at the apparent threshold in that work the energy per proton was thus 1.7 keV below the threshold measured with an atomic beam. The deuterons, of the same magnetic rigidity as the diatomic beam, which were scattered to give a calibration point near the polonium alpha group, were thus of lower energy than supposed and hence the energy of the alpha group was actually lower. When this correction

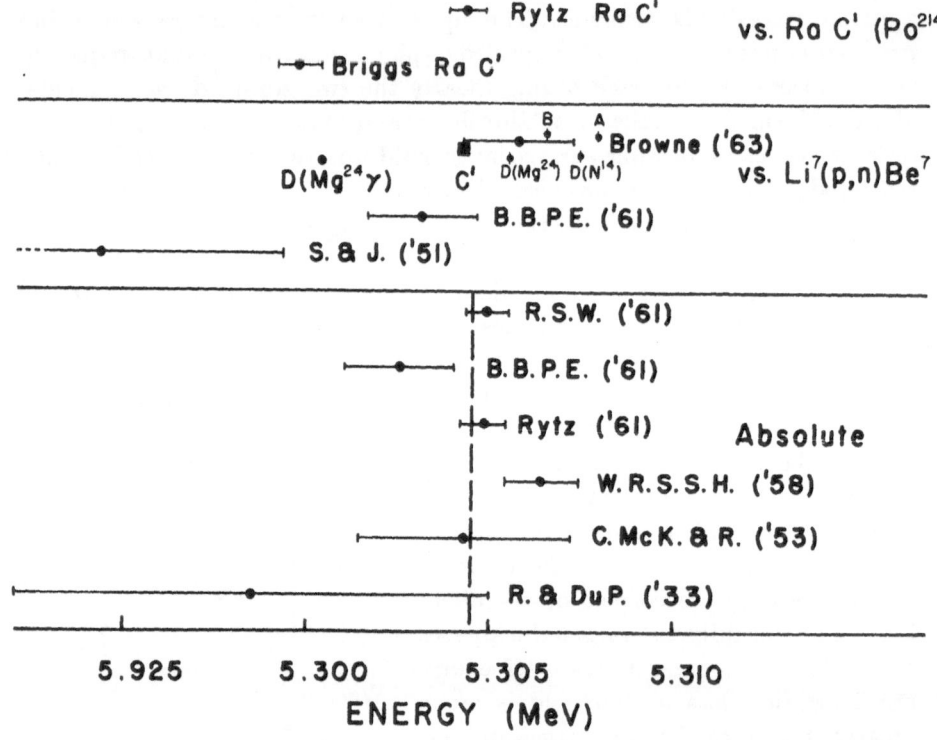

Fig. 4. Summary of determinations made up to 1963, of the energy of alpha particles from ^{210}Po. The lower part of the figure shows values from six absolute measurements, the middle part three comparisons with the $Li^7(p, n)Be^7$ reaction threshold energy, and the top third two values based on different numbers for the energy of alpha particles from RaC'. Labels are initials or names of authors reporting the values

is applied and when a $Li^7(p, n)Be^7$ threshold energy of 1.8807 MeV is used, a value of 5.3042 ± 0.0020 MeV is obtained for the alpha energy from ^{210}Po. This value is shown as point C' in Fig. 4. This figure summarizes all the data discussed in this paper. A comparison of Fig. 4 with Fig. 1 shows the great improvement in the situation since the first Conference on Nuclidic Masses. The points labeled A, B, D (Mg^{24}), and D (N^{14}) are based on the data of reference 2, adjusted to the threshold energy of 1.8807 MeV. Point A was given low weight in the original work because the spectrograph field was cycled in this method. Further experience with differential hysteresis in the spectrograph indicates this value should be given very low weight. Point B was also given low weight because the beam analyzer field was changed in this method. Point D (N^{14}) involved using the spectrograph calibration curve and the Q-values for $N^{14}(d, p)N^{15*}$ reaction measured against the $Li^7(p, n)$ threshold by the Wisconsin group. It is somewhat less reliable than the point D (Mg^{24}) which was obtained by using only energy ratios for scattering from various masses and the excitation energy of the Mg^{24} first excited state obtained by the Wisconsin group against the (p, n) threshold. The weighted average of ponts A, B, C' and D is seen to be in good agreement with the absolute determinations and the point C' which is given most weight, in excellent agreement.

An overall view of Fig. 4 shows that of all measurements of the ^{210}Po alpha energy based on absolute, $Li^7(p, n)$, or RaC' energies only the Sturm-Johnson measurement disagrees with the value of 5.3045 ± 0.0005 MeV. To obtain this agreement it is necessary to use the Rytz value for RaC' and discard the Briggs value. In summary, the absolute measurements are in excellent agreement, the ratio of ^{210}Po to ^{214}Po alpha energies is well known, and the ratio to the $Li^7(p, n)$ threshold seems to be in good agreement (but see below). A third independent absolute measurement of the RaC' energy is needed.

Lest the impression be gotten that everything is in order and all discrepancies removed, the point D (Mg^{24}) in Fig. 4 should be noted, and the uncertainties in the $Li^7(p, n)$ threshold measurement discussed in the following paper by R. G. HERB should be considered. Point D (Mg^{24}) is based on the recent determination[10] of the excitation energy of the first state in Mg^{24} in terms of the absolute γ-ray energy. The discrepancy between this point and point D (Mg^{24}) represents a discrepancy between the γ-ray energy scale and the value used here for the $Li^7(p, n)$ threshold.

In the light of the work reported in the next paper by R. G. HERB it seems that the uncertainties in the threshold measurements are, at this time, greater than those in the absolute measurements of the alpha energies. More work needs to be done in comparisons with the γ-ray energy scale. The value of 5.3045 ± 0.0005 MeV is suggested for the ^{210}Po alpha energy and it is suggested that this is at least as reliable a calibration point as the $Li^7(p, n)$ threshold.

Calculation of Threshold Shift with H_2^+ Beam

DAHL, COSTELLO, and WALTERS[9] calculated the thick target yield curve for a (p, γ) resonance in aluminum. They assumed that the molecular ion loses its binding electron at the instant it strikes the target. In the absence of the binding electron the two protons repel each other. The impulses received by the protons may have a large effect on their laboratory energy. The effect depends on the orientation of the internuclear axis with the beam direction. For ions with internuclear axes aligned with the beam, the front proton will receive energy from the Coulomb field faster than it loses energy through ionization for some distance into the target. This means that there will be protons in the target with energies above the (p, n) threshold energy when the beam energy is below threshold. At the suggestion of HERB and following the method given by DAHL[11],

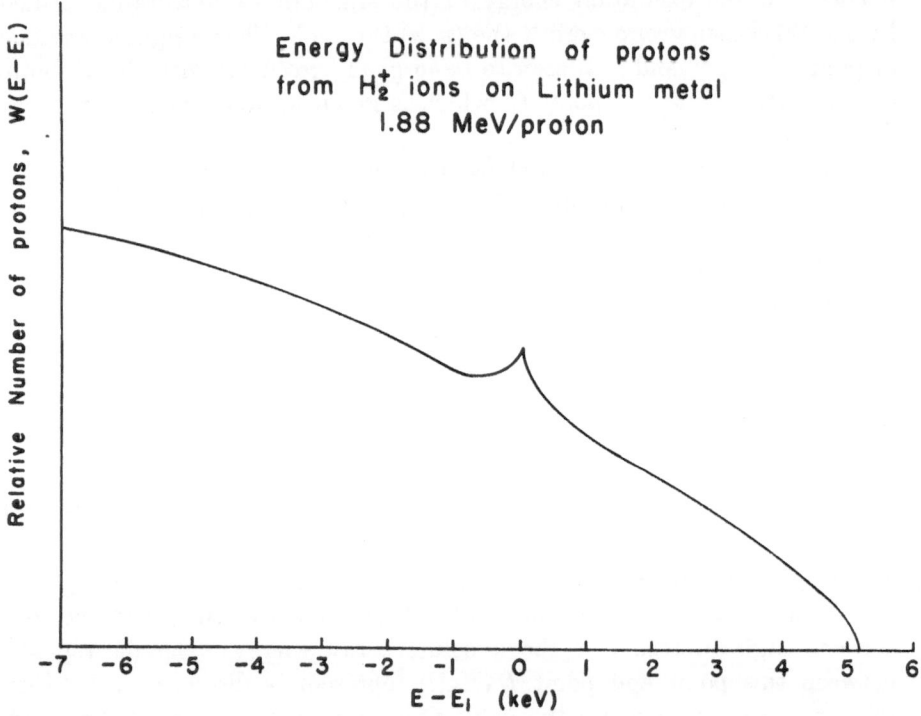

Fig. 5. Calculated distribution of protons as a function of energy difference from incident energy when lithium metal is bombarded with $H_2{}^+$ ions with an energy of about 1.88 MeV per proton

D. W. PALMER and the author calculated the energy distribution of protons in a lithium metal target bombarded with monoenergetic H_2 ions. Vibration effects were considered comparable to beam energy spread and were included at a latter point in the calculation. The resulting distribution is shown in Fig. 5. Some protons are found to have energies over 5 keV above the beam energy.

A plot of cross section versus energy above threshold was made using the expression given by NEWSON et al.[12] for the thin-target yield. This curve was folded into the energy distribution of Fig. 5 to give a thick target yield curve for monoenergetic ions. To derive the expected yield curve for our beam analyzer resolution the distribution observed in the elastically scattered protons from the diatomic beam was folded into the monoenergetic yield curve. The result is the dashed curve in Fig. 6. A Gaussian shaped beam distribution of half width 3.6 keV (considerably wider than the experimental value) was also tried. No appreciable change was made in the shape of the yield curve. If an ionization loss of 2 ev/A (about 3 times that for lithium metal) is used the solid curve of Fig. 6 is obtained. The crosses in the figure show experimental points from one of the runs and the dotted line indicates the extrapolation made from

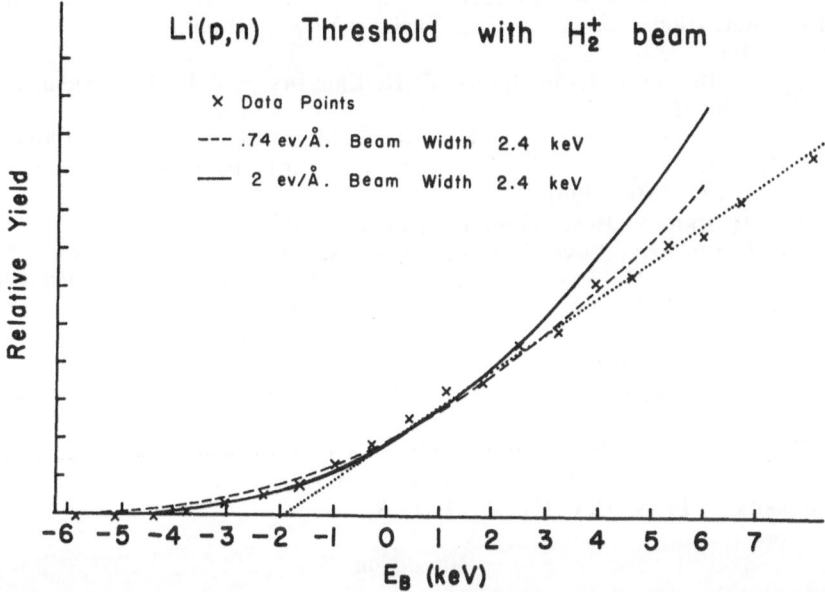

Fig. 6. Calculated and observed yield near threshold for the $Li^7(p, n)Be^7$ reaction induced with an H_2^+ beam. Crosses are experimental points, the dashed curve is calculated for an ionization loss of 0.74 ev/A and the solid curve for an ionization loss of 2 ev/A

these points to find an apparent threshold. Good agreement is found between theory and experiment up to an energy of about 4 keV above true threshold or 6 keV above the apparent threshold. One does not expect agreement much above this energy because, with the detector position used, some neutrons miss the detector as the "cone" opens. The theoretical curve may be too high because by the time this energy is reached some protons are over 10,000 A into the target and it is probably unreasonable to assume they are still moving under the influence of each others Coulomb field.

It is clear from the calculated curve that the value obtained for threshold with a diatomic beam depends strongly on the region of the yield curve above threshold used in the extrapolation. The displacement found in the present work applies only to our particular experimental arrangement. Work with other arrangements will be done.

The good fit to the threshold, giving the same displacement from true threshold as that measured assuming no atomic effects in the elastic scattering supports the latter assumption. A theoretical estimate of the effect of the field of the unscattered proton on the energy of the scattered proton is, however, needed.

References

[1] C. P. Browne, J. A. Galey, J. R. Erskine, and K. L. Warsh, Proceedings of the International Conference on Nuclidic Masses, Hamilton, Toronto Press, Toronto, 1960.

[2] C. P. Browne, J. A. Galey, J. R. Erskine, and K. L. Warsh, Phys. Rev. 120, 905 (1960).

[3] C. P. Browne and T. A. Eastwood, Phys. Rev. 124, 1494 (1961).

[4] E. H. Beckner, R. L. Bramblett, G. C. Phillips, and T. A. Eastwood, Phys. Rev. 123, 2100 (1961).

[5] G. H. Briggs, Revs. Modern Phys. 26, 1 (1954).

[6] C. P. Browne, Phys. Rev. 126, 1139 (1962).

[7] A. Rytz, H. H. Staub, and H. Winkler, Helv. Phys. Acta 34, 960 (1961).

[8] A. Rytz, Helv. Phys. Acta 34-3, 240 (1961).

[9] P. F. Dahl, D. G. Costello, and W. L. Watters, Nuclear Phys. 21, 106 (1960) and following paper.

[10] G. Murray, R. L. Graham, and J. S. Geiger, Bull. Amer. Phys. Soc. 7, 72 (1962).

[11] P. F. Dahl, Ph. D. thesis, University of Wisconsin 1960 (unpublished).

[12] H. W. Newson, R. M. Williamson, K. W. Jones, J. H. Gibbons, and H. Marshals, Phys. Rev. 108, 1294 (1957).

Discussion

W. W. Buechner: I would like to point out that our experience at MIT indicates that, as has been found at Notre Dame and at Rice, a ^{210}Po source on a silver wire gives an extremely reproducible calibrating standard. Over the past 15 years we have made perhaps thousands of such sources, some made with a great deal of care and others prepared in rather casual ways, by several generations of students. These sources have been used in connection with the calibration of several analyzing magnets and spectrographs. The excellent internal consistency of our Q-value measurements since 1946 indicates that such sources provide an extremely constant alpha energy source. The work of Browne shows that such sources emit particles with the same energy as those prepared for the work reported by Rytz. I do not believe that Q-values based on ^{210}Po suffer from appreciable systematic errors due to differing methods for preparing or using polonium sources.

D. M. van Patter: This question relates to Dr. Buechner's comments. Considering the procedure used at MIT for locating the high energy side of the Po-α peak, how old must a source be before this high energy point shifts down 1 keV from the position using a fresh source?

C. P. Browne: Our experience has been that in a period of the order of a week, one will see that the high energy edge no longer is a straight line and then you realize that you really can't say what the energy is. It is hard to say what a keV shift is. You would decide it is a bad source because you are not getting a straight line. Now if you just went ahead and said that I will just draw the best straight line through these scattered points, I would imagine you might be something like a keV off. I might make one other remark about source age. It is our practice not to leave the source in the vacuum system any longer than we have to, for two reasons: first, the contamination of the vacuum system that everyone finds with polonium. Second, we find that the source is likely to pick up surface layers from pump oils and contaminants in a vacuum system much faster than if it is just kept in a dust-free air atmosphere. We usually keep the source in the vacuum something like hours while we make a run and then we remove it and just keep it covered. When we do this we feel that a source that is less than a week old is safe. However, for the precision results we have reported, we have always made the source and used it either that day or the next day.

D. M. VAN PATTER: My question referred to the procedure used at MIT, which at the time I was there consisted of locating a point at $1/3$ of the maximum on the high energy side of the Po-α peak.

C. P. BROWNE: We have always used the one-third height. I have looked at this procedure in this work reported here. I could just as well have used the peak height or the half height. It doesn't matter what point you use.

D. M. VAN PATTER: One last comment. At the time of the Mass Conference in Mainz, I compared the existing MIT Q-values with other measurements, as well as comparing reaction cycles to the D_2—He^4 mass difference. This analysis revealed that the MIT Q-values should be increased systematically by 0.1%, which has now been accomplished by the increase in the standard Po-α energy of approximately 0.1% in the intervening time. In addition, comparison of MIT Q-values made with much different geometries shows excellent agreement. Therefore, in opposition to Dr. Rytz's remarks concerning possible systematic errors in nuclear reaction energies due to the use of Po-α calibrations, I do not believe that any major systematic error is present in the MIT Q-value measurements which could originate from the use of Po-α calibrations (i. e. in excess of about 0.05% of the Q-value). Of course, for isolated measurements there could be systematic errors arising from other sources such as surface contamination target preparation or thickness and so forth.

A. SPERDUTO: At MIT we routinely make our calibration checks within 24 hours after preparation of a ^{210}Po source. To check the effect of aging and different methods of source preparation, we made a series of studies three years ago measuring the energy changes as a function of time, starting from one hour after preparation to one week. As Dr. BROWNE has observed, we too have found deviations only of the order of 1 keV over the period of one week. It is gratifying to add that a recent complete recalibration of our single gap broad range spectrograph has shown deviations only of the order of 1 keV over the original calibration made in 1955.

A. I. ABRAMOV: I would like to expand the question asked by Dr. VAN PATTER. (1) What recommendations can you give with regard to the preparation of sources, in particular, as far as it concerns the choice of a support and the technique of deposition of the layer? (2) What value of the energy can be attributed with certainty at the present time to Po-α? (3) What is the degree of precision of this figure?

C. P. BROWNE: We purchase PoCl in dry form from the Mound Laboratory of Monsanto Chemical Company. About 1 cc of distilled water is added and three drops of concentrated HCl is added. A clean, polished, pure silver wire is dipped in this solution for 2 to 10 minutes, depending on the age of the solution. I suggest a value of 5.3045 ± 0.0005 MeV for the energy of the alpha particle from ^{210}Po.

A. RYTZ: I should prefer a slightly higher ^{210}Po energy, namely 5304.8 keV. Further, I still think that polonium source preparation is critical.

B. KARLIK: I have studied the energy distribution of α-particles as influenced by the conditions of the preparation of the sources somewhat in detail in about 1935 and I shall be glad to send reprints of this work to the colleagues who have participated in this discussion. A factor that has not been mentioned here so far is the deposit of carbon from CO_2 contained in the atmosphere or the gas in which the sources are kept, due to the disintegration of the CO_2 by the strong ionization near the surface of the source. The investigations at the time were made with strong α-sources as we were interested in monoenergetic α-particles for the production of nuclear reactions but corresponding smaller effects will already play a role with the precisions now under discussion.

A. RYTZ: May I ask Dr. BROWNE which values of fundamental constants he used in calculating his energy value? Since the differences are now so small, this point is no longer negligible.

C. P. BROWNE: For some of the older work, perhaps we have not made the corrections to the newer values for fundamental constants, but we have used the same values as Dr. RYTZ in his latest work to well within the experimental accuracy. I think a realistic uncertainty on the ratio measurements is about 2 keV and I think there is an uncertainty of at least 0.5 keV in the overall average so the change in constants still makes a very small difference.

Atomic Effects on Nuclear Reaction Yield Curves

By

R. G. Herb

University of Wisconsin, Madison 6, Wisconsin

With 15 Figures

Introduction

Non-nuclear effects that influence the shape of nuclear reaction yield curves were first studied at Wisconsin in 1959 in an attempt to check the linearity of an electrostatic analyzer when further work was planned on proton-proton scattering. The first effect studied[1, 2] is a large asymmetric shift of a thick target resonance yield curve when hydrogen diatomic ions are used for bombardment. Results were explained by taking into account Coulomb repulsion between the two protons of the diatomic ion assuming that the electron is lost immediately when the ion strikes the target.

The work with diatomic ions will not be discussed further in this paper. It led to observation of phenomena termed the "Lewis Effect" and subsequent work in this area at Wisconsin has been devoted to clarification of this effect.

Nature of the Lewis Effect

Approximately three years ago H. W. Lewis explained to the author that a thick target yield curve over a nuclear reaction resonance should exhibit a maximum just above the resonance. This, he explained, is due to discrete energy losses suffered by charged particles in passing through matter, and should be expected if the various spreading factors such as resonance width and beam energy spread are substantially less than the maximum energy loss step. Protons incident on a target at an energy well above a resonance lose energy stepwise by ionization and excitation. As they approach the resonance energy, a few will experience large energy losses that carry them beyond the resonance region. These protons have no chance for a nuclear reaction. If incident beam energy is at the resonance energy all protons have a chance to interact and a peak is observed just above the mid-resonance energy. Early experimental work with the $Al^{27}(p, \gamma)Si^{28}$ resonance at 992 keV showed peaks that appeared and then

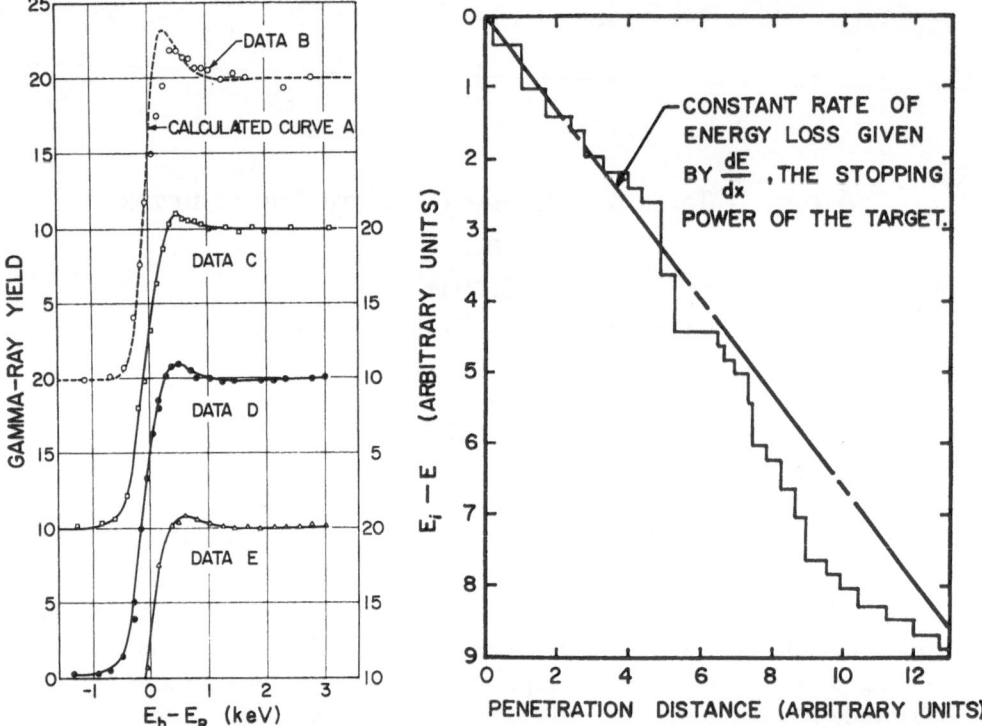

Fig. 1. Early experimental data (WALTERS et al., 1961) on four aluminum targets over the 992 keV resonance and calculated curve using $1/Q^2$ collision spectrum[3]

Fig. 2. Schematic diagram showing the energy of a proton as a function of penetration distance into a target[2]

disappeared. Intensive work over a period of six months finally gave sufficient reproducibility to where the effect was established. Fig. 1 shows these first results which were published as a Physical Review Letter[3].

Analysis

To explain their results the authors resorted to a Monte Carlo calculation. The probability $P(Q)$ of a proton losing energy in a step of magnitude Q was assumed to be given by $P(Q) \sim 1/Q^2$ where $Q_{min} < Q < Q_{max}$. Q_{max} is the energy loss in a head-on collision of a proton with a free electron, and Q_{min} is chosen to fit stopping power data. Fig. 2 shows the energy of a typical proton as a function of distance as it passes into a target.

The computer was programmed to pick a random number which determined the value of the first energy loss Q_1 suffered by proton No. 1

and this energy loss was subtracted from incident energy and was recorded. A second random number determined Q_2 and successive Q's were so determined until the total energy loss of proton No. 1 was 3 keV or greater. The program provided a distribution of Q values to conform with the $1/Q^2$ law. Ten thousand protons were programmed and results determined a function $\eta(E_i - E)$ which gave the probability density of protons with energy between E and $(E + dE)$ where E_i is incident energy. The curve

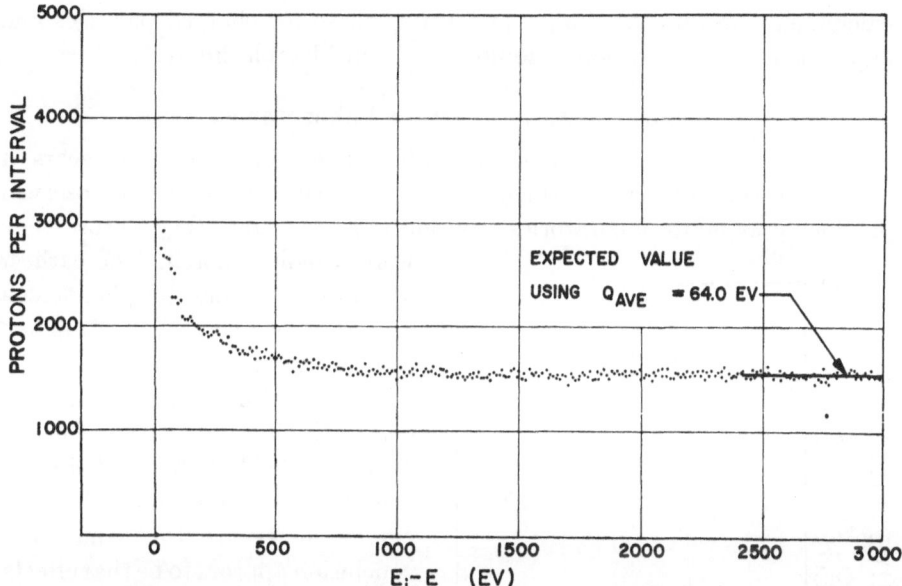

Fig. 3. Probability density of protons in a thick target as a function of energy from Monte Carlo calculation using $1/Q^2$ collision spectrum[2]

of Fig. 3 would give the thick-target yield curve predicted for the case where all energy spreading factors are zero.

The low energy portion of an experimental yield curve from a very thin aluminum target was reflected through its maximum to give a curve which was assumed to combine effects of all spreading factors, exclusive of discrete energy losses. This curve was folded into the curve of Fig. 3 to give the calculated curve of Fig. 1. The fit is qualitatively good but the peak height is greater than peaks exhibited by the experimental curves.

Other Methods of Analysis

H. W. Lewis worked out an analytic solution to this problem for a thick target[4]. His method has not as yet been used. It may save much computer time if resonances at many different energies and in different materials are to be programmed.

U. FANO discusses the general problem in a series of papers[5], and using his transport-equation approach JOHN BEAM working with C. H. BLANCHARD solved the thick target case utilizing a stepwise integration method[6]. This method can be adapted for any collision spectrum. In common with other solutions it is at present only applicable to thick target yield curves.

The author learned recently of the work of PLACZEK[7] where the effects of discrete energy losses of neutrons in passing through matter was accurately calculated. Results are similar in form to those for charged particles and experimental investigations should be of considerable interest.

Early Results at Other Laboratories

From communications and discussions over the past three years it is evident that humps on thick target yield curves had been observed previously at many laboratories. Because of their transitory nature they were usually ignored. The earliest attempt to explore the phenomena appears to be that of J. W. MUELLER working with H. H. STAUB at the University of Zurich. In a Master's degree thesis (University of Zurich, 1958) he considers effects of discrete energy losses assuming that all energy losses are the same. His conclusion appears to be that effects may not be observable experimentally. A. DEL CALLAR working with BONDELID did a Master's degree thesis (Catholic University of America, 1959) on resonance energies due to proton bombardment of aluminum. In his Master's thesis he has the curve of Fig. 4. This curve is excellent but it was not given journal publication and there appears to have been no attempt at explanation.

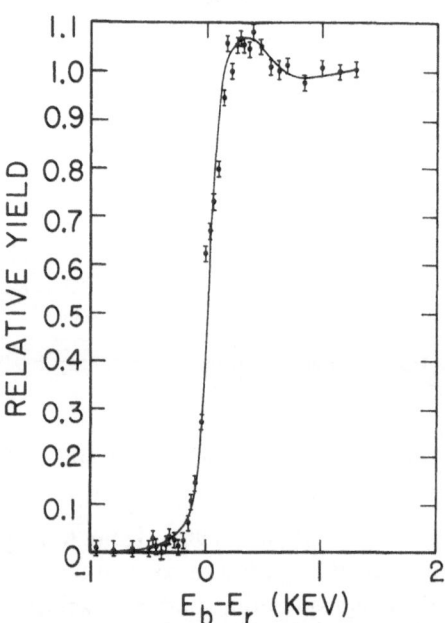

Fig. 4. Thick target curve over the 992 keV aluminum resonance obtained by A. DEL CALLAR. Master's degree thesis, Catholic University of America (1958)

Thin Targets and Targets of Intermediate Thickness

In 1961 the Monte Carlo program at Wisconsin was extended to where the computer kept a record of both energy and position of a proton as it passed through the target. Twenty thousand protons were programmed through the computer. From results of this program yield curves could

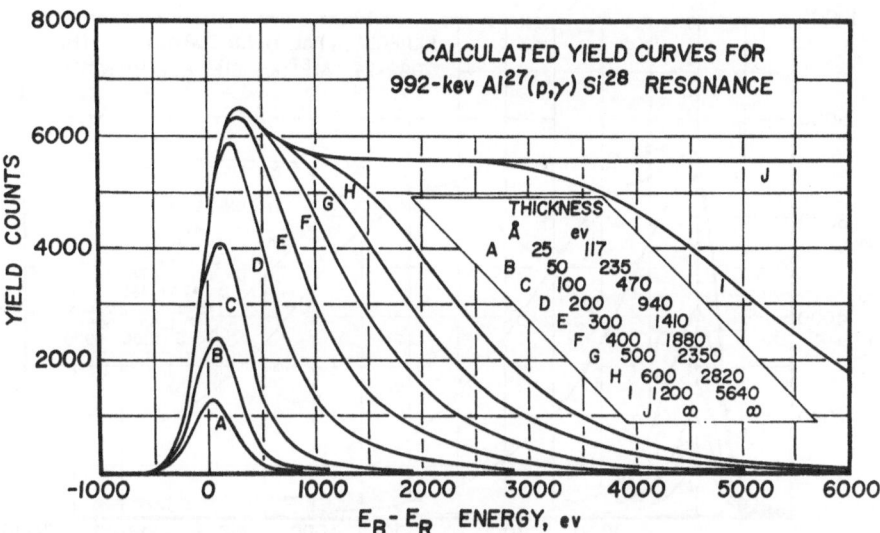

Fig. 5. Calculated yield curves for the $Al^{27}(p, \gamma)Si^{28}$ resonance at 992 keV assuming $1/Q^2$ collision spectrum[8]. Target thickness values are given in Å and in eV. To produce these curves a thin target yield curve corresponding to a resolution of about 4000 was folded into proton density distribution curves as determined by a computer

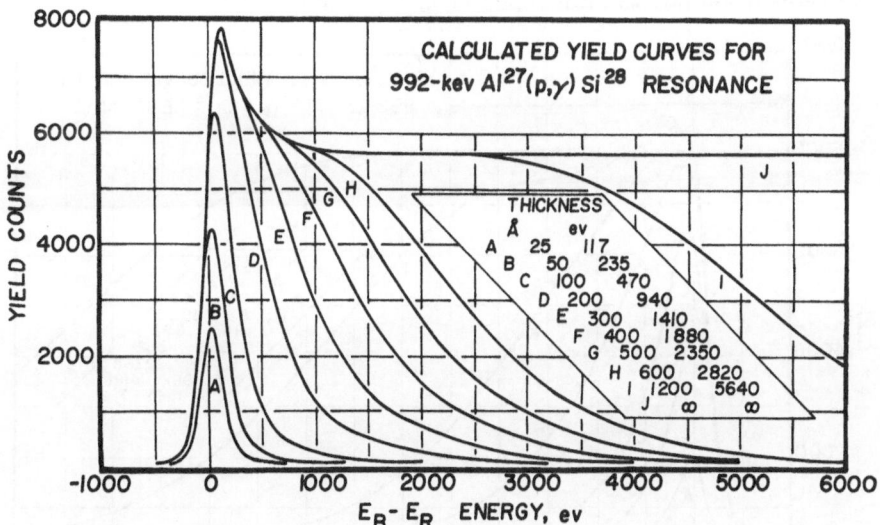

Fig. 6. Calculated yield curves for the $Al^{27}(p, \gamma)Si^{28}$ resonance at 992 keV assuming $1/Q^2$ collision spectrum[8]. Target thickness values are given in Å and in eV. To produce these curves a thin target yield curve corresponding to a resolution of about 8000 was folded into proton density distribution curves as determined by a computer

be prepared for a target of any thickness. Figs. 5 and 6 show two families of yield curves prepared from this data. Fig. 7 shows the first family of

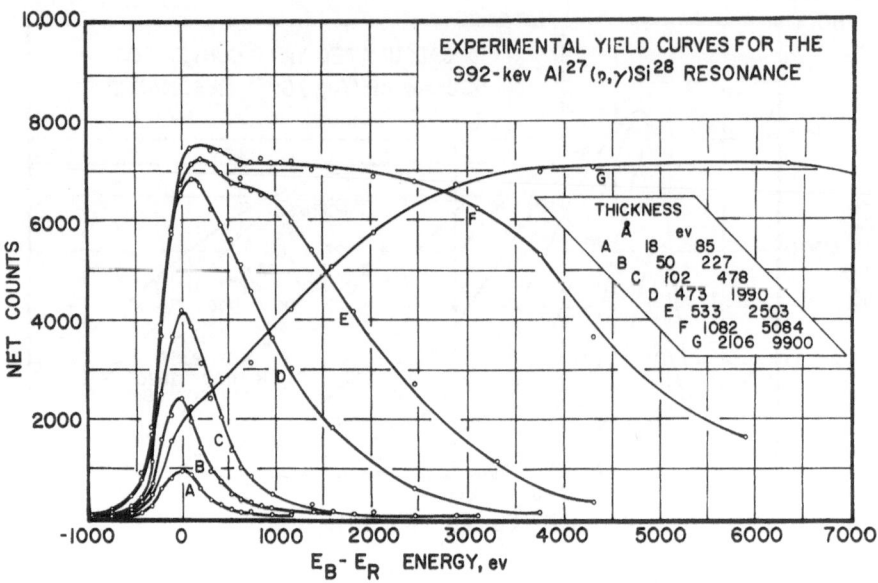

Fig. 7. Experimental yield curves over the $Al^{27}(p, \gamma)Si^{28}$ resonance at 992 keV[8]. Beam energy E_B is plotted relative to resonance energy E_R. Target thickness values are determined from areas under curves. Each target was produced by evaporation of Al onto previous target. A yield curve was obtained after each deposition. Results of curve G are due to evaporation of molybdenum from the oven evaporator

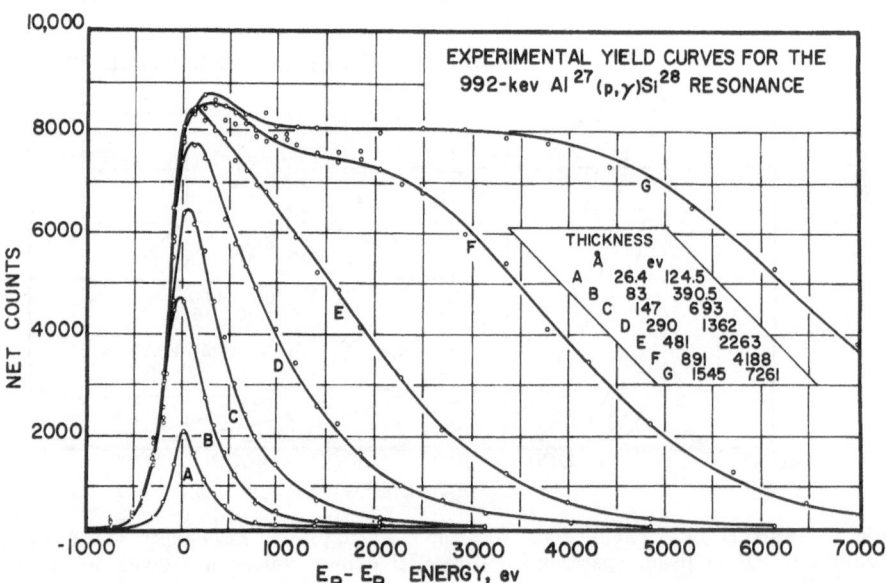

Fig. 8. Results from another group of targets[8], with conditions similar to those for the curves of Fig. 7

experimental curves obtained. Results were so striking and so unexpected that a second family of experimental curves were obtained (Fig. 8).

Agreement was excellent. These results[8] were presented at the Washington Physical Society meeting in April 1962. Subsequent work by BONDELID and BUTLER confirm these results[9].

The energy positions of peaks as a function of target thickness is far different than had been previously assumed. Procedures used previously for determining mid-resonance position are seen to be greatly in error for targets of intermediate thickness.

Contaminant Studies

The extended Monte Carlo program made possible the study of any proportion of inert contaminant at any position in a target. The

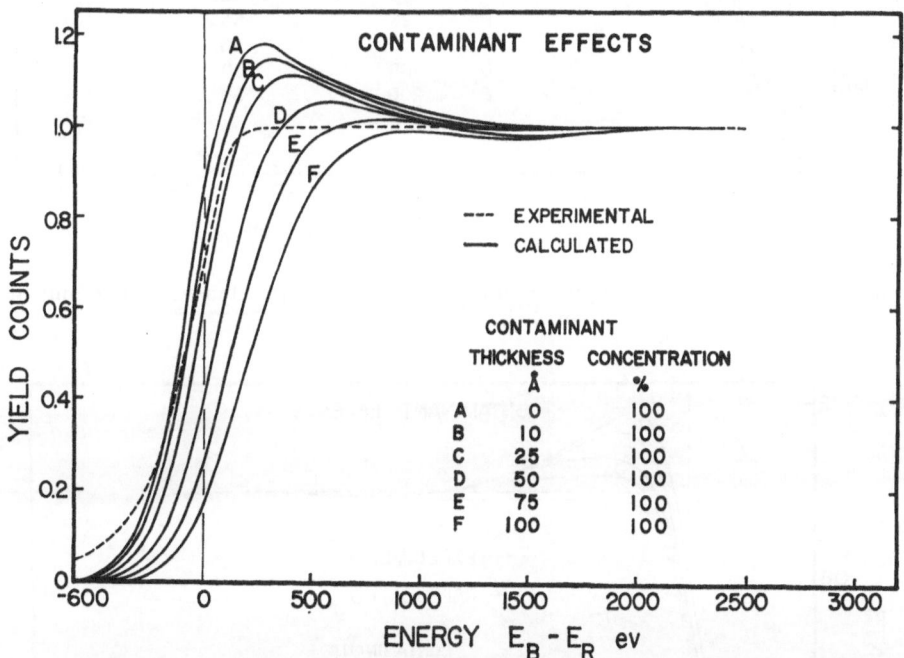

Fig. 9. Calculated thick target yield curves for $Al^{27}(p, \gamma)Si^{28}$ resonance at 992 keV assuming the $1/Q^2$ collision spectrum[8] with "aluminum equivalent" contaminant layers of 100% concentration. Dashed curve is experimental

approximation must be made that the inert contaminant is identical to aluminum in all energy loss processes.

Fig. 9 shows the form of curves to be expected if the contaminant consists of a uniform layer on the target surface. As the thickness of contaminant layer increases the Lewis peak decreases in height but the half value energy shifts rapidly toward higher energy. Experimental curves observed at Wisconsin behaved far differently. In Fig. 10 also, where the contaminant was assumed to have a 50% concentration

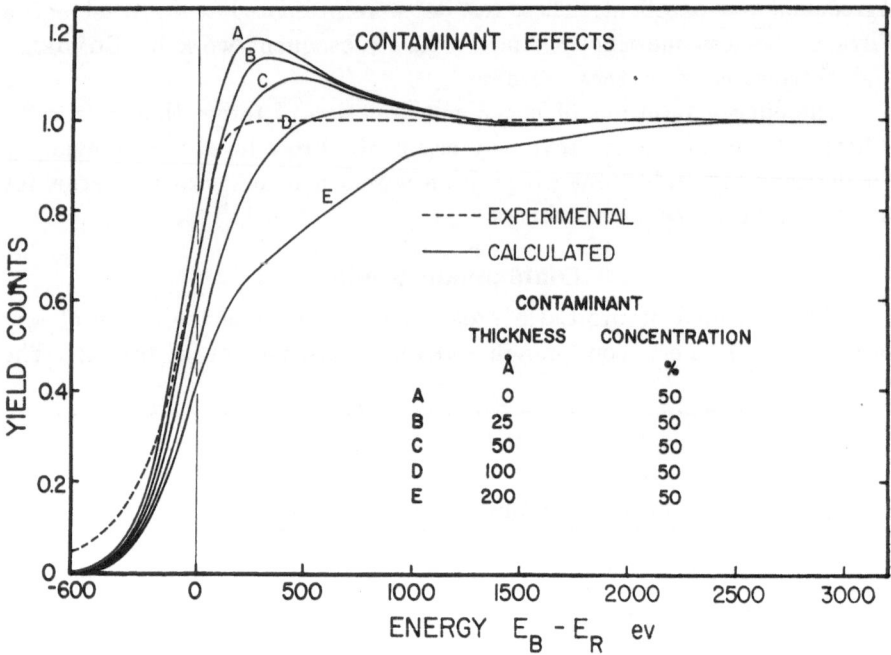

Fig. 10. Curves calculated with conditions of curves of Fig. 9 but with contaminant of 50% concentration[8]

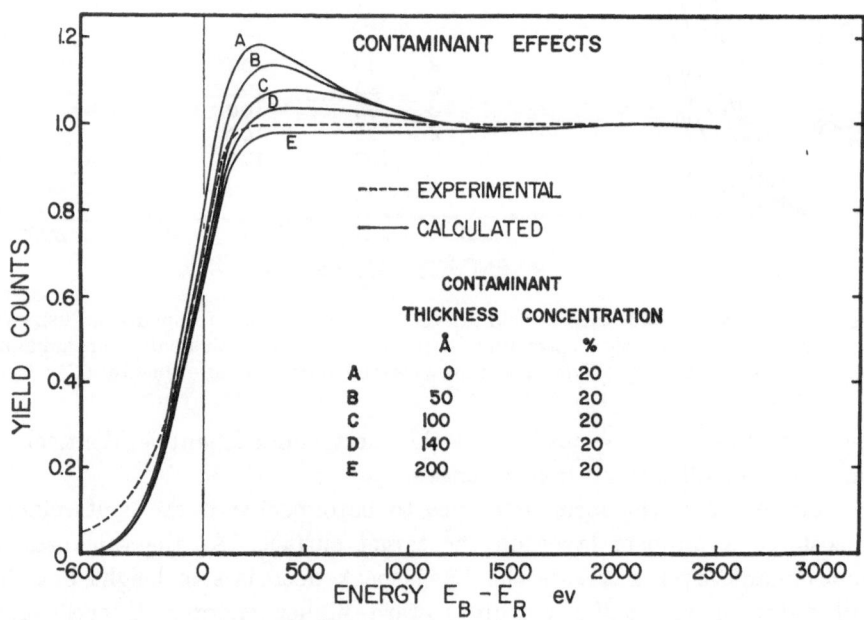

Fig. 11. Curves calculated with conditions of curves of Figs. 9 and 10, but with contaminant of 20% concentration[8]

agreement with experimental curves was poor. The 20% contaminant concentration used for the curves of Fig. 11 gave excellent agreement at a 200 Angstrom thickness with a number of experimental curves from which the Lewis peak had just disappeared. Lumps of contaminant 200 Å

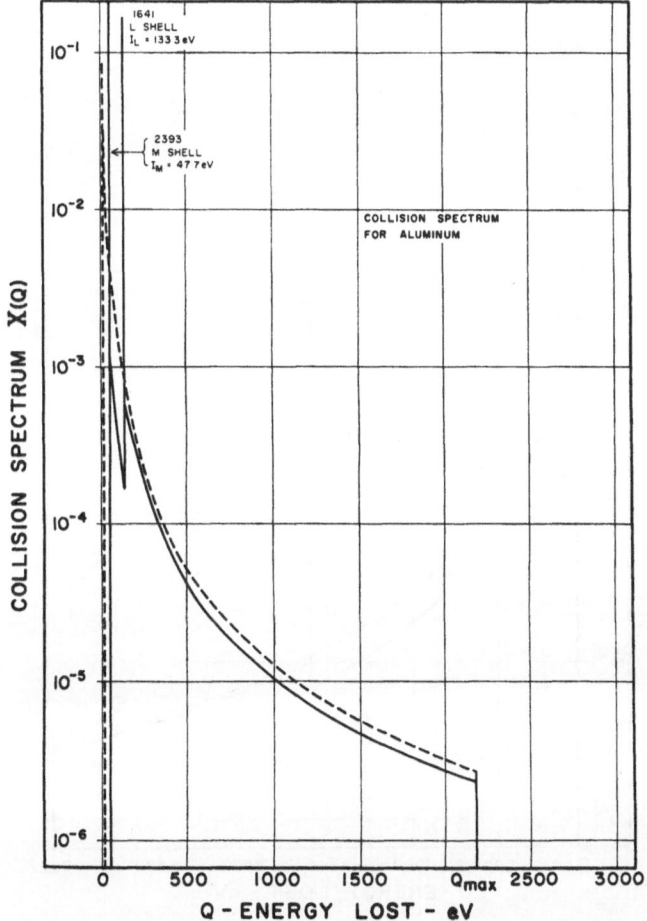

Fig. 12. Collision spectrum for 992 keV protons in aluminum including electron shell effects[11]

thick and covering 20% of the surface cannot be distinguished from interspersed contaminant with a 20% concentration extending to a depth of 200 Å.

A wide variety of other contaminant distributions were tried. They showed that experimental curves could be fitted with a relatively narrow range of contaminant concentrations and distributions.

The Collision Spectrum

In analyses discussed previously of the energy losses suffered by a charged particle in passing through matter all electrons are assumed to be free. The probability of an energy loss Q is proportional to $1/Q^2$. For

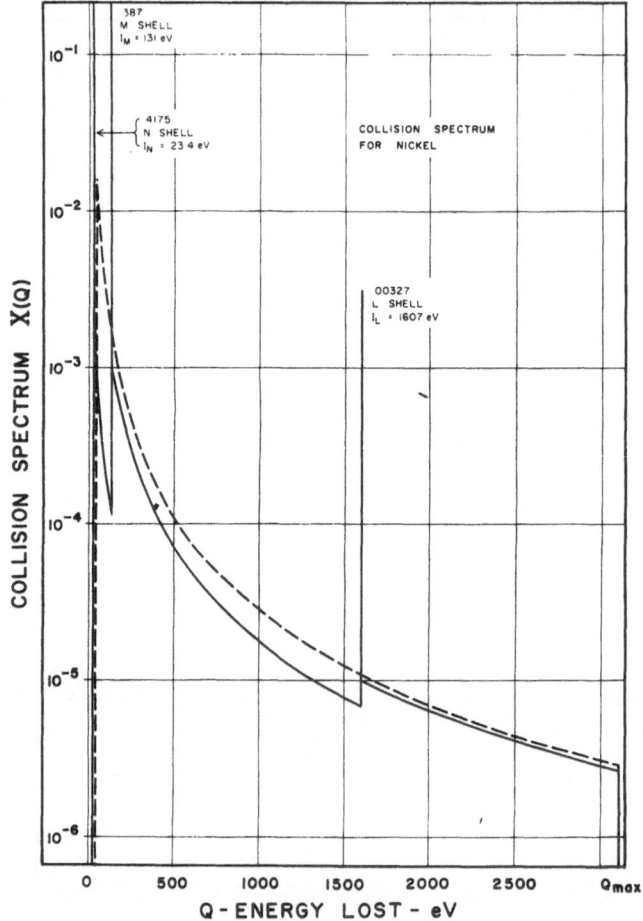

Fig. 13. Collision spectrum for 1424 keV protons in nickel including electron shell effects[11]

many applications this crude approximation has worked out satisfactorily. A minimum energy loss, Q_{min}, is assumed and its value is adjusted to fit stopping power measurements.

BETHE worked out the quantum mechanical expression for energy losses. BOHR, in his 1948 review article[10], points out that energy losses can be classified into two catagories, resonant contributions and free electron, non-resonant contributions. The free electron contributions are assumed to vary as $1/Q^2$ for electrons from a given shell down to an energy near the mean binding energy of these electrons. The resonant contributions of each electron to energy loss are assumed to be approximately equal to its non-resonant contributions.

J. G. SKOFRONICK and D. G. COSTELLO used results of R. M. STERN-HEIMER to estimate a mean ionization energy I_n for all electrons in the nth shell, and they worked out collision spectra for aluminum and nickel[11]. Their results for aluminum and nickel are shown in Figs. 12 and 13 respectively. Resonance contributions from electrons of one n value will be distributed over a relatively narrow energy range, and they were lumped into one δ function located in energy position following Sternheimer's method. Contributions of the K electrons of aluminum were computed using the treatment given by LIVINGSTON and BETHE and were found to be negligible. Thus in aluminum only 11 electrons per atom contribute to energy losses. A lower Lewis peak should be therefore expected than in previous calculations.

The collision spectrum shown in Fig. 13 for nickel is for 1424 keV protons. In this case Q_{max} is less than the K shell ionization potential and therefore K shell electron contributions can be ignored. L shell electrons make a substantial contribution to energy loss and the resonance term is at a sufficiently high energy to have an appreciable effect on the shape of an experimental yield curve.

Recent Experimental Results[11]

In all previous experimental work the Lewis peaks showed considerable variation in height and in the case of aluminum they usually disappeared after extended bombardment. In an attempt to improve target environment, changes were made as follows:

(1) Getter-ion pumps on the accelerator and electrostatic analyzer were modified to utilize sublimation of titanium rods. Pressures thereafter were usually in the region of 3×10^{-8} mm Hg.

(2) The target chamber of metal and ceramic was isolated from the electrostatic analyzer by a high impedance capillary and a sublimation type pump was provided for the target chamber. Here the pressure after baking was usually maintained at about 3×10^{-9} mm Hg except during evaporation of target material.

(3) Targets materials were heated by electron bombardment and relatively fast evaporation rates (from 1 to 5 keV per minute) appeared to give aluminum targets of higher purity than were given with slow rates.

(4) A stairstep voltage was applied to the target during bombardment to give a yield curve without changing the accelerator voltage. Effects due to changing background and variation in detector circuits were thereby largely eliminated.

Fig. 14 shows experimental curves over the 992 keV Al resonance
for eight different targets. The experimental curves were taken under a
great variety of conditions. Each is fitted to the calculated curve shown
by the solid line utilizing the collision spectrum of Fig. 12. In the

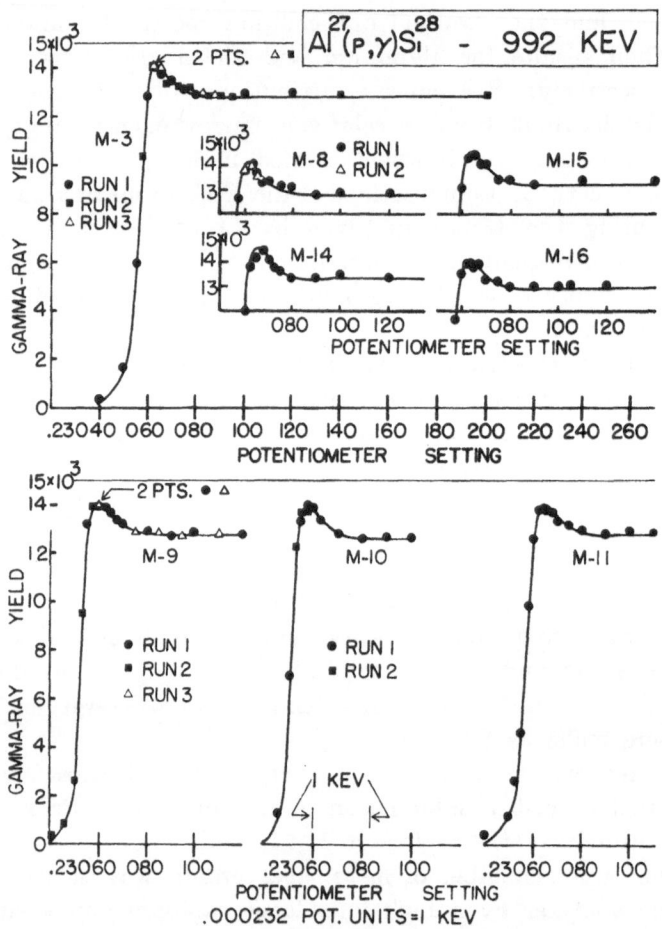

Fig. 14. Experimental curves from eight thick aluminum targets with improved
vacuum conditions[11]. The solid curve is calculated utilizing the collision spectrum
of Fig. 12, and is identical for all eight cases

determination of the calculated curve there are no adjustable parameters.
The eight calculated curves shown are identical. In each case agreement
of experimental data to the calculated curve is within statistical accuracy.

Curves M-3, M-8, M-9, and M-10 show repeat data after a
bombardment time of 1 to 2 hours. No target deterioration can be detected.
Targets yielding the curves of M-8, M-14, M-15, and M-16 were first
bombarded at the energy region of the peak and the peak heights were

therefore determined after bombardment times of no more than 15 minutes. The target yielding curve *M*-11 was prepared one day and the yield curve was taken the following day.

No change in peak height was noticed when CO or CH_4 gas was admitted to the target chamber to where pressure was 1 to 4×10^{-6} mm Hg and the target was bombarded for two hours.

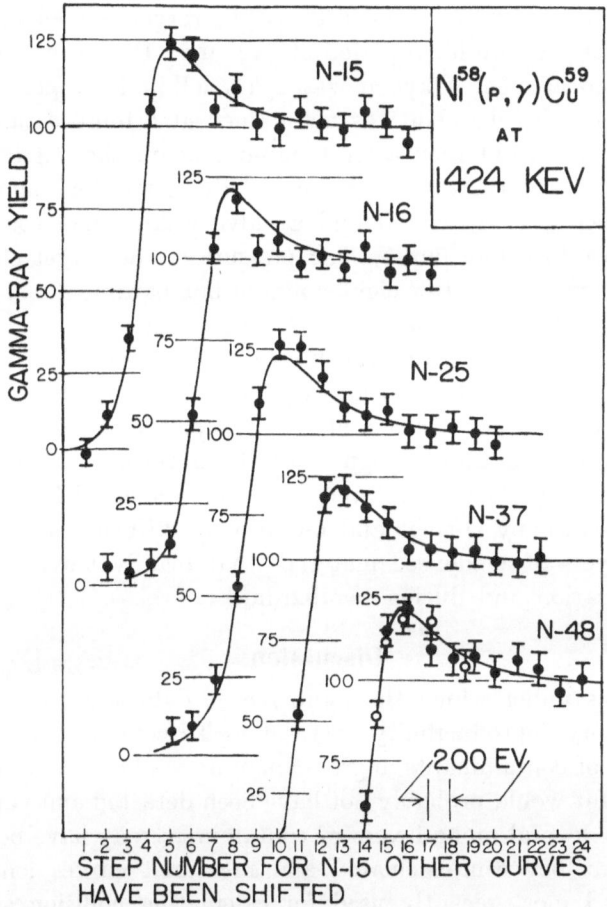

Fig. 15. Experimental curves from five nickel targets[11]. The solid curve is identical for all five cases and is calculated utilizing the collision spectrum of Fig. 13

All targets for the curves of Fig. 14 were prepared when the pressure during evaporation was 4×10^{-7} mm Hg or lower. Targets prepared when the chamber pressure was above 1×10^{-6} mm Hg showed smaller peaks.

Results for thick nickel targets over the 1424 keV resonance in Ni[58] are shown in Fig. 15 and the solid curves which are identical for all five cases shown were calculated utilizing the collision spectrum of Fig. 13.

Again, agreement in all cases appears to be within statistical accuracy. Nickel targets appeared to be less susceptible to contaminants during preparation than the aluminum targets. The target yielding curve N-16 was prepared at the rate of 0.24 keV per minute with the pressure at 7×10^{-6} mm Hg and the target for N-25 was prepared at the rate of 2.4 keV/min with the pressure at 1×10^{-6} mm Hg. No difference is detectable.

After obtaining the N-15 yield curve the target was let up to oxygen at atmospheric pressure for a period of 64 hours. Data taken after pump-out showed no change. Oxygen was again let in to a pressure of one atmosphere and the chamber was maintained at a temperature of 100° C for 12 hours. Data obtained after pump-out again showed no change.

After obtaining the N-25 yield curve the getter-ion pump attached to the chamber was turned off and a valve was opened connecting the target chamber to a fore pump. The pressure in the target chamber was then 2×10^{-4} mm Hg. After eight hours of bombardment the yield curve showed no detectable change.

More work is needed to establish conditions under which yield curves do not repeat. The baked out surfaces of this target chamber and the clean gettering surface of the pump probably trapped very effectively vapors that under other conditions might have been trapped by the target.

The unbaked surfaces of a target chamber connected for a long period to a system sealed by O-rings and pumped by diffusion pumps probably has heavy layers of condensed materials that readily move to the target during preparation and during bombardment.

Discussion

Conditions under which the more recent data was taken appear to give satisfactory reproducibility of results. Targets were in these cases probably free of contaminants in gross amounts. Volume contamination up to a few percent would probably not have been detected and surface layers amounting to several mono-layers of contaminant may have been present.

Agreement between calculated and measured curves lends support to the method most recently used for calculating collision spectra.

Work is now under way at Wisconsin on a number of extensions of these studies. In one extension an attempt is being made to improve energy resolution substantially to where the contaminant problem can be further clarified and to give a more sensitive check on calculated collision spectra. In another extension direct measurements of proton energy losses in thin foils are being attempted using two electrostatic analyzers in series, each with a resolution of about 30,000. In a third extension the determination of threshold energies is being investigated taking into account discrete energy losses.

Recently work by D. G. COSTELLO showed that many previous determinations of spreading factors such as beam energy spread and resonance width are in error. If the step in a thick target yield curve is used to determine beam energy spread or resonance width neglecting discrete energy losses the value determined may be much too small. In a recent case in the literature the energy spread calculated by the authors neglecting discrete energy losses was 23% below the value calculated when effects of discrete energy losses were included.

Conditions that must be satisfied for accurate determination of yield curves appear to be rapidly clarifying. These conditions do not appear to be difficult to meet and failure to meet them is very easily recognized from a thick target yield curve.

Corrections to old data will be difficult to apply because of uncertainty in regard to experimental conditions.

Remeasurements are needed but as the criteria for accuracy become better established intercomparisons of energy standards between laboratories will become much easier and much more dependable.

References

[1] P. F. DAHL, D. G. COSTELLO, and W. L. WALTERS, Nuclear Phys. 21, 106 (1960).

[2] W. L. WALTERS, D. G. COSTELLO, J. G. SKOFRONICK, D. W. PALMER, W. E. KANE, and R. G. HERB, Phys. Rev. 125, 2012 (1962).

[3] W. L. WALTERS, D. G. COSTELLO, J. G. SKOFRONICK, D. W. PALMER, W. E. KANE, and R. G. HERB, Phys. Rev. Letters 7, 284 (1961).

[4] H. W. LEWIS, Phys. Rev. 125, 937 (1962).

[5] U. FANO, Phys. Rev. 92, 328 (1953); 93, 1172 (1954).

[6] J. E. BEAM, and C. H. BLANCHARD, to be submitted for publication.

[7] G. PLACZEK, Phys. Rev. 69, 423 (1946).

[8] D. W. PALMER, J. G. SKOFRONICK, D. G. COSTELLO, W. E. KANE, and R. G. HERB, Bull. Amer. Phys. Soc. 7, 301 (1962). — J. G. SKOFRONICK, D. W. PALMER, D. G. COSTELLO, A. L. MORSELL, W. E. KANE, and R. G. HERB, Bull. Amer. Phys. Soc. 7, 301 (1962). — D. G. COSTELLO, W. E. KANE, A. L. MORSELL, D. W. PALMER, J. G. SKOFRONICK, and R. G. HERB, Bull. Amer. Phys. Soc. 7, 301 (1962). — D. W. PALMER, J. G. SKOFRONICK, D. G. COSTELLO, A. L. MORSELL, W. E. KANE, and R. G. HERB, Phys. Rev. 130, 1153 (1963).

[9] R. O. BONDELID, and J. W. BUTLER, Phys. Rev. 130, 1078 (1963).

[10] N. BOHR, Kgl. Danske Videnskab Selskab, Mat.-Fys. Medd. 18, 89 (1948).

[11] J. G. SKOFRONICK, and D. G. COSTELLO, Bull. Amer. Phys. Soc. 8, 320 (1963). — D. G. COSTELLO (introduced by R. G. HERB), A. L. MORSELL, J. G. SKOFRONICK, and D. W. PALMER, Bull. Amer. Phys. Soc. 8, 320 (1963); A paper covering this work to be submitted to Nuclear Physics.

Discussion

H. STAUB: There is certainly no doubt about the reality of the Lewis effect, but I would like to draw attention to the fact that a number of other mechanisms can lead to the appearance of a hump on a thick target yield curve. Among such mechanisms are inhomogeneity in density and grain structure of targets. Also, since the theory of discrete energy losses is certainly not perfect in every detail, it seems to me that the interpretation might be quite difficult. In particular, the magnitude of the minimum energy loss can alter the size of the hump. Assuming a probability of a loss Q to be $P(Q) = A/Q^2$, we get from

$$\int_{Q_{min}}^{Q_{max}} P(Q)\, dQ = 1 \quad \text{that } P(Q) = \frac{Q_{min}}{Q^2}$$

since $Q_{max} \gg Q_{min}$. In a metal, Q_{min} might be quite small for the conduction electrons and consequently, the ratio of probability of small to large energy losses increases, thereby decreasing the Lewis hump. In any case, I feel that with present accuracy of 0.1 keV the shift of the midpoint is just not yet significant. Possibly gaseous targets might be suitable for observing an undistorted Lewis effect.

R. G. Herb: It is a bit difficult to give a short answer. The theory is really in quite good shape. The determining factor is not the small energy losses, the big contribution is from the large ones. Here the theory is in quite good shape except for minute details and so I think you are wrong. The experimental work will soon be repeated, I am sure, in many places. It exists; there is no question about it. The effects of a target non-uniformities you speak of are not so hard to take care of and so it is an effect that should not be ignored. I think it will be relatively easy to get reproducible results if you are just careful about going over the target techniques where contaminents are held down to fairly low levels and you use a baked-out target chamber. Then I think both the experimental results and the calculated results are quite close to correct. I am sure they are not quite the last word. We certainly have improvements to make. I should also point out the shifts in energy here are not exceedingly great, say 125 or 150 eV is about as high as you would normally get one way but if you stay with your old system of standards you will get these variations. I don't think that you can use standard targets that are covered with carbonized pump oil. I think that is a bad standard but it is quite easy to improve these standards and I think it is important that we do it.

J. B. Marion: I would like to add a remark following Prof. Staub's comment on the possibility that targets may be composed of both thick and thin layers. We have at the University of Maryland made some measurements which are quite sensitive to such effects. We bombard a LiF target with protons of energy slightly (say, 20 keV) above the threshold for the $Li^7(p, n)$ threshold. At such an energy the neutrons are confined to a forward cone. We place one counter in the forward direction and another counter at the edge of the cone. Since the cross section is a reasonably flat function of energy, the counting rate in the forward detector is proportional to the volume of target material, whereas the edge counter is sensitive only the neutrons produced at the front surface of the target and the counting rate is therefore proportional to the surface area of target material. We observe the ratio of the counting rates, edge/forward, and find that as a function of time the ratio stays constant for a short while and then begins to decrease. We interpret this to be due to the formation of crystals or spikes of target material, with portions of the target backing being completely exposed. Now, this effect may occur only in alkali halides but it is possible that other types of target material will also exhibit the effect.

Nuclear Resonance Energies

By

H. H. Staub

Physik-Institut der Universität Zürich, Zürich, Switzerland

In the determination of mass values or rather mass differences through the Q-value of a nuclear reaction, two energies of particles, the energy of the primary particle and the energy of the reaction product must be determined. For intensity reasons doubly focussing sector magnets are usually used for both measurements. This type of spectrometer must be calibrated since it is impossible to compute the exact shape of particle orbits. Moreover ordinary and so called "differential" hysteresis affect the performance of the spectrometer and make it necessary that a rather closely spaced set of accurately known calibration energies is available. For calibration energies, resonance energies of isolated single narrow resonances whose width is of the order of the desired accuracy and threshold energies are suitable. Resonance energies are therefore mostly those of particle capture reactions with subsequent γ de-excitation. Threshold energies are commonly those of (p, n) reactions.

For the accurate and absolute determination of resonance energies three methods are currently in use.

1) Determination of the absolute particle momentum with the 180° homogeneous magnetic field spectrometer (Zurich and Rice Institute).

2) Energy determination with cylindrical electrostatic deflectors (A. E. I., Wisconsin and Naval Research Laboratory).

3) Velocity determination by time-of-flight method (Wisconsin).

Since the last Nuclidic Mass Conference in 1960 considerable progress in experimental techniques has been accomplished in all three methods. As a consequence the rather disturbing discrepancies between the values from different laboratories have now been greatly reduced.

a) Magnetic deflectors: At the laboratory of the University of Zurich a new greatly improved 180° magnetic spectrometer had been constructed in 1960[1]. This magnet is mechanically much more stable and its field, before any correction is applied, is more homogeneous than that of the previously used magnet although the ratio of pole width to gap is the same.

The radial field gradient is much smaller and the azimuthal fluctuations, which in the old magnet were caused by the periodic structure of the yoke are absent. The most significant improvement however was the use of a narrow aperture slit at 90° whose position relative to entrance and exit slits is known with high accuracy. Consequently the particle orbits throughout the magnet are known and the uncertainty in the radial field configuration is greatly reduced. Numerous other improvements are listed[2].

b) Targets: It is now generally recognized that the most reliable results are obtained by using thick targets and monoatomic ion beams, although it seems that the rather large shift which the thick target yield mid-point shows when molecular ions are used can be reasonably accounted for and corrected[3]. This is of course important if one wishes to extend a given standard to higher energy values.

The increased accuracy of present day measurements has made it necessary to consider seriously effects caused by the target surface and by the statistical nature of the energy loss as the bombarding particle passes through the target material. Effects due to this process are discussed in the paper of R. G. HERB at this conference. However it should be pointed out that even very small structural irregularities of the target may cause similar effects and of similar or even much larger magnitude. In general shifts of the midpoint energy of thick target yield curves associated with the "hump" at the high energy side will amount to about 100 eV at an energy of 1 MeV or 1 part in 10^4 if a resolution of this order of magnitude is used. But deviations of this magnitude are also very easily caused by target surface conditions. We therefore believe that resonance energy determination with accuracy higher than that reached at present can only be accomplished with gaseous targets where surface conditions are absent and where the atomic effects can be calculated with high accuracy. Finally it is worth mentioning that contamination of the targets and the energy shifts produced thereby have been greatly reduced partly due to improvements in high vacuum techniques and by carefully designing the targets in such a manner that its surroundings are kept at temperatures well below that of the target itself.

c) Recently GASTEN[4] reported measurements with the time of flight method previously used by SHOUPP, JENNINGS, and JONES[5]. The result for the $^{27}Al(p, \gamma)$ resonance is given as 991.4 keV. Unfortunately no errors are stated which makes a comparison with other values impossible at present.

The range of resonance energies has been considerably pushed towards higher energies in recent years. Narrow resonances of (p, γ) reactions suitable for energy calibration have been accurately determined up to 1843 keV for $^{58}Ni(p, \gamma)^{59}Cu$[6, 7]. Undoubtedly there are narrow (p, γ) resonances at still higher energies. Their usefulness however might be impaired by strong background radiation. However reactions of the type (α, γ) prove to be satisfactory as evidenced by the measurements of the Zurich group[8] on the 3200 keV narrow resonance of $^{24}\text{Mg}(\alpha, \gamma)^{28}\text{Si}$.

Table 1. Summary of resonance energy measurements

Reaction	A. E. I.	N. R. L.	Rice	Zürich
$^{19}\text{F}(p, \alpha, \gamma)^{16}\text{O}$	873.9 ± 0.8; 4.2[9]	872.4 ± 0.4; 4.5 ± 0.3[10]	872.3 ± 0.5[11]	871.8 ± 0.25; 4.8 ± 0.2[12]
$^{27}\text{Al}(p, \gamma)^{28}\text{Si}$	994.0 ± 1.0; < 0.4[9]	991.91 ± 0.3; 0.1 ± 0.05[6]	992.2 ± 0.5[11]	991.83 ± 0.10; 0.10 ± 0.02[12]
$^{27}\text{Al}(p, \gamma)^{28}\text{Si}$		1317.19 ± 0.4[6]		
$^{58}\text{Ni}(p, \gamma)^{59}\text{Cu}$		1423.64 ± 0.43; 0.05 ± 0.05[6]		
$^{13}\text{C}(p, \gamma)^{14}\text{N}$		1747.06 ± 0.53; 0.075 ± 0.05[6]		
$^{58}\text{Ni}(p, \gamma)^{59}\text{Cu}$		1843.45 ± 0.56; 0.10 ± 0.05[6]		1842.9 ± 0.45[7]
$^{24}\text{Mg}(\alpha, \gamma)^{28}\text{Si}$				3199.8 ± 1.0; 1.8 ± 0.3[8]

Summary: In Table 1 the results of resonance energy measurements made during the past few years are summarized. Under the heading of the laboratory where the work was performed the first number gives the absolute resonance energy in abs. keV with its stated uncertainty, the second number the observed resonance width Γ. The last number refers to the publication as listed under references. The low energy resonances (< 800 keV) have not been listed since the agreement of the various values has been quite good and no very recent measurements have been published.

References

[1] H. WINKLER, and W. ZYCH, Helv. Phys. Acta 34, 449 (1961).
[2] A. RYTZ, H. H. STAUB, and H. WINKLER, Helv. Phys. Acta 34, 960 (1961).
[3] R. O. BONDELID, and J. W. BUTLER, N. R. L. Report 5897 (1963).
[4] B. R. GASTEN, Bull. Aner. Phys. Soc. 7, 549 (1962).
[5] W. E. SHOUPP, B. JENNINGS, and W. JONES, Phys. Rev. 76, 502 (1949).
[6] R. O. BONDELID, and J. W. BUTLER, Phys. Rev. 130, 1078 (1963).
[7] H. H. STAUB, and H. WINKLER, Helv. Phys. Acta 33, 526 (1960).
[8] A. RYTZ, H. H. STAUB, H. WINKLER, and F. ZAMBONI, Nuclear Phys. 43, 229 (1963).
[9] S. E. HUNT, R. A. POPE, D. V. FRECK, and W. W. EVANS, Phys. Rev. 120, 1740 (1960).
[10] R. O. BONDELID, and C. A. KENNEDY, Phys. Rev. 115, 1601 (1959).
[11] E. H. BECKNER, R. L. BRAMBLETT, G. C. PHILIPS, and T. A. EASTWOOD, Phys. Rev. 123, 2100 (1961).
[12] A. RYTZ, H. H. STAUB, H. WINKLER, and W. ZYCH, Helv. Phys. Acta 35, 341 (1962).

Discussion

R. G. HERB: You made a statement that low energy losses are quite influential in determining what we call the Lewis effect. This is not right. The high energy losses cause all or almost all of the effects.

H. H. STAUB: That is very true. Of course it is the higher energy losses which really cause the Lewis effect but they will become relatively infrequent if you have very probable low energy losses because, after all, your average energy loss is given from the experimental results.

R. G. HERB: This is not correct. The high energy losses are computed independently and fairly accurately. Low energy losses are relatively unimportant. The situation is not as you stated it.

H. H. STAUB: I think I have to take exception because of the calculation of MUELLER where he has assumed the relatively large but uniform loss. There the Lewis effect becomes very, very small. If a proton has a large chance of losing its energy in very small steps then of course this will wash out the Lewis effect to a certain extent.

R. G. HERB: His calculation was a very approximate one which I don't think bears on the present subject. I would also like to point out that when you fit your curve with a certain resolution you will find that the apparent widths of the resonance, what we would like to call spreading factors, is likely to be off by a fairly substantial amount if you ignore the discrete energy losses. When you assume your resolution is a certain figure and the resonant width another figure and you try to fit the curve, ignoring discrete energy losses, then you will be quite far off. This is another way in which the discrete energy losses enter. Also the form of your curves is quite typical of targets which are rather badly contaminated. We got these curves for months and months before we cleaned up the targets.

H. H. STAUB: It would also be typical for clean targets if the discrete energy losses had a different form.

R. G. HERB: No, this is not right.

H. H. STAUB: They are calculated on this basis.

R. G. HERB: The calculations are getting pretty good. You are quite wrong in saying that the low energy losses influence them greatly.

H. H. STAUB: These curves are calculated taking no atomic effects into account. They give a shape which you would simply blame on contaminated targets. If contamination bears an important effect we would have to see a gradual shift of the mid-point towards higher energies. We have never observed such a thing.

R. G. HERB: This is because the contaminent tends to go on in apparently a spotty condition. Our curve shows very clearly that you can have a fairly large change before you get a great shift of the mid-point. I would guess in looking at your curve that they are probably low by approximately 100 volts.

H. H. STAUB: I certainly would not like to put a higher accuracy on any one of these results taken on evaporated targets than 100 volts for the reason I pointed out. Certainly the Lewis effect exists. Certainly surface conditions are of importance. As I said, I don't think we should attach a higher accuracy on present day results than 100 eV for the aluminum (p, γ) resonance.

Absolute Energy Standards for Van de Graaff Accelerators[*]

By

G. C. Phillips

Department of Physics, Rice University, Houston, Texas, USA

With 16 Figures

I. Introduction

The Van de Graaff accelerator provides a useful instrument for the determination of nuclear reaction Q-values and thus for the measurement of nuclidic mass-differences. This unique accelerator is usually operated so as to have beam energy-homogeneity of the order of one part in several thousand. To use particle beams of such precisely defined energy to determine nuclidic mass-differences however, it is necessary to ascertain, *in an absolute manner*, the energy of the particles. This paper will present the work done at Rice University on establishing such an energy standard. All of the Rice work has been on simple energy standards. It is our view that the calibration measurement of energy should be a simple one; they should all be laboratory standards that can be easily *reproduced* anywhere with a minimum of effort. Nevertheless, target preparation techniques are very important and it is unfortunate that space will not allow this subject to be discussed here. Professor HERB has discussed some aspects of this problem in a preceeding paper.

Similar work to the Rice work has been done in a number of other laboratories to establish an absolute energy scale. Historically the first of such measurements used natural alpha particle emitters as calibration sources[1]. Of additional interest has been the determination of neutron-threshold[2] and gamma ray resonance energies[3]. This session of the International Conference on Nuclidic Masses contains new contributions to various of these subjects and Professor MARION will summarize the present status of all the results. Thus this paper will only consider the Rice University work. Absolute calibration will be presented that covers the proton energy range from 0.87 MeV to 9.5 MeV and will include all the methods in current laboratory use: the use of conventional and Tandem

* Work supported in part by the United States Atomic Energy Commission.

Van de Graaffs, the calibration of neutron thresholds, gamma ray resonances, and natural alpha particles. All of these absolute energy determinations at Rice have employed a common measuring instrument: a 180° homogeneous-field, magnetic spectrometer. Thus the measurements reported here comprise a rather wide energy-range of absolute measurement, and were all measured with a common technique.

II. Experimental Methods

A. *The* 180° *magnetic spectrometer.* The magnetic spectrometer employed in all the Rice University absolute energy measurements is of

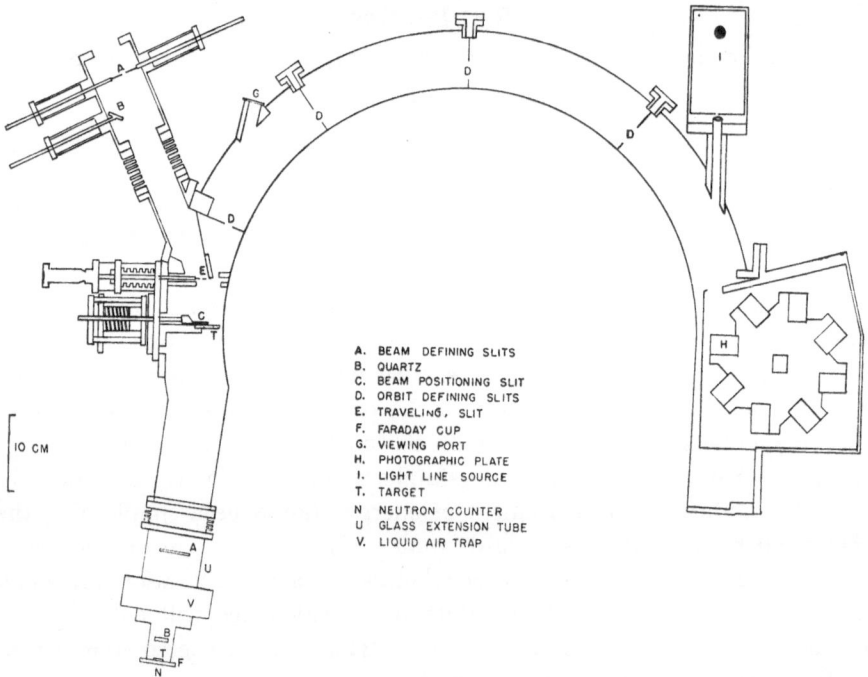

A. BEAM DEFINING SLITS
B. QUARTZ
C. BEAM POSITIONING SLIT
D. ORBIT DEFINING SLITS
E. TRAVELING, SLIT
F. FARADAY CUP
G. VIEWING PORT
H. PHOTOGRAPHIC PLATE
I. LIGHT LINE SOURCE
T. TARGET
N NEUTRON COUNTER
U GLASS EXTENSION TUBE
V. LIQUID AIR TRAP

Fig. 1. The vacuum tube of the Rice University 180° magnetic spectrometer, including the extension tube added for these measurements. The radius of curvature of the system is approximately 35 cm. The solid angle defined by the slit system is about 2×10^{-4} steradian

the annular-type[4]. The use of this instrument for such measurements has been described in detail in a number of publications[5]. Basically the measurement is very simple: One measures the radius of curvature, ϱ, of the charge particles being deflected in the field B and one also measures the field, B. The absolute measurement of $B\varrho$ comprises a measurement of the momentum and energy of the particles. If this is done for certain, easily reproduced experiments, then the measurements form an energy standard. In the Rice measurements the radius of curvature is established by the following details (see Fig. 1): The VAN DE GRAAFF beam is first magnetically analyzed by 90° deflection and then the beam is carefully positioned upon the target by means of a traveling slit which reads the beam position and then is withdrawn. The magnetically deflected reaction products from the target are accepted at an angle of about 180° to the beam direction. Another traveling slit, located by a micrometer, measures the angular deviation from 180°. A reaction angle of 180° is ideal for all reaction studies because the variation of Q-value determination with angle is a minimum at this angle.

The detected particles are allowed to strike a photoplate after traversing a half-circle. The photo-plates are each indexed with a light focussed to a fine line. The distance between the slit that determines the beam position and the light-line that is inscribed upon each photo-plate is measured in comparison to a National Bureau of Standards calibrated standard-meter bar.

The frequency of proton-moment magnetometer resonance signals are used to determine the magnetic fields in the 90° beam analyzer and in the 180° magnetic spectrometer.

Thus the basic measurements necessary to deduce an absolute energy are just a pair of frequencies and a distance. However, a correction must be made for small variations in B around the orbit of the particles[6]. For this purpose the field is measured at each 15° interval about the orbit and the correction applied.

To make absolute measurements one must assume some absolute criterion of spectrometer line-shape analysis. The one we have assumed is quite simple[7]: The particles of the greatest radius of curvature, when moving in a uniform field, will be those that have the greatest energy on emerging from the target or the source. This rather simple assumption coupled with the measured focal properties of the Rice spectrometer seems to be borne out by the consistency of all the results and by detailed line-shape analyses[7].

Fig. 1 shows the magnetic spectrometer and the details discussed above.

B. *Neutron counter techniques.* At proton or deuteron energies of about 5 MeV or less the problem of detecting neutron thresholds is rather simple; MARION and BONNER have described these methods[8]. Fig. 1 shows the method used at Rice to make such absolute energy measurements: The accelerator beam could either bombard the neutron (or gamma ray) target or could bombard the spectrometer target. This method allowed a measure of beam energy, and results using the 5.5 MeV Van de Graaff accelerator are reported below. However, at higher beam energies other problems arise that are due to the strong backgrounds of neutrons and gamma rays produced by the higher energy beams of projectiles. The method that we have developed at Rice to minimize the problem is shown in Fig. 2[9]. The beam of the Van de Graaff accelerator is magnetically analyzed by 90° deflection and the field measured by a proton moment magnetometer. The beam is then passed through a thin target, which produces the neutrons, and a neutron counter is arranged to intercept the almost forward-going neutrons and yet allow the beam to pass through. The beam of the Van de Graaff accelerator is then allowed to pass through a large vacuum tube and escape from the experimental area. Upon striking matter, the intense neutron fluxes that are produced are absorbed in a neutron cave (Fig. 3).

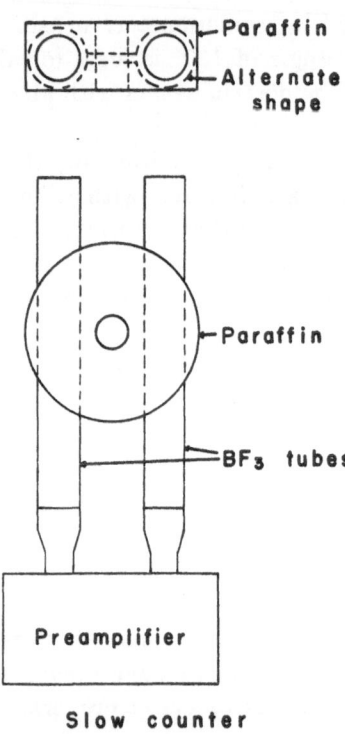

Fig. 2. The neutron counter used to detect threshold neutrons on the tandem accelerator. The beam passes through the central hole. Threshold neutrons are moderated by the paraffin cylinders that surround each of the two $B^{10}F_3$ counters

The absolute energy measurements are made using the 180° spectrometer; however, for this purpose the beam was deflected into another drift tube by a switching magnet located after the beam had been 90° analyzed.

In this arrangement it is very important to ascertain that a minimum amount of beam strikes anything except the desired thin target material when a high energy neutron threshold is to be determined. This purpose has been accomplished at Rice by an arrangement wherein the beam is positioned and focussed by electrostatic deflecting plates and alternating-gradient-lenses and is measured by means of two electrodes: The first electrode is called the washer (Fig. 4) and a minimum current ($< 5\%$) is allowed to strike it. The second electrode will accept all the beam when the beam is properly positioned. Because of alliteration, and hope, this was called the "wishy-washer". These measurements ascertain that the beam strikes on the center of the target; when the positioning has been determined the "wishy-washer" is withdrawn. This method is simple and has had some success as is shown in Fig. 5 where neutron backgrounds are plotted. These methods are recommended for neutron threshold measurements at tandem energies because they provide a low level method of measuring neutron-threshold calibration energies. In addition, these methods have

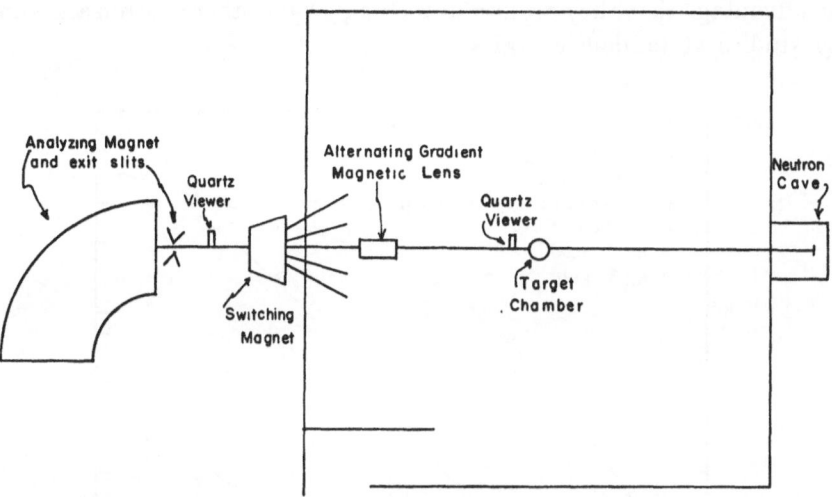

Fig. 3. Schematic plan view of the low-background method of measuring neutron threshold using the Rice University tandem Van de Graaff accelerator. The beam is stopped outside the experimental room and the neutrons produced are absorbed in a six-foot cube of concrete and 2000 pounds of borax

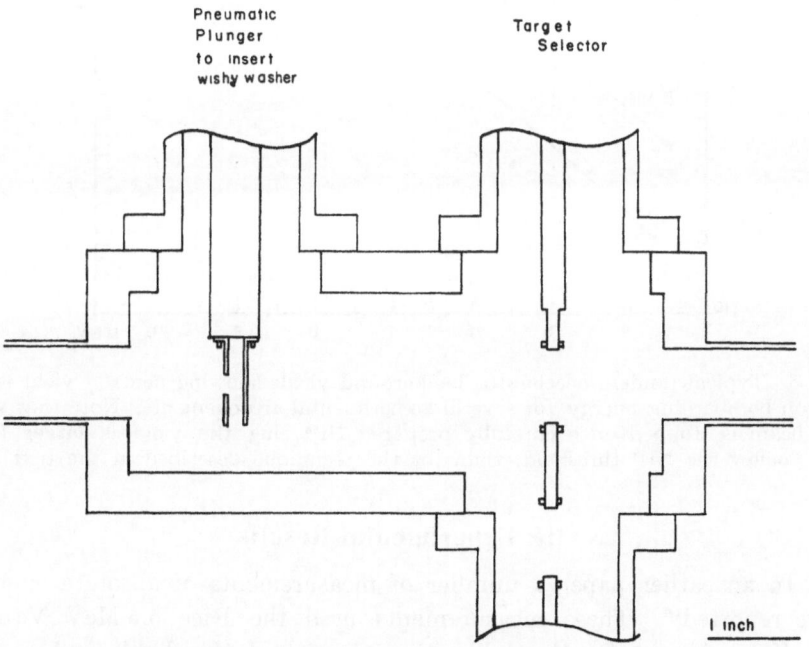

Fig. 4. The beam-positioning washer arrangement used on the tandem accelerator. Beam current may be measured on either the plate with the hole (the washer) or on the slug back of it. The whole arrangement may be withdrawn and the positioned beam will then pass through the thin foil target

the advantage that they are useful techniques for other neutron and gamma ray studies at tandem energies.

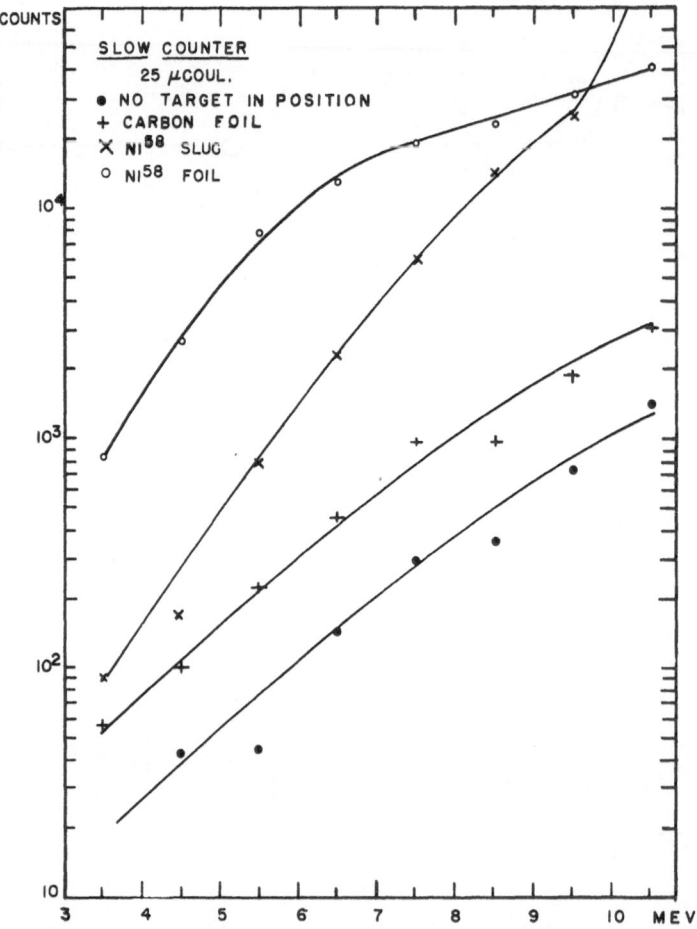

Fig. 5. Typical tandem accelerator background yields showing neutron yield *versus* proton bombarding energy for several experimental arrangements. Note that when the beam is stopped on a carefully prepared Ni⁵⁸ slug the yield is larger (even below the Ni⁵⁸ threshold) than for the technique described in the text

III. Experimental Results

In an earlier paper a number of measurements of absolute energies were reported[10]. These measurements used the Rice 5.5 MeV Van de Graaff accelerator for the calibration of neutron thresholds and gamma ray resonances and also measured the energies of the natural ^{210}Po alpha particles. These results will be repeated here, for completeness, and in addition recent results, employing the Rice tandem, will be reported.

A. Measurements with the Rice 5.5 MeV Van de Graaff accelerator

1) *The Li⁷(p, n) Be⁸ threshold.* This threshold has become one of the
standards for accelerator determinations of nuclear Q-value energies and

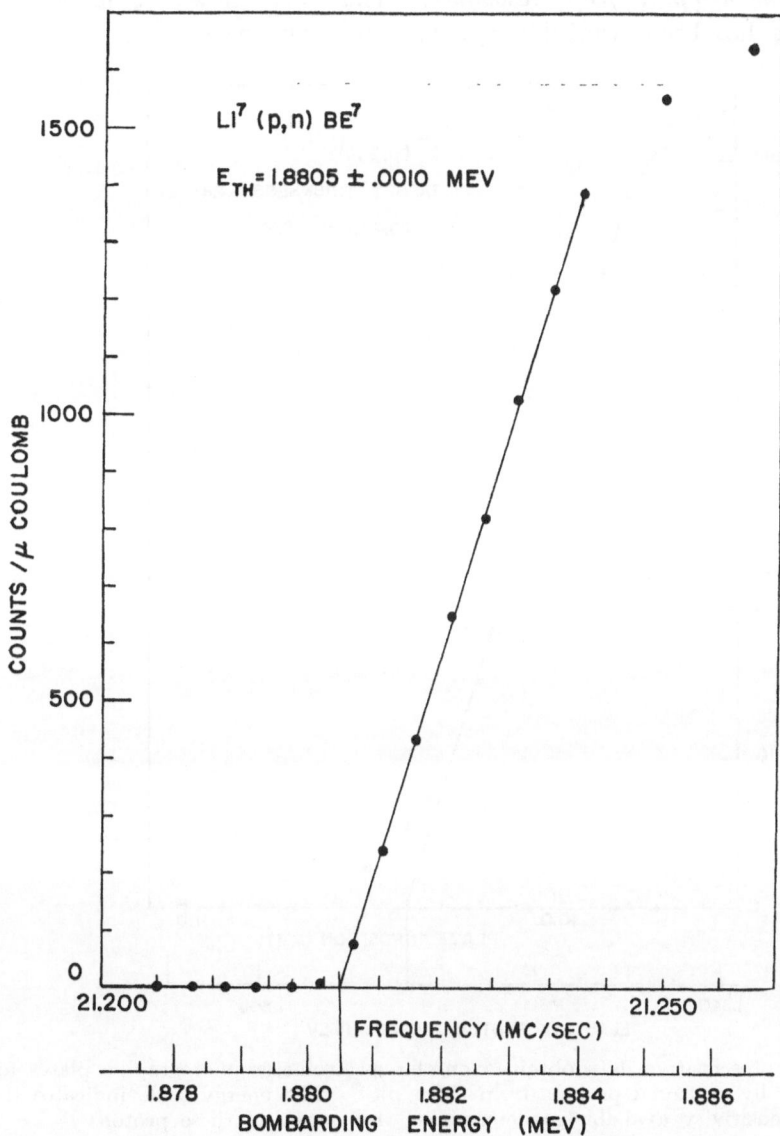

Fig. 6. Neutron counting data obtained from $Li^7(p, n)Be^7$. The energy sensitivity
of this threshold determination is the highest of all the thresholds studied

nuclidic mass-differences. A typical threshold is shown in Fig. 6 and the
determination of the proton energy is shown in Fig. 7[10]. For Fig. 7 and
for all the energy determinations reported here a C^{12} foil was used in

the 180° spectrometer as a scattering material to determine the energy of the beam particles. This threshold was measured to be 1.8805 ± ± 0.0008 MeV.

2) *The F¹⁹(p, α γ)O¹⁶ resonance.* This well-known gamma ray resonance has been studied and a typical resonance curve is shown in

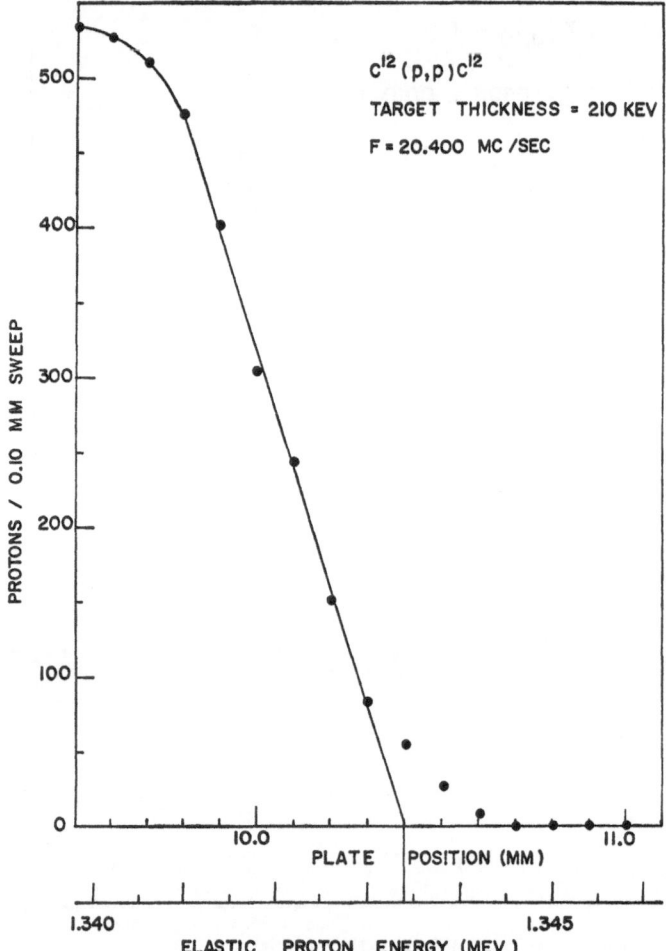

Fig. 7. Representative data obtained on the spectrometer photographic plates for the elastically scattered protons from $C^{12}(p, p)C^{12}$. The energy scale indicates the sensitivity available in determining the energy of these protons

Fig. 8[10]. Again the proton energy was determined by scattering from a C^{12} foil by using the 180° magnetic spectrometer. This resonance was determined to be at 0.8723 ± 0.0005 MeV.

3) *Other measurements below 5.5 MeV.* The gamma ray resonance of the reaction $Al^{27}(p, γ)Si^{28}$ has been of some importance for calibration

purposes. A typical thick target resonance curve is shown in Fig. 9. The resonance was measured to occur at $0.9922 \pm 0.0005\,\mathrm{MeV}$.

An important, and perhaps the most important, neutron threshold below 5.5 MeV is that of the $C^{13}(p, n)N^{13}$ reaction. Fig. 10 shows the strong threshold that is obtained for this reaction. This reaction and the $Li^7(p, n)Be^7$ threshold comprise the best calibration points below 5.5 MeV. It is my opinion that the $C^{13}(p, n)N^{13}$ reaction is a better calibration

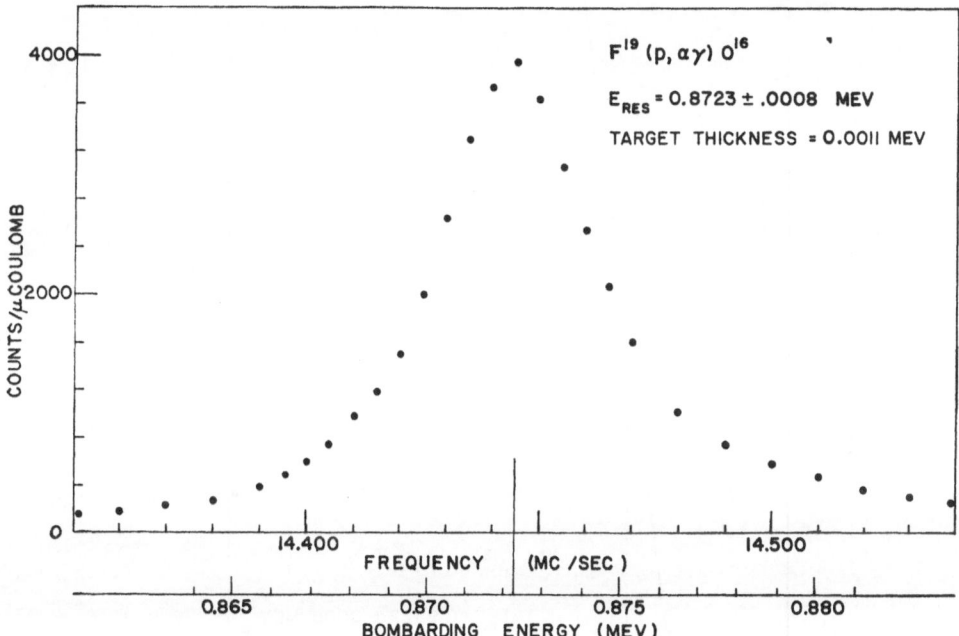

Fig. 8. Gamma ray counting data obtained from $F^{19}(p, \alpha \gamma)O^{16}$, with a 1.13 keV thick LiF target. The energy scale has been offset 0.57 keV so that the center of the resonance corresponds to the true resonant energy

point than $Li^7(p, n)Be^8$ because of the former's freedom from many of the Li^7 target problems. I thus recommend $C^{13}(p, n)N^{13}$ as a better energy calibration point than $Li^7(p, n)Be^7$ for Van de Graaff accelerators. This threshold was measured as $3.2353 \pm 0.0015\,\mathrm{MeV}$ with the Rice 5.5 MeV Van de Graaff accelerator.

B. The Po²¹⁰ particle groups

The alpha particles of Po^{210} have been a laboratory standard of energies for many years. There has been some controversy, however, about the variable results due to differing methods of preparation of sources; nevertheless, it has been shown[10] that these sources are easily prepared in reproducible fashion by various techniques. It is necessary, however, that they be used quickly. Thus for historical reasons, and for practical ones, Po²¹⁰ is an important calibration standard. Fig. 11 shows some of the spectra obtained with the Rice spectrometer for this source. Table 2 shows the results.

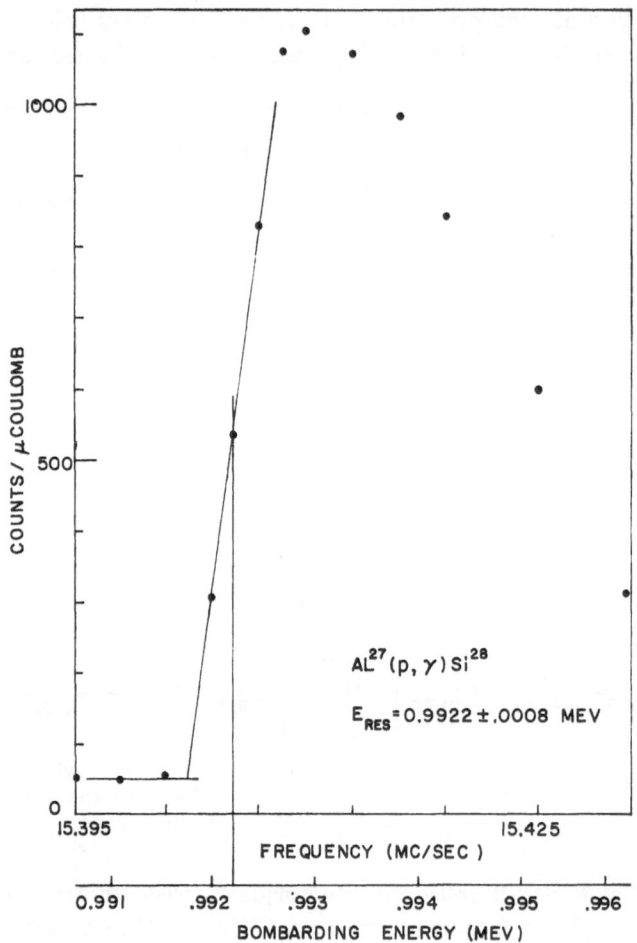

Fig. 9. Gamma ray counting data obtained from $Al^{27}(p, \gamma)Si^{28}$. The assigned resonant energy is the energy found at half-height of the peak since $T \gg \Gamma_{nat}$ where T is the target thickness and Γ is the natural width of the resonance. This experiment also gives an independent determination of the energy resolution of the bombarding protons, and is seen to yield $R \approx 2000$ ($R = E/\Delta E$)

C. Measurements with the Rice tandem Van de Graaff accelerator

1) *The $C^{13}(p, n)N^{13}$ reaction.* This reaction has also been measured with the Rice tandem accelerator[9]. The techniques of the neutron threshold determination have been described above and apply to all those measured

with the tandem accelerator. The beam energy was measured with the same 180° spectrometer and the method was the same as the earlier measurement using the 5.5 MeV accelerator except that the beam was deflected from the neutron target, by a switching magnet, into the spectro-

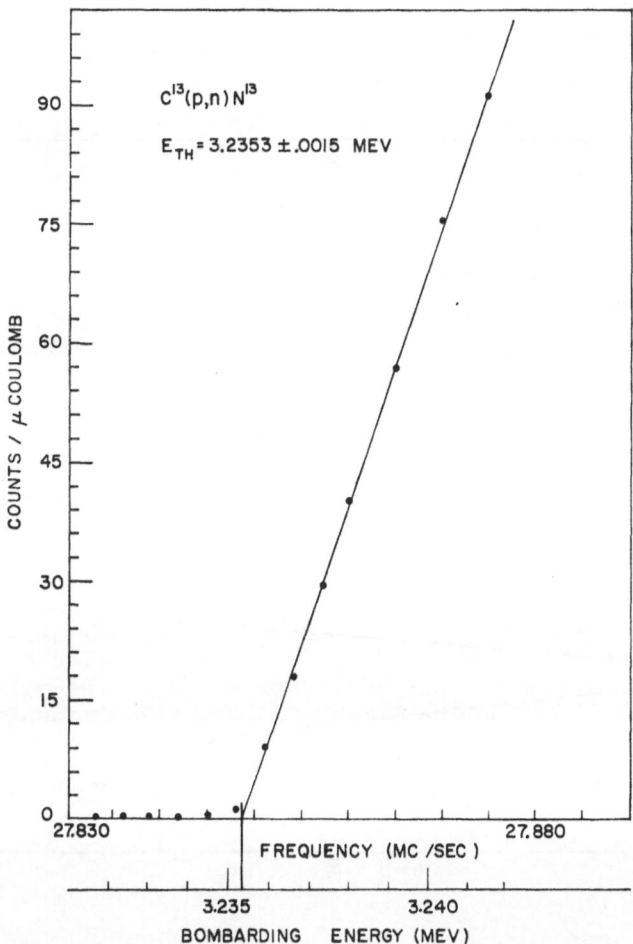

Fig. 10. Neutron threshold for the $C^{13}(p, n)N^{13}$ reaction measured on the Rice 5.5 MeV Van de Graaff accelerator. This reaction is recommended as a primary energy standard

meter. Fig. 12 shows a typical threshold curve. This curve demonstrates the power of the methods described because the excited states of N^{13} can be easily studied with good signal-to-noise.

2) *Higher energy measurements*. The techniques of measuring absolute energies of neutron-thresholds have been applied to various studies at tandem energies[9]:

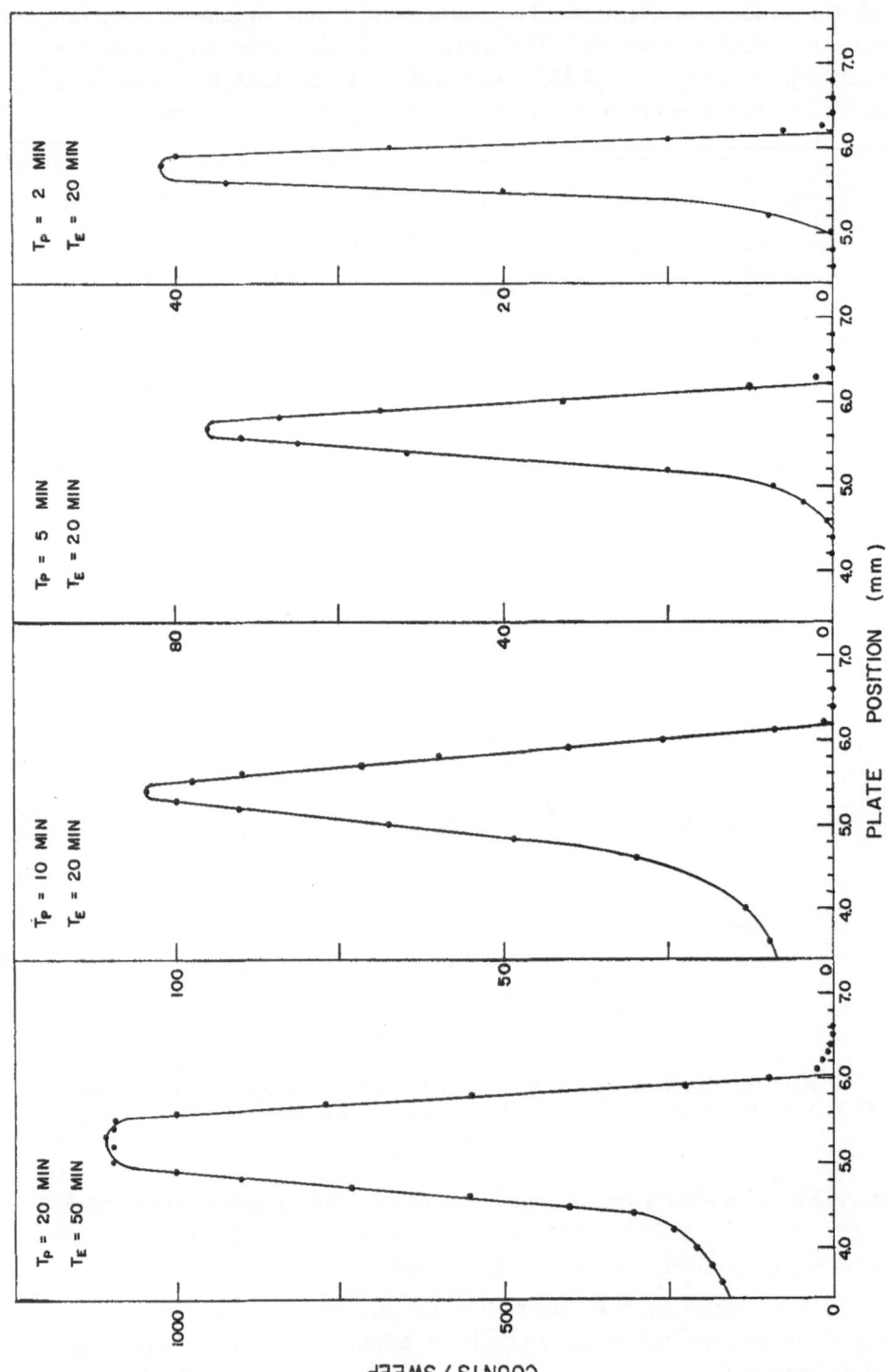

Fig. 11. Alpha-particle counting data obtained from several different sources. T_p is the time which the source backing was allowed to remain in the polonium solution. T_E is the exposure time of the spectrogram

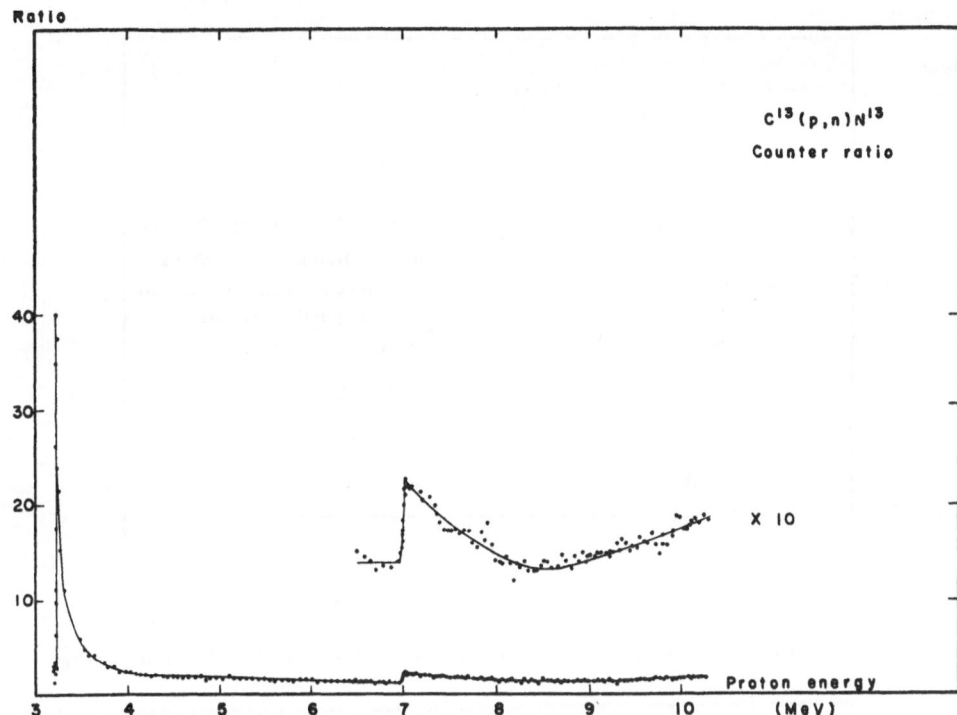

Fig. 12. Neutron threshold for the reaction $C^{13}(p, n)N^{13}$ as measured on the Rice tandem accelerator. The sensitivity of the technique allows the threshold for an excited state in N^{13} to be easily observed

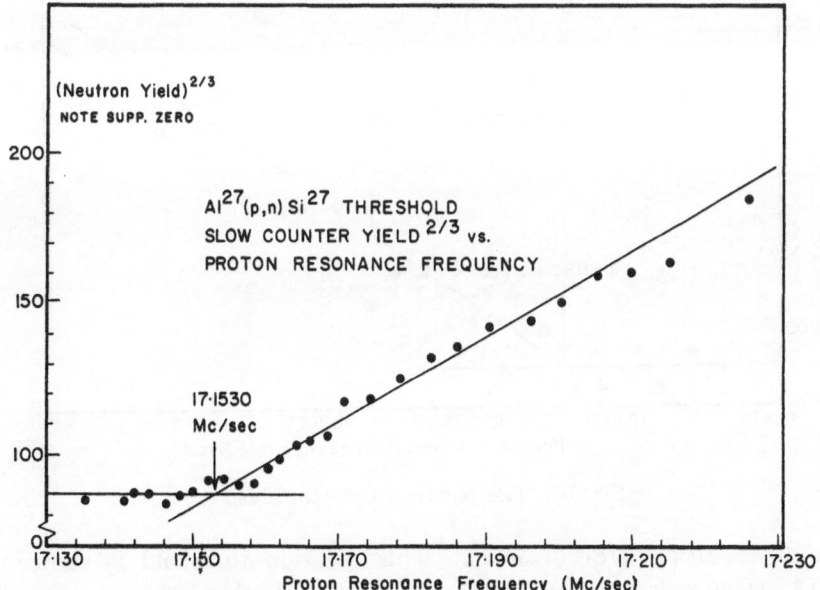

Fig. 13. The $Al^{27}(p, n)Si^{27}$ threshold

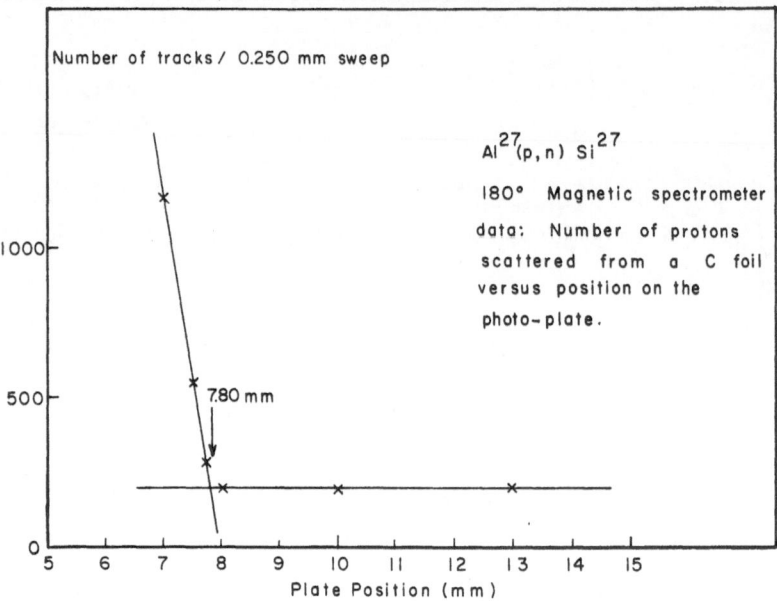

Fig. 14. Spectrometer data for determining the $Al^{27}(p, n)Si^{28}$ threshold energy

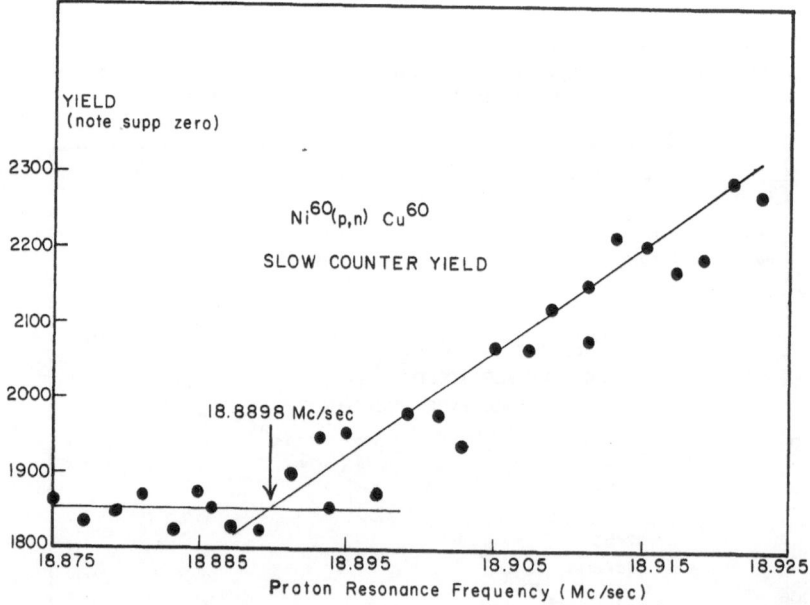

Fig. 15. The $Ni^{60}(p, n)Cu^{60}$ threshold

a) *The $Al^{27}(p, n)Si^{27}$ reaction.* This reaction threshold is shown in Fig. 13. The proton energy was measured with the 180° spectrometer and typical data are shown in Fig. 14. The results are shown in Table 2.

b) *The $Ni^{60}(p, n)Cu^{60}$ reaction.* This threshold has been measured and the results are shown in Fig. 15 and Table 2.

c) *The $Ni^{58}(p, n)Cu^{58}$ reaction.* This reaction provides one of the best calibrations at higher tandem energies. The threshold is very easy to measure if a thin foil is used or if a thick target is used. A typical thick target threshold curve is shown in Fig. 16.

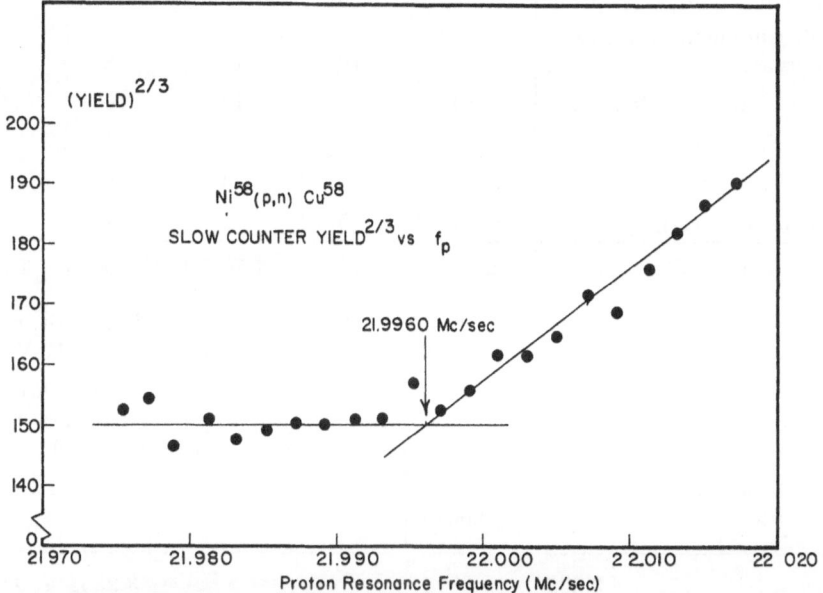

Fig. 16. The $Ni^{58}(p, n)Cu^{58}$ threshold

IV. Summary of Results and Conclusions

A. *Assignment of errors.* A number of sources of error contribute to obtaining an assigned error for each absolute energy measurement. These are summarized in Table 1.

B. *Summary of results.* Table 2 shows the set of absolute calibration energy standards determined at Rice University along with their assigned errors. The primary purpose of these experiments was the performance of a number of calibration measurements under a set of consistent experimental conditions. Most of the measurements had been made previously (some with higher accuracy than that obtained herein), but they had been conducted at a wide variety of laboratories and, consequently, employed various experimental procedures. It was not possible to know if consistent results could be expected to follow from these several measurements. Every possible effort has been made to employ consistent techniques in the present work, and it is believed that the results are as internally consistent as they could be made. Table 2 shows the results

18*

Table 1. A summary of the sources of error and the assigned value of these errors in each of the absolute energy measurements. Each error is given as parts per 10^5

Error Source	$C^{13}(p, n)$	$C^{13}(p, n)$	$Al^{27}(p, n)$	$Ni^{60}(p, n)$	$Ni^{58}(p, n)$
1. Measurement of particle group on photo plate ..	7	6	8	10	7
2. Measurement of radius of curvature	8	10	10	8	8
3. 180° Magnetic field	10	10	10	10	10
4. 90° Magnetic field	4	4	4	4	4
5. Hartree correction	6	5	3	8	9
6. Threshold assignment ..	3	5	12	21	5.5
Sum Error (keV)	2.36	2.43	4.76	7.10	7.75

Table 2. A summary of all Rice University absolute energy measurements with assigned errors. All energies are in MeV

Beckner et al., Phys. Rev. 123, 2100 (1961) (5.5 MeV Van de Graaff)	Rickards, Bonner, and Phillips, Bull. Amer. Phys. Soc. 8, 114 (1963) (Tandem accelerator)
$Li^7(p, n)Be^7$ .... 1.8805 ± 0.0008	
$B^{11}(p, n)C^{11}$ .... 3.0164 ± 0.0015	
$C^{13}(p, n)N^{13}$ .... 3.2353 ± 0.0015	3.2354 ± 0.0024
$F^{19}(p, n)Ne^{19}$ .... 4.2332 ± 0.0020	$Al^{27}(p, n)Si^{27}$ .... 5.7943 ± 0.0047
$F^{19}(p, \alpha \gamma)O^{16}$ .... 0.8723 ± 0.0005	$Ni^{60}(p, n)Cu^{60}$ .... 7.0236 ± 0.0071
$Al^{27}(p, \gamma)Si^{28}$ .... 0.9922 ± 0.0005	$Ni^{58}(p, n)Cu^{58}$ .... 9.5160 ± 0.0078
Po^{210} α-particles .... 5.3023 ± 0.0015	

of all the experiments performed. All the measurements are absolute insofar as the masses of the proton, alpha particle, and C^{12} nucleus are known absolutely. The measurements also require knowledge of the value of the gyromagnetic ratio and the charge-to-mass ratio of the proton, but these constants are known to accuracies of about 1 part to 10^5. The methods employed and the constants used in computing these energies are contained in the Appendix of ref. 10. The error assignments have been made on the basis of actual measurements whenever possible, with consideration given to the number of measurements made and the average deviation present in the results obtained. The error assigned to each reaction measurement has been in all cases larger than the calculated root-mean-square error but smaller than a simple arithmetic sum of the errors present.

V. Acknowledgments

The work summarized in this paper was carried out at Rice University over the last thirteen years in collaboration with a number of my former students and my colleagues. The development of the spectrometer as an absolute energy determining instrument is largely due to the work of Drs. K. F. FAMULARO, C. R. GOSSETT, and R. R. SPENCER. The development of the techniques of neutron and resonance energy determinations are due largely to Dr. E. H. BECKNER while the recent development of the new techniques for the tandem accelerator have been carried out by Mr. B. E. BONNER and Mr. JORGE RICKARDS (on leave from the University of Mexico). I gratefully acknowledge their help.

I also wish to acknowledge the subcommittee on Nuclear Constants of the National Academy of Sciences — National Research Council, and especially subcommittee chairmen, D. M. VAN PATTER and J. B. MARION, for encouragement to perform this work.

References

[1] W. B. LEWIS, and B. V. BOWDEN, Proc. Roy. Soc. (London) A 145, 235 (1934).
[2] H. B. WILLARD, and W. M. PRESTON, Phys. Rev. 81, 480 (1951).
[3] R. O. BONDELID, and C. A. KENNEDY, Phys. Rev. 115, 1601 (1959).
[4] J. D. COCKCROFT, J. Sci. Instr. 10, 71 (1933). — BUECHNER, VAN DE GRAAFF, STRAIT, STERGIOPOULOS, and SPERDUTO, Bull. Amer. Phys. Soc. 23, 30 (1948).
[5] E. KLEMA, and G. C. PHILLIPS, Phys. Rev. 86, 951 (1952). — K. FAMULARO, and G. C. PHILLIPS, Phys. Rev. 91, 1195 (1953). — C. R. GOSSETT, G. C. PHILLIPS, and J. T. EISINGER, Phys. Rev. 98, 724 (1955). — T. E. YOUNG, G. C. PHILLIPS, and R. R. SPENCER, Phys. Rev. 108, 72 (1957).
[6] D. R. HARTREE, Proc. Cambridge Phil. Soc. 21, 746 (1923).
[7] K. FAMULARO, and G. C. PHILLIPS, Phys. Rev. 91, 1195 (1953).
[8] J. B. MARION, and T. W. BONNER, in Fast Neutron Physics, J. B. MARION, and J. L. FOWLER, ed., Part II, p. 1865. New York: Interscience. 1963.
[9] J. RICKARDS, B. E. BONNER, and G. C. PHILLIPS, Bull. Amer. Phys. Soc. 8, 114 (1963).
[10] E. H. BECKNER, R. L. BRAMBLETT, G. C. PHILLIPS, and T. A. EASTWOOD, Phys. Rev. 123, 2100 (1961).

Discussion

H. H. STAUB: Do you also consider the threshold energy of $F^{19}(p, n)$ as a good medium energy standard?

G. C. PHILLIPS: Yes, the $F^{19}(p, n)$ threshold is a good one and it is one of those that we, at Rice, have measured and that we believe one can have considerable confidence in. However, since the $C^{13}(p, n)$ threshold is nearby in energy, and since it has a rather better signal-to-noise than the $F^{19}(p, n)$ I would recommend that the C^{13} threshold be used whenever possible.

R. G. HERB: I didn't get a chance to talk about our work on threshold reactions. Work on the analysis of $Li^7(p, n)Be^7$ threshold curves shows that inclusion of the discrete energy losses may make appreciable shifts. In one case investigated, a 200 eV shift was obtained. The magnitude of the shift depends on the extrapolation method. As you go higher in energy the shifts will undoubtedly be greater in magnitude. I don't know how they will change relatively.

G. C. Phillips: I certainly think that this is a very interesting point that Professor Herb is emphasizing here today about the importance of target preparation, the cleanliness of the target chamber and the potential importance of the Lewis effect in making these determinations. I believe, however, at the present time there is very little evidence that this effect is sufficiently large to change any of the absolute energies, say below 5 MeV. Now as to what it would do above that, I would appreciate perhaps your amplifying why you think it is going to get bigger as one goes to higher bombarding energy.

R. G. Herb: I think the absolute effect will get greater, the relative effect is very hard to guess until you go through the calculations. I think that you will find that as soon as you include the effects of discrete energy losses, your extrapolation methods improve. You will find you can fit the curve better to a straight line. I think you will generally appreciate including this correction rather than ignoring it because it can be included quite accurately.

G. C. Phillips: I think all of us who have done such measurements have always found that if you take the errors from an error table such as I have shown (Table 1) and if you compound the estimates from a root-mean-square viewpoint that it frequently gives a number that can be somewhat smaller, not very much smaller, but somewhat smaller than the sort of deviations that you find in your repeated measurements using different targets and slightly different techniques. For that reason at Rice we have always taken a conservative opinion of our results and included somewhat larger errors than anyone else I believe does for similar measurements. So I think that what Dr. Herb is reporting is certainly important and I believe that we should all take it very seriously and do something about it.

J. Freeman: The Rice measurement for Po^{210} is a little lower than the value recommended earlier today as a "best value". For those who are using the latter as a standard, should one accept the Rice values as they stand or should one consider the possibility of a systematic error and normalize the values appropriately?

G. C. Phillips: The Rice measurements are all absolute measurements and thus they each provide an entirely independent absolute determination: each stands on its own feet, so to say. Yet, since all the Rice measurements were made with techniques as nearly the same as possible then the set of all the determinations provides a relative energy scale as well. I would not care to recommend any change in our values at all. It seems to me that they agree, within quoted errors, with all the other determinations. We can, however, await Professor Marion's recommendation of "best values".

J. B. Marion: I would like to express agreement with the comment that C^{13} should be used for calibration purposes. There are two reasons for this. First, in addition to the $C^{13}(p, n)$ threshold at 3.2 MeV, there is a sharp (p, γ) resonance at 1.75 MeV which is quite useful. Second, the stability of the targets is high. Dr. Bondelid of NRL informs me that he has used C^{13} targets over a period of several years and finds no changes in the values of the measured energies.

H. H. Staub: One would certainly expect that the shift of a resonance or threshold energy due to the Lewis effect would increase with energy since the maximum energy loss increased proportional to the energy of the incident particle.

R. G. Herb: Yes, as I said, the shift should go up in absolute value, percentagewise, I don't know. It it difficult to predict because so far we don't have an analytic formula.

Recommended Values of Calibration Energies

By

Jerry B. Marion*

Department of Physics and Astronomy, University of Maryland, College Park, Maryland, USA

Introduction

Since the time of the last International Conference on Nuclidic Masses, there has been considerable activity in several different phases of precision beam energy measurements. First, several new absolute measurements have been made of the energies of standard calibration points, two of these using an altogether new technique. Second, the beams from tandem accelerators have been used to establish reference energies up to 14.6 MeV. Finally, detailed investigations have been made concerning the influence of energy loss effects and of the structure of target materials on the determination of resonance and threshold energies.

In this report, rather than attempt to summarize all of the resonance and threshold energies that have been used as calibration points, I shall confine myself to a discussion of only the most important reactions. In the low energy region these are, of course, the $Al^{27} (p, \gamma)$ resonance at 0.99 MeV and the $Li^{7}(p, n)$ threshold at 1.88 MeV. In addition, we now have sufficient information to be able to list recommended values of calibration points for proton energies of approximately 3, 4, 6, 7, and 9 MeV.

Procedures for the Treatment of Data

During the last four or five years, experimenters have become increasingly aware of the subtleties that are involved in performing precision energy measurements. Also, better equipment has become available and techniques have improved significantly. These facts coupled with the large number (relatively speaking!) of experiments that have been performed recently, make it reasonable that in determining recommended values for the primary calibration energies only results should be used which are of recent origin. The measurements which I

---

* Supported in part by the U. S. Atomic Energy Commission.

have used in the summary of the Al^{27} and Li^7 energies have all been
published since 1959. (There is one exception, which I shall mention
later.) In the tabulations of the higher energy (and less precisely known)
points, I have included some of the older measurements for purposes of
comparison; these results are, however, not given appreciable weights in
arriving at recommended values.

In analyzing the Al^{27} and Li^7 measurements, I have used the
procedure outlined in the 1961 review article[1]. That is, the means have
been calculated by giving to the individual results weights which are
inversely proportional to the stated uncertainties; the uncertainty in the
mean is taken to be the larger of the internal or external error. For the
remaining measurements, the recommended values have been arbitrarily
selected, but by giving the greatest weights to recent and to absolute
determinations.

It is apparent that these procedures involve a great deal of arbitrariness.
However, I believe that the present state of affairs does not permit a
completely objective analysis to be made. For example, there is considerable
variation in the methods by which various experimenters arrive at their
uncertainties. A set of measurements of the same quantity made at different
laboratories therefore does not constitute a statistical sample and there
is then no reason to choose to weight the results by the inverse squares
of the stated uncertainties. In constructing a mass table, for example,
there are mathematical reasons for adopting such a weighting procedure,
but such a choice is, in itself, arbitrary, and it is made even more so by
the assignment of factors by which the stated uncertainties of the
measurements are multiplied. When one deals with a small set of
measurements it seems more reasonable to adopt a procedure that does
not eliminate otherwise good measurements which have been analyzed in
such a way as to produce a rather generous uncertainty. The $(\Delta E)^{-1}$
weighting is therefore a compromise between a $(\Delta E)^{-2}$ weighting and no
weighting at all. (An unweighted average has, in fact, been used for some
of the secondary calibration points.)

I have emphasized this point concerning weighting procedures since
I believe that it is good to be reminded that the rules of statistics do not
strictly apply to small sets of data from different laboratories. In practice,
however, the distinction is not very important—at least, not with the
measurements presently available. For example, the mean value for the
$Li^7(p, n)$ threshold energy would be raised by only 80 eV if the inverse-
square weighting were used, and would be lowered by the same amount
for no weighting at all.

I shall therefore assume that my procedures are either acceptable
or that they can be tolerated since they do not significantly influence the
results and shall proceed to a discussion of the various measurements.

The $Al^{27}(p, \gamma)$ Resonance

Table 1 shows the four determinations of the Al^{27} (p, γ) resonance energy which have been used to obtain the recommended value of 991.82 ± 0.10 keV*. These data are all from recent, absolute measurements. The agreement is exceptionally good; the greatest deviation from the mean is 350 eV. The older, absolute measurements from Wisconsin[6] and from Associated Electrical Industries[7] have not been included; both of these values are near 993.5 keV and are unrealistically high. Perhaps the target conditions and/or the systematic effects peculiar to electrostatic analyzers were not as carefully controlled as in the more recent experiments. The Zürich measurement[3] which is listed is taken to supercede an older value[8] of 991.1 ± 0.2 keV which was obtained before their improved equipment was installed. The listed NRL value[4] replaces a previously published[9] result of 992.4 ± 0.5 keV; the new value was obtained from a more sophisticated analysis of the detailed shape of the resonance curve measured under carefully controlled conditions.

Table 1. $Al^{27}(p, \gamma)Si^{28}$ Resonance Energies

Laboratory	Resonance Energy (keV)	Method	Reference
Rice (1961)	992.2 ± 0.5	Absolute magnetic	2
Zürich (1962)...........	991.83 ± 0.10	Absolute magnetic	3
Naval Research Lab (1963)	991.9 ± 0.3	Absolute electrostatic	4
Wisconsin (1963)	991.6 ± 0.2	Absolute velocity	5

Recommended value (weighted mean): 991.82 ± 0.10 keV.

The most recent addition to the list, from Wisconsin[5], deserves special mention since it was obtained by using a unique device. Essentially, the method consists of modulating the beam from the accelerator and then passing the bursts through two pick-up coils separated by a known distance. The measurement of the time of flight between the pick-up coils can be converted to a frequency measurement and can therefore be made with high precision. The great advantage of a velocity measurement of this type is that it is not subject to some of the bothersome effects that plague users of electrostatic and magnetic analyzers—end effects, stray fields, plate charging, non-uniform magnetic fields, etc. This instrument has also been used to determine the $Li^7(p, n)$ threshold energy, and it is to be hoped that the Wisconsin group will extend their measurements to other calibration points as well.

---

* This represents a decrease of 180 eV from the previously recommended value[1].

The Li⁷(p, n) Threshold

Table·2 shows the seven values of the Li⁷(p, n) threshold energy which have been used to obtain the recommended value of 1880.36 ± 0.22 keV*. Three measurements have been eliminated from the list: the 1949 Wisconsin result[6] of 1882.2 ± 1.9 keV, obtained with an electrostatic analyzer; the 1949 Westinghouse velocity measurement[15] which gave 1881.2 ± 1.9 keV; and the 1951 Wisconsin measurement[16] which compared the threshold energy with the energy of the α particles from RaC′.

Table 2. Li⁷(p, n)Be⁷ Threshold Energies

Laboratory	Threshold Energy (keV)	Method	Reference
Wisconsin (1954)........	1879.4 ± 1.0	Electrostatic; comparison of threshold energy with energy of Mg²⁴*; Chalk River value used for Mg²⁴*	10 11
Naval Research Lab (1959)	1881.2 ± 0.9	Absolute electrostatic analysis of beam	9
Notre Dame (1960)	1880.8 ± 1.0	Magnetic; comparison of threshold energy with Po α energy; Rytz's value used for Po α	12 14
Zürich (1960)..........	1880.3 ± 0.5	Absolute magnetic analysis of beam	13
Rice (1961)	1880.5 ± 0.8	Absolute magnetic analysis of scattered protons	2
Zürich (1961)..........	1880.48 ± 0.25	Absolute magnetic analysis of beam	14
Wisconsin (1963)........	1879.8 ± 0.6	Absolute velocity measurement of beam	5

Recommended value (weighted mean): 1880.36 ± 0.22 keV.

The first value on the list is from the only relatively old measurement[10] retained in the summary of primary calibration points. In this experiment a comparison was made of the threshold energy with the energy of the first excited state of Mg²⁴ which was reached by inelastic scattering. The energy of Mg²⁴* was, at that time, known best in terms of the absolute measurements of the energies of the Au¹⁹⁸ and RaC′ γ rays which were related to the Mg²⁴ energy via measurements of the Co⁶⁰ γ rays. There was a difference of 1.7 keV in the threshold energy, depending upon whether the gold or the radium γ-ray energy was used. But now

---

* This result is 340 eV lower than the previously recommended value[1].

there is a quite precise result from Chalk River[11] which gives the Mg^{24} energy directly in terms of the accurately known Au^{198} γ ray. Thus, a single value can now be given for the Wisconsin threshold measurement, and this result turns out to be slightly lower than either of the two previously listed values[10]. It should be noted that of the values listed in Table 2 this measurement is the only one in which a linear extrapolation of the neutron yield was used rather than the now-recommended $(yield)^{2/3}$ method[1]. However, in attempting to apply the $(yield)^{2/3}$ procedure to these data, I was unable to shift the intercept by more than 100 eV. Since the stated uncertainty in this measurement is 1.0 keV, it was decided to retain this result, unaltered, in the list.

The Notre Dame group[12] has measured the ratio of the energy of Po α particles to the $Li^{7}(p, n)$ threshold energy. This result has been coupled with the value of 5304.79 ± 0.4 keV for Po α, as recommended by RYTZ[14], to obtain a threshold value of 1880.8 ± 1.0 keV.

The first absolute determination of the $Li^{7}(p, n)$ threshold energy[6] yielded a result of 1882.2 keV. Since that time new measurements have steadily driven the mean value downwards until it now stands almost 2 keV lower than the original figure. This trend is also apparent in the case of the $Al^{27}(p, \gamma)$ resonance. One might be inclined to say that we are slowly learning how to prevent pump oil from covering the surfaces of our targets. The real reason (as yet unknown) is probably more subtle than this, however, since the effects of carbon buildup on targets has been known for some time and everyone now takes precautions against this type of systematic error. On the other hand, the detailed structure of even carbon-free targets is now known[4, 17] to influence the positions of resonances and thresholds. These effects, due to residual dirt, oxides, and even the crystalline structure of the target material, become significant as higher and higher precision is sought.

Professor HERB has reported in this session on the work at Wisconsin regarding the effects of discrete energy loss in targets and the problems of target contaminants. In view of these newly studied effects, it is probable that the uncertainties which have been assigned to results of measurements of the primary calibration energies should be increased somewhat. It seems unlikely, however, that corrections for such effects will shift either of the weighted mean values which I have stated by amounts greater than the assigned uncertainties. Because of the desire to continually improve the precision of energy measurements, it is hoped that further experiments will be undertaken which will be analyzed in such a way as to take account of these small but important effects.

I would also like to draw attention to the fact that in the set of $Al^{27}(p, \gamma)$ resonance measurements and among the *direct* determinations of the $Li^{7}(p, n)$ threshold energy, the absolute velocity experiments have produced the lowest of the reported values. This point may not be significant, but certainly an effort should be made to determine whether velocity measurements tend to yield systematically low energy values. An investigation of this type is to be made at the University of Maryland in the near future.

(p, n) Thresholds above 3 MeV

Table 3 shows the available data (since 1955) for six (p, n) reactions which have thresholds in the range from 3.2 to 9.5 MeV. It is fortunate that the energy spacing of these thresholds is such as to provide a useful range of calibration points up to 10 MeV.

Table 3. Higher-Energy (p, n) Thresholds

Reaction	Threshold Energy (keV)	Calibration	Reference	Recommended Value (keV)
$C^{13}(p, n)$	3237.2 ± 1.6	Absolute electrostatic (NRL)	9	
	3235.3 ± 1.5	Absolute magnetic (Rice)...	2	$3236.0 \pm 1.0*$
	3235.4 ± 2.4	Absolute magnetic (Rice)...	18	
$F^{19}(p, n)$	$4240. \pm 8$	Several reactions (ORNL)..	19	
	$4235. \pm 5$	Several reactions (Rice)....	20	
	$4240. \pm 5$	Several reactions (Rice)....	21	$4234.2 \pm 1.5**$
	$4227. \pm 6$	Several reactions (CRL)....	22	
	4233.2 ± 2.0	Absolute magnetic (Rice)...	2	
	4234.7 ± 1.0	Absolute magnetic (Zürich).	23	
$Al^{27}(p, n)$	$5792. \pm 10$	Several reactions (ORNL)..	19	
	$5798. \pm 5$	Several reactions (Rice)....	20	
	$5803. \pm 4$	ThC α (6089.7 keV) (AERE)	24	$5798.6 \pm 5*$
	5794.3 ± 4.7	Absolute magnetic (Rice)...	18	
	5806.0 ± 6.3	Several reactions (Japan)..	25	
$Ni^{60}(p, n)$	$7028. \pm 20$	Several reactions (CRL)....	26	$7024. \pm 7**$
	7023.6 ± 7.1	Absolute magnetic (Rice)...	18	
$Fe^{54}(p, n)$	$9203. \pm 5$	ThC' α (8786.4 keV) (AERE)	24	9203 ± 5
$Ni^{58}(p, n)$	$9459. \pm 70$	Several reactions (CRL)....	26	
	9516.0 ± 7.8	Absolute magnetic (Rice)...	18	$9514. \pm 5**$
	$9513. \pm 5$	ThC' α (8786.4 keV) (AERE)	24	

* On the basis of the unweighted average; uncertainty arbitrarily chosen.
** Arbitrarily chosen.

Several of the older measurements, which were made on a relative basis, are included for comparison. For the $F^{19}(p, n)$ and $Al^{27}(p, n)$ reactions it is quite comforting to note the excellent agreement between the relative and the absolute measurements.

Hopefully, we can look forward to more absolute measurements in this energy region so that these, and perhaps other thresholds, can be firmly established as calibration energies.

Helium-Ion Measurements

With tandem accelerators now operating above 10 MeV and in the near future probably up to 30 MeV, it becomes increasingly important to establish some accurately known resonance and threshold energies employing helium ions. A known (α, n) threshold which occurs at 7 MeV, for example, will, if measured with He^{4+} ions, provide a calibration point for a magnetic analyzer at an equivalent proton energy of 28 MeV. Unfortunately, too little attention has been given to precision measurements of α-particle energies and the available data is meager. The same is true of He^3-induced reactions, due in part to the fact that (He^3, n) thresholds are the only processes that lead to effects that are well-defined in terms of the bombarding energy. In spite of the obvious importance of helium-ion reactions, I am aware of the publication of only two precision, absolute measurements of (He^3, n) threshold energies and of only one such experiment concerning α-particle resonance energies. There exists a list of at least ten candidates for calibration points which involve helium ions as the inducing agent. Table 4 shows this list. In this table I have included only one result for each of the points; in some cases this is the most precise value that is available, in the rest it is the *only* value that is available.

Table 4. Energies of Helium-Ion Resonances and Thresholds

Reaction	Energy (keV)	Reference
$C^{12}(He^3, n)$	$E_{th} = 1436.2 \pm 0.9$	29
$Li^6(\alpha, \gamma)$	$E_R = 2605;\ \Gamma \leqslant 1.5$	30, 31
$C^{13}(\alpha, n)$	$E_R = 2800 \pm 3;\ \Gamma \sim 4$	32
$Li^6(He^3, n)$	$E_{th} = 2966.1 \pm 1.7$	33
$Mg^{24}(\alpha, \gamma)$	$E_R = 3199.8 \pm 1.0;\ \Gamma = 1.8 \pm 0.3$	34
$Na^{23}(\alpha, n)$	$E_R = 3492 \pm 3;\ \Gamma < 1$	32
$Li^7(\alpha, n)$	$E_{th} = 4379 \pm 6$	35
$Li^6(\alpha, n)$	$E_{th} = 6630 \pm 20$	36
$C^{12}(\alpha, n)$	$E_{th} = 11341 \pm 15$	37
$C^{12}(\alpha, n)$	$E_R = 14600 \pm 40;\ \Gamma = 220 \pm 60$	37

Clearly, if the energies of these resonances and thresholds were all established with precision, this would be an impressive list of useful calibration points. I can only implore those of you who have suitable instruments and appropriate accelerators to pursue these measurements. Perhaps by the time of the next International Conference a rather complete list of energies suitable for calibration purposes will be available covering the energy range up to equivalent proton energies of 26 MeV with $Li^6(\alpha, n)$ or even up to 45 MeV with $C^{12}(\alpha, n)$.

Conclusion

In summary, there now seems to be an increasing tendency for agreement among the various laboratories for reaction energies that involve protons of less than 2 MeV, and even up to 9 MeV the energy scale is in reasonably acceptable condition. This is not to say that further work is unnecessary, however, since increased precision is useful not only for its own sake but also for the fact that it almost always uncovers some new, interesting and important effect—for example, the so-called Lewis effect.

On the other hand, there is a definite need for increased effort in the area of helium-ion reactions. These reactions serve not only to provide calibration points for high energy protons, but (α, n) threshold measurements, for example, are probably the most precise method currently available for determining mass differences over a gap of three mass units.

The high rigidity of helium ions in magnetic analyzers allows the possibility of some interesting inter-comparisons of energy standards. For example, the $C^{12}(He^3, n)$ and $F^{19}(p, n)$ thresholds require magnetic field settings that differ by only about 60 keV in equivalent proton energy if the bombarding particles are He^{3+} ions. Also, if doubly ionized helium is used, then the field settings differ by only about 80 keV between the $C^{12}(He^3, n)$ threshold and the $Al^{27}(p, \gamma)$ resonance. Thus, a precise comparison could be made between $Al^{27}(p, \gamma)$ and $F^{19}(p, n)$ even without an absolute energy measuring device. Also it might be easier to compare the $Fe^{54}(p, n)$ threshold with that for the $Li^6(He^3, n)$ reaction than with a radioactive α-particle energy as has been done. In this case the field settings are only about 360 keV apart. For He^{3++} ions the $Li^6(He^3, n)$ threshold occurs at a field difference of only 340 keV from that for the $Li^7(p, n)$ threshold. Other similar possibilities exist and it would seem important that some of these checks on the internal consistency of the energy scale be made. The use of helium ions in two different charge states to alter the magnetic rigidity is subject to considerably less uncertainty than measurements made with atomic and molecular hydrogen beams.

References

[1] J. B. MARION, Revs. Modern Phys. **33**, 139 (1961).

[2] E. H. BECKNER, R. L. BRAMBLETT, G. C. PHILLIPS, and T. A. EASTWOOD, Phys. Rev. **123**, 2100 (1961).

[3] A. RYTZ, H. H. STAUB, and W. ZYCH, Helv. Phys. Acta **35**, 341 (1962).

[4] R. O. BONDELID and J. W. BUTLER, Phys. Rev. **130**, 1078 (1963).

[5] B. R. GASTEN, to be published.

[6] R. G. HERB, S. C. SNOWDEN, and O. SALA, Phys. Rev. **75**, 246 (1949).

[7] S. E. HUNT, R. A. POPE, D. V. FRECK, and W. W. EVANS, Phys. Rev. **120**, 1740 (1960).

[8] H. H. STAUB and H. WINKLER, Nuclear Phys. 17, 271 (1960).

[9] R. O. BONDELID and C. A. KENNEDY, Phys. Rev. 115, 1601 (1959).

[10] K. W. JONES, R. A. DOUGLAS, M. T. McELLISTREM, and H. T. RICHARDS, Phys. Rev. 94, 947 (1954).

[11] G. MURRAY, R. L. GRAHAM, and J. S. GEIGER, Bull. Amer. Phys. Soc. 7, 72 (1962), and private communication.

[12] C. P. BROWNE, Paper No. 2 of this session.

[13] H. H. STAUB and H. WINKLER, Helv. Phys. Acta 33, 526 (1960).

[14] A. RYTZ, Paper No. 1 of this session.

[15] W. E. SHOUPP, B. JENNINGS, and W. JONES, Phys. Rev. 76, 502 (1949).

[16] W. J. STURM and V. JOHNSON, Phys. Rev. 83, 542 (1951).

[17] D. W. PALMER, J. G. SKOFRONICK, D. G. COSTELLO, A. L. MORSELL, W. E. KANE, and R. G. HERB, Phys. Rev. 130, 1153 (1963).

[18] G. C. PHILLIPS, J. RICKARDS, and B. E. BONNER, private communication.

[19] J. D. KINGTON, J. K. BAIR, H. O. COHN, and H. B. WILLARD, Phys. Rev. 99, 1393 (1955).

[20] J. B. MARION, T. W. BONNER, and C. F. COOK, Phys. Rev. 100, 91 (1955). The value for $Al^{27}(p, n)$ has been corrected and differs slightly from the published result.

[21] R. A. CHAPMAN and H. BICHSEL, unpublished results (1957).

[22] D. A. BROMLEY, A. J. FERGUSON, H. E. GOVE, J. A. KUEHNER, A. E. LITHERLAND, A. ALMQVIST, and R. BATCHELOR, Canad. Journ. Physics 37, 1514 (1959).

[23] A. RYTZ, H. WINKLER, F. ZAMBONI, and W. ZYCH, Helv. Phys. Acta 34, 819 (1961).

[24] J. M. FREEMAN, R. E. WHITE, J. H. MONTAGUE, G. MURRAY, and W. E. BURCHAM, private communication of preliminary results.

[25] K. OKANO and K. NISHIMURA, Journ. Phys. Soc. Japan 18, 1563 (1963).

[26] H. E. GOVE, J. A. KUEHNER, A. E. LITHERLAND, E. ALMQVIST, D. A. BROMLEY, A. J. FERGUSON, P. H. ROSE, R. P. BASTIDE, N. BROOKS, and R. J. CONNOR, Phys. Rev. Letters 1, 251 (1958).

[27] A. RYTZ, H. H. STAUB, H. WINKLER, and F. ZAMBONI, Nuclear Phys. 43, 229 (1963).

[28] J. W. NELSON, E. B. CARTER, G. E. MITCHELL, and R. H. DAVIS, Phys. Rev. 129, 1723 (1963).

[29] J. W. BUTLER and R. O. BONDELID, Phys. Rev. 121, 1770 (1961).

[30] L. MEYER-SCHÜTZMEISTER, and S. S. HANNA, Phys. Rev. 108, 1506 (1957).

[31] W. E. MEYERHOF and L. F. CHASE, Jr., Phys. Rev. 111, 1348 (1958).

[32] R. M. WILLIAMSON, T. KATMAN, and B. S. BURTON, Phys. Rev. 117, 1325 (1960).

[33] K. L. DUNNING, J. W. BUTLER, and R. O. BONDELID, Phys. Rev. 110, 1076 (1958).

[34] A. RYTZ, H. H. STAUB, H. WINKLER, and F. ZAMBONI, Nuclear Phys. 43, 229 (1963).

[35] H. BICHSEL and T. W. BONNER, Phys. Rev. 108, 1025 (1957).

[36] M. K. MEHTA, W. E. HUNT, H. S. PLENDL, and R. H. DAVIS, Bull. Amer. Phys. Soc. 6, 226 (1961).

[37] J. W. NELSON, E. B. CARTER, G. E. MITCHELL, and R. H. DAVIS, Phys. Rev. 129, 1723 (1963).

Discussion

J. Mattauch: I would be very much interested in what Dr. Breitenberger or Dr. Cohen would say to the statement that weighting $1/E$ is intermediate between no weighting and the weighting according to $1/E^2$. I am not clear about how the error of the average is computed and what it means.

E. R. Cohen: I think that one cannot really argue with the statement that $1/E$ weighting is intermediate between unit weight and $1/E^2$ weight. The question is what is the significance of this. The important aspect of the $1/E^2$ weight is that it is the only way that you can really make statistical analysis between internal and external consistency. If you use the $1/E$ weight or any other weight, you can only calculate the external consistency measure, you do not have χ^2 tests or other statistical tests to make. In addition, the $1/E^2$ weight is, in the strictest statistical meaning of the word, the most efficient weighting that you can use. It allows you to assign the smallest error to the average you then obtain, but it certainly requires that the variations of the data from the true value are due only to the statistical fluctuations that are properly represented by the weighting factors that you have assigned to them. In such a case as we have here, where there is quite an uncertainty as to just what the statistics mean, the $1/E$ weighting is not as bad as it might sound.

C. P. Browne: I would like to comment just once more on the statement about the extrapolation for the Li(p, n) threshold, whether one uses a linear or a $2/3$ power extrapolation. I think Professor Herb could do this better than I since I want to point out results of Palmer and Herb at Wisconsin. Using finite energy loss and the resolution of the analyzer and the distance above threshold over which one extrapolated as variables, they calculated sets of curves showing how far from the true threshold you will be under various conditions. We tried to use these curves with our data and got remarkable agreement. In going from a linear to a $2/3$ power extrapolation, we got a total shift of 0.4 keV. In our particular case one was 0.2 keV above and the one 0.2 keV below the true threshold. If you change the resolution however, this does not change in a simple fashion. In view of these rather large differences, I believe the uncertainty of ± 0.2 keV on the recommended threshold value is optimistic.

H. H. Staub: I think it is a comforting thought that it makes so little difference whether one takes a linear or a $2/3$ power extrapolation, although from a purely scientific point of view it really should be a $2/3$ power extrapolation.

Q-Value Measurements at M. I. T.*

By

A. Sperduto and W. W. Buechner

Department of Physics and Laboratory for Nuclear Science,
Massachusetts Institute of Technology, Cambridge 39, Massachusetts

(Presented by A. SPERDUTO)

With 7 Figures

The original planning of the program for making precision energy measurements of nuclear reaction products at M. I. T. can perhaps be traced back to 1939. At that time plans were made for the construction of a magnetic spectrograph of the type built by COCKCROFT[1] and used by RUTHERFORD and his collaborators[2] in their early measurements on the deflection of radioactive alpha emitters. Two one-ton cylindrical sections of Armco iron were procured in 1940 but, because of the interruption brought about by the Second World War, the design and construction of the 180-degree focusing magnet were delayed until 1945.

In 1947, the first measurements of reaction Q-values were made at the High Voltage Laboratory, and at the same time we initiated a program primarily directed toward measuring energy levels in light nuclei[3].

During the past 16 years we have used fundamentally the same method in making these measurements; that is, the method of employing a uniform magnetic field for deflecting charged reaction particles and of measuring the magnetic field and trajectory radius, thereby obtaining a measure of the magnetic rigidity, or $B\varrho$ of the particle. The principal measurements involve the accurate determination of both the energy of the incident bombarding particles and the emitted reaction particles. The incident energy is usually directly determined by measuring the elastically scattered particles at a particular angle from the nucleus under study. Using the reaction kinematics involving the laws of conservation of energy and momentum, together with these energy measurements, leads to a

---

* This work has been supported in part through an AEC contract, with funds provided by the U. S. Atomic Energy Commission, by the Office of Naval Research, and by the Air Force Office of Scientific Research.

determination of the mass difference between target and residual nucleus; the so-called *Q-Value.*

The M. I. T. measurements over the past 16 years were made with two different particle accelerators and three different uniform-field magnetic spectrographs. In some instances the same reaction was studied with different combinations of accelerator and spectrograph. From 1947 to 1951, the source of high-energy protons and deuterons was provided by the original 35-foot high, 2.5 MeV air-insulated Van de Graaff generator[4]. From 1951 to the present, the pressure-insulated ONR accelerator[5] has provided nearly mono-energetic beams of protons, deuterons, and alpha particles with energies ranging up to 8.5 MeV.

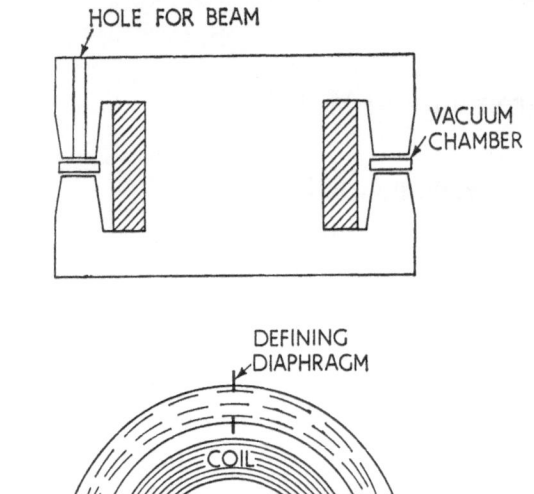

Fig. 1. Schematic drawing of 180-degree uniform-field magnetic spectrograph

Only the Cockcroft type uniform-field spectrograph was used with both accelerators. Fig. 1 is a schematic diagram showing the 180-degree focusing properties of this analyzer. Note that both the target and detector are located within the region of the field. Only the incident particles are affected by the fringing field on entering the target region through a slot normal to the gap. Scattered and reaction particles remain within the uniform annular field region for those deflections focused on the detector. The M. I. T. magnet has a mean radius of 35 cm, a $^1/_2$ in. gap, a useable width in the annular region of about 5 cm, and an energy resolution $E/\Delta E$ of approximately 1000. The nuclear emulsion detector covers an energy range of about 10%. At present, two broad-range magnetic spectrographs are in routine use with the 8.5 MeV ONR accelerator; the single-gap magnet since 1954, and the multiple-gap instrument since 1961.

Fig. 2 shows a schematic drawing illustrating the main features of a broad-range spectrograph, as first proposed by Bainbridge in 1947[6]. The uniform magnetic field is bound by circular entrance and exit faces. The chief advantages of this type of spectrograph are (1) that both the object and image points (i. e., target and detector) are located outside the region of the magnetic field; and (2) that simultaneous recording of particle groups with a range of energy of approximately 2.4 can be realized. In the M. I. T. design[7], particles whose radii of curvature in the uniform field range between 36.6 and 56.6 cm are focused at different points along a hyperbolic surface 75 cm in length. The resolution attainable in practice is between 1000 and 1500.

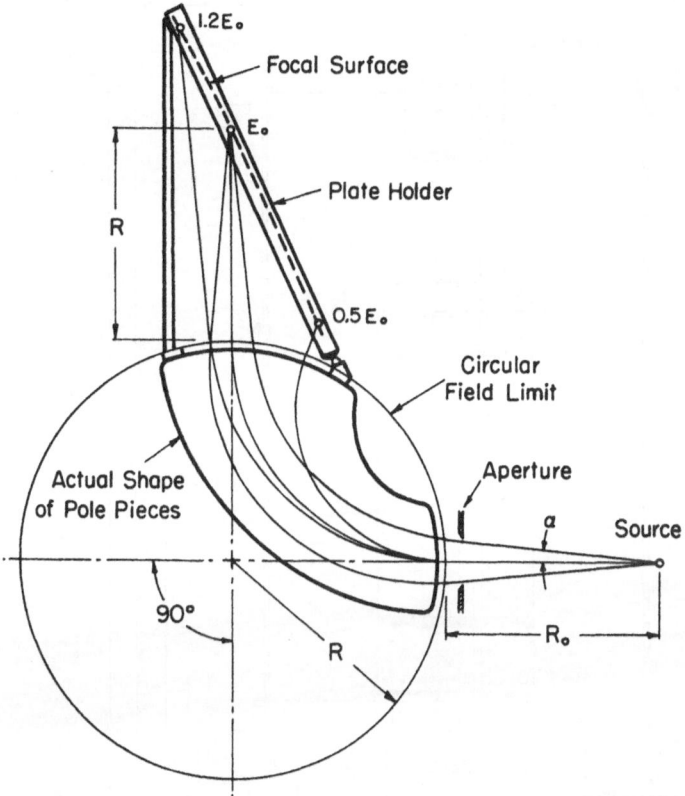

Fig. 2. Geometry of the broad-range magnetic spectrograph

The pole face and focal surface geometries for the M. I. T. multiple-gap spectrograph[8] are identical for all gaps and are also very nearly the same as for our single-gap spectrograph. Fig. 3 shows the present experimental arrangement at M. I. T. Here are shown schematically and to scale the ONR accelerator, the beam-analyzing magnet, and the tank enclosing the multiple-gap spectrograph. The incident beam from the accelerator is directed through a hole in the toroid and focused by an electrostatic lens onto a target in the center of the tank.

Fig. 4 shows in more detail a cross-sectional view of the magnet structure and the vacuum tank, which is independently supported. The only mechanical linkages between the vacuum vessel and magnet base

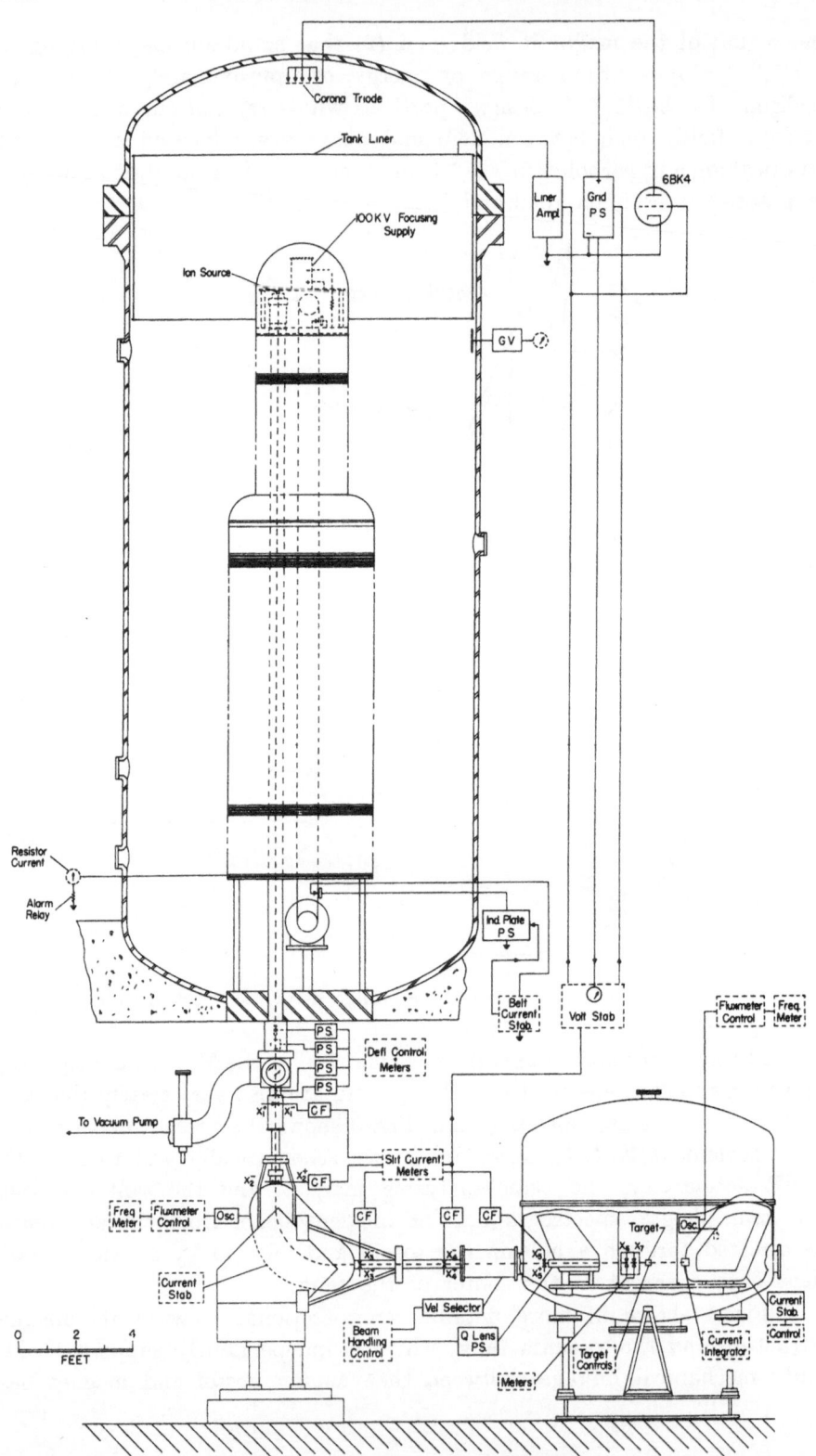

Fig. 3. Present experimental arrangement at M. I. T. showing 8.5 MeV ONR accelerator and multiple-gap spectrograph

are sylphon bellows, used to provide the shock proof and vacuum-tight seal between the tank and each of the six magnet supporting legs. The trajectories of particles corresponding to minimum and maximum energies

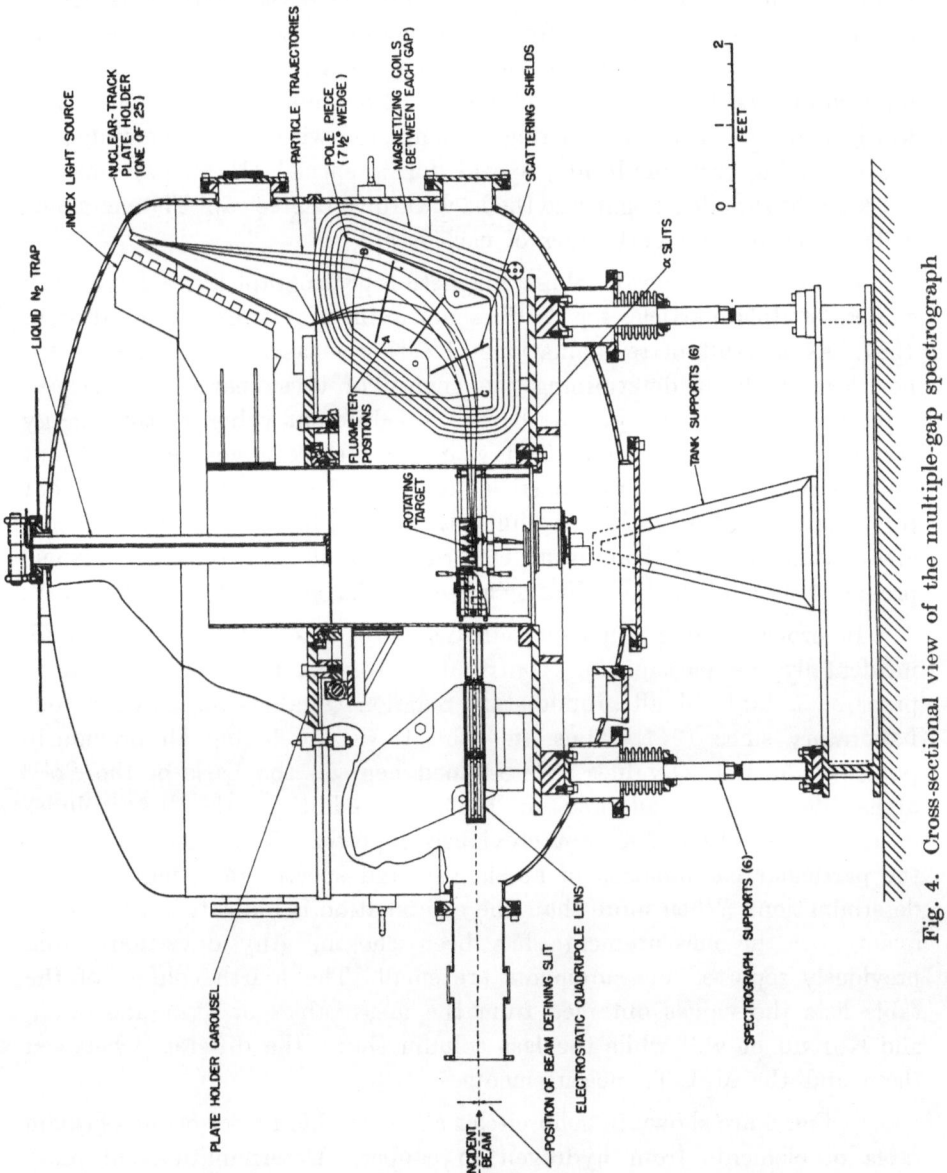

Fig. 4. Cross-sectional view of the multiple-gap spectrograph

for a given field setting are shown focused on the nuclear emulsion detector surface. A cylindrical section of pipe 2 feet in diameter encloses the target region, with slots provided opposite the entrance pole faces to permit reaction particles to enter the gaps. This cylindrical pipe also serves to support the holders for the nuclear track plates on a rotatable platform mounted on a large bearing.

Fig. 5 shows an enlarged plan view of the target region. The incident beam entering from the left is focused by the electrostatic lens and is collimated by a slit system before impinging on a target. The gaps, $^3/_8$ in. wide, are spaced at 7.5-degree intervals from zero to 90 degrees in the forward quadrant and between 90 and 172.5 degrees in the opposite backward quadrant. The toroidal ring is completed with solid iron wedges in the remaining two quadrants, except for a special $1^1/_2$ in. gap halfway between the zero-degree gap and the backward 90-degree gap. The energizing coils are located on both sides of each gap.

In order to calibrate these magnetic spectrographs, we have used sources of alpha particles for Po^{210} as our primary energy standard since 1947. As a result of re-evaluations of earlier measurements[9] and recent new attempts[10] at determining the energy of these particles, we have reported Q-value measurements based on a different value for this energy standard from time to time. Today, in 1963, there is general agreement that the value of the alpha particles from Po^{210} is significantly different from the value of 5.298 MeV we initially adopted in 1947. At the Nuclidic Mass Conference at Hamilton, Ontario in 1960, we reported[10] some previously unpublished Q-values on a Po^{210} alpha standard of 5.3042 MeV.

In order to maintain a certain degree of internal consistency and, incidentally, as part of our contribution to this Conference, we have prepared a table of all ground-state reaction Q-values measured in our Laboratory since 1947. Thus, in Table 1, we are listing all previously published M. I. T. Q-values redetermined here on the basis of the Po^{210} alpha standard we adopted in 1960. In addition, the list includes measurements of Q-values not previously reported. The table also shows the particular combination of accelerator and spectrograph used in each determination. When more than one combination is indicated, a weighted average of the measurements has been chosen. Any deviations from previously reported measurements are small. The fourth column of the table lists the values obtained from the mass tables of Everling et al., and Koenig et al.[11] while the last column shows the difference between these and the M. I. T. measurements.

In Fig. 6 are shown in solid circles all the stable nuclei in the periodic table of elements from hydrogen to copper. Determinations of mass differences in this region from Q-value measurements made at M. I. T. are shown in the diagram by lines connecting the circles and the reaction mechanism involved in each case is indicated by the code in the upper left hand corner of the figure. This diagram is a combination of the mass-link diagrams presented at the Conference in Mainz[12] in 1956 and

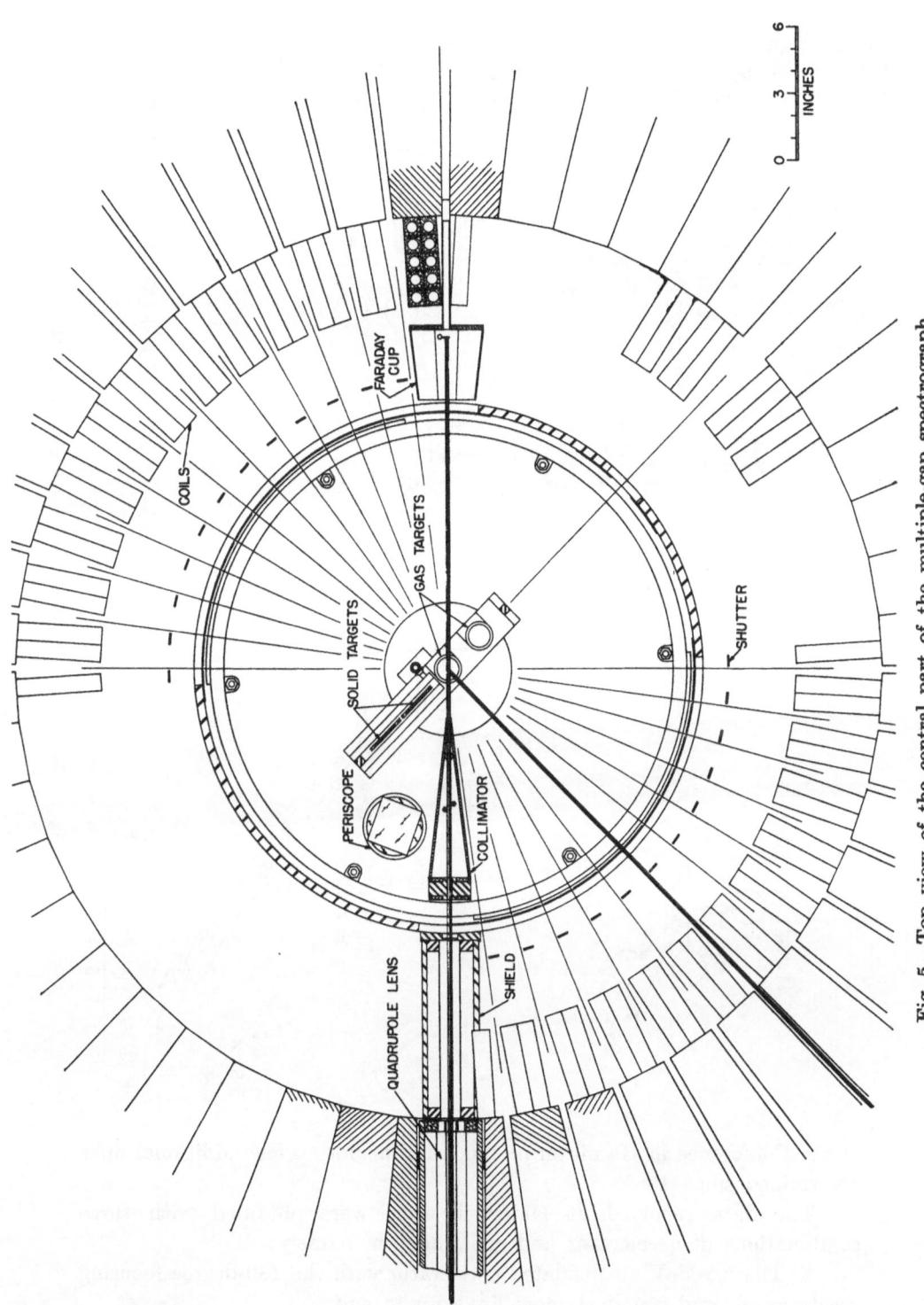

Fig. 5. Top view of the central part of the multiple-gap spectrograph

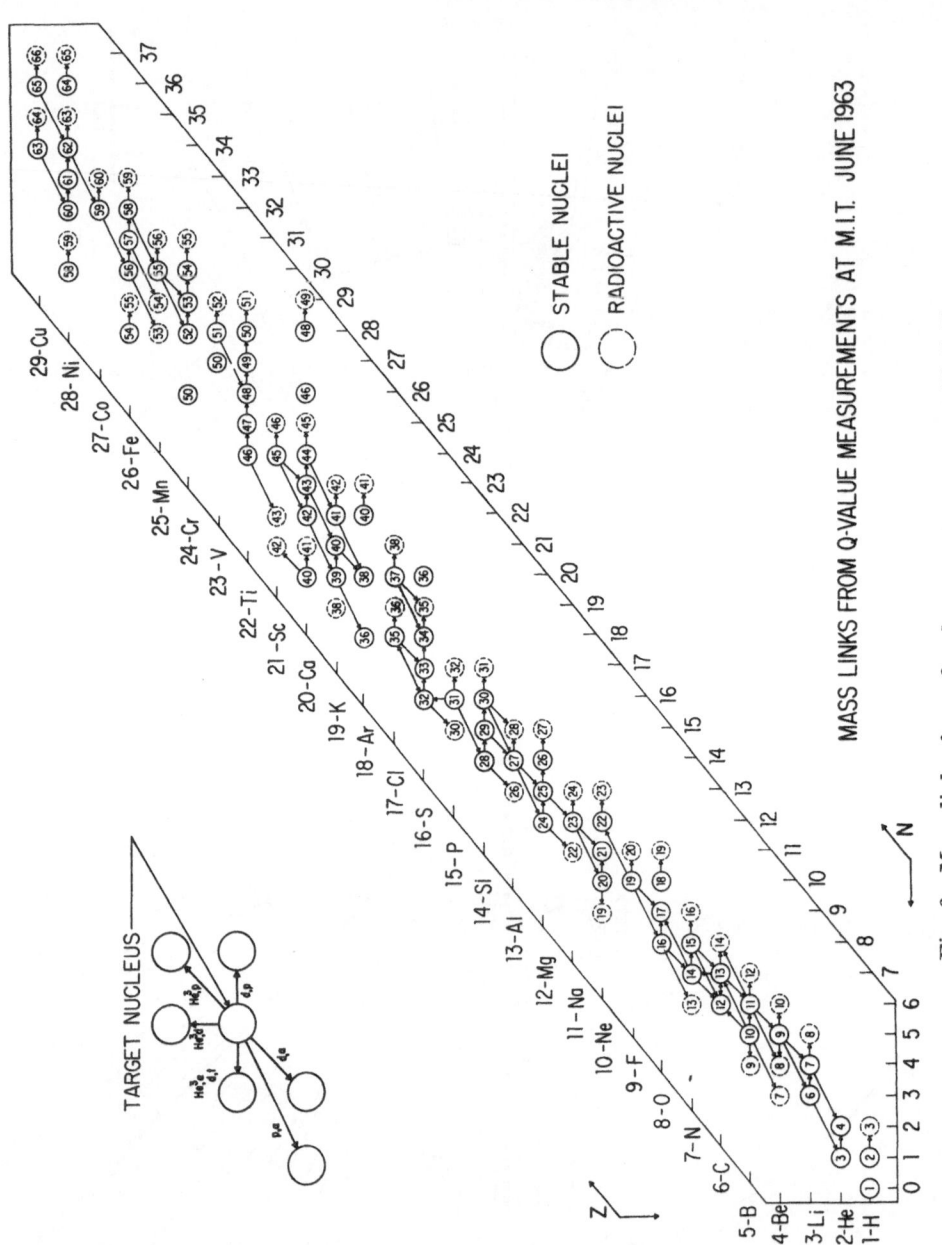

Fig. 6. Mass links from Q-value measurements at M. I. T.

at the Conference in Hamilton in 1960 and includes a few additional links determined since 1960.

The data reported at Mainz in 1956 were obtained with three combinations of accelerator and spectrograph, namely:

1. The 2.5-MeV air-insulated accelerator with the 180-degree focusing spectrograph and involved mass links up to sulfur.

Table 1. Ground State Q-Values Measured at M. I. T.[1]

Reaction	$Q_{\text{M. I. T.}}$ in keV	Source and Spectro-graph[2]	Q_M[3] in keV	$Q_{\text{M. I. T.}} - Q_M$
		(d, p) reactions		
$H^2(d, p)H^3$	4034 ± 6	a	4033 ± 0.2	$-\ 1$
$He^3(d, p)He^4$	18377	c	18352 ± 0.4	$+\ 25$
$Li^6(d, p)Li^7$	5024 ± 7	a	5028 ± 1	$-\ 4$
$Li^7(d, p)Li^8$	-188 ± 7	a	-192 ± 1	$+\ 4$
$Be^9(d, p)Be^{10}$	4590 ± 8	a, b	4590 ± 2	0
$B^{10}(d, p)B^{11}$	9244 ± 11	a	9231 ± 1	$+\ 13$
$B^{11}(d, p)B^{12}$	1137 ± 5	a	1145 ± 1	$-\ 8$
$C^{12}(d, p)C^{13}$	2722 ± 4	a, b, c, d	2722 ± 1	0
$C^{13}(d, p)C^{14}$	5951 ± 8	a, b	5951 ± 1	0
$N^{14}(d, p)N^{15}$	8623 ± 3	a, c	8610 ± 1	$+\ 13$
* $N^{15}(d, p)N^{16}$	267 ± 8	d	275 ± 6	$-\ 8$
$O^{16}(d, p)O^{17}$	1920 ± 3	a, b, c, d	1917 ± 1	$+\ 3$
* $O^{18}(d, p)O^{19}$	1727 ± 8	d	1732 ± 4	$-\ 5$
$F^{19}(d, p)F^{20}$	4377 ± 7	a	4374 ± 4	$+\ 3$
$Ne^{20}(d, p)Ne^{21}$	4534 ± 7	a, c	4534 ± 1	0
$Ne^{22}(d, p)Ne^{23}$	2968 ± 7	a, c	2968 ± 5	0
$Na^{23}(d, p)Na^{24}$	4736 ± 7	a, c	4734 ± 3	$+\ 2$
$Mg^{24}(d, p)Mg^{25}$	5102 ± 7	a	5106 ± 2	$-\ 4$
$Mg^{25}(d, p)Mg^{26}$	8889 ± 12	a	8873 ± 3	$+\ 16$
$Mg^{26}(d, p)Mg^{27}$	4211 ± 6	a	4212 ± 3	$-\ 1$

[1] This table includes all reactions measured at M. I. T. during the period 1947 to 1963. Practically all previously published Q-values were based on a calibration standard using 5.299 MeV ($B \varrho = 331.59$ kg/cm) for the energy of Po^{210} alpha particles. In 1960 we adopted the new value of 5.3042 MeV ($B \varrho = 331.750$ kg/cm) for the energy of these alpha particles. The Q-values tabulated here have been adjusted to conform with this new calibration standard.

[2] In this column is indicated both the accelerator used as the source of bombarding particles and the spectrograph used for analyzing the reaction products. When more than one source and spectrograph are indicated, the value quoted is a weighted average of the measurements obtained. The letter code refers to the following:

a — 2.5 MeV air-insulated Van de Graaff accelerator and 180° focusing spectrograph.

b — 8.5 MeV ONR Van de Graaff accelerator and 180° focusing spectrograph.

c — 8.5 MeV ONR Van de Graaff accelerator and single gap broad-range spectrograph.

d — 8.5 MeV ONR Van de Graaff accelerator and multiple gap broad-range spectrograph.

[3] F. EVERLING, L. A. KOENIG, J. H. E. MATTAUCH, and A. H. WAPSTRA, Consistent Set of Q-values, Nuclear Data Tables Part I and II National Academy of Sciences — National Research Council Feb. 1961 — also Nuclear Phys. 15, 342 (1960).

* Q-values not previously reported.

** Assignment uncertain; may be due to excited state.

*** Isotopic assignment uncertain.

(Table 1, continued)

Reaction	$Q_{M.I.T.}$ in keV	Source and Spectrograph[2]	$Q_M{}^3$ in keV	$Q_{M.I.T.} - Q_M$
$Al^{27}(d, p)Al^{28}$	5503 ± 10	a, c	5499 ± 3	$+ 4$
$Si^{28}(d, p)Si^{29}$	6252 ± 10	a	6253 ± 3	$- 1$
$Si^{29}(d, p)Si^{30}$	8396 ± 13	a	8390 ± 4	$+ 6$
$Si^{30}(d, p)Si^{31}$	4368 ± 7	a	4367 ± 4	$+ 1$
$P^{31}(d, p)P^{32}$	5712 ± 8	a, c	5712 ± 2	0
$S^{32}(d, p)S^{33}$	6420 ± 6	a, c	6418 ± 3	$+ 2$
$S^{33}(d, p)S^{34}$	9202 ± 10	c	9196 ± 4	$- 6$
$S^{34}(d, p)S^{35}$	4762 ± 10	c	4757 ± 4	$+ 5$
$Cl^{35}(d, p)Cl^{36}$	6360 ± 8	c	6352 ± 5	$+ 8$
$Cl^{37}(d, p)Cl^{38}$	3885 ± 8	c	3885 ± 8	0
$Ar^{40}(d, p)Ar^{41}$	3878 ± 6	c	3868 ± 11	$+ 10$
$K^{39}(d, p)K^{40}$	5579 ± 10	b, c	5573 ± 3	$+ 6$
$K^{41}(d, p)K^{42}$	5314 ± 12	c	5304 ± 21	$+ 10$
$Ca^{40}(d, p)Ca^{41}$	6146 ± 9	b, c	6136 ± 8	$+ 10$
$Ca^{42}(d, p)Ca^{43}$	5716 ± 10	b, c	5705 ± 5	$+ 11$
$Ca^{43}(d, p)Ca^{44}$	8922 ± 14	b, c	8911 ± 6	$+ 11$
$Ca^{44}(d, p)Ca^{45}$	5193 ± 10	b, c	5195 ± 6	$- 2$
$Ca^{48}(d, p)Ca^{49}$	2919 ± 6	c	2919 ± 6	0
$Sc^{45}(d, p)Sc^{46}$	6541 ± 8	c, d	6542 ± 5	$- 1$
* $Ti^{46}(d, p)Ti^{47}$	6675 ± 8	c, d	6662 ± 8	$+ 13$
* $Ti^{47}(d, p)Ti^{48}$	9409 ± 8	c, d	9396 ± 8	$+ 13$
* $Ti^{48}(d, p)Ti^{49}$	5930 ± 8	c	5922 ± 4	$+ 8$
* $Ti^{49}(d, p)Ti^{50}$	8741 ± 8	c	8713 ± 6	$+ 28$
* $Ti^{50}(d, p)Ti^{51}$	4157 ± 8	c	4137 ± 21	$+ 20$
$V^{51}(d, p)V^{52}$	5098 ± 9	b, c	5079 ± 6	$+ 19$
$Cr^{52}(d, p)Cr^{53}$	5725 ± 6	c	5719 ± 4	$+ 6$
* $Cr^{53}(d, p)Cr^{54}$	7480 ± 12	d	7497 ± 5	$- 17$
* $Cr^{54}(d, p)Cr^{55}$	4027 ± 8	c	3800 ± 140	$+ 227$
$Mn^{55}(d, p)Mn^{56}$	5052 ± 5	c	5046 ± 4	$+ 6$
$Fe^{54}(d, p)Fe^{55}$	7084 ± 8	c	7075 ± 5	$+ 9$
$Fe^{56}(d, p)Fe^{57}$	5425 ± 8	c	5416 ± 3	$+ 9$
$Fe^{57}(d, p)Fe^{58}$	7815 ± 8	c	7823 ± 6	$- 8$
$Fe^{58}(d, p)Fe^{59}$	4357 ± 8	c	4361 ± 6	$- 4$
$Co^{59}(d, p)Co^{60}$	5267 ± 11	b, c	5272 ± 4	$- 5$
$Ni^{58}(d, p)Ni^{59}$	6785 ± 8	c, d	6776 ± 4	$+ 9$
$Ni^{60}(d, p)Ni^{61}$	5604 ± 8	c	5598 ± 6	$+ 6$
$Ni^{61}(d, p)Ni^{62}$	8379 ± 8	c	8365 ± 9	$+ 14$
$Ni^{62}(d, p)Ni^{63}$	4623 ± 8	c, d	4617 ± 5	$+ 6$
$Ni^{64}(d, p)Ni^{65}$	3876 ± 6	c, d	3910 ± 200	$- 24$
$Cu^{63}(d, p)Cu^{64}$	5697 ± 8	c	5691 ± 4	$+ 6$
$Cu^{65}(d, p)Cu^{66}$	4837 ± 8	c	4836 ± 7	$+ 1$
* $Se^{78}(d, p)Se^{79}$	** 4660 ± 6	c	4754 ± 45	$- 94$
* $Se^{80}(d, p)Se^{81}$	** 4490 ± 6	c	4590 ± 60	$- 100$
* $Zr^{90}(d, p)Zr^{91}$	** 4929 ± 13	d	4981 ± 49	$- 52$
* $Rh^{103}(d, p)Rh^{104}$	** 4735 ± 10	d	4567 ± 14	$- 32$
* $Ag^{107}(d, p)Ag^{108}$	** 4973 ± 10	c	4988 ± 9	$- 15$

(Table 1, continued)

Reaction	$Q_{\text{M. I. T.}}$ in keV	Source and Spectro-graph[2]	$Q_M{}^3$ in keV	$Q_{\text{M. I. T.}} - Q_M$
* $Ag^{109}(d, p)Ag^{110}$	** 4590 ± 5	c	4591 ± 5	$- 1$
$Ba^{138}(d, p)Ba^{139}$	2495 ± 10	c	2495 ± 10	0
* $Ce^{140}(d, p)Ce^{141}$	3210 ± 10	d	3304 ± 27	$- 96$
* $Pr^{141}(d, p)Pr^{142}$	3626 ± 10	d	3611 ± 28	$- 14$
* $Tb^{159}(d, p)Tb^{160}$	4165 ± 20	d	4280 ± 400	$- 115$
* $Ta^{181}(d, p)Ta^{182}$	3835 ± 8	d	3847 ± 30	$- 12$
* $W^{182}(d, p)W^{183}$	3912 ± 5	d	3960 ± 8	$- 48$
* $Pb^{208}(d, p)Pb^{209}$	1705 ± 15	d	1695 ± 26	$+ 7$
$Bi^{209}(d, p)Bi^{210}$	2369 ± 10	d	2412 ± 28	$- 43$

(p, α) reactions

Reaction	$Q_{\text{M. I. T.}}$ in keV	Source and Spectro-graph[2]	$Q_M{}^3$ in keV	$Q_{\text{M. I. T.}} - Q_M$
$Li^6(p, \alpha)He^3$	4025 ± 6	a	4022 ± 1	$+ 3$
$Li^7(p, \alpha)He^4$	17357 ± 14	a	17347 ± 1	$+ 10$
$Be^9(p, \alpha)Li^6$	2144 ± 6	a	2125 ± 1	$+ 19$
$B^{10}(p, \alpha)Be^7$	1153 ± 4	a	1147 ± 1	$+ 6$
$B^{11}(p, \alpha)Be^8$	8575 ± 11	a	8586 ± 1	$- 11$
$N^{15}(p, \alpha)C^{12}$	4965 ± 7	a	4964 ± 1	$+ 1$
$O^{16}(p, \alpha)N^{13}$	$- 5211 \pm 10$	c	$- 5218 \pm 1$	$+ 7$
$F^{19}(p, \alpha)O^{16}$	8122 ± 9	a, c	8114 ± 1	$+ 8$
$Na^{23}(p, \alpha)Ne^{20}$	2373 ± 8	a, c	2379 ± 2	$- 6$
$Al^{27}(p, \alpha)Mg^{24}$	1596 ± 7	a	1595 ± 1	$+ 1$
$Si^{30}(p, \alpha)Al^{27}$	$- 2368 \pm 10$	c	$- 2378 \pm 4$	$+ 10$
$P^{31}(p, \alpha)Si^{28}$	1911 ± 10	a, c	1917 ± 3	$- 6$
$Cl^{35}(p, \alpha)S^{32}$	1865 ± 8	c	1865 ± 3	0
$Cl^{37}(p, \alpha)S^{34}$	3029 ± 8	c	3030 ± 3	$- 1$
$K^{39}(p, \alpha)A^{36}$	1287 ± 7	c	1292 ± 4	$- 5$
* $K^{41}(p, \alpha)A^{38}$	4018 ± 10	c	4035 ± 5	$- 17$
$Ca^{42}(p, \alpha)K^{39}$	118 ± 7	b, c	126 ± 4	$- 8$
$Ca^{43}(p, \alpha)K^{40}$	$- 14 \pm 8$	b, c	$- 6 \pm 5$	$- 8$
$Ca^{44}(p, \alpha)K^{41}$	$- 1058 \pm 10$	b, c	$- 1047 \pm 6$	$- 11$
$Sc^{45}(p, \alpha)Ca^{42}$	2343 ± 8	c	2341 ± 6	$+ 2$
*** $Ti^{46}(p, \alpha)Sc^{43}$	$- 3217 \pm 14$	c	$- 3082 \pm 11$	$- 135$
$V^{51}(p, \alpha)Ti^{48}$	1162 ± 10	c	1166 ± 5	$- 4$
$Mn^{55}(p, \alpha)Cr^{52}$	2570 ± 8	c	2572 ± 4	$- 2$
$Fe^{56}(p, \alpha)Mn^{53}$	$- 1060 \pm 9$	c	$- 1061 \pm 10$	$+ 1$
$Fe^{57}(p, \alpha)Mn^{54}$	237 ± 9	c	238 ± 7	$- 1$
$Fe^{58}(p, \alpha)Mn^{55}$	420 ± 9	c	410 ± 7	$+ 10$
$Co^{59}(p, \alpha)Fe^{56}$	3245 ± 8	c	3240 ± 5	$+ 5$
$Ni^{62}(p, \alpha)Co^{59}$	342 ± 10	c	352 ± 7	$- 10$
$Cu^{63}(p, \alpha)Ni^{60}$	3757 ± 8	c	3756 ± 6	$+ 1$
$Cu^{65}(p, \alpha)Ni^{62}$	4345 ± 8	c	4343 ± 6	$+ 2$

(d, α) reactions

Reaction	$Q_{\text{M. I. T.}}$ in keV	Source and Spectro-graph[2]	$Q_M{}^3$ in keV	$Q_{\text{M. I. T.}} - Q_M$
$Be^9(d, \alpha)Li^7$	7157 ± 8	a	7153 ± 1	$+ 4$
$B^{11}(d, \alpha)Be^9$	8024 ± 7	a, c	8027 ± 1	$- 3$
$C^{13}(d, \alpha)B^{11}$	5165 ± 10	a	5167 ± 1	$- 2$

(Table 1, continued)

Reaction	$Q_{M.I.T.}$ in keV	Source and Spectrograph[2]	$Q_M{}^3$ in keV	$Q_{M.I.T.} - Q_M$
$N^{14}(d, \alpha)C^{12}$	13588 ± 6	c	13574 ± 0.4	$+ 14$
$N^{15}(d, \alpha)C^{13}$	7689 ± 6	a	7687 ± 1	$+ 2$
$O^{16}(d, \alpha)N^{14}$	3113 ± 6	a	3110 ± 0.4	$+ 2$
$F^{19}(d, \alpha)O^{17}$	10060 ± 10	a	10031 ± 1	$+ 29$
$Na^{23}(d, \alpha)Ne^{21}$	6909 ± 10	a	6913 ± 2	$- 4$
$Mg^{24}(d, \alpha)Na^{22}$	1955 ± 12	c	1964 ± 5	$- 9$
$Mg^{25}(d, \alpha)Na^{23}$	7026 ± 13	a	7047 ± 2	$- 21$
$Al^{27}(d, \alpha)Mg^{25}$	6700 ± 10	a	6700 ± 2	0
$Si^{28}(d, \alpha)Al^{26}$	1429 ± 4	c	1421 ± 4	$+ 8$
$Si^{29}(d, \alpha)Al^{27}$	6000 ± 11	a	6012 ± 4	$- 12$
$Si^{30}(d, \alpha)Al^{28}$	3123 ± 10	a	3121 ± 4	$+ 2$
$P^{31}(d, \alpha)Si^{28}$	8166 ± 11	a	8170 ± 3	$- 4$
$S^{32}(d, \alpha)P^{30}$	4892 ± 10	c	4892 ± 10	0
$Cl^{35}(d, \alpha)S^{33}$	8285 ± 10	c	8183 ± 4	$+ 2$
$Cl^{37}(d, \alpha)S^{35}$	7791 ± 12	c	7787 ± 3	$+ 4$
$Ca^{40}(d, \alpha)K^{38}$	4655 ± 10	a, c	4655 ± 10	0
* $Sc^{45}(d, \alpha)Ca^{43}$	8028 ± 12	d	8045 ± 6	$- 17$
$Mn^{55}(d, \alpha)Cr^{53}$	8283 ± 8	c	8291 ± 4	$- 8$
(d, t) reactions				
$Be^9(d, t)Be^8$	4602 ± 13	a	4592 ± 1	$+ 10$
$B^{10}(d, t)B^9$	$- 2189 \pm 10$	c	$- 2182 \pm 2$	$- 7$
$C^{13}(d, t)C^{12}$	1311 ± 6	a	1311 ± 1	0
(p, d) reactions				
$Be^9(p, d)Be^8$	562 ± 4	a	559 ± 1	$+ 3$
$C^{13}(p, d)C^{12}$	$- 2722 \pm 7$	c	$- 2722 \pm 1$	0
(α, p) reactions				
$B^{10}(\alpha, p)C^{13}$	4068 ± 12	b, c	4064 ± 1	$+ 4$
$B^{11}(\alpha, p)C^{14}$	789 ± 17	b, c	784 ± 1	$+ 5$
$N^{14}(\alpha, p)O^{17}$	$- 1200 \pm 17$	b, c	$- 1193 \pm 1$	$- 7$
$F^{19}(\alpha, p)Ne^{22}$	1674 ± 11	b, c	1675 ± 1	$- 1$
$S^{32}(\alpha, p)Cl^{35}$	$- 1862 \pm 17$	b, c	$- 1865 \pm 3$	$+ 3$
(α, d) reactions				
$B^{10}(\alpha, d)C^{12}$	1341 ± 15	b, c	1342 ± 1	$- 1$
(He³, d) reactions				
* $C^{13}(He^3, d)N^{14}$	2048 ± 14	d	2056 ± 1	$- 8$
* $P^{31}(He^3, d)S^{32}$	3356 ± 13	d	3369 ± 2	$- 13$
(He³, α) reactions				
* $Ne^{20}(He^3, \alpha)Ne^{19}$	3750 ± 13	d	3702 ± 5	$+ 48$
(He³, p) reactions				
* $Ca^{40}(He^3, p)Sc^{42}$	4966 ± 20	d	5400 ± 600	$- 434$

2. The 8.5 MeV ONR accelerator with the 180-degree focusing spectrograph.

3. The 8.5 MeV ONR accelerator with the single-gap broad-range spectrograph. Together (2) and (3) provided some mass links up to scandium.

At the Hamilton Conference, all of the new measurements reported involved the use of the broad-range spectrograph and the 8.5 MeV accelerator. These include most of the reactions shown linking the masses from scandium to copper. Since 1961, the emphasis in our research program has been in nuclear spectroscopy, with most of our efforts directed to the use of the multiple-gap spectrograph for obtaining intensity measurements of reaction products as a function of angle. A few of the links shown in Fig. 6 have been obtained with this spectrograph, particularly those in the titanium and chromium region. We are hoping to add more links in this region through the use of He^3-induced reactions, which will help to complete some cycles and especially to fill in some gaps near argon and titanium. A few recent determinations of mass differences between neighboring nuclei above copper in the Periodic Table are listed in Table 1 and all involve (d, p) reactions on target nuclei ranging from selenium to bismuth.

Although most of the activities of the Laboratory during the past three years have been concentrated on angular-distribution studies, we have nonetheless been concerned about the problems and limitations associated with the use of the multiple-gap magnet for making accurate Q-value measurements. There are two chief difficulties requiring special precautions in dealing with the energy measurements; difficulties not encountered in the case of our single-gap magnet. Measurements of the homogeneity of the field in the trajectory region of the single-gap spectrograph showed deviations of the order of 0.1% only above 10,000 gauss. In our multiple-gap instrument, these deviations within a single gap and from one gap to another approach more nearly 1%, even with field strengths below 10,000 gauss. This degree of inhomogeneity was not unexpected, since, after completion of machining of the pole faces, it was discovered that the indicated tolerance on flatness was not maintained, and thus variations of several mils in the gap width were noted in some areas along the trajectory path.

The second difficulty can best be seen by noting that the flux path in a circle around the toroidal ring through point A (see Fig. 4) is less than the flux path through point B. This difference in flux path accentuates the effect of hysteresis in the iron, and this in turn bears on the reproducibility of our measurements. We should like, of course, to make accurate energy measurements at any and all angles. However, we do not have flux-meter probes in all gaps. Ideally, if we can achieve reproducibility, then a single probe can be used as a reference to obtain a measure of the field at a single point (say, point A, Fig. 4). The field in other regions of the same gap, or in different gaps, can be expected to be different, but the important requirement is that the field at any point bear a fixed relationship to that where the reference probe is located, and particularly that this relationship be maintained over a range of field settings and also for the calibration measurements.

That these requirements have not been fully realized to date is illustrated in Fig. 7. Here are shown two sets of data obtained with the

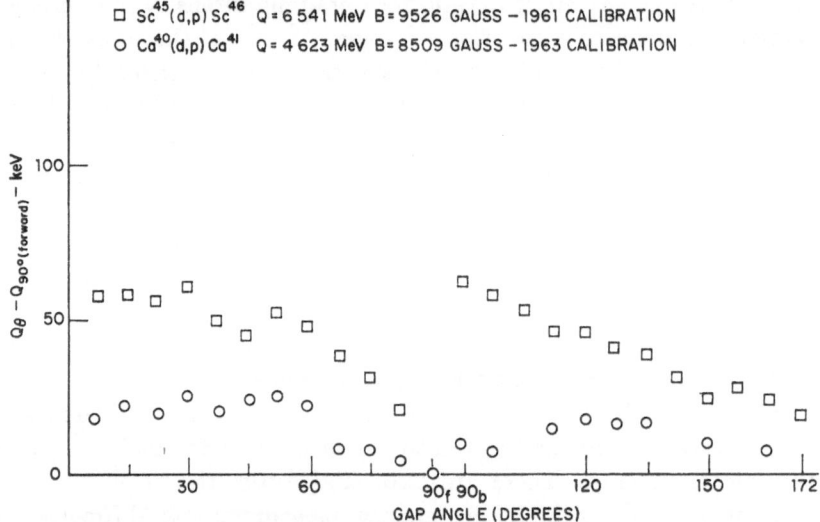

Fig. 7. Q-value measurements as a function of angle with multiple-gap spectrograph under two conditions of operation

multiple-gap spectrograph. For a given incident energy, the Q-value of the group corresponding to the ground-state transition of the $Sc^{45}(d, p)Sc^{46}$ reaction was measured as a function of angle. The deviations observed from the value measured in the 90-degree gap in the forward quadrant are plotted as a function of angle. From a more recent experiment and for a reaction group corresponding to the first excited state in Ca^{41}, the deviations from the forward 90-degree gap are again shown as a function of angle. The significant improvement noted is associated with two specific changes made in the use of the spectrograph; one is related to the effect of hysteresis and the other, to probe location. For the Sc^{46} data, the reference probe was located at position B (see Fig. 4) on the outer edge of the toroidal ring. Furthermore, the field settings were made without special care in applying the energizing current, both for the calibration and the reaction runs.

In the case of the Ca⁴¹ data, the reference probe was located at position *A* on the inside edge of the toroidal ring. In addition, a special procedure was initiated for energizing the magnet. In an attempt to reduce the effect of hysteresis in the iron and thereby perhaps to achieve greater reproducibility, we have adopted a cycling routine for making field settings. This simply involves, first, slowly increasing the current in the energizing coils to that corresponding to the maximum field setting, then decreasing the current to zero and finally raising the current to approach the desired field from zero.

With continuing studies on the characteristics of the multiple-gap magnet, we hope to improve the accuracy of these measurements. Meanwhile, we are taking precautions in the measurement of any *Q*-value by referring each measurement to a secondary standard, such as a contaminant group observed near the same trajectory region as the group in question.

During the past 16 years, the overall precision realized with the use of the uniform-field magnetic spectrographs has been of the order of 0.1% in energy. Any substantial improvement over this precision would require greater homogeneity of the magnetic field, as well as greater precision in the measurement of the radius of curvature of the particle trajectory.

In connection with plans for a new accelerator at M. I. T., we are working on the design of an improved beam analyzer and a new precision magnetic spectrograph. Considerations are being directed at possibly improving the precision of energy measurements by another order of magnitude. It is hoped that, by the time of the next Mass Conference, a report on the use of these new facilities can be made.

References

[1] J. D. COCKCROFT, J. Sci. Instr. 10, 71 (1933).

[2] Lord RUTHERFORD, C. C. WYNN-WILLIAMS, W. B. LEWIS, and B. V. BOWDEN, Proc. Roy. Soc. (London) A 139, 617 (1933).

[3] W. W. BUECHNER, E. N. STRAIT, C. G. STERGIOPOULOS, and A. SPERDUTO, Phys. Rev. 74, 1569 (1948).

[4] L. C. VAN ATTA, D. L. NORTHRUP, C. M. VAN ATTA, and R. J. VAN DE GRAAFF, Phys. Rev. 49, 761 (1936).

[5] W. W. BUECHNER, A. SPERDUTO, C. P. BROWNE, and C. K. BOCKELMAN, Phys. Rev. 91, 1502 (1953).

[6] K. T. BAINBRIDGE, Solvay Report, 7th Congress, Chem. R. Stoopes, Brussels (1947).

[7] C. P. BROWNE, and W. W. BUECHNER, Rev, Sci. Instr. 27, 899 (1956).

[8] H. A. ENGE, and W. W. BUECHNER, Rev. Sci. Instr. 34, 155 (1963).

[9] G. H. BRIGGS, Rev. Modern Phys. 26, 1 (1954).

[10] H. E. DUCKWORTH, Editor, Proceedings of the International Conference on Nuclidic Masses, Toronto, Canada; University of Toronto Press, 1960.

[11] F. EVERLING, L. A. KOENIG, J. H. E. MATTAUCH, and A. H. WAPSTRA, "Consistent set of energies liberated in nuclear reactions I." Targets in mass region A ⩽ 66; Nuclear Data Tables, Part 1. — L. A. KOENIG, J. H. E. MATTAUCH,

304   A. Sperduto and W. W. Buechner: Q-Value Measurements at M. I. T.

and A. H. Wapstra, Targets in Mass Region $67 \leqslant A \leqslant 199$, Part 2. U. S. Government Printing Office, Washington 25, D. C.

[12] H. Hintenberger, "Nuclear Masses and Their Determination". London: Pergamon Press. 1957.

Discussion

G. C. Phillips: I certainly wish to commend you on the beautiful and accurate work you have done through the years at M. I. T. Frankly, however, I doubt that your new techniques will gain you the order-of-magnitude accuracy increase that you hope for. Nevertheless, I certainly wish you luck in this venture.

Nuclear Reaction Measurements
at the National University of Mexico

By

M. Mazari*, A. Jáidar, G. López, A. Tejera*, J. Careaga, R. Domínguez, and F. Alba*

(Presented by M. MAZARI)

Laboratorios Van de Graaff, Instituto de Física. Universidad Nacional Autónoma de México. México, D. F.

With 8 Figures

I. Introduction

The following paper will cover three aspects on ground state Q-value measurement of nuclear reactions, which have been or are under development at the National University of Mexico, since the first International Conference on Nuclidic Masses.

The first part consists of a set of remeasurements of Q-values already presented in Canada by our group[1], but which were questioned by Dr. W. W. BUECHNER, who suggested that the observed differences with respect to very accurately known masses might be attributed to saturation effects of the magnetic spectrograph field[2], and to the size of the beam hitting the target (personal communication).

The second part includes Q-values determined by (p, α), (d, p), and (d, α) reactions using gas embedded targets on metallic backings and reactions induced by bombardment with 1.51 MeV He[3] particles, mostly on evaporated solid targets.

Finally, the third part consists of a brief description of a toroidal multiple gap broad-range magnetic spectrograph (stereo-spectrograph) under construction at the National University of Mexico (UNAM), similar to that developed initially at the Massachusetts Institute of Technology (M. I. T.)[3], with which a large number of stable and radioactive isotopes can be accurately linked by means of charged particle reactions. Due to the interest in accurate absolute Q-value determinations, some

---

* Consultants to the Comisión Nacional de Energía Nuclear.

characteristics of a possible iron free 180°, 200 cm radius spectrograph are also included.

II. Remeasurements of Q-Values

a) Source Geometry

As mentioned by Dr. BUECHNER, one source of possible error for some Q-values presented at the last meeting, could be the difference in size of the beam spot (1.2 mm in height by 2.5 mm wide), when comparing

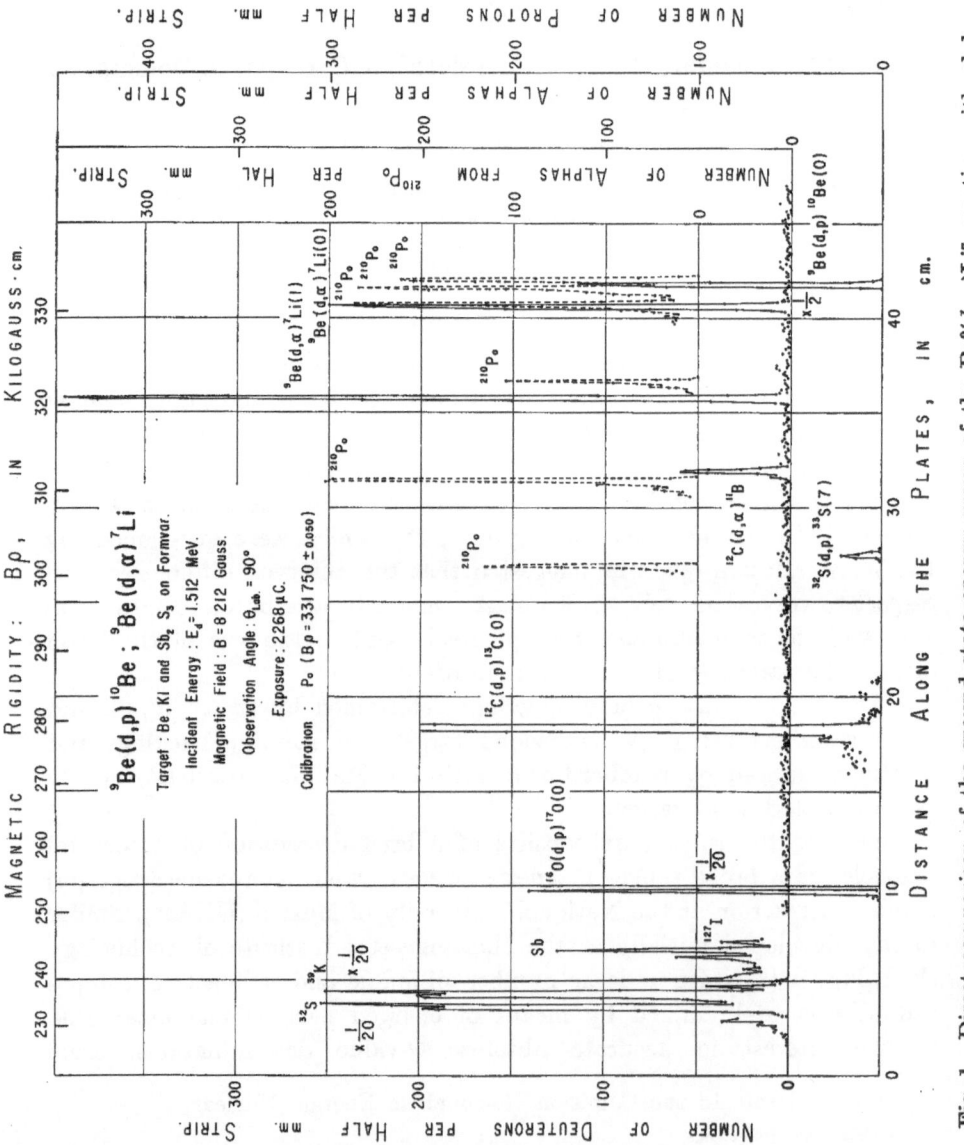

Fig. 1. Direct comparison of the ground state energy group of the Be⁹(d, α)Li⁷ reaction with alpha particles emitted from Po²¹⁰ under identical field conditions

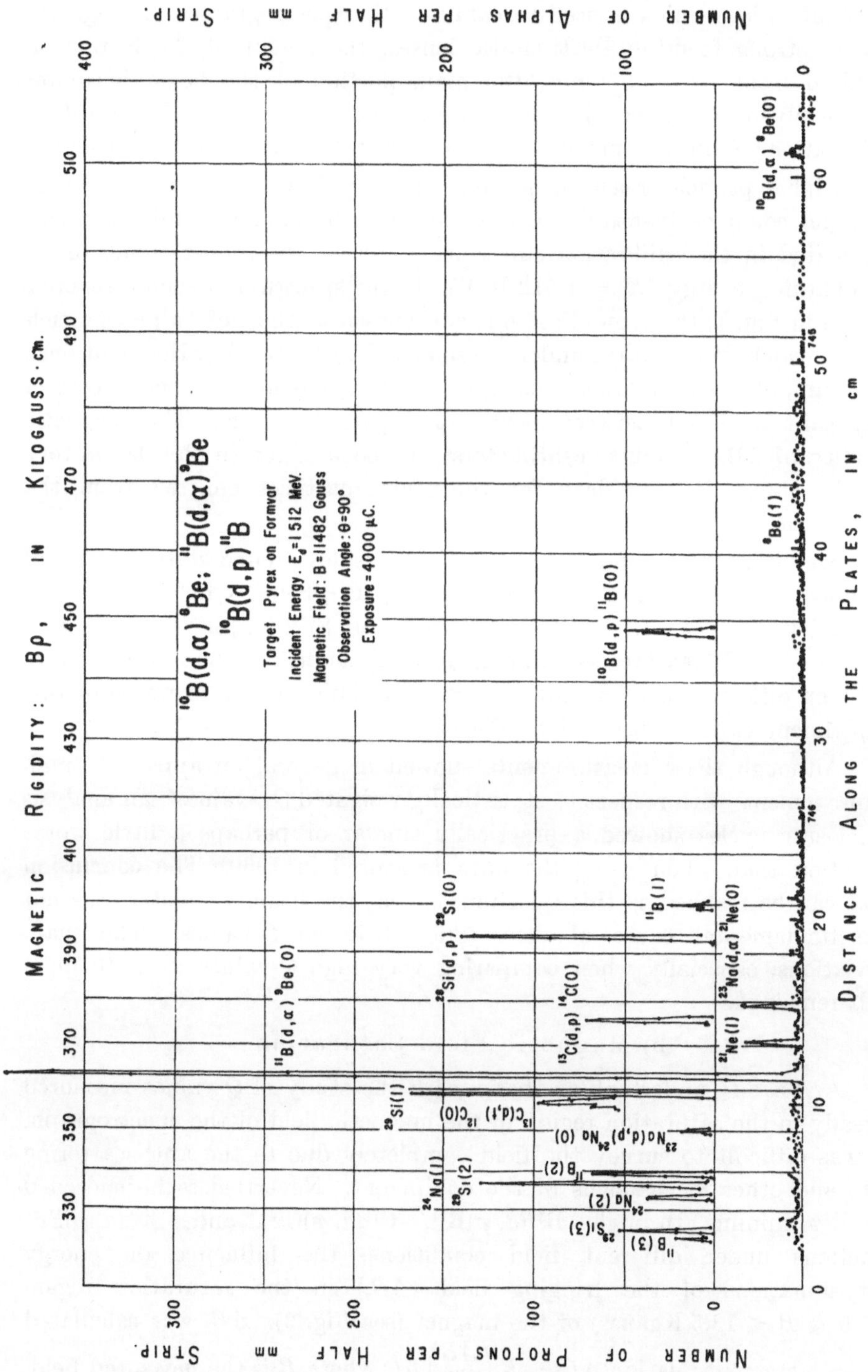

Fig. 2. Simultaneous reactions on various substances contained in Pyrex leading to ground states of the residual nuclei Be[8], Be[9], B[11], C[12], C[14], Ne[21], Na[24] and Si[29]

it with the silver wire dipped in a Po²¹⁰ solution (0.25 mm in diameter by 3 mm wide) which is used to calibrate the spectrograph. Through the use of strong focusing electrostatic lenses, the height of the beam was reduced from 1.2 to 0.3 mm; the main portion of the beam becoming practically of the same dimensions as the calibrator wire. A careful re-orientation of the equipment was done prior to this set of experiments.

Alpha particle reaction groups emitted from the $Be^9(d, \alpha)Li^7(O)$ reaction could be identically compared with the 5.3 MeV alpha particles originated in the calibration substance Po²¹⁰, by the proper choice of the bombarding energy $(E_d = 1.512 \text{ MeV})$. In the spectrum of such a reaction shown in Fig. 1, the ground state is superposed to the Po²¹⁰ alpha particle group, which was obtained under the same field setting; after the mentioned exposure, others were taken with small field increments in order to correct any shift observed between both alpha particle peaks. The magnetic rigidity of 331750 Gauss · centimeters⁴ has been taken in this laboratory as the calibration standard for the alpha particles emitted from the Po²¹⁰ source.

Simultaneous ground state transitions besides excited states of a good number of reactions like $B^{10}(d, \alpha)Be^8$, $B^{10}(d, p)B^{11}$, $B^{11}(d, \alpha)Be^9$, $C^{13}(d, p)C^{14}$, $C^{13}(d, t)C^{12}$, $Na^{23}(d, \alpha)Ne^{21}$, $Na^{23}(d, p)Na^{24}$, and $Si^{28}(d, p)Si^{29}$ were obtained by the use of Pyrex targets. Under identical conditions the relative errors between different reactions are expected to diminish. An example of this experiment can be observed in Fig. 2.

Although these measurements showed in general an apparent small improvement[5] with respect to statistically weighted Q_m-values[6], an analysis of closure cycles showed a practically similar or perhaps a little worse situation than when using the data presented in 1960. The conclusion that can be reached to this question, is that the beam geometry was not the fundamental cause of discrepancy between Q-values. The main deviations, especially when comparing very high Q-values $(Q > 10 \text{ MeV})$ still remained.

b) Magnetic Field Saturation

A. TEJERA undertook as thesis work the study of Q-values measured already in the saturation region of the magnetic field of the spectrograph. It was difficult to survey the field completely due to the anti-scattering slits and other mechanisms of the equipment. Nevertheless he succeeded in determining through $B^{10}(d, p)B^{11}$, $C^{13}(d, p)C^{14}$, and $Si^{28}(d, p)Si^{29}$ reactions under different field conditions, the influence on energy determinations of the fringing field (ΔB_s) in the saturation region $(10.5 \leqslant B \leqslant 13.5 \text{ Kgauss})$ of the magnet (see Fig. 3). ΔB_s was calculated by an increment relation $\Delta B_s = K \left(\dfrac{\Delta B}{\Delta I} \right) \Delta I$, where B is the measured field,

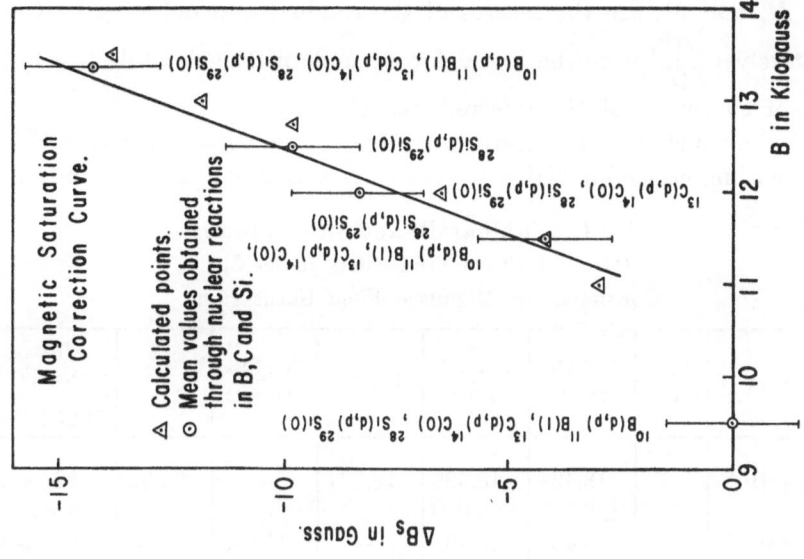

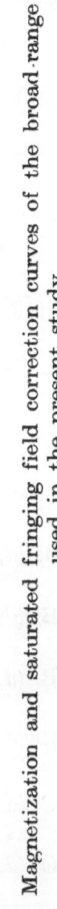

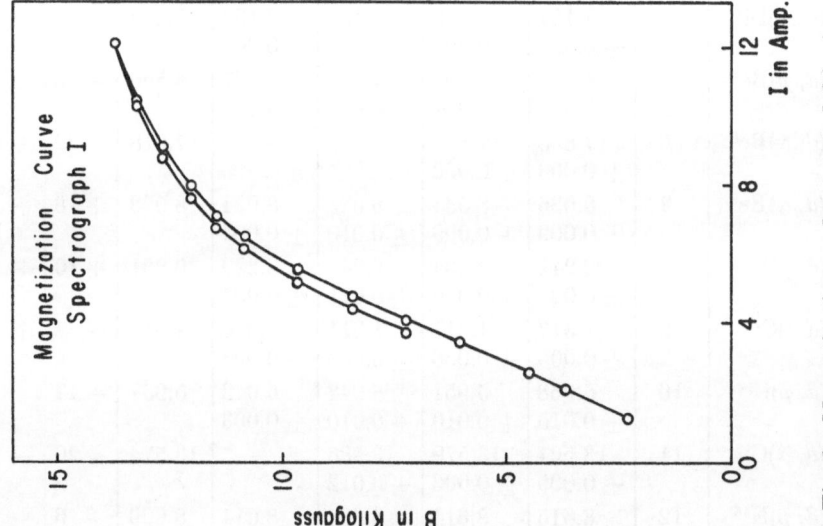

Fig. 3. Magnetization and saturated fringing field correction curves of the broad-range spectrograph used in the present study

I the current in the coils and K a constant determined experimentally ($K = -1.26 \times 10^{-3}$).

By using the practically linear relation between B and ΔB_s, it was possible to correct Q-value measurements due to saturation effects[7], by the use of the following expression derived from the Q equation:

$$\Delta Q = \frac{M_r + M_s}{M_r} \frac{\Delta E_s}{\Delta(B\varrho)} \varrho \, \Delta B_s, \tag{1}$$

M. Mazari et al.:

where M_r and M_s are the masses of the residual and outgoing particles, respectively; $\dfrac{\Delta E_s}{\Delta(B\varrho)}$ can be found for example in Enge's tables[8], ϱ is the radius of curvature of the referred particle.

Q-values including the corrections due to magnetic saturation effects have been summarized in Table 1, where a comparison against old UNAM's

Table 1. Nuclear Reaction Q-Values
(Ground State Transitions in MeV)
Corrected for Magnetic Field Saturation

Reaction	Fig. Ident.	UNAM 1960[1]	UNAM 1963	M. I. T.[1]	Weighted Average[1]	$Q_m{}^6$	$Q_{\text{UNAM}} - Q_m$ (keV) 1961	$Q_{\text{UNAM}} - Q_m$ (keV) 1963
$He^3(d, p)He^4$	1	18.434 ± 0.010	18.380 ± 0.010	18.377	—	18.352	+ 82	+ 28
$Li^6(d, \alpha)He^4$	2	22.431 ± 0.010	22.403 ± 0.012	—	22.386 ± 0.011	22.375	+ 56	+ 28
$Li^7(p, \alpha)He^4$	3	17.406 ± 0.030	17.373 ± 0.006	17.357 ± 0.014	17.346 ± 0.010	17.347	+ 59	+ 26
$Be^9(d, \alpha)Li^7$	4	7.164 ± 0.012	7.162 ± 0.004	7.157 ± 0.008	7.153 ± 0.003	7.153	+ 11	+ 9
$Be^9(d, p)Be^{10}$	5	4.598 ± 0.012	4.595 ± 0.004	4.590 ± 0.009	4.587 ± 0.005	4.590	+ 8	+ 5
$B^{10}(d, \alpha)Be^8$	6	17.850 ± 0.006	17.830 ± 0.006	—	—	17.818	+ 32	+ 12
$B^{11}(d, \alpha)Be^9$	7	8.036 ± 0.009	8.035 ± 0.009	8.023 ± 0.010	8.024 ± 0.004	8.028	+ 8	+ 7
$B^{10}(d, p)B^{11}$	8	9.241 ± 0.006	9.234 ± 0.006	9.244 ± 0.011	9.229 ± 0.005	9.231	+ 10	+ 3
$C^{13}(d, t)C^{12}$	9	1.317 ± 0.005	1.311 ± 0.006	1.311 ± 0.006	1.310 ± 0.003	1.310	+ 7	+ 1
$C^{13}(d, p)C^{14}$	10	5.968 ± 0.015	5.951 ± 0.010	5.942 ± 0.010	5.943 ± 0.003	5.951	+ 17	0
$N^{14}(d, \alpha)C^{12}$	11	13.594 ± 0.005	13.579 ± 0.006	13.588 ± 0.012	—	13.574	+ 20	+ 5
$N^{14}(d, p)N^{15}$	12	8.615 ± 0.010	8.614 ± 0.006	8.623 ± 0.010	8.614 ± 0.007	8.609	+ 6	+ 4
$O^{16}(d, \alpha)N^{14}$	13	3.108 ± 0.008	3.110 ± 0.006	3.111 ± 0.006	3.115 ± 0.002	3.111	— 3	— 1
$Na^{23}(d, \alpha)Ne^{21}$	14	6.907 ± 0.012	6.911 ± 0.009	6.908 ± 0.010	—	6.913	— 6	— 2
$Na^{23}(d, p)Na^{24}$	15	4.731 ± 0.009	4.736 ± 0.005	4.736 ± 0.007	4.727 ± 0.005	4.734	— 3	+ 2
$Si^{28}(d, p)Si^{29}$	16	6.254 ± 0.015	6.252 ± 0.010	6.252 ± 0.010	—	6.253	+ 1	— 1

measurements, M. I. T.'s and Q_m-values can be made. When analyzing these numbers in closed cycles, an important improvement is observed, the

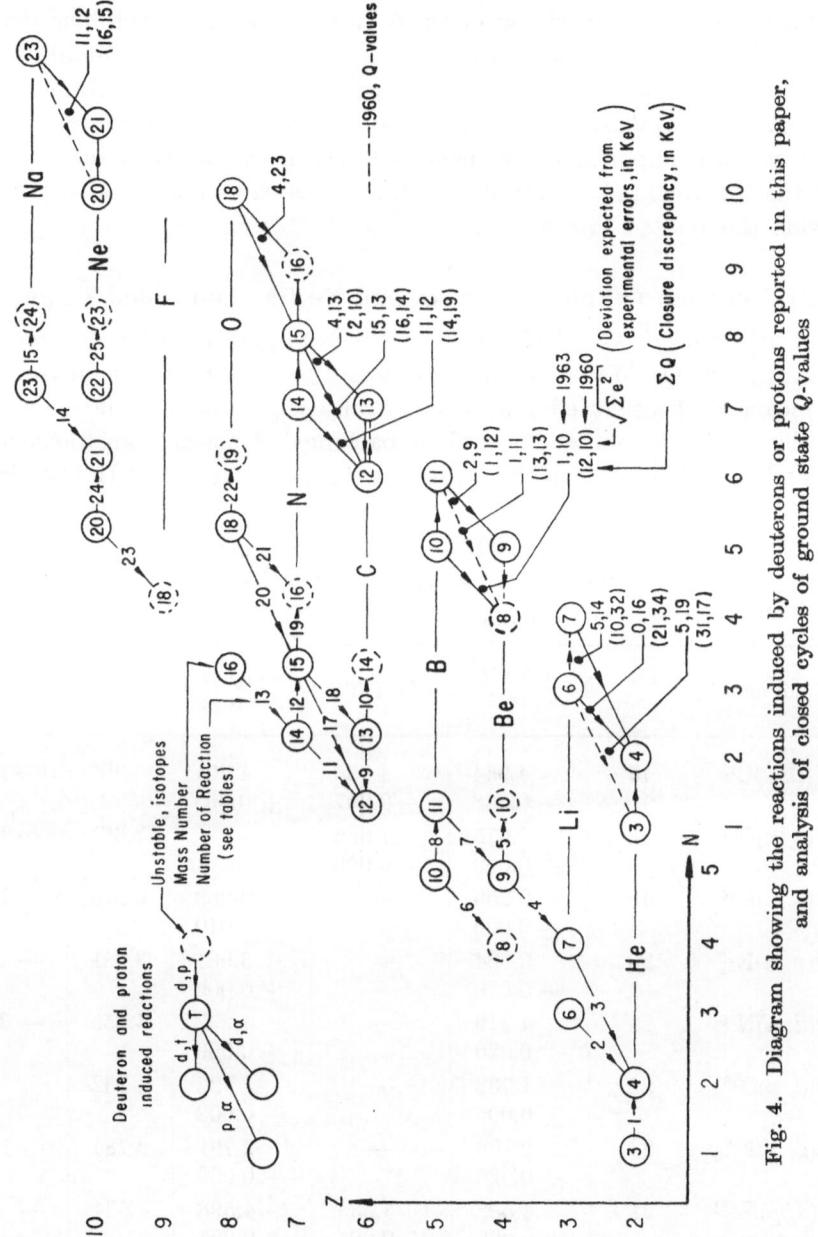

Fig. 4. Diagram showing the reactions induced by deuterons or protons reported in this paper, and analysis of closed cycles of ground state Q-values

internal differences ($\delta = \sum Q_i$) are in general much lower than the external proposed errors ($\varepsilon = \sqrt{\sum e_i^2}$). Such cycles are shown in the lower part of Fig. 4.

The conclusion of this set of corrected remeasurements is as follows: the definitely smaller errors resulting in the cycles as compared with those quoted when adding statistically independent determinations, gives confidence on the correctness of the Q-values, and it is even tempting to reduce the assigned errors noticeably. Even for the highest numbers like reactions 1, 3 and 11 of Table 1, the agreement between UNAM's results and those of M. I. T. is excellent (overlap of experimental errors). Nevertheless there are still important discrepancies between the actual determinations and the Q_m weighted numbers[6], which in some cases remain several times outside the quoted errors.

III. Reaction Q-Values on N, O, and Ne Gas Embedded Targets

Enriched N^{15}, O^{18}, Ne^{20}, Ne^{22}, and Ar^{40} targets were kindly prepared for us by Dr. O. ALMÉN at the Chalmers Tekniska Högskola in Sweden, by means of accelerated ions magnetically separated and injected into metallic molybdenum and tantalum backings. Nuclear transitions have been tried by bombarding them with protons, deuterons and He^3 particles.

Table 2. Nuclear Reaction Q-Values
(Ground State Transitions in MeV)
Further Measurements

Reaction	Fig. Ident.	UNAM 1961[9]	M. I. T.	Weighted Average[1]	Q_m[6]	$Q_{\text{UNAM}} - Q_m$ (keV)
$N^{15}(p, \alpha)C^{12}$	17	4.954 ± 0.008	4.965 ± 0.007	4.961 ± 0.003	4.965	— 11
$N^{15}(d, \alpha)C^{13}$	18	7.675 ± 0.009	7.688 ± 0.006	—	7.687	— 12
$N^{15}(d, p)N^{16}$	19	0.259 ± 0.006	—	0.269 ± 0.010	0.276	— 17
$O^{18}(p, \alpha)N^{15}$	20	3.964 ± 0.010	—	3.961 ± 0.009	3.981	— 17
$O^{18}(d, \alpha)N^{16}$	21	4.219 ± 0.020	—	4.237 ± 0.009	4.255	— 36
$O^{18}(d, p)O^{19}$	22	1.733 ± 0.006	—	1.732 ± 0.008	1.732	+ 1
$Ne^{20}(d, \alpha)F^{18}$	23	2.766 ± 0.020	—	2.791 ± 0.009	2.784	— 18
$Ne^{20}(d, p)Ne^{21}$	24	4.532 ± 0.006	4.533 ± 0.007	4.528 ± 0.006	4.534	— 2
$Ne^{22}(d, p)Ne^{23}$	25	2.974 ± 0.006	2.967 ± 0.007	2.966 ± 0.005	2.968	+ 6

Po^{210} alpha particle magnetic rigidity standard = 331750 Gauss · cm.[4]

The argon targets did not give any yield when bombarded with 1.8 MeV protons or deuterons, and none of them did when 1.5 MeV He³ projectiles were used.

Table 2 shows the list of the measured reactions[9]. It can be observed that about half of them reached a precision comparable with others mentioned in section II, larger errors ($\pm$ 20 keV in two cases) have been assigned to the rest, which all involved the emission of an alpha particle. This is understandable since alphas and deuterons lose more energy than protons, the targets appearing to be thicker in such processes. The possibility of having the main part of the gas at a given depth below the surface might be another cause of a shift in the determinations.

Closed cycle errors based on these results and some old measurements made at UNAM are drawn in the upper part of the same Fig. 4, as it can be considered an extension of the earlier work. The situation in this upper part of Fig. 4 is very similar to that at the first conference; in some cases the internal errors are lower than the external ones, but in others they are about equal. Therefore it still seems convenient to keep assigning errors in accordance to the individual experiments, where the thickness is considered, as well as other technical limitations of the method employed and not yet diminish them according to how close the discrepancy in the cycle appears.

IV. He³ Induced Reactions

The commercial availability of He³ gas permitted our group to start a program of Q-value measurements involving nuclear reactions induced by this type of particle. Such a study is being made using He³ particles of 1.51 MeV, maximum energy that our 90° deflector could bend. Although exposures were tried on nuclei as heavy as phosphorus, no yield was observed for substances above oxygen.

Unfortunately it was not possible to get very thin H³ and He³ targets, although accurate Q-value determinations originating in these substances would be desirable.

Similarly, reactions on deuterium were performed on thick targets as well as on pure Formvar thin targets. A comparison made between thick and thin target measurements as shown in Table 3 (reaction 28) suggests that the usual criterion to define the energy of a given group at $1/3$ height of the high energy side of a peak is not valid for thick targets. $1/2$ of the height seems to give better agreement, although reaction 30 does not confirm this suggestion. Reactions on the very light nuclei have been included in Table 3, although the precision is too low for mass determination purposes.

He³ reactions on Li⁶ targets leading to Li⁵ as the residual nucleus produces an extremely wide peak for the ground state. It seems better

Table 3. Nuclear Reaction Q-Values
(Ground State Transitions in MeV)
He³ induced reactions

Reaction	Fig. Ident.	UNAM	Other measurements[10]	Q_m[6]	Q_{UNAM} $-Q_m$ (keV)
$H^2(He^3, \alpha)H^1$	26	18.44 [a] ± 0.24	—	18.353	—
$H^3(He^3, \alpha)H^2$	27	14.31 [a] ± 0.12	—	14.319	—
$H^2(He^3, p)He^4$	28	18.382 ± 0.015 18.39 [a] ± 0.06	—	18.353	+ 29 —
$H^3(He^3, d)He^4$	29	14.53 [a] ± 0.15	—	14.319	—
$H^3(He^3, p)He^5$	30	11.42 [a] ± 0.15	11.15 ± 0.07	11.137	—
$Li^6(He^3, \alpha)Li^5$	31	16.74 [b] ± 0.05	14.85 ± 0.15	14.915	—
$Li^7(He^3, \alpha)Li^6$	32	13.322 ± 0.016	—	13.325	— 3
$Li^6(He^3, d)Be^7$	33	0.136 ± 0.003	—	0.116	+ 20
$Li^6(He^3, p)Be^8$	34	16.824 ± 0.012	16.60	16.788	+ 36
$Li^7(He^3, d)Be^8$	35	11.795 ± 0.013	—	11.759	+ 36
$Li^7(He^3, p)Be^9$	36	11.215 ± 0.015	—	11.201	+ 14
$Be^9(He^3, \alpha)Be^8$	37	18.931 ± 0.013	—	18.912	+ 19
$Be^9(He^3, d)B^{10}$	38	1.123 ± 0.005	—	1.094	+ 29
$Be^9(He^3, p)B^{11}$	39	10.344 ± 0.013	—	10.325	+ 19
$B^{10}(He^3, \alpha)B^9$	40	12.171 ± 0.015	12.130 [c] 12.110 [d] ± 0.020	12.137	+ 34
$B^{10}(He^3, d)C^{11}$	41	3.226 ± 0.010	3.174 [d] ± 0.015	3.198	+ 28

[a] Thick targets; Q-values taken at half height.
[b] Q-values taken at maximum energy of continuum.
[c] Rice University (Ref. 11).
[d] University of Manchester (Ref. 12).
[e] AWRE, Aldermaston (Ref. 13).

(Table 3, continued)

Reaction	Fig. Ident.	UNAM	Other measurements[10]	Q_m[6]	$Q_{\text{UNAM}} - Q_m$ (keV)
$B^{11}(He^3, p)C^{13}$	42	13.221 ± 0.010	—	13.186	+ 35
$C^{12}(He^3, p)N^{14}$	43	4.806 ± 0.009	4.780 e ± 0.015	4.779	+ 27
$O^{16}(He^3, p)F^{18}$	44	2.055 ± 0.005	2.033 c 2.044 e ± 0.015	2.022	+ 33

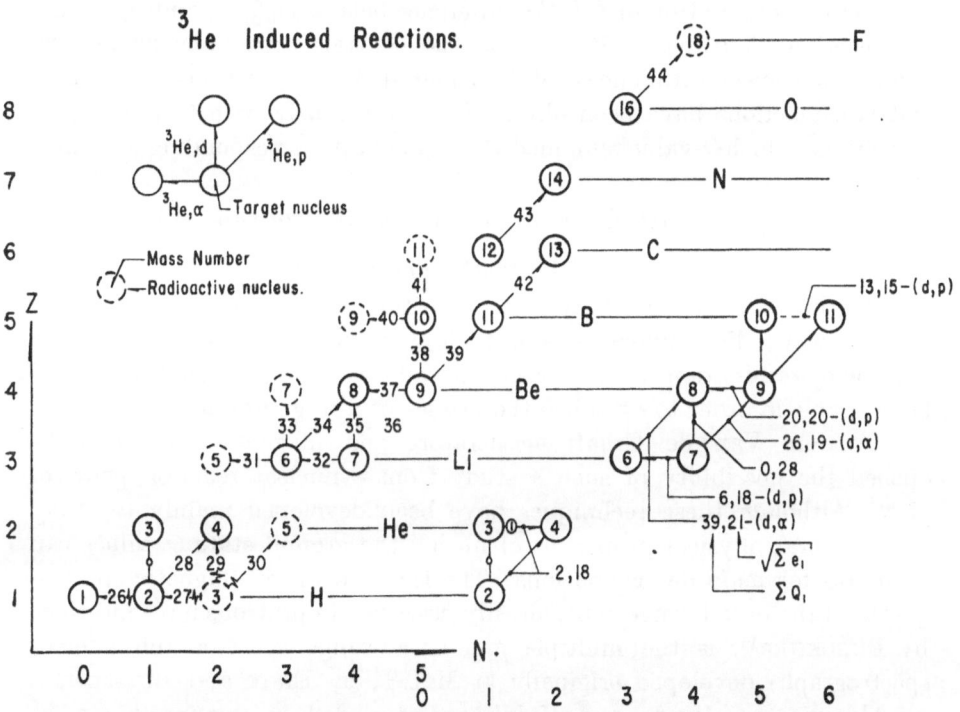

Fig. 5. He³ induced reactions and closed cycle analysis of Q-values as compared with (d, p) and (d, α) transitions

to give a lower limit to the mass of Li⁵ by the point where this continuum starts; which has been well observed in this reaction (31, Table 3).

Measurements on thin Formvar targets like $H^2(He^3, p)He^4$ (reaction 28) confirm Q-values obtained at M. I. T. and UNAM (reaction 1) by using accelerated deuterons; these Q-values are both 29 keV higher than the values calculated from mass tables.

Ground state Q-values of reactions induced by He³ particles in target nuclei like Li⁷, Be⁹, B¹⁰, B¹¹, C¹² and O¹⁶ have also been listed in

Table 3. It is surprising that, except for one case, the Q-values are systematically an average of 26 keV above Q_m-values, suggesting that the He³ mass might be in defect by this amount. Only a few references to He³ reactions studied by other laboratories have been obtained. Low precision work on reactions 30, 31, and 34 has been summarized in Ajzenberg-Lauritsen compilations[10]. High precision measurements for reactions 40[11, 12], 41[12], 43[13] and 44[11, 13] have been obtained at Rice University, at the University of Manchester and at Aldermaston as shown in Table 3.

When comparing Q-values in closed cycles (see Fig. 5) against (d, p) and (d, α) reactions, the closure errors do not seem to correlate too well, as was mentioned previously. The discrepancies can be attributed mainly to reactions originating in Li⁷, the difference between Q'_{UNAM} and Q_m values for reaction 35 being $+ 36$ keV; on the contrary, reactions 32 and 36 show much lower differences, of $- 3$ and $+ 14$ keV respectively. These last two reactions have been obtained simultaneously with $C^{12}(He^3, p)N^{14}$ transitions, which Q-value remained 27 keV above its corresponding Q_m-value.

V. Future Q-Value Measurement Program

1. Stereo-Spectrograph

As the charged particle nuclear transition energy measurement is an accurate way to link radioactive and stable nuclei masses, an extension to such Q-value determinations at higher energies all along the periodic table would be desirable in order to establish the masses of a large number of isotopes.

Tandem Van de Graaff accelerators and magnetic spectrographs opened the possibility of such a study from a nuclear reaction point of view. Although these techniques have been developed mainly to study more general physical properties of nuclei, the ground state Q-values can be useful for mass determinations. The last version of magnetic spectrographs of the broad-range type, already described in previous mass meetings by BUECHNER[14], is the multiple gap broad-range spectrograph (stereo-spectrograph) developed originally at M. I. T. by ENGE and BUECHNER[3]. At Aldermaston, Hinds and Middleton succeeded in putting a similar spectrograph into operation, even before M. I. T. did.

Due to the versatility of such an instrument, the construction of a similar one was adopted in Mexico and it is expected to be in operation by the end of the present year. This spectrograph is going to be used first in conjunction with a 3 MeV Dynamitron machine[15], although it has been designed to cover reactions induced at Tandem energies (15 MeV).

a) Characteristics of the Spectrograph

The design of the Mexican stereo-spectrograph[16] was based on the statistical regression Q-value curves for various types of reactions induced

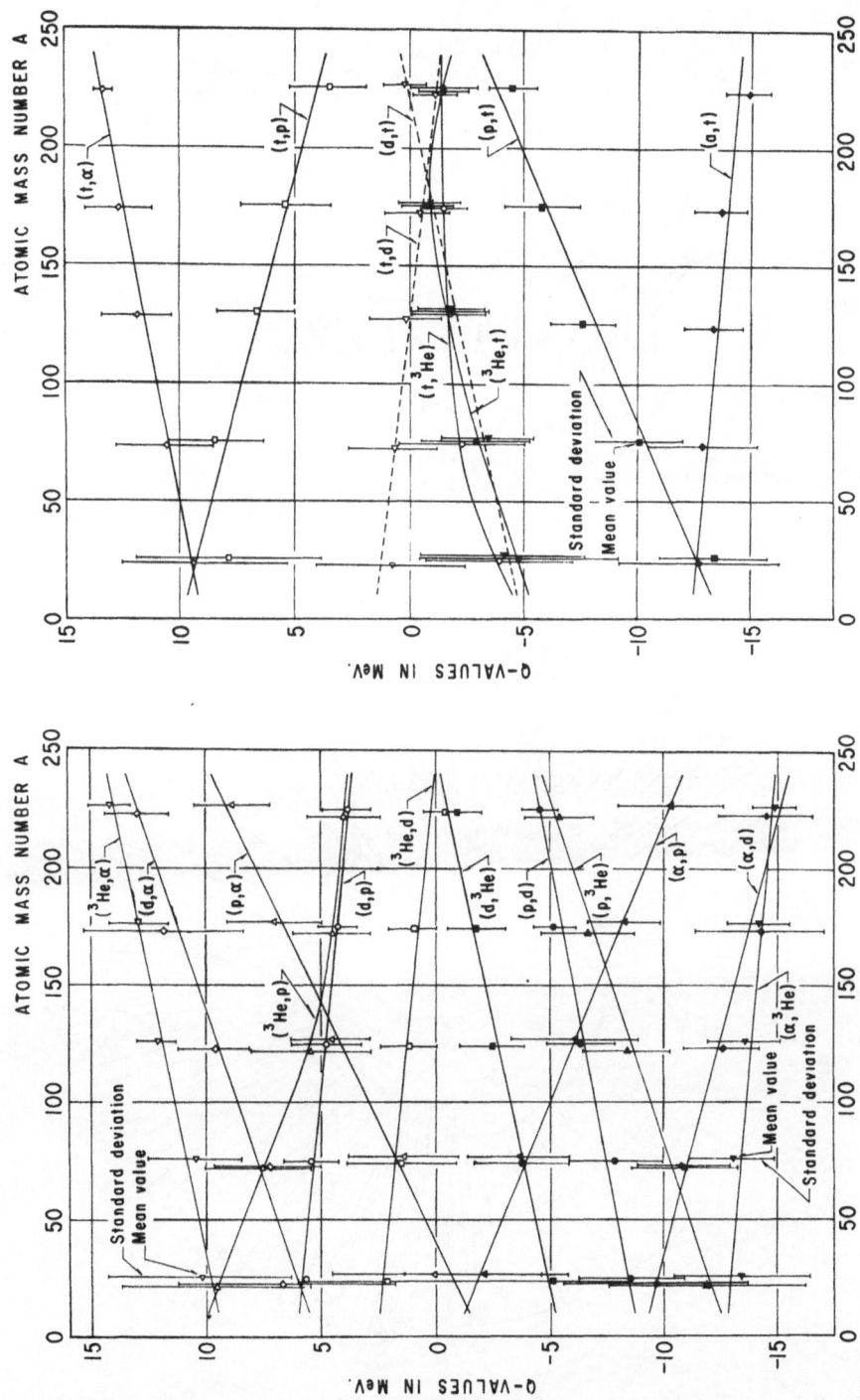

Fig. 6. Q-value vs mass number statistical regression curves for nuclear reactions calculated from spectroscopic mass determinations

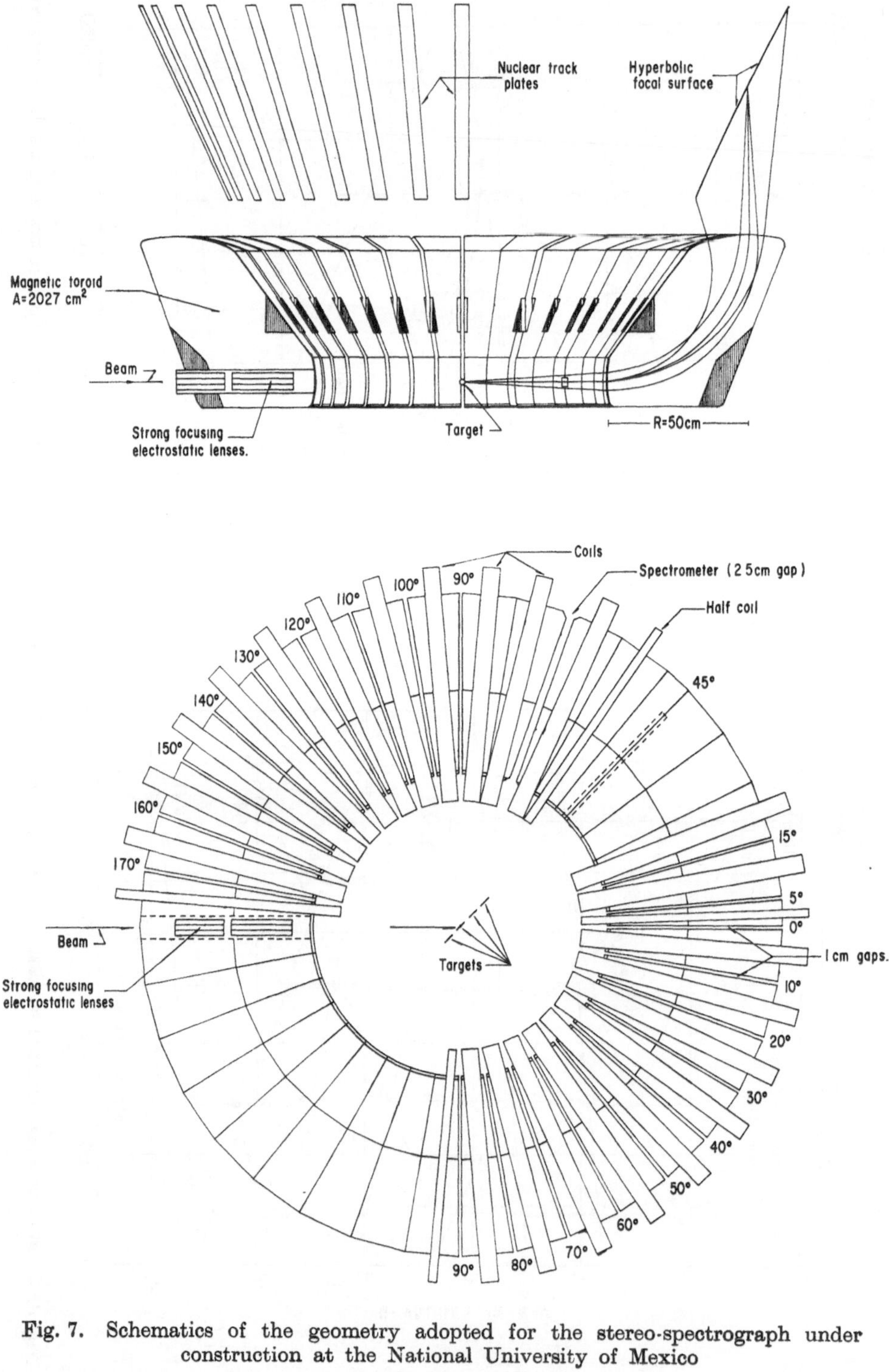

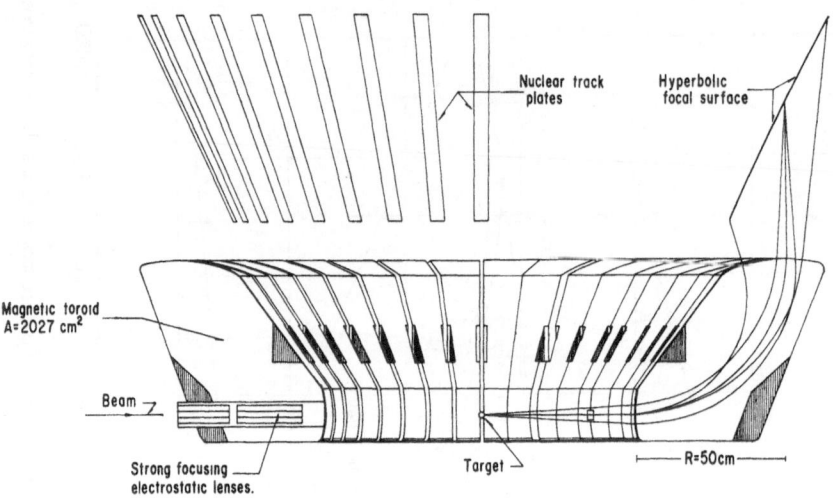

Fig. 7. Schematics of the geometry adopted for the stereo-spectrograph under construction at the National University of Mexico

by light nuclei, which are shown in Fig. 6; these were obtained from the available atomic masses.

From the expected particles and energies it was possible to select the radius (50 cm) and intensity of the magnetic field (16000 Gauss) in order to bend up to 38 MeV protons, with which practically any reaction can be analyzed. 21 one centimeter gaps have been located from 0° to 90° in one quadrant of the toroid and from 90° to 170° in the opposite quadrant, with the addition of two extra gaps at 5° and 15° in order to survey the lower angle group distributions in small steps. One extra 2.5 cm gap, which is located at 70° with respect to the beam direction, is planned to be used as a spectrometer and probably to initiate polarization experiments. The geometry adopted for the Mexican spectrograph, shown schematically in Fig. 7, permits the simultaneous analysis of particles by target reflection and transmission.

The whole instrument will be housed in a vacuum tank with flexible joints to the jacks which support the toroid. It can be levelled, translated and rotated by means of screws. 21 plate holders ($2'' \times 30''$) focussed to optimum position can rotate on a precision roller bearing in order to index the plates as well as to translate zones on the plates without breaking the vacuum.

The coils, each constructed with a 63×3.2 mm copper solid strip and with one central heat exchanger, will produce the 16 Kgauss magnetic field in the toroid, for which a stabilized power supply of 500 A, 70 V, 35 kW is needed. The overall weight of the instrument is 24 tons. For more detailed information on this instrument see reference 3.

b) Possibilities of a Tandem-Spectrograph Technique

From the 266 stable isotopes with enough abundance ($> 0.1\%$) to prepare thin targets by magnetic separation procedures, about 940 nuclear spectra can in principle be studied; there exists besides the possibility of producing around another 100 nuclei not yet known. About 270 isotopes (mainly radioactive) and fission fragments cannot be reached by this method. It is estimated that around 560 research man · years are needed to make a survey of nuclear properties along the periodic table. This refers only to reactions where the incident and outgoing particles are hydrogen or helium isotopes.

The masses of such a large number of radioactive isotopes which could be determined with relatively good precision through nuclear reactions, and which certainly would be very difficult to measure by direct mass spectroscopic methods, will rely basically on a central column of stable masses precisely established by mass spectroscopic techniques.

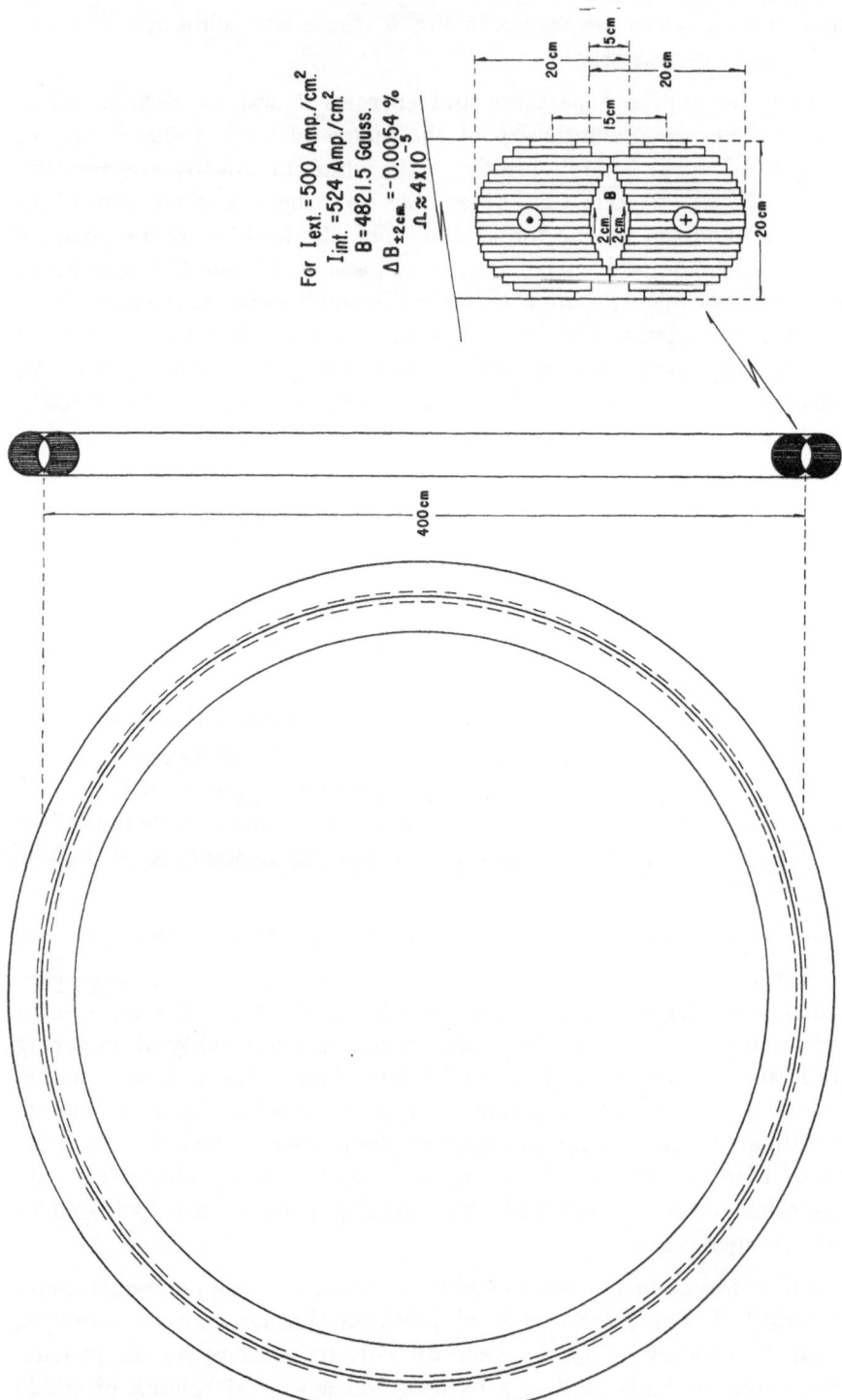

For $I_{ext.} = 500$ Amp./cm.^{2}
$I_{int.} = 524$ Amp./cm^2
$B = 4821.5$ Gauss.
$\Delta B_{\pm 2\,cm.} = -0.0054$ %
$\Lambda \approx 4 \times 10^{-5}$

Fig. 8. Iron free absolute 180° magnetic spectrograph for Q-value measurements

2. Absolute Q-value Spectrograph

H. A. ENGE suggested to us last year the possibility of employing an absolute homogeneous field iron free 180° spectrograph, based on the idea of J. W. BLAMEY[17], used for the orbital magnet of the 10 GeV proton synchrotron of the Australian National University.

According to theoretical calculations done for us at Oak Ridge by Dr. W. F. GAUSTER and his group[18], a geometry as the one shown schematically in Fig. 8 offers an attractive possibility for an accurate absolute Q-value spectrograph. The main result of these calculations can be summarized as follows. Assuming the cross section of the double torus as formed by 20 strips of one cm wide rectangles to fit the proposed solid cross section on each side as well as possible, and by increasing the inner torus magnet current by 4.88%, the field gradient is expected to be 0.0070% per cm. The field strength at ± 2 cm from the center becomes 0.0054% smaller than the central field, which is 4821.5 gauss when supplying a current of 500 Amp/cm² in the outer coil.

Under the proposed conditions this type of spectrograph could bend protons of energies up to 43 MeV. The solid angle for reactions would be of the order of 4×10^{-5} steradians, and the machine would have an energy resolution of 6000 for practical purposes. This corresponds to an improvement of a factor of five to ten in Q-value measurements as compared to those presented in this paper, besides having the advantage of not needing a calibrator substance to compare against and which involves another source of error.

VI. Acknowledgments

The authors are indepted to Drs. W. W. BUECHNER and H. A. ENGE of M. I. T. for valuable discussions and suggestions; to Dr. W. F. GAUSTER of Oak Ridge for his magnetic field calculations that saved our group a good amount of time; to LUIS and FRANCISCO VELÁZQUEZ who are in charge of the accelerator and made all the exposures and to E. OSALDE, B. GALVÁN, S. CASTILLO and G. MORENO for their patient scanning of the photographic plates.

References

[1] MAZARI, DOMÍNGUEZ, JÁIDAR, RICKARDS, ALBA, LÓPEZ, and O. DE LÓPEZ, Q-values of Nuclear Reactions in the Light Element Region. Proceedings of the International Conference on Nuclidic Masses, p. 276. H. E. Duckworth, Edit. University of Toronto Press. 1960.

[2] W. W. BUECHNER, Q-values from Charged-Particle Reactions. Proceedings of the International Conference on Nuclidic Masses, p. 271. H. E. Duckworth, Edit. University of Toronto Press. 1960.

[3] H. A. ENGE, and W. W. BUECHNER, Multiple-Gap Magnetic Spectrograph for Charged-Particle Studies. Rev. Sci. Instr. 34, 155 (1963).

[4] WAPSTRA, NIJGH, and VAN LIESHOUT, Nuclear Spectroscopy Tables, p. 128. North Holland Publishing Company. 1959.

[5] TEJERA, MAZARI y JÁIDAR, Remedición de Valores Q_0 de Reacciones Nucleares entre los Elementos Ligeros. Rev. Mex. Fís. 10, 229 (1961).

[6] EVERLING, KÖNIG, MATTAUCH, and WAPSTRA, Atomic Masses of Nuclides for $A \leqslant 70$. Nuclear Phys. 15, 342 (1960).

[7] A. TEJERA, PhD Thesis. Faculty of Science. UNAM. 1962.

[8] H. A. ENGE, Table of Charged Particle Energies vs Magnetic Field Strength × Orbit Radius. Bergen: A. S. John Griegs Boktrykkeri. 1954.

[9] G. LÓPEZ y O. ALMÉN, Valores Q_0 en la Región Carbono-Neón para Determinaciones Precisas de Masas Nucleares. Rev. Mex. Fís. 10, 239 (1961).

[10] F. AJZENBERG-SELOVE, and T. LAURITSEN, Energy Levels of Light Nuclei VI. North Holland Publishing Company. 1959.

[11] SPENCER, PHILLIPS, and YOUNG (Rice University), Nuclear Phys. 21, 310 (1960); Phys. Rev. 116, 962 (1959).

[12] FORSYTH, BARROS, JAFFE, TAYLOR, and RAMAVATARAM (University of Manchester), Proc. Phys. Soc. 75, 291, 772 (1960).

[13] S. HINDS, and R. MIDDLETON (Atomic Weapons Research Establishment, Aldermaston), Proc. Phys. Soc. 73, 721 (1959); 75, 745 (1960).

[14] W. W. BUECHNER, The Determination of Nuclear Q-values, p. 85. Nuclear Masses and Their Determination. Pergamon Press. 1957.

[15] 3 MeV Dynamitron Accelerator donated to UNAM by the U. S. Government.

[16] MAZARI, ALBA, LÓPEZ y MELLO, Proyecto de un Laboratorio Avanzado de Espectroscopía Nuclear, p. 223. IV Inter-American Symposium on the Peaceful Application of Nuclear Energy IANEC. 1962.

[17] J. W. BLAMEY, The Orbital Magnet and Power Supply of the 10 GeV Proton Synchroton at the Australian National University. Symposium du CERN. Comptes Rendus 1, 344 (1956).

[18] W. F. GAUSTER, Personal Communication.

Discussion

D. M. VAN PATTER: I wonder if you could give additional information concerning your procedure in obtaining the correction curve for saturation. If you obtain corrections using a fixed angle of observation and fixed bombarding energy, then the corrections for a given reaction group apply to different radii of curvature as the magnetic field is increased. Since the saturation might not be the same for all radii of curvature, your correction curve might be subject to systematic error.

M. MAZARI: The correction was obtained at 90° with respect to the beam direction, and at different bombarding energies. The entire saturation region was covered at all radii of curvature with different reactions as indicated in the paper and averaged. The results remained near the mean curve and it seems that this first order approximation is sufficient for the actual correction purpose.

D. M. VAN PATTER: Then I take it that your correction curve is not based on observations of a single reaction group.

M. MAZARI: That is correct, it is based on different ground and excited state reactions.

C. P. BROWNE: Concerning the measurement of He³ induced reactions and the uncertainty in the He³ mass, I shall comment on work at Notre Dame.

We have measured the $N^{14}(He^3, p)O^{16}$ reaction and are working on $B^{10}(He^3, p)C^{12}$. Here the mass differences are well known. If we use the masses from the Wapstra table, good agreement is obtained. If one attempts to find the He^3 mass directly from nuclear reactions, independent of mass spectrometer data one finds large discrepancies in published results. A preliminary Notre Dame result lies near the middle of the published values. More precise nuclear Q-value work is planned in the region of light nuclei.

K. T. BAINBRIDGE: Dr. MAZARI showed a slide of a very ingenious iron-free magnetic analysis. The suggestion for its design was attributed to Dr. BLAMEY. I believe that Professor I. I. RABI first suggested this general method for securing a uniform field in the *Reviews of Scientific Instruments* in the 1930's.

A. O. NIER: At the Hamilton conference it was pointed out that there was a systematic difference between a mass table constructed purely from mass spectroscopic results and one constructed purely from reaction energies. During the last three years there have been no systematic mass spectroscopic measurements in the region of disagreement. On the other hand, there are new Q-value measurements and changes in calibration energies. Perhaps someone can comment on the present state of the systematic discrepancy.

C. P. BROWNE: We are working on the question of possible systematic discrepancies between mass spectrometer results and nuclear reaction results. It is too early to answer the question with certainty but it appears that there is now no systematic discrepancy with mass spectrometer results.

M. MAZARI: I can make a comment here. I compared our Q-values, all the Q-values that we have, with all the values that Dr. SPERDUTO has presented just now and we have very, very close agreement. I think the maximum deviation in one or two cases is something like 20 or 25 keV as a maximum. Usually we are within, say, 5 to 10 keV. So the Q-values at M. I. T. and Mexico are very close together.

A. O. NIER: If the Mexican and M. I. T. values are in agreement, this would mean that the discrepancy to which I referred still exists.

A. H. WAPSTRA: I don't think there is, in fact, any serious and systematic discrepancy between mass spectroscopic results and Q-values. There was one, but as you might remember, we said at Kingston that the systematic discrepancy would disappear if you assumed the calibration energy (Po α) to be 5 keV higher. At the moment, the Po α energy is adopted to be quite a lot higher. It is 7 keV higher so you might say that instead of a -5 we have a $+2$ discrepancy.

Q-Value Measurements at the University of Pittsburgh

By

Roshan K. Patell and **Bernard L. Cohen**

University of Pittsburgh, Pittsburgh 13, Pa.

(Presented by B. L. Cohen)

The 15 MeV deuteron beam from the University of Pittsburgh cyclotron is being used to determine neutron binding energies by measuring *Q*-values for (d, p) and (d, t) reactions. Thin — and in many cases, isotopically enriched — targets are irradiated by the magnetically focussed and analysed beam; reaction products are magnetically analysed and detected by photographic plates. The absolute energy calibration is achieved by running a spectrum from ^{57}Fe (d, p) or (d, t) adjacent to each run on an unknown without making any changes other than shifting the plate and target holders. The nearest well known particle groups from the ^{57}Fe reactions were used as standards. The energy resolution was typically 60 keV.

Table 1. Summary of Sources of Error

Nature of Error	Error in (d, p) (keV)		Error in (d, t) (keV)	
	Average	Maximum	Average	Maximum
Standard *Q*-values..............	11	11	11	11
Target thickness of				
1) Standard	3	3	6	6
2) Unknown	9	25	13	37
Change in cyclotron energy from standard	13	25	13	25
Plate handling	9	15	5	9
Spectrograph calibration	10	15*	7	10*
Uncertainty in angle of detection (via center of mass correction) .	3	7	6	15
Uncertainty in cyclotron energy (via center of mass correction, dE/dx)	1	1	2	4
Error in location of peak center..	9	25	5	15
Standard error (average case)	15		25	

The sources of error are listed in Table 1: they may be explained as follows:

Value of Standard.

The Q-values for the ^{57}Fe reactions are known[1] with 8 keV accuracy, and the excitation energies of the excited states (which are usually used as standards) are known to $\pm$ 8 keV[2], so the standards are assumed to have an uncertainty of 11 keV.

Target Thickness.

The energy difference between standard and unknown must be corrected for differences in the energy loss in the respective targets, so that uncertainty in the target thickness is an important source of error. Average thicknesses were determined by area measurement and weighing, and in several cases uniformity scans were made with an alpha-particle source – solid state detector arrangement. In general, the thickness uncertainty was about 5%. In some cases, additional errors are encountered because targets are wrinkled, or on backings of uncertain thickness.

Variation in Cyclotron Energy.

The magnetic analysis of the incident beam holds its energy constant only to within about 80 keV, so that a change in energy between runs on the standard and unknown could cause very large errors. In order to check this, a study was made of the apparent shifts between successive runs on the standard. The errors thus determined were used although two unknowns were run and plates were changed between these, whereas unknowns are calibrated against a standard run immediately adjacent to it and on the same plate.

Plate Handling.

Errors due to locating the plate in its holder and on the microscope stage, and due to the marker system used are included here.

Spectrograph Calibration.

The spectrograph energy scale has been calibrated by use of known levels in light nuclei, but the errors in this process are not negligible. Differences in the magnetic field pattern due to hysteresis have been observed at times. Parts of the system are somewhat moveable, although alignment is frequently checked with plumb-bobs and levels. All of these effects would be much more serious were it not for the fact that an ^{57}Fe calibration run is made on each plate.

Uncertainty in Angle of Detection.

The angle at which the cyclotron beam enters the scattering chamber is subject to small variations, causing an uncertainty in the angle between incident and outgoing particles. There is an additional contribution from alignment problems. These cause an uncertainty in the center of mass correction that must be applied to the energy difference between the ^{57}Fe and unknown peaks.

Uncertainty in Cyclotron Energy.

The energy of the cyclotron beam is unknown (and subject to long term drifts) by as much as $\pm$ 200 keV. This has an effect on the center of mass corrections. It also affects the target thickness correction, but this is negligible. Other effects are cancelled in the use of the ^{57}Fe for calibration.

Error in Location of Peak Center.

This depends on the resolution to some extent. In many cases, the statistics are so poor that the peak profile is very uncertain.

Table 2. Neutron Binding Energies (MeV)

(Revised as of February 1, 1964)

Nucleus	This Work	Previous	Nucleus	This Work	Previous
Zr^{91}	7.185 $\pm$ .025	7.199 $\pm$ .007	Cd^{112}	9.392 $\pm$ .020	9.857 $\pm$ .006
Zr^{92}	8.622 $\pm$ .025	8.630 $\pm$ .006	Cd^{113}	6.557 $\pm$ .025	6.534 $\pm$ .004
Zr^{93}	6.716 $\pm$ .030	6.750 $\pm$ .010	Cd^{114}	9.045 $\pm$ .030	9.041 $\pm$ .004
Zr^{94}	8.222 $\pm$ .025	8.187 $\pm$ .010	Cd^{115}	6.141 $\pm$ .025	6.148 $\pm$ .012
Zr^{95}	6.448 $\pm$ .030	6.468 $\pm$ .013	Cd^{116}	8.714 $\pm$ .020	8.684 $\pm$ .012
Zr^{96}	7.852 $\pm$ .025	7.853 $\pm$ .013	Cd^{117}	5.763 $\pm$ .030	
Zr^{97}	5.563 $\pm$ .030	5.578 $\pm$ .032	In^{115}	9.047 $\pm$ .030	9.022 $\pm$ .013
Nb^{93}	8.838 $\pm$ .020	8.822 $\pm$ .044	In^{116}	6.719 $\pm$ .025	6.759 $\pm$ .061
Nb^{94}	7.190 $\pm$ .020	7.195 $\pm$ .025	Sn^{113}	7.729 $\pm$ .025	8.041 $\pm$ .012
Mo^{93}	8.077*$\pm$ .020	7.933 $\pm$ .065	Sn^{114}	10.310 $\pm$ .020	10.005 $\pm$ .012
Mo^{94}	9.699 $\pm$ .020	9.807 $\pm$ .065	Sn^{115}	7.558 $\pm$ .025	7.523 $\pm$ .012
Mo^{95}	7.362*$\pm$ .020	7.371 $\pm$ .005	Sn^{116}	9.575 $\pm$ .020	9.570 $\pm$ .010
Mo^{96}	9.181 $\pm$ .020	9.155 $\pm$ .005	Sn^{124}	8.518 $\pm$ .035	8.511 $\pm$ .012
Mo^{97}	6.807*$\pm$ .020	6.816 $\pm$ .004	Sn^{125}	5.755 $\pm$ .030	5.742 $\pm$ .015
Mo^{98}	8.637 $\pm$ .020	8.640 $\pm$ .004	Au^{197}	8.078 $\pm$ .030	8.067
Mo^{99}	5.912*$\pm$ .020	6.119 $\pm$ .300	Au^{198}	6.507 $\pm$ .030	6.497
Mo^{100}	8.296 $\pm$ .020	8.097 $\pm$ .300	Pt^{194}	8.384 $\pm$ .020	
Mo^{101}	5.391*$\pm$ .020		Pt^{195}	6.125 $\pm$ .020	6.178 $\pm$ .040
Pd^{104}	10.020 $\pm$ .025	10.047 $\pm$ .032	Pt^{196}	7.940 $\pm$ .020	7.922 $\pm$ .012
Pd^{105}	7.108 $\pm$ .030	7.063 $\pm$ .016	Pt^{197}	5.852 $\pm$ .020	
Pd^{106}	9.527 $\pm$ .025	9.543 $\pm$ .013	Pt^{198}	7.563 $\pm$ .020	
Pd^{107}	6.548 $\pm$ .030	6.542 $\pm$ .004	Pt^{199}	5.572 $\pm$ .020	
Pd^{108}	9.235 $\pm$ .030	9.223 $\pm$ .006	Pb^{206}	8.096 $\pm$ .030	8.124 $\pm$ .035
Pd^{109}	6.161 $\pm$ .030	6.147 $\pm$ .007	Pb^{207}	6.751 $\pm$ .030	6.731 $\pm$ .008
Ag^{109}	9.205 $\pm$ .030	9.182 $\pm$ .021	Pb^{208}	7.372 $\pm$ .025	7.376 $\pm$ .008
Cd^{110}	9.916 $\pm$ 0.25	9.857 $\pm$ .006	Bi^{209}	7.474 $\pm$ .030	7.432 $\pm$ .010
Cd^{111}	6.984 $\pm$ .020	6.965 $\pm$ .005			

Uncertainty in Peak Identification.

This could be a source of error in a very few cases. This problem is somewhat eased by our experience on relative cross sections to be expected. Known excited states are identified in almost all cases.

From Table 1, the average standard error expected from these effects is about 25 keV. Measurements were made for about 50 nuclei for which the *Q*-values are known from other data[1]; the distribution of discrepancies is in rough agreement with this result. Studies were made of the dependence of these discrepancies on distance between standard and unknown lines, and on the portion of the plates where the measurements were made; no very important correlations were noted.

The preliminary results for cases where there are discrepancies with previous data or where the estimated errors are much reduced are listed in Table 2. These data have been fully processed, but it is felt that an independent set of check runs should be made before the results are considered final.

References

[1] L. A. KÖNIG, J. H. E. MATTAUCH, and A. H. WAPSTRA, Nuclear Phys. 31, 1 (1962).

[2] Mass. Inst. Tech. Lab. Nuc. Science Progress Report, Nov. 1, 1957.

Discussion

H. E. DUCKWORTH: Professor COHEN has done us the courtesy of showing us his results in advance. Dr. BARBER and I shall present results tomorrow for several elements giving the mass difference between isotopes that are two mass units apart. Thus, they are equivalent to the sum of two (d, p) reactions or a (d, p) plus a (d, t) and, as such, can be compared to Professor COHEN's results. In zirconium two comparisons are possible and the agreement in one case is 5 keV and in the other case 3 keV. Good agreement is also found for tin isotopes. With molybdenum, cadmium and tellurium, however, where comparisons are possible, the two sets of data are systematically different by approximately 100 keV, or 50 keV per reaction. If our results turn out to be correct, they would suggest that Professor COHEN's (d, p) *Q*-values are too large by roughly this amount. I should add that Dr. JOHNSON and Dr. RIES, of the University of Minnesota, will also present results tomorrow which are in good agreement with ours.

B. L. COHEN: I'm willing to bet anyone here at least 10 to 1 that I am wrong and you are right but if anybody could give me any idea how we could be off 100 keV I would like to know.

G. C. PHILLIPS: It seems to me that this method of Dr. COHEN's is a very good one because it is a null-type of measurement and should allow accurate relative calibration of nuclear reaction energies. However, to use this method it seems to me than an absolute criteria of how the energy of a broad line is to be picked is needed. For example, one has different target thickness for unknown and for comparison spectra and one must determine how to compare them. When targets are as thick as several hundred kilovolts, and when there are strong fringing field effects in the spectrometer, the corrections can be quite large and rather complicated. I would suggest these effects should be investigated so that the accuracy of this interesting method can be extended.

B. L. COHEN: We always assume that the reaction takes place at the center of the target. I should say that these discrepancies are in cases where the targets are particularly thick so that this might be a cause.

The Relative Masses of Some Uranium Isotopes Determined from Nuclear Reaction Studies

By

R. Middleton and **H. Marchant**

Atomic Weapons Research Establishment, Aldermaston, Berkshire, England

With 8 Figures

1. Introduction

During the past few years the Aldermaston multi-angle spectrograph[1] and Tandem accelerator have been used to study a wide range of nuclear reactions with relatively light nuclei. Recently, however, some charged particle reaction studies have been performed on the uranium isotopes and presented here is a summary of the results so far obtained. Since measurements and analysis are still in progress these results are preliminary and may be subject to slight revision.

The main purposes of the investigation were (1) to measure accurate ground state Q-values, (2) to obtain detailed information about the low lying energy levels and (3) to generally explore the possibility of interpreting the angular distributions and making spin and parity assignments.

The principal experimental difficulty in the study of heavy nuclei is to obtain sufficient intensity while preserving good energy resolution conditions. To this end a multi-angle spectrograph is an enormous advantage particularly if it is desired to measure angular distributions. However, even a single exposure is fairly lengthy and the present exposures of between 4000 and 7000 micro-coulombs required 12 to 15 hours running time. This was with targets of about $500 \ \mu\mathrm{gm} \cdot \mathrm{cm}^{-2}$ which were sufficiently thick to have an adverse effect on the energy resolution. Usually the latter was about 20 keV for the forward angles but for angles greater than $90°$ which view the target by reflection the resolution was frequently reduced to about 30 keV.

2. Results

$U^{238}(d, p)U^{239}$.

The $U^{238}(d, p)U^{239}$ reaction has been studied in most detail and a typical proton energy spectrum is shown in Fig. 1. This was measured

R. MIDDLETON and H. MARCHANT:

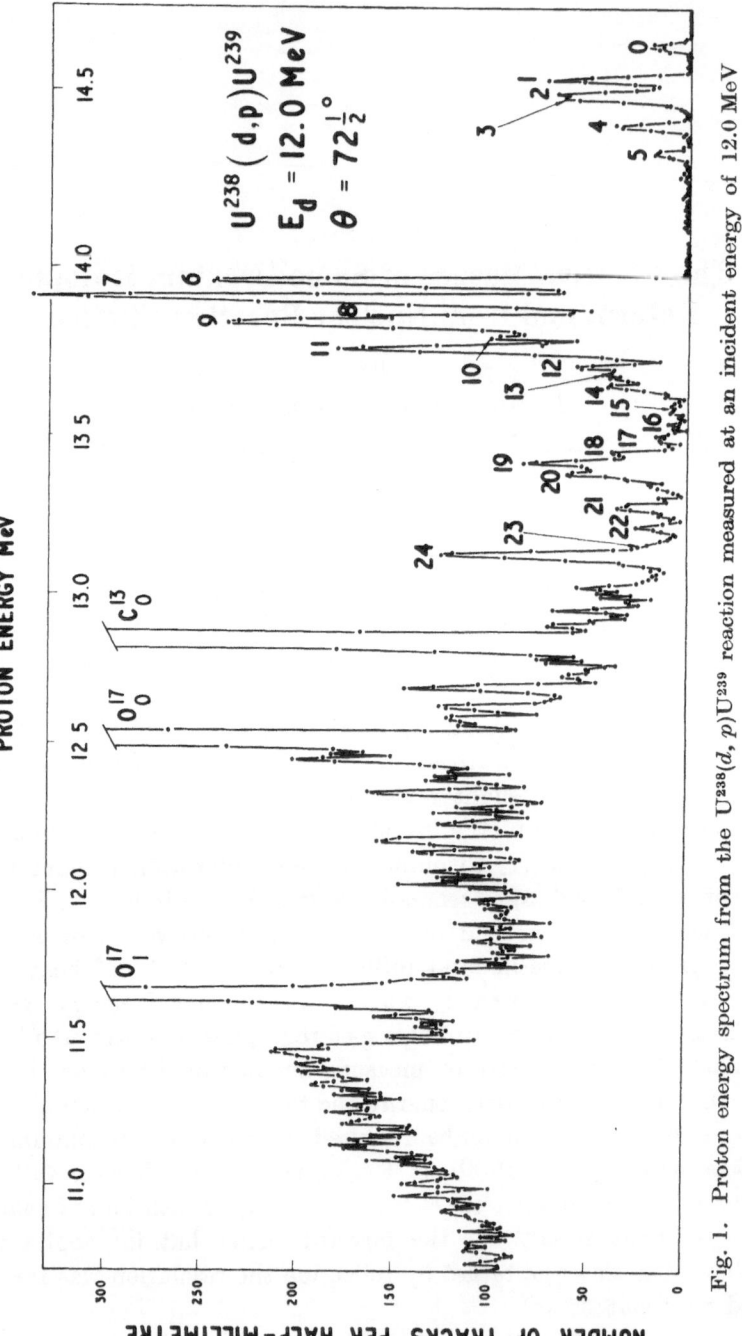

Fig. 1. Proton energy spectrum from the $U^{238}(d, p)U^{239}$ reaction measured at an incident energy of 12.0 MeV

at an angle of $72\frac{1}{2}°$ using deuterons of 12.0 MeV. The groups identified as arising from U^{238} are labelled numerically with O referring to the ground state. Those arising from target impurities are labelled with the symbol for the final nucleus with a subscript to indicate the appropriate state. The excitation energies of 24 excited states are given in Table 1.

To minimise the errors in the spectrograph energy calibration the ground state Q-value was determined in terms of the accurately known Q-value of the $C^{12}(d, p)C^{13}$ reaction (2.722 MeV). This was a particularly convenient and accurate procedure since it was possible to find an angle at which the two ground state groups almost coincided. The Q-value determined in this way is 2.588 $\pm$ 0.020 MeV which may be compared with the accepted mass Q-value of 2.537 MeV (KÖNIG, MATTAUCH and WAPSTRA[2]).

Table 1. Energy Levels of U^{239} Determined from the $U^{238}(d, p)U^{239}$ Reaction

Group Number	Energy (MeV)	l	Group Number	Energy (MeV)
0	0	2	13	0.96 $\pm$ 0.015
1	0.092 $\pm$ 0.005	4	14	0.99 $\pm$ 0.015
2	0.131 $\pm$ 0.005	1	15	1.06 $\pm$ 0.015
3	0.138 $\pm$ 0.005		16	1.11 $\pm$ 0.015
4	0.222 $\pm$ 0.005	4	17	1.15 $\pm$ 0.015
5	0.301 $\pm$ 0.005	5	18	1.19 $\pm$ 0.015
6	0.688 $\pm$ 0.005	2	19	1.22 $\pm$ 0.015
7	0.738 $\pm$ 0.010	2	20	1.25 $\pm$ 0.015
8	0.789 $\pm$ 0.010		21	1.36 $\pm$ 0.015
9	0.809 $\pm$ 0.010	2	22	1.43 $\pm$ 0.015
10	0.848 $\pm$ 0.010		23	1.49 $\pm$ 0.015
11	0.883 $\pm$ 0.010	3	24	1.52 $\pm$ 0.015
12	0.933 $\pm$ 0.010	—		

Angular distributions of the ground and several excited state groups have been measured and are shown in Fig. 2. Perhaps the most striking feature of these is their similarity which arises due to the dominance of coulomb effects. At first sight it would appear doubtful if reliable l-values could be assigned but it was possible because of the excellent agreement with distorted wave stripping calculations.

The full line curves in the figure are the results of D. W. B. A. calculations performed by MACEFIELD at Aldermaston. Since at the outset only the l-value for the ground state angular distribution was known with certainty the procedure was to find a set of distorting parameters to fit this. The first calculations were made using proton and deuteron parameters taken from experimental measurements on near by nuclei at roughly the appropriate energies. Very good agreement was obtained almost immediately and the improved fit shown in Fig. 2 was obtained

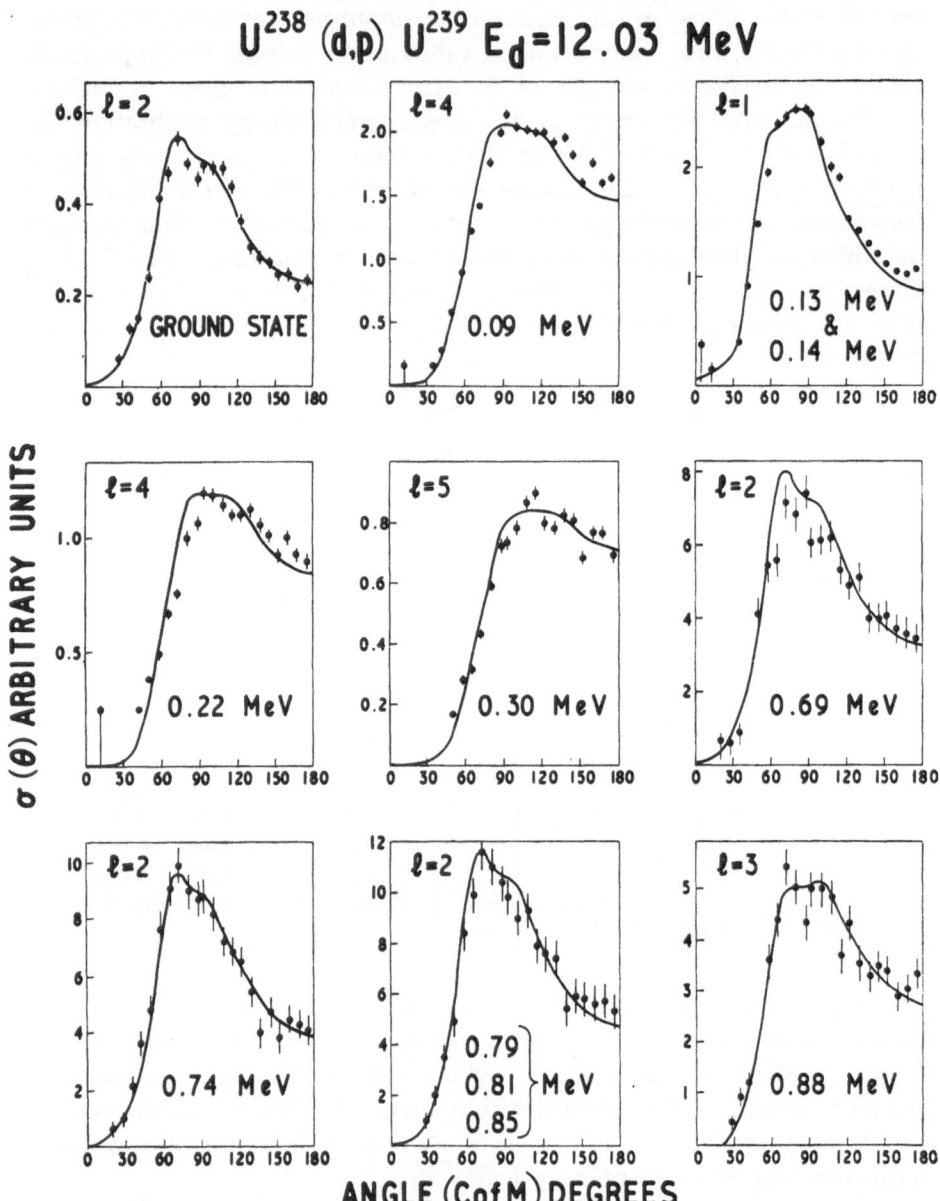

Fig. 2. Angular distributions of the ground and several excited state groups from the $U^{238}(d, p)U^{239}$ reaction. The full line curves were calculated from D. W. B. A. theory

by slightly increasing the imaginary part of the proton potential. Having thus established a set of parameters for the ground state these were used to calculate the curves for the excited states. A detailed account of the fitting procedure will be published elsewhere but it can be concluded that we are fairly confident that the l-value assignments for $l \leqslant 4$ are correct. For $l > 4$ there is an uncertainty of about one unit of orbital angular momentum.

From comparisons with Pu^{241} it is expected that the first member of the ground state rotational band should occur at about 45 keV while the first observed level is at 92 keV. However the Nilsson model[3] predicts the reduced width of this state to be less than that of the ground state

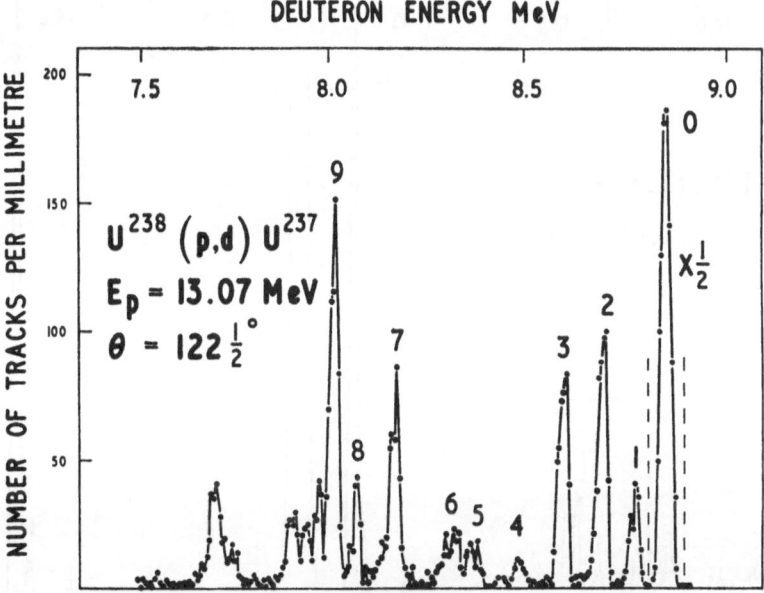

Fig. 3. Deuteron energy spectrum from the $U^{238}(p, d)U^{237}$ reaction measured at an incident energy of 13.0 MeV

by at least an order of magnitude and therefore it is not surprising that it was not observed. The observed state at 92 keV is almost certainly the $9/2^+$ second member of the ground state rotational band. Supporting this is the approximately correct excitation energy, the $l = 4$ assignment to the angular distribution and the ratio of the peak differential cross-section to that of the ground state is 3.7 ± 0.1 compared with 3.5 predicted by the Nilsson model with D. W. B. A. theory.

The third and fourth members of the ground state rotational band are expected to occur at about 150 and 225 keV respectively. Although fairly intense groups are observed at not very different energies it is unlikely that these are they since the Nilsson model predicts both states to have small reduced widths. Transitions to these states would also be expected to be weak because of the high angular momentum transfer ($l = 6$).

$U^{238}(p, d)U^{237}$.

Fig. 3 shows a typical spectrum from the $U^{238}(p, d)U^{237}$ measured with protons of 13.0 MeV and at an angle of $122^1/_2°$. Following the procedure for the previous reaction groups identified arising from U^{238} are labelled numerically with O referring to the ground state. Note that the strongest transition is to the ground state and that this is shown

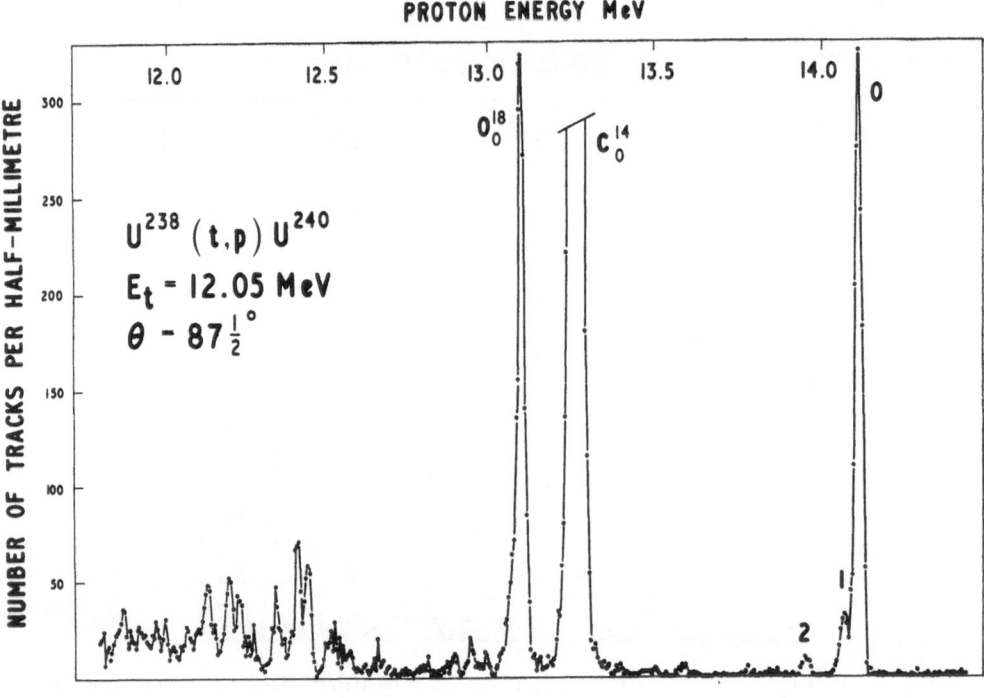

Fig. 4. A typical proton energy spectrum from the $U^{238}(t, p)U^{240}$ reaction

times $^1/_2$ in the figure. Also that the group attributed to the ground state is probably a doublet since the first excited state of U^{237} is known to be at about 10 keV excitation energy[4].

The ground state Q-value was determined in terms of the Q-value of the $C^{13}(p, d)C^{12}$ reaction i. e. using the same Q-value as was used to determine the Q-value of the $U^{238}(d, p)U^{239}$ reaction. The value obtained is -3.951 ± 0.020 MeV which may be compared with the accepted mass Q-value of -3.817 MeV.

$U^{238}(t, p)U^{240}$.

The third reaction to be studied with U^{238} is the (t, p) reaction proceeding to U^{240}. This is of particular interest since as predicted by YOSHIDA[5] the ground state transition is enhanced. This effect is well illustrated in Fig. 4 which shows a typical proton energy spectrum measured at $57^{1}/_{2}°$. It will be noticed that in addition to the strong transition to the ground state relatively weak transitions are observed to the first and second members of the ground state rotational band. There then appears to be a gap in the spectrum extending to about 1 MeV excitation which is followed by a region containing a large number of weakly excited levels.

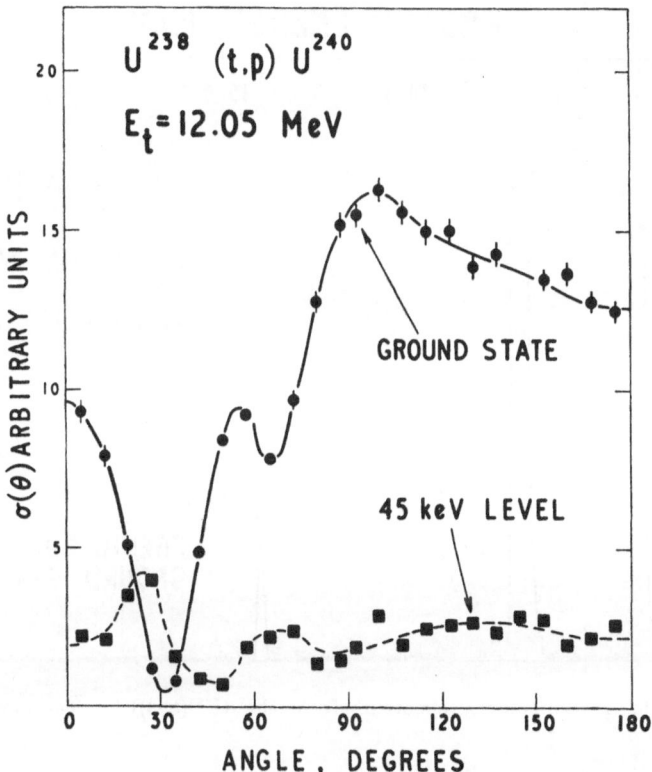

Fig. 5. Angular distributions of the ground and first excited state proton groups from the $U^{238}(t, p)U^{240}$ reaction

This spectrum is probably typical of a (t, p) reaction with a heavy even-even target nucleus. The enhanced transition to the ground state and the relative certainty with which it can be identified clearly make this a very useful technique for determining nuclear mass differences. It is also noteworthy that the inverse (p, t) reaction should behave similarly and might prove equally useful.

The ground state Q-value has been determined in terms of the well known Q-value of the $O^{16}(t, p)O^{18}$ reaction (3.706 MeV) and is 2.242 ± 0.020 MeV. This may be compared with the accepted mass Q-value of 2.130 MeV.

Angular distributions have been measured for the ground and first excited state groups and are shown in Fig. 5. Since these were obtained very recently no distorted wave calculations have yet been performed. It is gratifying, however, that the ground state distribution closely resembles the predicted shape for an $l = 0$ (d, p) transition and little difficulty is anticipated in obtaining agreement with double stripping theory.

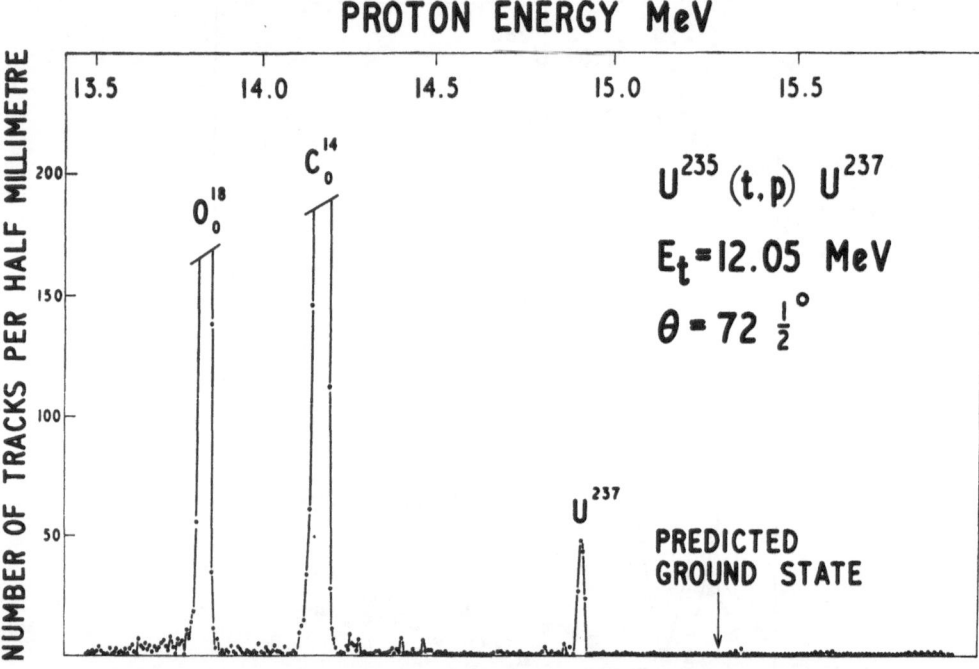

Fig. 6. Proton energy spectrum from the $U^{235}(t, p)U^{237}$ reaction

$U^{236}(t, p)U^{238}$.

In order to establish a link between U^{238} and U^{236} the $U^{236}(t, p)U^{238}$ reaction has been studied. Unfortunately only a very limited amount of separated U^{236} isotope was available and a complicated painting technique had to be used to prepare the target. This resulted in a rather unsatisfactory target containing a large number of light element impurities but largely due to the U^{236} ground state transition being enhanced this was readily identified. The reaction Q-value was determined in terms of the $C^{12}(t, p)C^{14}$ reaction Q-value (4.614 MeV) and is 2.900 ± 0.020 MeV. This may be compared with the accepted mass Q-value of 2.978 MeV.

$U^{235}(t, p)U^{237}$.

Another (t, p) reaction studied is the $U^{235}(t, p)U^{237}$ reaction and Fig. 6 shows a typical proton energy spectrum measured at an angle of $72^{1}/_{2}°$.

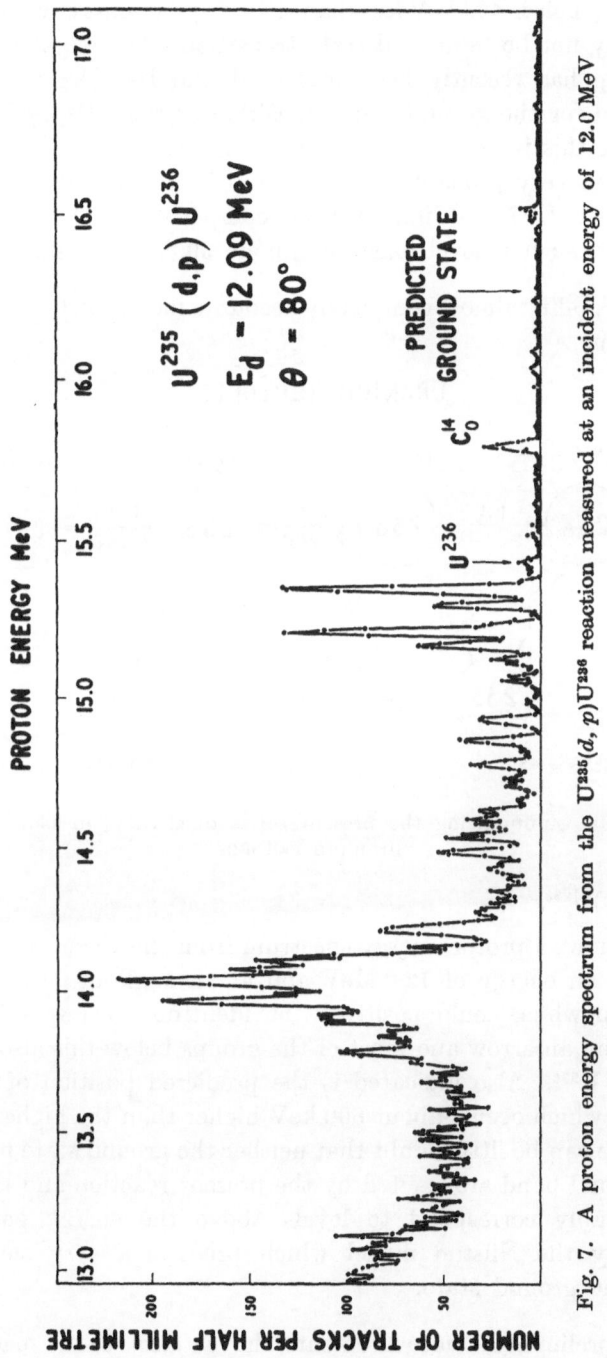

Fig. 7. A proton energy spectrum from the $U^{235}(d, p)U^{236}$ reaction measured at an incident energy of 12.0 MeV

Only the one group of protons indicated could be identified corresponding to U^{237} and the Q-value for this group was determined to be 2.980 ±

$\pm$ 0.020 MeV. This is 423 keV less than the accepted mass Q-value suggesting that this may not be te ground state transition. The angular distribution of this group has recently been measured and its close resemblance to that obtained for the ground state transition of the $U^{238}(t, p)U^{240}$ reaction suggests that this is an $L = 0$ transition. Since the ground state of U^{235} is $7/2 -$ it is very probable that the transition proceeds to a low lying $7/2 -$ level in U^{237} and indeed from comparison with Pu^{239} it is very probable that such a level exists at about 390 keV excitation.

This possibility almost completely accounts for the 423 keV discrepancy in the Q-value.

URANIUM ISOTOPES

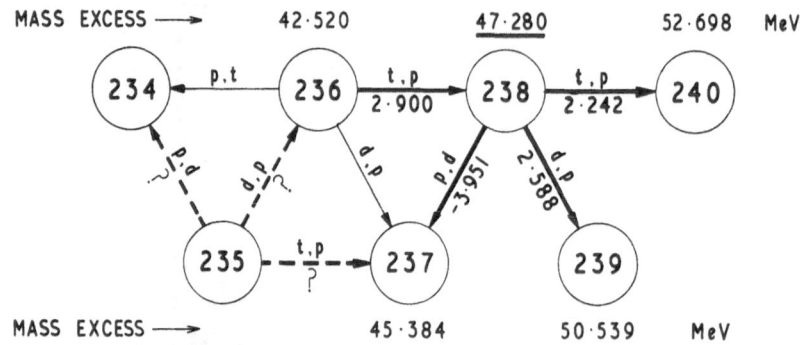

Fig. 8. Diagram summarising the present series of Q-value measurements on the uranium isotopes

$U^{235}(d, p)U^{236}$.

Fig. 7 shows a proton energy spectrum from the $U^{235}(d, p)U^{236}$ reaction measured at an energy of 12.0 MeV and at an angle of 80°. The highest energy group which could positively be identified corresponding to U^{236} is indicated by an arrow and most of the groups below this also correspond to states in U^{236}. Also indicated is the predicted position of the ground state groups which occurs about 800 keV higher than the highest identified group. There can be little doubt that neither the ground state nor members of its rotational band are excited by the present reaction and the observed groups probably correspond to levels above the energy gap. This is confirmed by the Nilsson model which predicts a very small reduced width for the ground state.

Some preliminary measurements have also been made on the $U^{235}(p, d)U^{234}$ reaction but again there is some doubt whether the ground state is being excited. However, these results have not yet been fully analysed.

3. Conclusions

Fig. 8 summarises the present series of measurements. Starting from U^{238} the relative mass of U^{239} has been determined from the (d, p) reaction, of U^{240} from the (t, p) reaction and of U^{237} from the (p, d) reaction. Also linked to U^{238} is the mass of U^{236} from measurements made on the $U^{236}(t, p)U^{238}$ reaction.

The reaction studies with U^{235} are less satisfactory and it is doubtful if the ground state transitions were observed for the (d, p), (t, p) or (p, d) reactions. Some of the difficulties associated with U^{235} might, however, be overcome by studying the (d, p), (p, d) and (p, t) reactions with U^{236} and it is hoped shortly to study these reactions.

References

[1] R. MIDDLETON, and S. HINDS, Angular Distributions of the Protons from the $Mg^{24}(d, p)Mg^{25}$ Reaction measured with a Multi-Channel Magnetic Spectrograph. Nuclear Phys. 34, 404 (1962).

[2] L. A. KÖNIG, J. H. E. MATTAUCH, and A. H. WAPSTRA, 1961 Nuclidic Mass Table. Nuclear Phys. 31, 18 (1961).

[3] S. G. NILSSON, Bound States of Individual Nucleons in Strongly Deformed Nuclei. Dan. Mat. Fys. Medd. 29, 16 (1955).

[4] J. SCHEER, Energy Levels of Nuclei $A = 213$ to $A = 257$. Landolt-Börnstein Tables.

[5] S. YOSHIDA, Note on the Two Nucleon Stripping Reaction. Nuclear Phys. 33, 685 (1962).

Discussion

A. H. WAPSTRA: Dr. MIDDLETON did not mention the errors assigned to the mass differences in the 1960 mass table with which he compared his result. The differences are all within these errors with the sole exception of $U^{238}(p, d)$. The second thing is that I am still a little worried about the (d, p) reaction on U^{238} for the following reason. MIDDLETON finds a Q-value of 2588 keV. There is also a (d, p) Q-value by HOLM of 2515 keV, about 70 keV less with a stated accuracy of 60 keV and also an (n, γ) reaction which gives a Q-value of 2526 ± 30 keV. I would like to hear your comments about these values.

R. MIDDLETON: The present work was performed with considerably improved energy resolution and since HOLM et al. did not resolve the ground and first four or five excited states, I am not surprised that there is a Q-value discrepancy. With regard to the (n, γ) measurements, I heard of these for the first time yesterday and am in no position to make any comments. However, I believe that Dr. FIEBIGER who made the measurements is at the conference and perhaps he would like to comment.

N. FIEBIGER: γ—γ coincidence measurements have been made for the reaction $U^{238}(n, \gamma)U^{239}$. A value for the binding energy of the neutron is derived from the sum of cascades. The value is relative to the reported strong high energy γ-line of 4.062 MeV (CAMPION et al., Canad. Journ. Physics 1959) and is 4.755 ± 0.02 MeV. NaI crystals were used for the experiment.

R. MIDDLETON: I would like to say that our error on this Q-value of 20 keV is, I think, a little pessimistic. I hope that shortly we might be able to revise this and bring the error down a bit.

D. M. VAN PATTER: With regard to the beautiful fit of the theoretical DWBA curve to the angular distribution of the ground-state $U^{238}(d, p)U^{239}$ groups which you showed on your first slide, was this a fit to the absolute experimental cross sections or was it a relative fit?

R. MIDDLETON: It was a relative fit.

G. C. PHILLIPS: I have two remarks to make on these interesting measurements. The first is that we have seen demonstrated the importance of nuclear structure effects and of nuclear reaction mechanism effects. Clearly until we know the results of these effects it may be difficult to use (d, p), (p, d), (t, p), etc., Q-value determination to ascertain ground state mass values for heavy elements in the way one can in lighter elements. The second remark is that I was impressed with the very small variation of calculated DWBA angular distribution shapes versus transferred l-value and I thus wonder what certainty one can have of the determination of angular momentum for these reactions in heavy elements.

R. MIDDLETON: It is largely because of considerations similar to your first remark that we undertook angular distribution measurements. These frequently help in the identification of the ground state transition. With regard to your second remark, I agree that the angular distributions are very similar. I have not time to give a full account of the fitting procedure. I can only repeat that we are fairly confident that the l-value assignments for l less than or equal to 4 are correct. We will be writing up this work in more detail presently. It is also interesting to note that the DWBA calculations made by MACEFIELD at Aldermaston suggest 25 MeV to be best the energy at which to make such measurements. At 25 MeV you get the familiar stripping curve and do get strong differences between l-values.

Accurate Measurements of (p, n) Thresholds in the Proton Energy Range 5 to 10 MeV

By

Joan M. Freeman, R. E. White*, J. H. Montague, G. Murray, and W. E. Burcham**

Atomic Energy Research Establishment, Harwell, Didcot, Berks, England

(Presented by JOAN M. FREEMAN)

With 7 Figures

I. Introduction

A number of precise (p, n) threshold measurements have been made in the proton energy range up to about 4 MeV, providing some suitable calibration points for electrostatic generators (see for example the review by MARION[1]). However, now that tandem accelerators have extended the proton energy range in which high-precision nuclear physics experiments can be carried out, the establishment of calibration points for higher proton energies is very desirable. Some (p, n) threshold studies in this region have already been made by BROMLEY et al.[2] and more recently by BONNER et al.[3]. In this work neutrons were detected by the counter-ratio technique, which has been so successful at proton energies up to about 6 MeV. However this method encounters some difficulties at higher energies because of an increasingly large neutron background produced when the direct or scattered protons react with various parts of the apparatus.

In order to avoid the difficulties inherent in neutron detection in the presence of a large background, we have adopted a method which has proved to have several advantages and is applicable to the measurement of most (p, n) ground-state (or, in one or two cases, near ground-state) thresholds in the energy region of interest to us. This is an activation technique which depends on the fact that in these cases the residual nucleus formed in the (p, n) reaction has a half-life somewhere in the region

---

* Now at the University of Auckland, New Zealand.
** On leave from the University of Birmingham.

from a few tenths of a second to a few seconds, and emits positrons with energies up to several MeV.

II. Experimental Method

Our measurements have made use of the system illustrated in Fig. 1. The proton beam from the Harwell tandem accelerator can be interrupted mechanically by means of a tantalum flap inserted a short distance beyond the control slits which follow the analyzing magnet. When the flap is out the beam passes through a 4.8 mm diameter aperture to the target for which the (p, n) threshold is to be measured. The target is irradiated

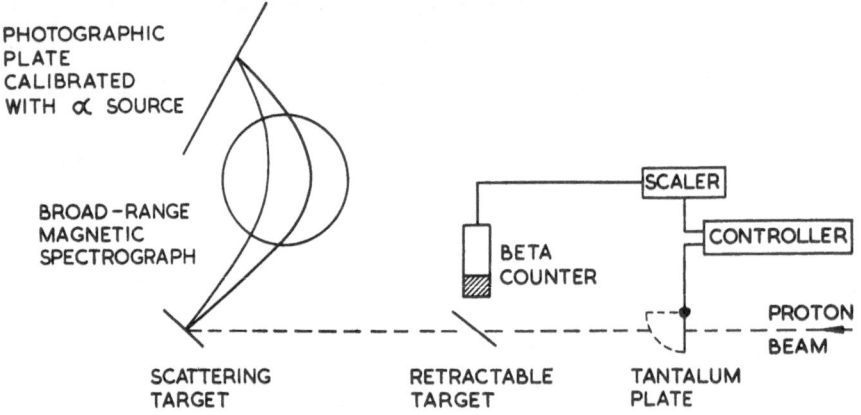

Fig. 1. Schematic diagram of the apparatus used for (p, n) threshold measurements and calibrations

for a period of about one half-life of the residual nucleus. The proton beam is then interrupted and, after a short delay, a scaler is automatically switched on to count positron activity from the target for a fixed time interval equal to one or two half-lives. The positron detector consists of a plastic scintillator covered by a thin aluminium light shield and backed by a photomultiplier. Amplified pulses from the detector pass through a discriminator, the bias of which is set so that positrons with energies down to about two-thirds of the end-point value for the decay can be counted, while positrons and gamma-rays from most contaminants, particularly N^{13} from $C^{13}(p, n)$, are rejected. In this way a very low background below theshold is obtained. The irradiation and counting sequence is repeated for a preset number of cycles (10 or 20 for a bombarding current of about 1 μA). The target current is integrated and the total number of counts from the positron detector is normalized to the total charge. The positron yield is measured in this way as a function of proton bombarding energy, the latter being varied in small steps, measured in terms of the frequency of the proton resonance magnetometer which defines the magnetic field at a fixed point in the analyzing magnet of the accelerator. A threshold curve can usually be obtained thus in about half an hour.

The positron counts recorded at each proton energy above threshold are directly proportional to the total (p, n) reaction yield and thus measure a more definite quantity than that obtained by observation of neutron emission with a moderated BF_3 counter; the latter not only has a sensitivity which may vary with energy and angle of neutron emission, but moreover, if set at $0°$, it receives the total neutron flux only as long as the cone of emitted neutrons has a half-angle less than that subtended by the counter. For medium-weight nuclei this cone spreads out very rapidly above threshold.

These considerations are relevant to the question of correct extrapolation of a threshold curve. If one measures a quantity directly proportional to the total (p, n) reaction yield and the target is effectively thick compared with the maximum energy range above threshold covered in the experiment, then the two-thirds power law for s-wave neutron emission, as discussed for example by MARION[1], can be used. This law follows from the fact that the total cross section σ, immediately above threshold, varies as $(\Delta E)^{1/2}$, where ΔE is the difference between the bombarding energy and the threshold energy. Therefore the total yield Y (with background subtracted) varies as $(\Delta E)^{3/2}$ and a plot of $Y^{2/3}$ versus energy should extrapolate linearly to the true threshold energy. In practice this procedure requires that the beam energy spread be small; in our measurements the spread was one part in 2000 or less.

III. Threshold Measurements

In the present work we have obtained threshold curves for the following reactions (the half-life and positron end-point for each residual nucleus are indicated in parenthesis): $Ni^{58}(p, n)Cu^{58}$ (3.2 sec, 7.5 MeV), with a target of 1 mgm/cm² pure Ni^{58} electroplated onto a gold backing; $Fe^{54}(p, n)Co^{54}$ (0.2 sec, 7.2 MeV) with a 0.1 mm thick disc of separated Fe^{54}; $S^{34}(p, n)Cl^{34}$ (1.5 sec, 4.5 MeV), using 500 and 200 μgm/cm² thick targets of CdS^{34} on gold backings; $Al^{27}(p, n)Si^{27}$ (4.2 sec, 3.8 MeV), with a 0.2 mm thick aluminium foil; $Mg^{26}(p, n)Al^{26m}$ (metastable first excited state, 6.4 sec, 3.2 MeV) using a 300 μgm/cm² layer of the separated isotope on gold.

Fig. 2 shows a typical threshold curve obtained for the $Al^{27}(p, n)Si^{27}$ reaction, which we studied because it has previously been investigated by a number of experimenters using the more usual techniques, and it served as a check on our method. The full circles represent the observed positron yields, normalised to a constant integrated beam current, and the open circles a plot of the two-thirds power of the yield with the mean back-

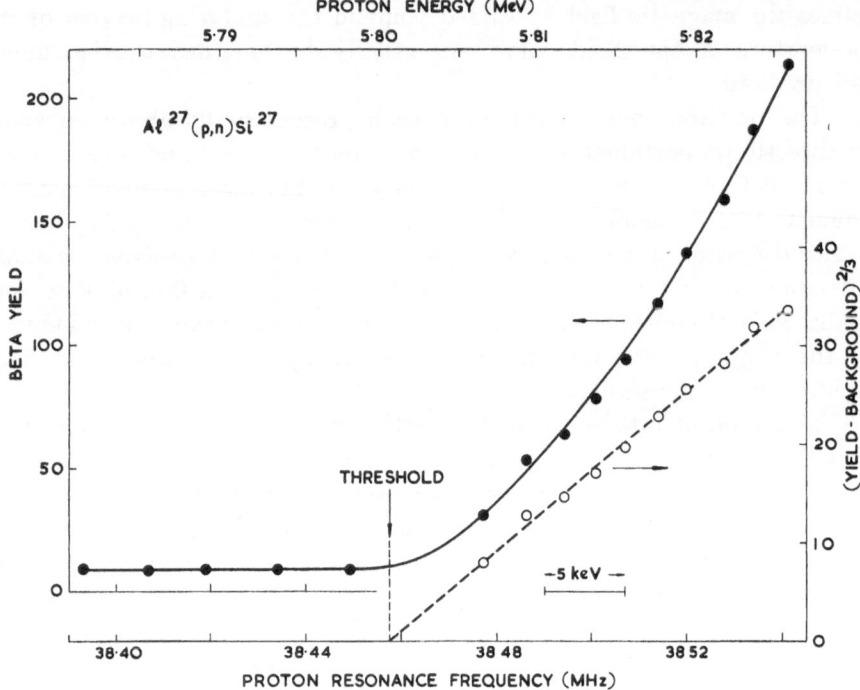

Fig. 2. Threshold curve for the reaction $Al^{27}(p, n)Si^{27}$. Full circles (left-hand ordinates): positron yield, normalised to a constant integrated beam current, plotted as a function of the frequency of the proton resonance magnetometer of the analysing magnet (note the displaced zero). Open circles (right-hand ordinates): two-thirds power of the yield with mean background subtracted

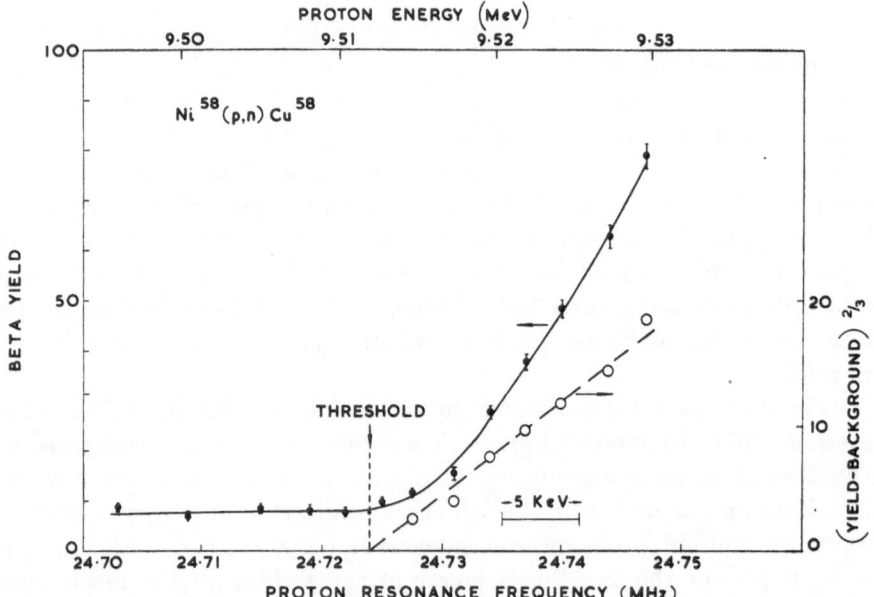

Fig. 3. Threshold curve for the reaction $Ni^{58}(p, n)Cu^{58}$. Details as given in the caption to Fig. 2

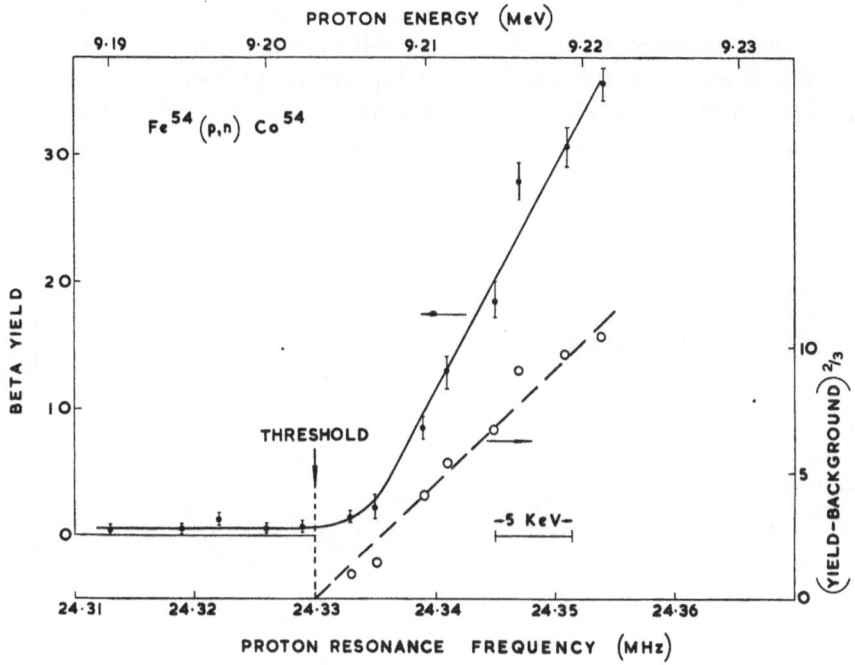

Fig. 4. Threshold curve for the reaction $Fe^{54}(p, n)Co^{54}$

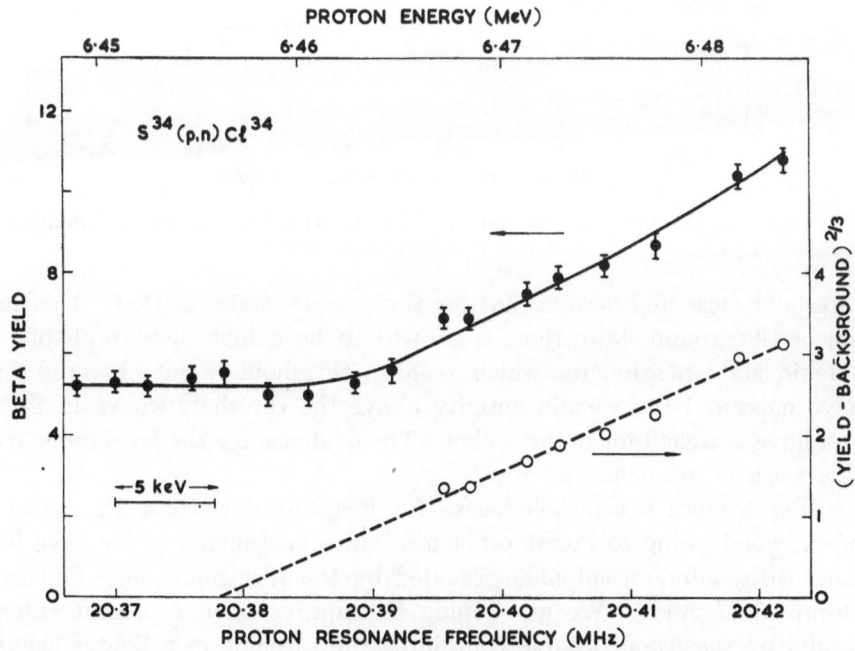

Fig. 5. Threshold curve for the reaction $S^{34}(p, n)Cl^{34}$

ground subtracted. The proton energy scale shown at the top of the graph was obtained from the energy calibration to be discussed below.

Fig. 3 shows a threshold curve for $Ni^{58}(p, n)Cu^{58}$. Note the low background obtained below threshold, even at proton bombarding energies of around 9.5 MeV. Fig. 4 shows a result for the Fe^{54} reaction, again with an almost negligible background. It should be noted that the

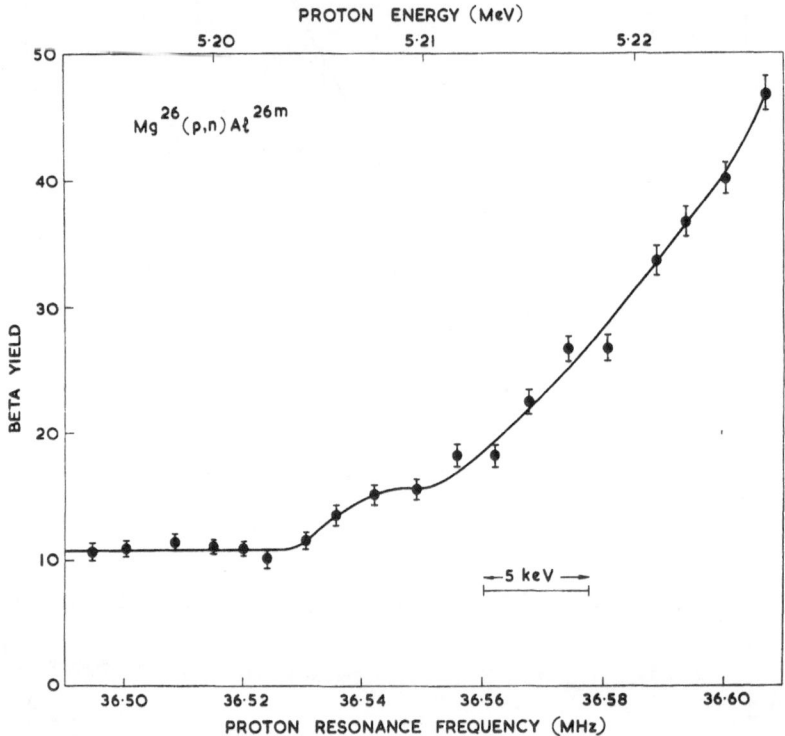

Fig. 6. Threshold curve for the reaction $Mg^{26}(p, n)Al^{26m}$. The proton energy scale shown is approximate only

threshold measured here is that for the $J = 0^+$ state in Co^{54}. This may not be the ground state; there is known[4] to be a high spin (probably 7^+) 1.5 min state nearby, for which a sharp threshold is not observed. We have measured the 1.5 min activity above the threshold shown in Fig. 4, and have searched for it just below. The evidence for the level order from this work is inconclusive.

Fig. 5 gives a threshold curve for $S^{34}(p, n)Cl^{34}$. Here the yield is not so good owing to target problems. Our experiments so far have been done with cadmium sulphide provided by the Electromagnetic Separator Group at Harwell[5]. We are hoping to improve on our present sulphur results by the use of targets containing the sulphide of a lighter element such as zinc. Because of the poorer statistics and the fact that S^{34} is a fairly light element, for which resonance effects might occur, one can question whether the apparent threshold inferred from Fig. 5 is in fact the true one. With this in mind we studied the Cl^{34} decay in detail, with a 40 keV thick target, at various points along the yield curve, using a multi-channel time-analyser. No evidence was found for the 1.5 sec activity from Cl^{34} below the threshold indicated in Fig. 5, so it appears unlikely that the true threshold lies below the apparent one.

Fig. 6 shows a threshold curve obtained for $Mg^{26}(p, n)Al^{26m}$. In this reaction it is known that the $J = 0^+$ state (6.4 sec Al^{26m}), for which an s-wave threshold is expected, lies above the long-lived $J = 5^+$ ground-state. This case appears to be an example of the way in which resonance structure may affect the shape of a threshold curve. The low level step below the steeper rise was reproduced in a number of runs with different targets and is believed to be genuine. A simultaneous measurement using the counter-ratio method confirmed the presence of resonance effects. A two-thirds power law plot was not therefore justified in this case, and a precise threshold measurement has not been made.

IV. Energy Calibration

To determine accurately the proton energy scale for a (p, n) threshold curve measured as described above we compared the proton energy at a selected point close to the threshold with the energy of an alpha-particle group from a suitably chosen standard source. For this calibration use was made of the Harwell broad-range magnetic spectrograph as illustrated in Fig. 1. Where possible the alpha-source was chosen to have an energy just slightly less than the proton energy to be measured.

After a threshold measurement the (p, n) target was withdrawn, and the proton beam was allowed to be scattered by reflection from a thin foil placed at the center of the spectrograph target chamber. The spectrograph angle was set so that scattered protons entering the spectrograph followed closely the same radius of curvature for the same magnetic field as had the alpha-particles from the source previously placed at the point where the protons were scattered. Use was made of a graduated quartz plate, which could be inserted immediately in front of the scattering foil, to locate exactly the position of the proton beam spot. The latter, and also the alpha source, were restricted by slits to an area $3/_4$ mm by 2 or 3 mm. The small lateral displacement of the proton and alpha groups obtained on the photographic plate at the spectrograph focal plane was accurately determined and the corresponding energy difference calculated from the spectrograph calibration. The incident proton energy was then obtained from this measurement, the known scattering target mass, and the scattering angle. The required (p, n) threshold energy thus followed using the proton resonance frequency readings for threshold and for the proton energy measurement.

For calibration of the Ni^{58} and Fe^{54} thresholds (9.5 and 9.2 MeV respectively) use was made of the alpha-particles from a ThC' (Po^{212}) source, prepared by collecting the recoil products over radiothorium. For the alpha-particle energy we adopted the value 8786.4 $\pm$ 2.8 keV given by Wapstra[6] as the average of a number of measurements normalized

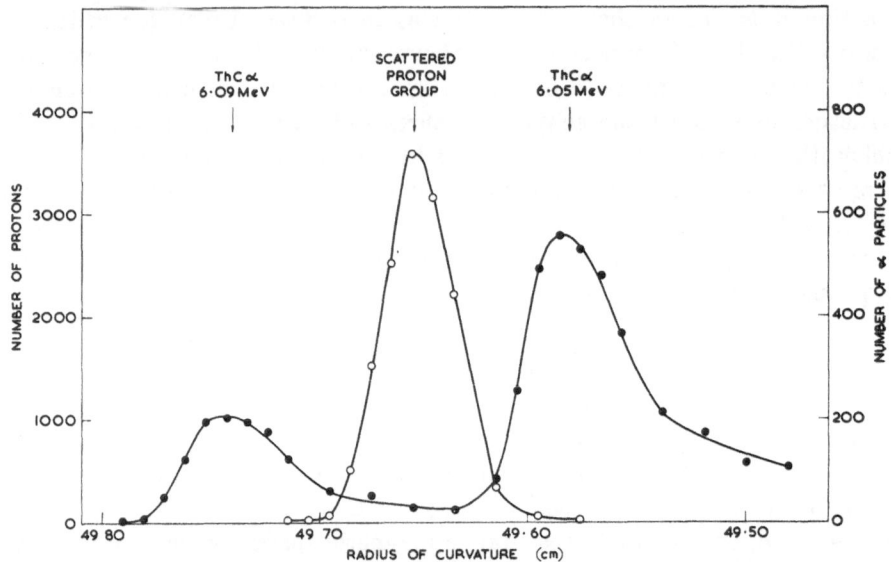

Fig. 7. Alpha-particle groups from a ThC source (full circles), obtained during a calibration run with the broad-range spectrograph, and a proton group (open circles) resulting from the scattering of the proton beam into the spectrograph after a measurement of the threshold for $S^{34}(p, n)Cl^{34}$

to a value of 5304.2 keV for Po^{210}. This value is a mean between the result 8785.4 $\pm$ 0.8 keV obtained by Rytz[7], whose measurement for Po^{210} was 5304.8 keV, and the value 8787.4 keV which would follow from Wapstra's ratio for ThC'/Po^{210} and the Rytz value for Po^{210}.

For the S^{34} threshold calibration (6.5 MeV) it was convenient to use the alpha-particle group from ThC (Bi^{212}), for which the value given by Wapstra relative to 5304.2 for Po^{210} was again used, namely 6089.7 $\pm$ 1.8 keV. This is very close to the result (6089.8 $\pm$ 0.7 keV) obtained by Rytz[7]. Fig. 7 shows an example of this ThC alpha-particle calibration group (and also the group at 6049.8 keV) together with a proton group obtained, during a $S^{34}(p, n)$ threshold run, by scattering the proton beam from a 30 μgm/cm² Al foil at an angle of 87.8° into the spectrograph. The spectrograph field was close to 7000 gauss for these exposures.

In the case of $Al^{27}(p, n)Si^{27}$ (threshold 5.8 MeV) we adopted initially for the alpha-particle calibration an Am^{241} source (5.48 MeV). However it became clear first, that the energy value for Am^{241} is not very well established, and secondly that our source appeared to have an inactive layer on top of it. This was demonstrated by rotating the source through $60°$ in the spectrograph chamber so that such a layer would appear twice as thick to alpha-particles entering the spectrograph from the source. A reduction of about 6 keV in the apparent alpha-particle energy was observed, whereas no reduction was found when the same experiment was performed with ThC and ThC′ sources. We then calibrated our Am^{241} source against the 6.0897 MeV alpha-particles from ThC, changing the spectrograph field by the appropriate amount to obtain the correct radius of curvature. As a check that hysteresis effects were not influencing the result, we compared the ThC′ and ThC alpha-particle groups in the same way and obtained ratios agreeing closely with those implied by Wapstra's and Rytz's values for the respective energies. We were thus able confidently to calculate the $Al^{27}(p, n)$ threshold results in terms of the ThC source for which the value quoted above was again used.

V. Results

Table 1 shows our results for a number of Ni^{58} runs, and the allowances made for possible systematic errors in the estimation of the scattering angle, the angle of the beam relative to the $0°$ direction, and the relative heights of the alpha-particle source and the beam spot on the scattering target. Including also the quoted uncertainty in the energy of the ThC′ alpha-particles we arrive at an error of ± 5 keV in our final mean threshold value.

Table 1. $Ni^{58}(p, n)Cu^{58}$ Threshold Measurements

Results (keV)	Sources of Error (keV)	
9511.6	Statistical...............................	± 1.3
9515.1	α-particle energy	± 3.0
9508.8	Angular scale............................	± 1.2
9513.6	Proton beam direction	± 2.0
9513.0	Relative positions proton beam and α-source....	± 3.0
9518.6		
9509.7	Total root mean square	± 5.0
Mean 9512.9 $\pm$ 1.3		

Final Result: 9513 ± 5 keV

Table 2 gives our results for Fe^{54} where the various errors are similar. Again we emphasise that this result may well not be the ground-state threshold; it refers to excitation of the 0^+ state of Co^{54}. However this result is of some interest, not only as an energy calibration point for tandem generators, but also because it enables one to calculate precisely the equivalent positron end-point for the Co^{54} decay, the ft-value of which is of significance in the theory of weak interactions.

Table 2. $Fe^{54}(p, n)Co^{54}$ Threshold Measurements

Results (keV)	Sources of Error (keV)	
9201.0	Statistical	± 0.6
9202.1	α-particle energy	± 2.9
9203.3	Angular scale	± 1.2
9204.7	Proton beam direction	± 1.9
9202.3	Relative positions proton beam and α-source	± 3.0
Mean 9202.7 $\pm$ 0.6	Total root mean square	± 4.8

Final Result: 9203 $\pm$ 5 keV

Table 3 shows the result obtained for Al^{27}. This agrees reasonably well with other measurements of this threshold[8].

Table 3. $Al^{27}(p, n)Si^{27}$ Threshold Measurements

Results (keV)	Sources of Error (keV)	
5806.0	Statistical	± 1.4
5809.3	α-particle energy	± 2.5
5800.3	Angular scale	± 0.6
5799.9	Proton beam direction	± 1.2
5804.5	Relative positions proton beam and α-source	± 2.0
5799.7	Total root mean square	± 3.8
5800.3		
Mean 5802.9 $\pm$ 1.4		

Final Result: 5803 $\pm$ 4 keV

Work on the $S^{34}(p, n)Cl^{34}$ reaction is still going on so we quote only a preliminary value for the threshold. With an allowance of an additional

Table 4. Summary of Results

Reaction	Threshold* (keV)	Q-value (keV)	Error (keV)	State of residual nucleus
$Al^{27}(p, n)Si^{27}$	5803	-5593	± 4	ground
$S^{34}(p, n)Cl^{34}$	6451	-6264	$^{+4}_{-7}$	ground
$Ni^{58}(p, n)Cu^{58}$	9513	-9350	± 5	ground
$Fe^{54}(p, n)Co^{54}$	9203	-9033	± 5	$J = 0^+$

* Referred to the energy of ThC and ThC′ alpha-particle groups given by WAPSTRA who adopts the value $E_\alpha = 5304.2 \pm 1.6$ keV for Po^{210} (WAPSTRA, NIJGH, and VAN LIESHOUT, Nuclear Spectroscopy Tables, North-Holland Publishing Co., 1959).

error of $\pm\, ^{0}_{6}$ keV for the effect of a possible surface layer of Cd on the target, our present result is $6451\,^{+\,4}_{-\,7}$ keV. Like the Fe^{54} case, this threshold is of particular interest in connection with beta-decay theory.

In conclusion, Table 4 summarizes our present threshold results and shows the corresponding Q-values.

References

[1] J. B. MARION, Accelerator Energy Calibrations. Revs. Modern Phys. 33, 139 (1961).

[2] D. A. BROMLEY, A. J. FERGUSON, H. E. GOVE, J. A. KUEHNER, A. E. LITHERLAND, E. ALMQVIST, and R. BATCHELOR, Canad. Journ. Physics 37, 1514 (1959).

[3] B. E. BONNER, J. RICKARDS, and G. C. PHILLIPS, Neutron Thresholds from the Proton Bombardment of C^{13}, Al^{27}, Cr^{52}, Ni^{58}, Ni^{60} and Ni^{62}. Bull. Amer. Phys. Soc., Series II, 8, 114 (1963); also G. C. PHILLIPS (this conference).

[4] D. C. SUTTON, Half-lives of and Isomerism in V^{46}, Mn^{50}, Co^{54} and Cu^{58}. Thesis, Princeton University, 1961.

[5] We are grateful to Mr. McILROY and Mr. REYNOLDS for providing these and the other separated targets for us.

[6] A. H. WAPSTRA, G. J. NIJGH, and R. VAN LIESHOUT, Nuclear Spectroscopy Tables, North Holland Publishing Co., 1959.

[7] A. RYTZ, Absolutmessung der Energie der wichtigsten natürlichen Alpha-Strahler. Helv. Phys. Acta 34, 240 (1961).

[8] J. B. MARION (this conference).

Discussion

J. B. MARION: Dr. FREEMAN has shown that if one plots the $^2/_3$ power of the net yield above background, then a linear curve results. In fact, the curve remains linear for 20 keV or so above threshold. I believe that this fact alone is sufficient to recommend the $^2/_3$ power procedure for those who wish to make intercomparisons of results from different laboratories. It is well to remember, however, that the calculation which states that the yield above threshold should be proportional to the velocity of the outgoing neutrons is the result of a zero-order R-matrix theory. The approximations which are made in such a calculation for high energy protons bombarding a medium-weight element are probably valid only for 100–200 eV above threshold. The reason why the curve is linear for such an extended energy region is, I believe, not known yet.

G. C. PHILLIPS: I would like to concur in Professor MARION's remarks there and also to emphasize the interesting resonance structure that Dr. FREEMAN has just shown us just above these thresholds. We have observed similar things at Rice. I think it can't be over-emphasized that in the medium-weight elements one has very complicated (p, n) and (p, γ) resonance structure. Any simple interpretations must be looked at very critically to ascertain what you mean by the threshold; you may be measuring the first resonance 3 keV above the threshold, so that one must be very careful about this and perhaps distinguish between a laboratory standard and the threshold.

J. M. FREEMAN: By the time one gets up to Ni^{58} at this level of excitation, I would have thought that one was getting out of the region of separate resonances so that it would be unlikely that one would miss completely the beginnings of a threshold if one can measure yields to a few percent of the intensity observed a few kilovolts above one's threshold.

G. C. PHILLIPS: We have studied all of the separated nickel isotopes and we see (p, n) resonance structure above threshold in all cases.

J. M. FREEMAN: There might be structure, but the cross-section would not go down to zero. You would expect to see something just above threshold. The activation technique is a sensitive method of checking just below an apparent threshold to see if there is in fact a small yield there. With a thick target I would have thought that one was reasonably sure of not missing the true threshold.

Q-Values as Measured Through Resonance Energies

By

H. H. Staub

Physik-Institut der Universität Zürich, Zürich, Switzerland

The usual determination of Q-values of a nuclear reaction involves the measurement of the energies of the primary particle and of one of the reaction products at a fixed angle. The Q-value then follows from the reaction kinematics. For intensity reasons one is usually forced, at least for the reaction product, to use large aperture magnetic spectrometers which must be calibrated with some particle of known energy, which should be as close to the unknown as possible. Moreover, ordinary and differential hysteresis add to the difficulties encountered in calibration.

A large number of nuclear reactions proceed by formation of the same level C of the compound nucleus, by different particles $A + B$ or $A' + B'$. If E_A and $E_{A'}$ are the kinetic energies of the primary particles A and A' projected at the target nuclei B and B' at rest forming the same compound nucleus C the energy equation is:

$$m_A + m_B + E_A/c^2 = m_C + E_C/c^2,$$
$$m_{A'} + m_{B'} + E_{A'}/c^2 = m_C + E_{C'}/c^2. \qquad (1)$$

and the momentum relation:

$$p_A = p_C,$$
$$p_{A'} = p_{C'}. \qquad (2)$$

The Q-value of the reaction $A + B \to A' + B'$ is

$$Q = m_A + m_B - (m_{A'} + m_{B'}). \qquad (3)$$

Expanding to the first order relativistic term one obtains

$$Q = E_{A'} \left(1 - \frac{m_{A'}}{m_C} \right) \left[1 - \frac{E_{A'}}{2\, m_C\, c^2} \left(1 + \frac{m_{A'}}{m_C} \right) \right] -$$
$$- E_A \left(1 - \frac{m_A}{m_C} \right) \left[1 - \frac{E_A}{2\, m_C\, c^2} \left(1 + \frac{m_A}{m_C} \right) \right], \qquad (4)$$

where m_A, $m_{A'}$, m_C are the masses of the bare nuclei and have to be known with moderate accuracy. An accurate determination of the Q-value of the reaction is therefore possible by simply measuring the energies of the primary particles E_A and $E_{A'}$ leading to the same resonance state of nucleus C. The resonance must of course be reasonably narrow and well isolated from others. The absolute determination of the energies can be done for instance with an absolute 180° magnetic spectrometer since usually the primary particles are in a well focussed beam and no consideration has to be given to the small size of the angle of acceptance of the spectrometer.

This method has been used to determine the Q-value of the reaction ^{27}Al$(p, \alpha)^{24}$Mg [1]. Levels with even spin positive parity or odd spin negative parity can be reached either by Al (p) or Mg (α). Among the several levels having this property there are however only 2 which for intensity reasons are suitable. They have both $I = 2^+$ with excitation energies of about 12.08 and 12.73 MeV[2]. The corresponding proton energies are 505 and 1183 keV and the α energies 2437 and 3200 keV.

Resonance energies were measured with an absolute 180° homogenous magnetic field spectrometer[3] using thick targets of Mg and Al. The resonance energies of protons and He$^+$ ions were taken at the midpoint of the thick target step and determined from the field values measured by nuclear magnetic resonance. The probe consisted of water in the case of protons and of ^{7}Li in the case of α particles. The energy values are given by the following relations including first order relativistic corrections. Protons:

$$E_P = \frac{\pi^2}{2\,\gamma_P} \frac{\nu_c}{\nu_n} D_0{}^2 \nu_0{}^2 \left(1 - \frac{E_P}{2\,m_P\,c^2}\right) \times 10^{-8} \text{ eV}, \tag{5}$$

α particles:

$$E_\alpha = \frac{\pi^2}{2\,\gamma_P} \frac{\nu_c}{\nu_n} D_0{}^2 \nu_0{}^2 \frac{m_H}{m_{He}} \left(1 - \frac{m_0}{m_{He}}\right) \left(1 - \frac{E_\alpha}{2\,m_{He^+}\,c^2}\right) \times 10^{-8} \text{ eV}, \tag{6}$$

where: $\gamma_P = 26751.2 \pm 0.1$ sec^{-1} Γ^{-1} gyromagnetic ratio of the proton[4].

$\dfrac{\nu_n}{\nu_c} = 2.79268 \pm 0.00001$ the proton magnetic moment[4].

$D_0 =$ distance of midpoints of slits.

$\nu_0 =$ proton magnetic resonance frequency corresponding to midpoint of yield.

$m_0 =$ electron rest mass.

$m_P =$ proton rest mass.

$m_H =$ mass of H atom.

$m_{He} =$ mass of He atom.

$m_{He^+} =$ mass of singly charged helium ion.

In the case of He^+ particles the proton magnetic resonance value was found from the observed frequency for ^7Li through the relations:

$$\nu_0 = \nu_{Li} \frac{\gamma_P}{\gamma_{Li}}; \qquad \frac{\gamma_P}{\gamma_{Li}} = 0.388645 \pm 0.000005^5.$$

The results are:

$$E_x = 12.08 \text{ MeV}: \; E_P = (504.88 \pm 0.15) \text{ keV},$$

$$E\alpha = (2437.4 \pm 1.0) \text{ keV}.$$

$$E_x = 12.73 \text{ MeV}: \; E_P = (1183.25 \pm 0.25) \text{ keV},$$

$$E\alpha = (3199.8 \pm 1.0) \text{ keV}$$

using the mass values for helium and hydrogen from the 1961 nuclidic mass tables[6] we get for the $^{27}Al(p, \alpha)^{24}Mg$ Q-value

$$Q = (1601.4 \pm 1.0) \text{ keV from the lower resonance,}$$

$$Q = (1602.1 \pm 1.0) \text{ keV from the higher resonance,}$$

or an average of

$$Q = (1601.7 \pm 0.7) \text{ keV,}$$

whereas the adjusted Q-value given in the nuclidic mass table[6] is

$$Q = (1594.5 \pm 1.1) \text{ keV.}$$

As a consequence the mass difference Δ of ^{27}Al and ^{24}Mg would be

$$\Delta = 3 \text{ mu} - (3261.9 \pm 0.8) \text{ keV}$$

from the present measurement as compared to:

$$\Delta = 3 \text{ mu} - (3269.1 \pm 2.7) \text{ keV}$$

from the mass tables.

There are a number of (p, γ) and (α, γ) resonances in other pairs of nuclei and preparations for a survey of several of these pairs are under way.

References

[1] A. Rytz, H. H. Staub, H. Winkler, and F. Zamboni, Nuclear Phys. 43, 229 (1963).

[2] P. J. M. Smulders, and P. M. Endt: Physica 28, 1093 (1962). — S. L. Andersen, H. Bö, T. Holtebekk, O. Lönsjö, and R. Tangen, Nuclear Phys. 9, 509 (1958).

[3] H. Winkler, and W. Zych, Helv. Phys. Acta 34, 449 (1961).

[4] E. R. Cohen, Most probable values of the physical constants, 1962.

[5] H. E. Walchli, O. R. N. L. — 1469 (1953), Suppl. II.

[6] L. A. König, J. H. E. Mattauch, and A. H. Wapstra, Nuclear Phys. 31, 18 (1962). — F. Everling, L. A. König, J. H. E. Mattauch, and A. H. Wapstra, Nuclear Phys. 18, 529 (1960).

Discussion

H. T. RICHARDS: The use of resonant energies to infer Q-values is not a new procedure. Almost a dozen years ago at Wisconsin, KAUFMANN et al. published in the *Physical Review* 88, 673 (1952) a Q-value of 1.613 ± 0.010 MeV for this same reaction by matching resonances and energy scales for the $Mg^{24}(x, p)Al^{27}$ and $Al^{27}(p, \alpha)Mg^{24}$ reactions. Direct Q-value measurements also at Wisconsin by DONAHUE et al. (1953) using electrostatic deflection gave a lower value of 1.594 ± .002 MeV. Changes in calibration energies are insufficient to account for the difference. Hence, there appears to remain a systematic difference between the direct deflection measurements and those inferred from matching resonant energies. We have re-examined the Wisconsin deflection measurement for possible neglected systematic effects. One such effect is that, for intensity reasons, the deflection measurement was made at a sharp resonance. This might affect the line shape but is difficult to believe that the resulting midpoint shift could be sufficient to account for the difference.

H. H. STAUB: I was not aware of this former measurement. Apparently it is not quoted any more and not used any more in the compilations.

H. T. RICHARDS: It was rejected from the mass table of WAPSTRA and MATTAUCH because it differed considerably from the Q-values measurement.

A. H. WAPSTRA: I want to deny the remark by Dr. RICHARDS that the value 1613 ± 10 keV for the $Al^{27}(p, \alpha)$ reaction was not used because it disagreed with other data. The only reason was that there was another value 1594 ± 2 keV with a so much higher statistical weight that it made no sense to retain the first result.

J. B. MARION: I think it is interesting to point out that it is possible to use a method similar to that described by Prof. STAUB in order to obtain a precision measurement of the binding energy of the deuteron. If one observes a (p, γ) resonance and then measures the (d, n) threshold to the same state, then a combination of the two energies yields a value for the binding energy of the deuteron. Drs. BUTLER and BONDELID plan such a measurement at the U. S. Naval Research Lab, using ^{40}Ca as the target for both reactions.

Some Atomic Masses in the Region from Gallium Through Molybdenum*

By

R. R. Ries**, R. A. Damerow, and W. H. Johnson

University of Minnesota, Minneapolis 14, Minnesota, USA

(Presented by R. R. RIES)

With 14 Figures

I. Introduction

In 1961, operational difficulties in the 16-inch double focusing mass spectrometer at the University of Minnesota became progressively more apparent, especially in measurements of heavier isotopes where maximum resolution is required. These difficulties necessitated the movement and reconstruction of the instrument. Some of the modifications made at that time will be discussed in this report. This improved spectrometer was then employed to measure a number of mass doublets in the region from gallium through xenon. Some of the atomic masses of the stable isotopes between $A = 69$ and $A = 100$ will be reported in this paper, while some masses between $A = 100$ and $A = 130$ will be reported in the following paper. These mass results are compared with other mass spectroscopic results and with nuclear reaction results wherever possible, and in addition, a partial mass table of radioactive atoms has also been calculated for this region by combining the stable mass results with available disintegration energies. Finally, one can then use these mass values to study the nuclear binding energy systematics in the region around the neutron shell closure at $N = 50$, as well as the proposed sub-shell at $N = 40$ and $Z = 40$.

II. The Instrument

The instrument employed for all previous measurements has been described in detail elsewhere[1, 2], but it might be worthwhile to briefly review the system once again in this report. A schematic of the spectro-

* Supported in part by Contract Nonr 710 (18) with the Office of Naval Research.
** Now at Max-Planck-Institut für Chemie (Otto-Hahn-Institut), Mainz, Germany.

meter locating the major components is shown in Fig. 1. The series combination of the 90° electrostatic analyzer and the 60° magnetic analyzer provides first and second order angle focusing and first order energy focusing at the fixed collector slit S_4. The radii are approximately 20 inches and 16 inches for the electric and magnetic fields, respectively, while the total path length is about 12 feet. The ion source is of the

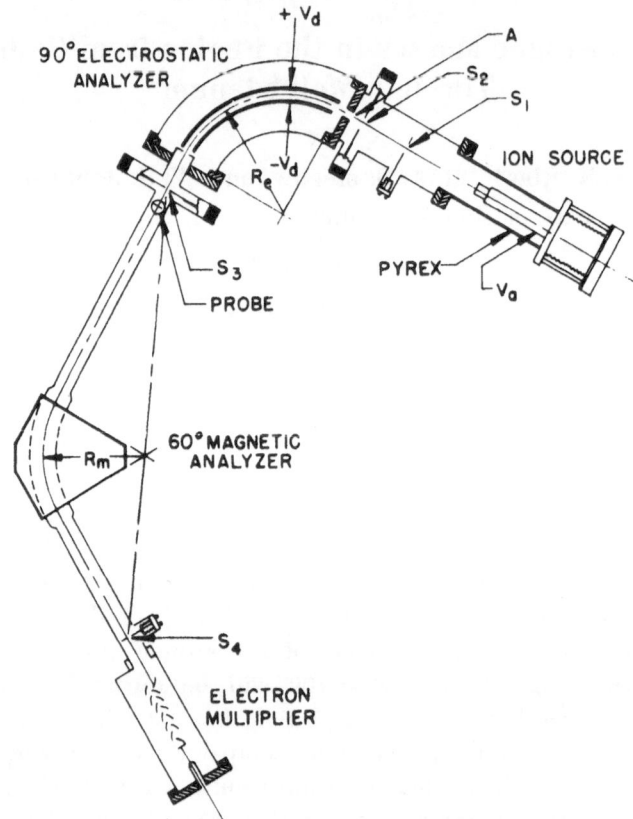

Fig. 1. Schematic of instrument

electron impact type, and a cylindrical lens focuses the ions onto the object slit S_1. Slit S_2 is positioned behind the object slit and thus defines the angular spread of the ion beam, while S_3 is located at the exit focal point of the electrostatic analyzer and is used to define the energy spread of the ion beam. Both of these slits are newly constructed adjustable systems which consist of two independent cutting edges, one on either side of the central ion beam. The new design permits a more precise control of the ion beam parameters. A typical value of α and β determined by these slits is about $1/2000$.

    The ion currents which pass through the collector slit are detected by a ten-stage electron multiplier followed by a fast electrometer amplifier, and finally the vertical amplifier of an oscilloscope. These currents are then observed as a vertical deflection on the oscilloscope screen.

The power supplies are of an elementary nature. The magnetic field is powered by a bank of 96 lead storage batteries, while the electric field is provided by a series-connected stack of 45-volt heavy duty "B" batteries. The 40-kilovolt ion accelerating potential is derived from a feedback regulated radio frequency DC power supply.

III. Method of Measurement

The method of measurement is based on the following principle: For a fixed magnetic field, the mass of an ion collected at the rear slit of

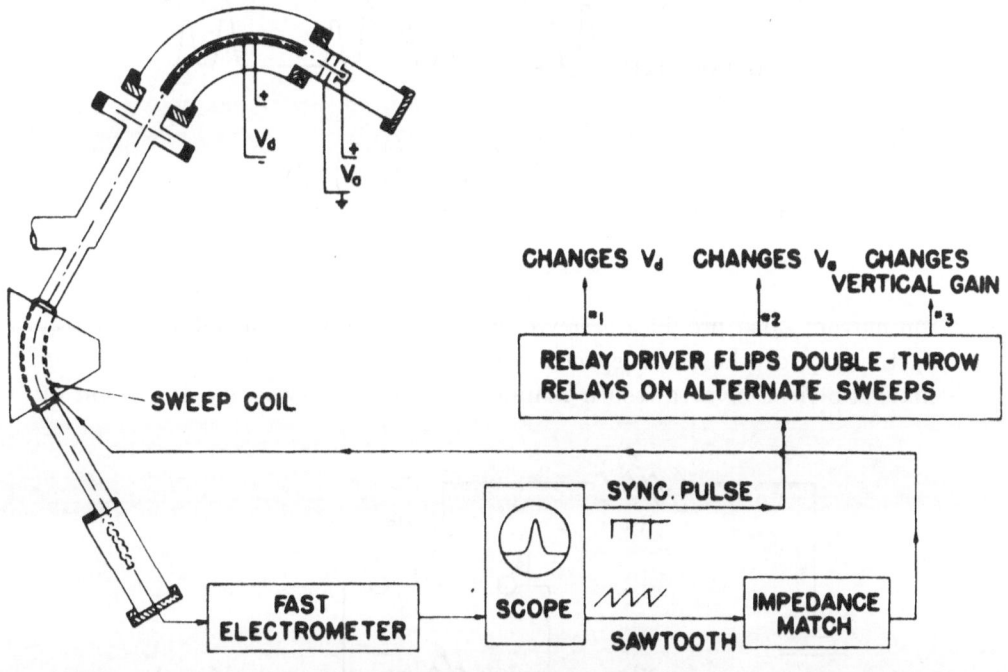

Fig. 2. Peak matching circuitry

this spectrometer is inversely proportional to all the electric fields in the instrument. One can then show that the mass difference between the two members of a mass doublet is related to the electrostatic deflection voltage according to the following equation:

$$\Delta M/M = \Delta V/V, \tag{1}$$

where ΔM is the doublet width, M is the mass of the lighter ion collected with the electrostatic deflection voltage V, and ΔV is the change in deflection voltage necessary to cause the heavier ion of the mass doublet, $M + \Delta M$, to also be focused at the fixed collector slit.

The actual measurement is carried out using the peak-matching technique. Fig. 2 shows a schematic view of the peak-matching circuitry. In this method, the oscilloscope which amplifies and displays the collected

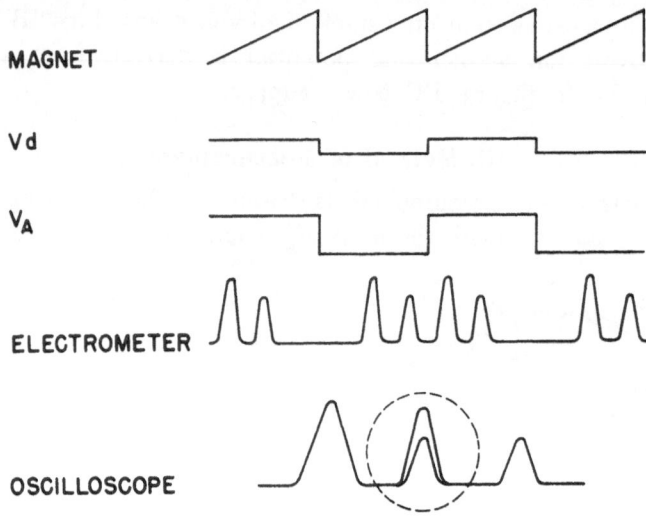

Fig. 3. Peak matching wave forms

ion current, also provides a sawtooth sweep voltage to a flat magnet coil in the gap of the main spectrometer magnet. This coil modulates the magnetic field in a sawtooth manner, periodically sweeping the beam of

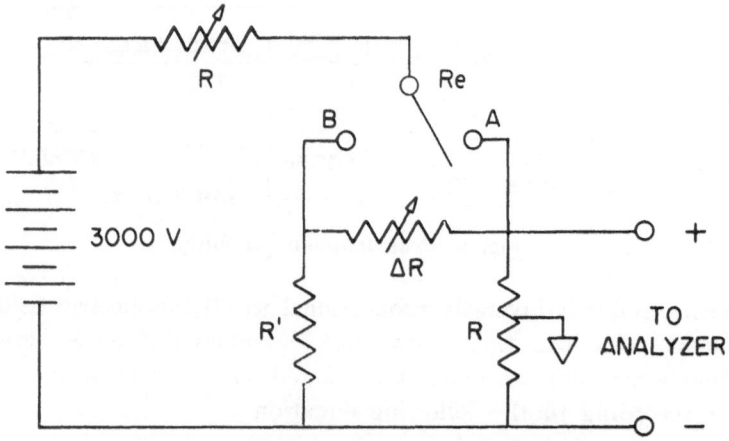

Fig. 4. Deflection plate switching circuit

ions across the collector slit, thus forming an ion peak. In addition, the oscilloscope also provides a synchronized pulse after each sweep activating the relays which change the voltages in the spectrometer. This is illustrated schematically in Fig. 3. If one changes the voltages V_a and V_d on alternate sweeps of the sawtooth voltage, the relative positions of the mass peaks in a doublet may be visually compared on the oscilloscope screen. If the change in voltage is chosen properly, the high mass peak on one sweep may be superimposed on the low mass peak of the following sweep, and the dispersion equation may then be applied.

In actuality, however, the voltage ratio $\Delta V/V$ is not directly measured. Instead, Fig. 4 shows a constant current network made up of precision resistors which are used for the measurement. This circuit determines the electrostatic deflection field in the instrument. The voltage ratio is then directly related to a resistance ratio, and the dispersion equation becomes:

$$\Delta M/M = \Delta R/R, \qquad (2)$$

where ΔR is the value of the variable resistance required to superimpose the two constituents of the mass doublet, and R is the value of the total fixed resistance as shown in the figure. By measuring known doublets, the relationship given by this equation may be shown to be correct to a very high accuracy. Thus the precision of a mass doublet measurement depends to a large extent on the precision with which one is able to reliably determine resistance ratios. In the report which follows, some of the characteristics of an improved resistor network at the University of Minnesota will be described in detail.

IV. Modifications

Several modifications in the instrument will now be described. One of the recurring difficulties in the original instrument was the random modulation of the ion beam resulting from building vibrations and from time-varying magnetic fields from nearby AC power lines. This modulation limited both the maximum usable resolution of the spectrometer, as well as the range of sweep frequencies which could be employed.

To resolve these difficulties, the instrument was moved to a sub-basement room which, first of all, had a factor of about five lower stray magnetic fields. In addition, the instrument was rebuilt and mounted rigidly onto a two ton cast iron surface plate. This iron table was in turn supported on four spring units. The weight of the table and spectrometer loaded the springs such that the natural frequency of the entire system was between one and two cps. All of this finally rests on a concrete slab which is independent of the building. These measures have proved to be very effective in isolating the spectrometer from the higher building vibration frequencies.

Another change made at this time sought to simplify, yet speed up the focusing time of the instrument. Energy and angle focusing are now accomplished by moving the entire magnet in either one of two nearly perpendicular directions in the horizontal plane of the spectrometer. The magnet is moved on a dual track arrangement by means of two small motors which are geared down to provide very small motions to the magnet. In operation, the magnet poles move approximately 0.001 inches per second, thus allowing a continuous observation of the peak shape as the magnetic field is oriented to achieve focus. This new magnet support has speeded up the focusing time immensely.

A new Tektronix oscilloscope which provides a continuously variable sweep rate from as low as 0.02 cps has also been installed. Though

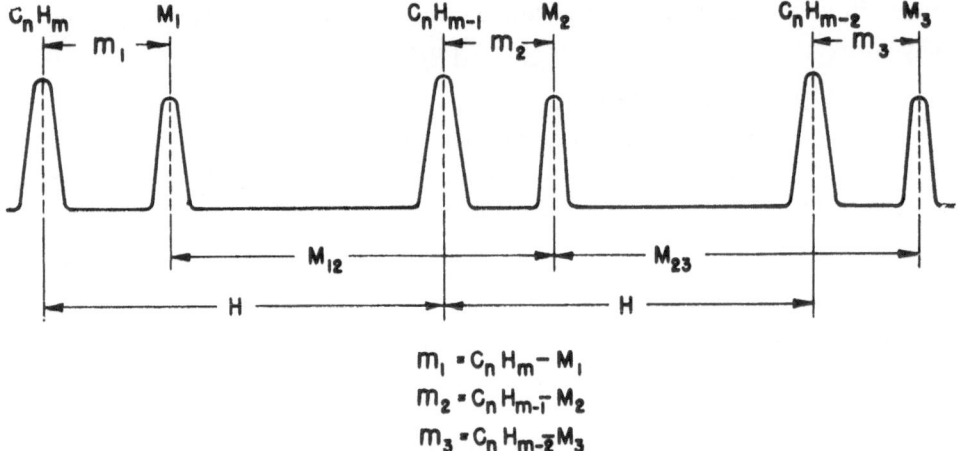

$$m_1 = C_n H_m - M_1$$
$$m_2 = C_n H_{m-1} - M_2$$
$$m_3 = C_n H_{m-2} - M_3$$

Fig. 5. Measurement scheme

the usual operational sweep rate is either about 2 cps or about 30 cps, the very slow sweep has been extremely useful for studying peak shapes, and for investigating the possible existence of close-lying contaminant peaks.

All of these modifications have resulted in improved performance of the instrument. The maximum usable resolution is now two or three times that of the original spectrometer, i. e., resolutions of about one part in 300000 have now been attained. We define resolution in terms of the width of the ion peak at one-half the maximum intensity. These changes have also decreased considerably the time necessary for focusing the instrument.

This spectrometer has also utilized more fully a new measurement scheme which is illustrated in Fig. 5. In the past, this instrument was employed mainly to measure narrow doublets made up of a hydrocarbon comparison ion of known mass and an ion of unknown mass at the same mass number. In the mass region reported here, these doublets have $\Delta M/M$ values of approximately 10^{-3}. Tests have shown that the instrument is capable of measuring wider doublets with sufficient accuracy to be worthwhile. In particular, this figure illustrates the wide doublets which we are now measuring. These wider doublets are composed of two adjacent isotopes of the same element. Doublets of this type are referred to as isotopic doublets, and these are now used to supplement the hydrocarbon doublet data in a weighted least squares fit. This will be discussed further in the following paper.

One of the difficulties that arises when hydrocarbon comparison ions are employed is caused by the rare isotope of carbon, ^{13}C. When the hydrocarbon C_mH_n is used as a comparison ion, a satellite ion is also present, $^{13}CC_{m-1}H_{n-1}$. In the mass region considered in this paper, a resolution (full width at half-height) of about $^1/_{50\,000}$ is required to completely resolve the ^{13}C satellite. The resolution values ranged from $^1/_{60\,000}$ to $^1/_{200\,000}$ during the period. Thus the ^{13}C satellite was at all times completely resolved.

V. Results

Mass doublets have been measured for all of the stable isotopes between gallium and molybdenum, but not all of these measurements can be presented in this report. A paper which includes all of the measurements has been submitted to the Physical Review. The doublet differences for strontium, zirconium, and molybdenum are found in Table 1. These values illustrate the precision of the measurements, as well as the typical size of the errors. These particular doublets have been chosen because of other recent work on these elements. The errors, composed of resistor uncertainties and statistical variations in the data, refer to the last quoted figure. These particular results carry one more significant figure than is

Table 1. Mass Doublets

Doublet	Mass Difference u	Error	Doublet	Mass Difference u	Error
C_6H_{12}—^{84}Sr	0.180 470 8	26	C_7H_8—^{92}Mo	0.155 790 0	32
C_6H_{14}—^{86}Sr	0.200 264 9	36	C_7H_{10}—^{94}Mo	0.173 159 6	32
$C_4H_7O_2$—^{87}Sr	0.135 722 2	35	C_7H_{11}—^{95}Mo	0.180 236 5	35
$C_4H_8O_2$—^{88}Sr	0.146 789 1	47	C_7H_{12}—^{96}Mo	0.189 226 9	30
^{87}Sr—^{86}Sr	0.999 618 1	115	$C_5H_5O_2$—^{97}Mo	0.122 937 6	23
^{88}Sr—^{87}Sr	0.996 739 6	116	$C_5H_6O_2$—^{98}Mo	0.131 375 4	28
			C_7H_{16}—^{100}Mo	0.217 730 3	42
$C_4H_{10}O_2$—^{90}Zr	0.163 377 1	55	^{95}Mo—^{94}Mo	1.000 757 2	122
C_7H_7—^{91}Zr	0.149 143 1	44	^{96}Mo—^{95}Mo	0.998 838 5	124
C_7H_8—^{92}Zr	0.157 569 4	38	^{97}Mo—^{96}Mo	1.001 346 3	123
C_7H_{10}—^{94}Zr	0.171 929 4	39	^{98}Mo—^{97}Mo	0.999 386 0	121
C_7H_{12}—^{96}Zr	0.185 628 0	57			
^{91}Zr—^{90}Zr	1.000 942 0	116			
^{92}Zr—^{91}Zr	0.999 397 2	117			

warranted by the magnitude of the error. Both the narrow hydrocarbon-isotope doublets and the wide isotopic doublets are representative of the other measurements in this region. The report which follows this one contains a detailed discussion and analysis of our error assignments, so nothing more will be mentioned about errors in this paper.

I would like to make a short remark about a new correction we have applied to our mass doublets at Minnesota. In order to check our dispersion relation, $\Delta M/M = \Delta R/R$, we measure a number of so-called hydrogen mass unit doublets of the form $C_m H_n - C_m H_{n-1}$ in the mass region of interest. This experimentally determined measurement of the hydrogen mass is then compared with the accepted value of the hydrogen mass, and the measure of agreement or disagreement serves as a test of our dispersion equation. The results of these measurements are shown in Table 2. In particular, the measure of disagreement given by the ratio of the accepted value to the measured value of hydrogen is employed to correct all the mass doublets which were measured concurrently with these hydrogen doublets. The measured results are to be compared with the known hydrogen mass, $1.0078247 \pm 2\,u$. The resulting dispersion constant, K, for each element is then used to multiply all mass doublets of this element. The values of K are always close to unity, and the average value over the

Table 2. Hydrogen Mass Unit Doublets

Element Studied	Mass Unit Measured	Average Result		Dispersion Constant K
		u	Error	
Sr	$C_6H_{13}-C_6H_{12}$	1.007833	2	0.999992
Zr	$C_7H_{12}-C_7H_{11}$	1.007820	2	1.000005
Mo..........	$C_7H_{12}-C_7H_{11}$ $C_7H_9-C_7H_8$	1.007832	3	0.999993

Hydrogen Mass 1.0078247 ± 2
Average Dispersion Constant for All Elements from Ga to Mo 0.999997 ± 2

Table 3. Atomic Masses

Isotope	Atomic Mass u	Error	Isotope	Atomic Mass u	Error
^{84}Sr	83.913425	3	^{92}Mo	91.906807	4
^{86}Sr	85.909278	4	^{94}Mo	93.905086	4
^{87}Sr	86.908882	4	^{95}Mo	94.905835	4
^{88}Sr	87.905634	5	^{96}Mo	95.904670	3
			^{97}Mo	96.906014	3
^{90}Zr	89.904696	5	^{98}Mo	97.905401	3
^{91}Zr	90.905631	4	^{100}Mo	99.907464	5
^{92}Zr	91.905028	4			
^{94}Zr	93.906317	4			
^{96}Zr	95.908268	6			

entire block of data was 0.999 997 ± 2. We feel this correction is now necessary because of the greater reliance we place on the wide isotopic doublets mentioned earlier.

The atomic masses of strontium, zirconium, and molybdenum have been calculated from the doublet values shown earlier, and these results are presented in Table 3. Where masses have been over-determined with both wide and narrow doublets, our final adopted value is just the result of the least squares fit. The error on these particular masses is then the

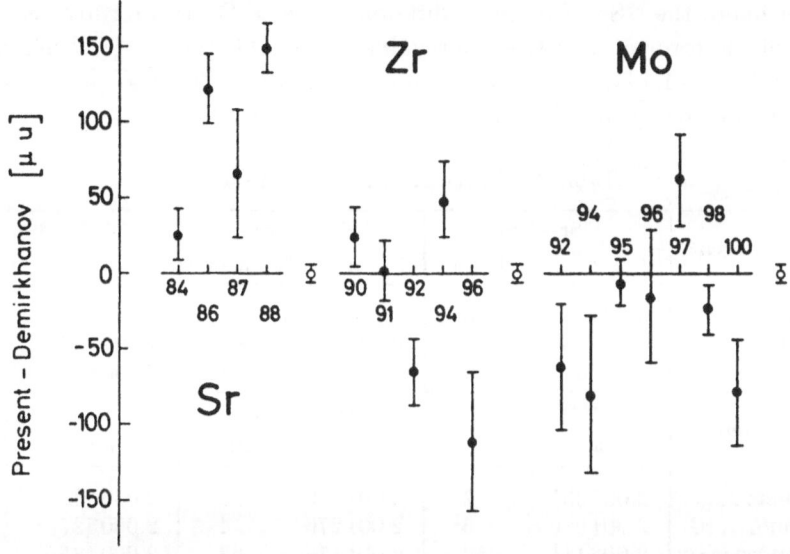

Fig. 6. Comparison of present mass results with results of DEMIRKHANOV

least squares adjusted error. The magnitude of these errors is typical for the entire mass region under investigation.

These particular elements have also been measured in 1960 by DEMIRKHANOV, DOROKHOV, and co-workers[3]. Fig. 6 presents a comparison between their results and the present work. The DEMIRKHANOV results have errors approximately 5–6 times larger than ours, a typical size of their error being about 30 μu compared to our 5 μu. The average overall disagreement for these three elements amounts to 60 ± 30 μu, or roughly twice the size of their errors. All four of their strontium masses are lighter than our present results, while 6 of their 7 molybdenum masses are greater than our values. There are particularly large differences for the isotopes 86Sr, 88Sr, 96Zr, and 100Mo.

A similar comparison was made with the former Minnesota results measured by COLLINS and co-workers[4] in 1954 on a smaller, less precise spectrometer. While the agreement for the entire mass region was quite reasonable, the comparison is not presented here because of the larger errors on these former values.

Another very significant comparison with other mass spectroscopic work can be made by considering mass differences rather than absolute masses. This comparison plays the same role as relative Q-values to absolute Q-values. Table 4 shows the present isotopic and double isotopic mass differences placed alongside other measurements. In the case of strontium, one can see that the ^{87}Sr$-^{86}$Sr results of both Demirkhanov and Collins compare quite favorably with the present work. On the other hand, the ^{88}Sr$-^{87}$Sr mass difference due to Demirkhanov lies about two of his errors lower than our value, while the ^{86}Sr$-^{84}$Sr difference is about three of his errors lower than our results. The less precise values of Collins lie within his error in two of the three cases.

Table 4. Isotopic Mass Differences

Mass Difference	^{88}Sr$-^{87}$Sr		^{87}Sr$-^{86}$Sr		^{86}Sr$-^{84}$Sr	
	u	Error	u	Error	u	Error
Present......	0.996752	6	0.999603	5	1.995853	5
Demirkhanov	0.996669	45	0.999660	48	1.995757	29
Collins	0.99702	19	0.99963	18	1.99611	27

Mass Difference	^{96}Zr$-^{94}$Zr		^{94}Zr$-^{92}$Zr		^{92}Zr$-^{90}$Zr	
	u	Error	u	Error	u	Error
Present......	2.001951	7	2.001289	5	2.000332	7
Barber	2.001970	5	2.001276	4	2.000327	4
Demirkhanov	2.002111	52	2.001175	33	2.000421	30

The bottom half of this table contains a comparison of the present zirconium mass differences with the recently published results of Barber, et al.[5] working under the direction of Professor Duckworth in Hamilton. The same quantities are also calculated from the doublet data of Demirkhanov. Even though the combined errors of Barber's and our work do not quite overlap in two of the three cases, the agreement is

Table 5. Molybdenum Isotopic Difference

Mass Difference	Present Results		Demirkhanov		Difference
	u	Error	u	Error	
^{94}Mo$-^{92}$Mo	1.998279	6	1.998297	67	-18 ± 67
^{95}Mo$-^{94}$Mo	1.000749	5	1.000675	54	$+74 \pm 54$
^{96}Mo$-^{95}$Mo	0.998834	5	0.998844	47	-10 ± 47
^{97}Mo$-^{96}$Mo	1.001345	4	1.001267	53	$+78 \pm 53$
^{98}Mo$-^{97}$Mo	0.999387	4	0.999473	34	-86 ± 34
^{100}Mo$-^{98}$Mo	2.002063	6	2.002118	39	-55 ± 39

clearly quite good, especially for the ^{92}Zr—^{90}Zr value. The results of
DEMIRKHANOV lie about three of his errors from our values, two of his
mass differences being smaller, and one larger than those reported here.

Table 5 presents a similar comparison for the molybdenum isotopic
differences. The comparison with DEMIRKHANOV is here somewhat better,
the agreement being very good for two cases, while the remaining mass
differences disagree by about $1^1/_2$ times DEMIRKHANOV's errors.

Another quite independent method of checking these mass differences
is afforded by nuclear reactions which connect two stable atoms. Table 6
illustrates such a comparison in this particular mass region. The present
strontium mass differences and the (n, γ) reactions of KINSEY and
BARTHOLOMEW[6] are truly in accord, and this kind of agreement gives
us considerably more confidence in our results, which differ by large amounts
from all other previous mass spectroscopic work. The (d, p) reaction due
to WALL[7] is also very acceptable, though it has a large error. The ^{90}Zr—^{88}Sr
mass difference is a rather controversial quantity because of conflicting
results for two nuclear reaction energies. The positron decay of ^{88}Y to
^{88}Sr has two distinct values, one represented by the work of RAMASWAMY
and JASTRAM[8], and the other by the results of RHODE[9], and the recent
(p, n) reaction by SHAFROTH[10]. These groups disagree by about 200 keV,
almost ten times larger than the error. In addition, there was a great
deal of uncertainty concerning the $Y^{89}(\gamma, n)Y^{88}$ reaction because of the
large spin change between the target and daughter nucleus. GELLER[11]
has just recently remeasured this threshold energy and has established
the ground state Q-value. GELLER has then chosen to combine the results
of SHAFROTH and RHODE along with his own Q-value to find the ^{90}Zr—^{88}Sr
mass difference as indicated. This value was not in accord with the mass

Table 6. Nuclear Reaction Mass Differences

Mass Difference	Present Result		Nuclear Reaction Result		Reaction	References
	u	Error	u	Error	·	
^{87}Sr—^{86}Sr	0.999 603	5	0.999 629	19	n, γ	K, B
			0.999 540	215	d, p	W
^{88}Sr—^{87}Sr	0.996 752	6	0.996 737	15	n, γ	K, B
^{90}Zr—^{88}Sr	1.999 062	7	1.999 046	29	$\gamma, n \begin{matrix}(p, n)\\(\beta^+)\end{matrix}$	$G \begin{matrix}(S)\\(R)\end{matrix}$
			1.999 246	21	MS	I, B, D
^{91}Zr—^{90}Zr	1.000 935	7	1.000 888	32	d, p	P, M, S
^{92}Zr—^{91}Zr	0.999 397	6	0.999 332	32	d, p	M, S, P
			0.999 368	43	n, γ	K, B
^{96}Mo—^{95}Mo	0.998 834	5	0.998 831	11	n, γ	K, B
^{98}Mo—^{97}Mo	0.999 387	4	0.999 772	110	d, p	W

spectroscopic value of ISENOR, BARBER, and DUCKWORTH[12]. It should, however, be mentioned that this mass spectroscopic result was a preliminary measurement on a single-focusing apparatus which, I believe, was then under construction. The present work clearly supports the calculation by GELLER.

The present zirconium isotopic differences disagree with the (d, p) reactions of PRESTON, MARTIN, and SAMPSON[13, 14] by about $1^1/_2$ and 2 times their error, respectively. Once again, the (n, γ) reactions of KINSEY and BARTHOLOMEW[6] support the present work to a high degree of precision.

Finally, the molybdenum (d, p) reaction of WALL[7] appears to lie approximately 5 of his errors from the present work.

VI. Nuclear Systematics

Atomic masses of all the stable isotopes between ^{69}Ga and ^{100}Mo have been determined. These stable isotopic masses were then combined

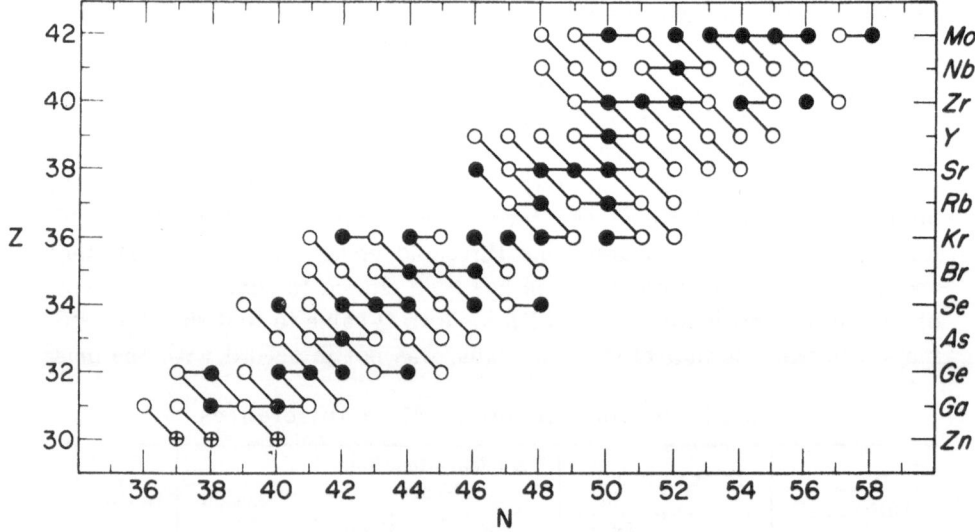

Fig. 7. Reaction scheme

with available nuclear reaction and disintegration energies as shown in Fig. 7 in order to calculate the atomic masses of radioactive atoms. From this group of stable and radioactive atoms one is able to study the nuclear systematics in this mass region.

Fig. 8 exhibits the general behavior of the average nuclear binding energy per nucleon. The even A isotopes of a given element are connected with a solid curve, while the odd A points for all the elements are connected by a dotted line. The characteristic parabolic shape in the even A curves appears in this region just as has been observed in other regions. The

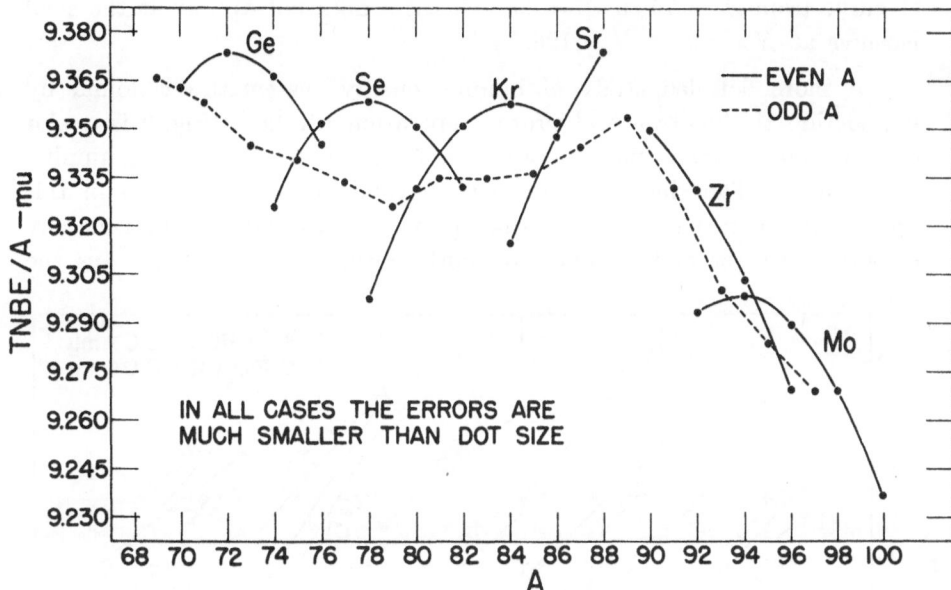

Fig. 8. Average nuclear binding energy per nucleon as a function of A

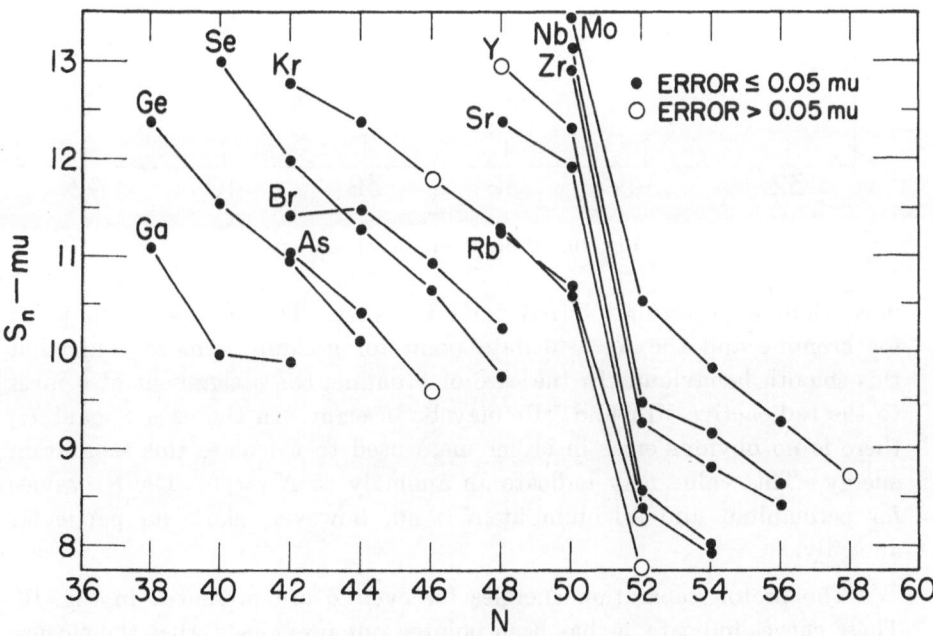

Fig. 9. Neutron separation energy

obvious break in the curve connecting odd A points near $A = 89$ is a
result of the shell closure at $N = 50$. There is an increase in the odd

A curve from $A = 79$ to $A = 89$. This rise does not appear at the shell closures at $N = 82$ or $N = 126$.

A more detailed study of binding energy systematics is found by considering the neutron and proton separation energies. Fig. 9 is a plot of the neutron separation energy as a function of the neutron number for even N. The sharp discontinuity beyond $N = 50$ is shown in this figure with a greater precision than previously available. The generally smooth character of the curves on both sides of $N = 50$ is perhaps the

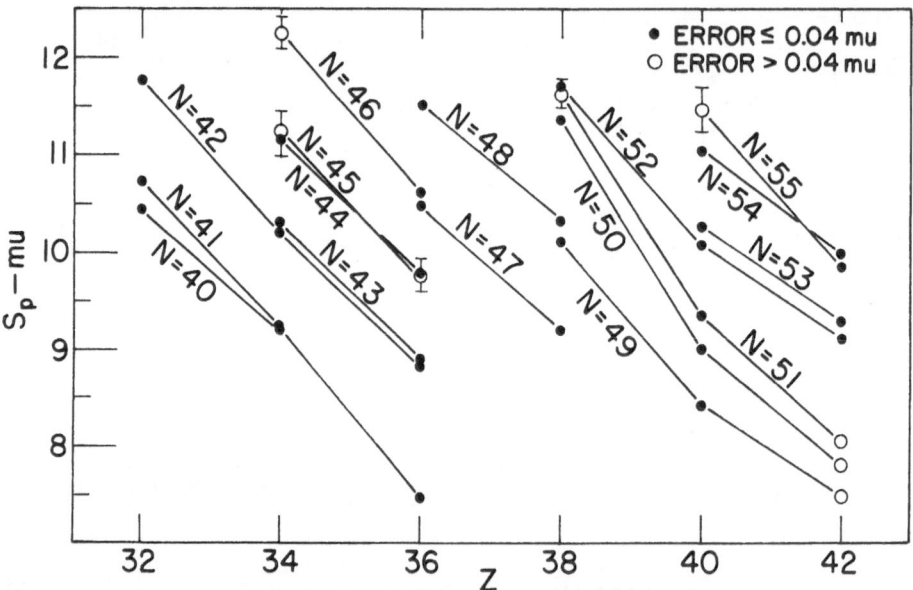

Fig. 10. Proton separation energy

most significant result inferred from the plot. The $N = 42$ data point for bromine and the $N = 40$ data point for gallium seem to contradict this smooth behaviour. In the case of bromine, the assignment of a mass to the radioactive ^{77}Br and ^{76}Br may be in error. In the case of gallium, there is no obvious error in either mass used to calculate this separation energy. This value may indicate an anomaly at $N = 40$. The S_n values for germanium and selenium at $N = 40$, however, show no particular anomaly.

The proton separation energies for even Z are presented in Fig. 10. These curves indicate, as has been pointed out previously, that the closure of a neutron shell seems to have no observable effect on the proton separation energies. The variation of S_p for a given N value as a function of Z is also smooth. There is a persistent change in slope at $Z = 40$ for the four curves that have data points at both $Z = 38$ and $Z = 42$.

As·in the neutron data, this may be an indication of a slight change in nuclear structure at nucleon number 40.

The study of the systematics of the binding energy of the last pair of nucleons is worthwhile because in most cases, only stable masses with small experimental errors are used in the calculation. Fig. 11 illustrates the variation of the binding energy of the last pair of neutrons as a function of neutron number, where the neutron number is even. Values for the same element are connected with a curve. Once again the shell

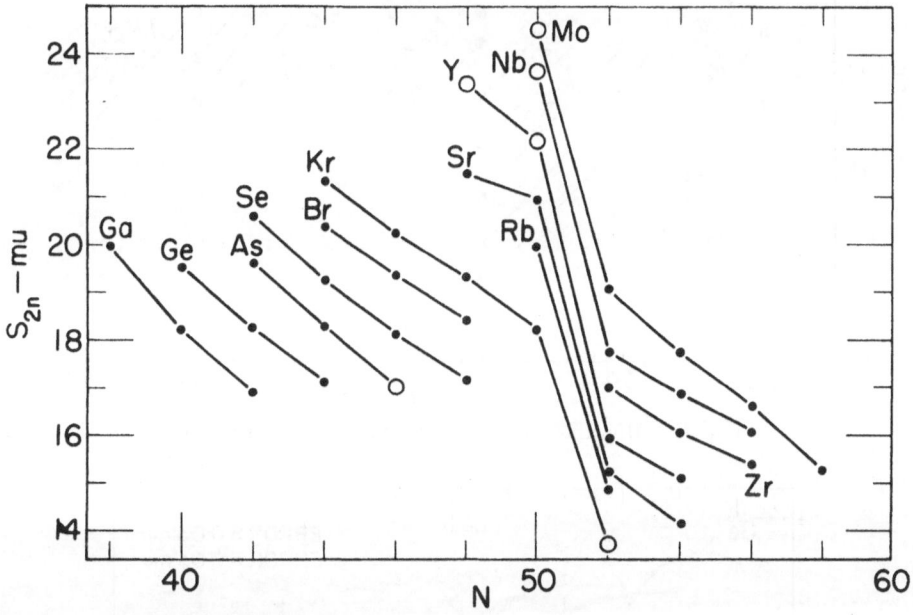

Fig. 11. Binding energy of the last pair of neutrons

closure is clearly visible at $N = 50$. The smooth behavior of these curves on either side of the closure is clearly evident, and it seems reasonable to suspect that an extrapolation of these curves, at least to the next point, should be quite reliable.

Fig. 12 shows the binding energy of the last pair of protons as a function of Z for even Z nuclei. For a given neutron number the smooth variation of these data with changes in Z is again evident. These results also indicate a slight change in slope at $Z = 40$. The consistency of slope for values of Z other than 40 is difficult to demonstrate because of the lack of three data points for a given N value.

The present mass data also permit a systematic study of the pairing energies for neutron and proton pairs in this mass region. For light nuclei, MAYER and JENSEN[15] have concluded that a correlation exists between the pairing energy and the j value of the odd nucleon of the pair. In this scheme, larger pairing energies are correlated with higher j values. In regions of high j values, one finds that it is energetically possible to have the pair occupy a high j value state rather than to pair in the lower

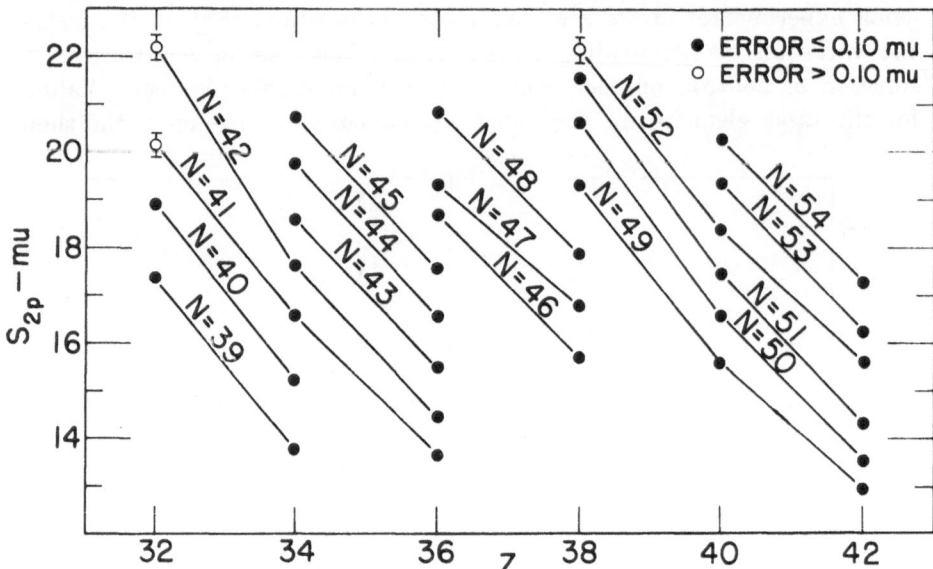

Fig. 12. Binding energy of the last pair of protons

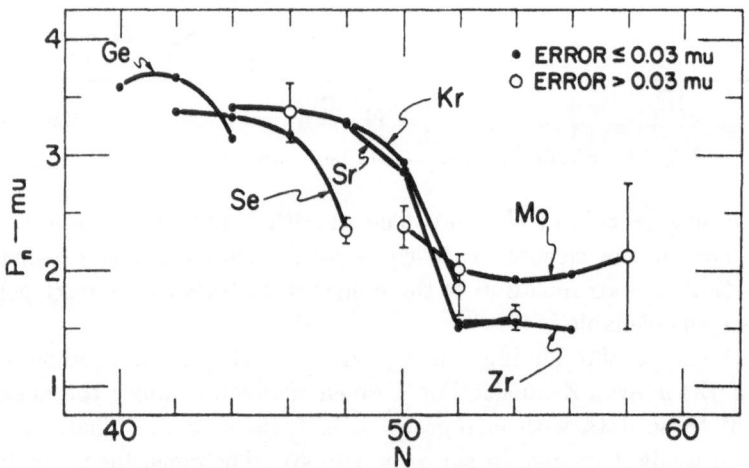

Fig. 13. Neutron pairing energy

spin state of the preceding odd nucleon. This mechanism is used to explain the absence of the highest j values from the ground state spins of odd nuclei.

We have plotted in Fig. 13 the neutron pairing energy P_n as a function of the neutron number. Values from the same element have been connected with a curved line. An attempt has been made to correlate the magnitude of these pairing energies with 1) the j value of the previous odd neutron, and 2) with the j value which the pair is assumed to have according to the filling scheme of MAYER and JENSEN[15]. Neither comparison produces very convincing results. The correlation between the value of j and the pairing energy is, in some cases, what MAYER and JENSEN have suggested, and in other cases, it is just the opposite. The one positive statement which may be made is that the value of the pairing energy

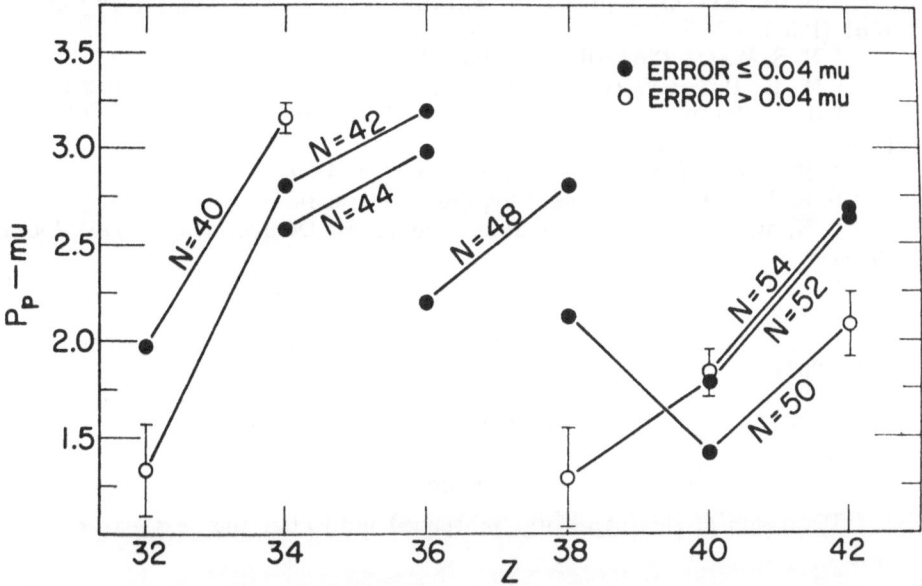

Fig. 14. Proton pairing energy

decreases, rather strikingly, following the shell closure at $N = 50$. Beyond $N = 50$ the P_n values are small, and for each element, are essentially constant in the region plotted. There does not seem to be any anomaly at the one point at $N = 40$.

The final plot, Fig. 14, shows the proton pairing energy as a function of Z for even Z nuclei. In this figure, points with the same N value are connected by lines. An attempt to correlate P_p with the j value of the pair is again not particularly fruitful. The proton pairing energies at $Z = 40$, with $j = 1/2$ for this pair, appear to be lower than practically all other values in this region. A j value of $1/2$ occurs only at $Z = 40$; values for other pairs in this region are all larger. Thus in this instance, low P_p is correlated with low j. There are instances, however, where this correlation is reversed. It is of interest to note that the value of P_p at $Z = 40$ is so small. This may indicate a structure change at $Z = 40$ that is not indicated at $N = 40$.

References

[1] K. S. Quisenberry, T. T. Scolman, and A. O. Nier, Phys. Rev. 102, 1071 (1956).

[2] E. G. Johnson, and A. O. Nier, Phys. Rev. 91, 10 (1953).

[3] R. A. Demirkhanov, V. V. Dorokhov, and M. I. Dzkuya, J. Exptl. Theoret. Phys. (USSR) 40, 1572 (1961); Soviet Phys. JETP 13, 1104 (1961).

[4] T. L. Collins, W. H. Johnson, and A. O. Nier, Phys. Rev. 94, 398 (1954).

[5] R. C. Barber, R. L. Bishop, W. McLatchie, P. van Rookhuyzen, and H. E. Duckworth, Canad. Journ. Physics 41, 696 (1963).

[6] B. B. Kinsey, and G. A. Bartholomew, Canad. Journ. Physics 31, 1051 (1953).

[7] N. S. Wall, Phys. Rev. 96, 664 (1954).

[8] M. H. Ramaswamy, and P. S. Jastram, Nuclear Phys. 19, 243 (1960).

[9] J. I. Rhode, O. E. Johnson, and W. G. Smith, Phys. Rev. 129, 815 (1963).

[10] S. M. Shafroth, Nuclear Phys. 28, 649 (1961).

[11] K. N. Geller, Nuclear Phys. 40, 177 (1963).

[12] N. R. Isenor, R. C. Barber, and H. E. Duckworth, Canad. Journ. Physics 38, 819 (1960).

[13] R. L. Preston, H. J. Martin, and M. B. Sampson, Phys. Rev. 121, 1741 (1961).

[14] H. J. Martin, M. B. Sampson, and R. L. Preston, Phys. Rev. 125, 942 (1962).

[15] M. G. Mayer, and J. H. D. Jensen, Elementary Theory of Nuclear Shell Structure. New York: J. Wiley and Sons, Inc. 1955.

Discussion

Discussion of this paper was postponed until after the next paper.

Atomic Masses from Ruthenium to Xenon*

By

Richard A. Damerow**, Richard R. Ries, and Walter H. Johnson

School of Physics, University of Minnesota, Minneapolis 14, Minnesota

(Presented by W. H. JOHNSON)

With 11 Figures

I wish to complete the report of progress in mass doublet measurements at the University of Minnesota. During the three years since the Hamilton

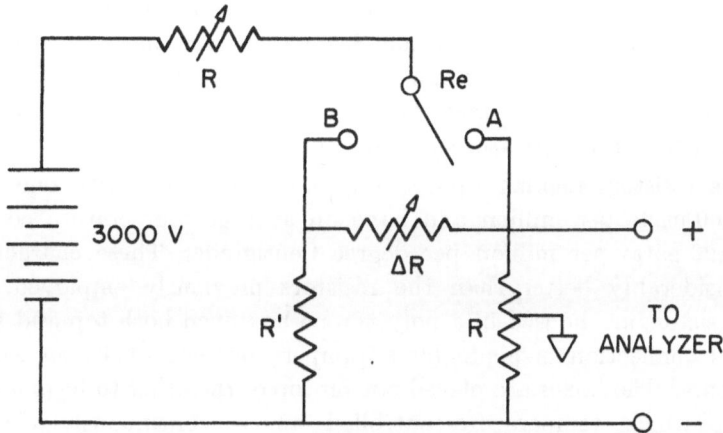

Fig. 1. The resistance network

Conference, we have rebuilt our spectrometer, and measured a number of doublets in the intermediate mass region. This paper will be divided into two parts, first a description of our new resistor circuitry and second, a report on results of some measurements in the region of ruthenium to xenon.

You will recall that the equation used to determine doublet differences ΔM with our instrument is

$$\Delta M = M \frac{\Delta V}{V}, \tag{1}$$

* Supported in part by Contract Nonr 710 (18) with the Office of Naval Research.

** Present Address: Sandia Corporation, Albuquerque, New Mexico.

where ΔV represents a change in electrostatic analyzer voltage, V the electrostatic analyzer voltage and M the mass of one member of the doublet. The voltages are derived from the resistance network shown in Fig. 1. One can show that $\Delta V/V$ is equal to $\Delta R/R$ for this circuit. Thus we can write

$$\Delta M = \frac{M \, \Delta R}{R} \qquad (2)$$

and the doublet measurement is reduced to determining the proper ΔR. For this purpose the peak matching method is employed[1]. Precise mass determinations in this system then require that ΔR and R be known accurately.

We have recently begun using a new resistance network that we feel is more accurate than that previously used. Although only a few tentative measurements have been made with this system, we expect that it will materially reduce our resistor errors and also substantially reduce the calibration time. The resistor R in Fig. 1 is constructed of one-hundred tenth-megohm resistors to form a total resistance of 10 megohms. These resistors, as well as those that form ΔR and R' were obtained from the General Instrument Corporation, Resistor Division. They are bobbinless resistors with the wire coils suspended in a viscous fluid inside a hermetically-sealed metal can. The bobbinless construction eliminates the thermal stresses on the wire coils and produces a resistor of higher stability and inherently lower temperature coefficient.

The resistors making up R were chosen to agree with one another to ± 100 parts per million and have an average temperature coefficient of $+ 1.20$ parts per million per degree Centigrade. These characteristics are considerably better than the resistors previously employed. A set of 10 resistors are housed in a polystyrene box open both top and bottom. A series connection is made by a jumper connector between each pair of resistors. The boxes are placed one on top of the other to form a passage through which temperature-controlled air is continuously circulated. Five of the resistor boxes contain the decade resistors which are part of the adjustable resistor ΔR. Each decade is switched by a paralleled pair of General Radio decade switches. These paralleled switches have reproducible contact resistances of less than 1 milliohm. The decades have values of 10 ohms, 100 ohms, 1 K ohms, 10 K ohms and 100 K ohms per step. In addition, a continuous variation in ΔR is accomplished by including in the circuit a 100 ohm, 10-turn helical potentiometer with a 25 milliohm resolution, located in the air stream with the resistors.

Calibration of the resistors is made by normal bridge technique using an internal standard. An absolute standard is not required because we are only interested in resistance ratios. The internal standard employed for these measurements is the parallel combination of the one-hundred tenth-megohm resistors. The series to parallel resistance ratio of a set of n resistors is given in equ. 3.

$$\frac{R_S}{R_P} = n^2 \left[1 + \frac{1}{n} \sum_{K=1}^{n} \beta_K{}^2 \right],$$ (3)

where

$$R_K = \overline{R} \left[1 + \beta_K \right].$$

With resistors matched to ± 100 ppm, the series parallel ratio for the present circuit is 10^4 to an accuracy of 1 part in 10^8. For our calibration, we assume the parallel combination of the 100 resistors to be exactly 1000 ohms; then the value of the series combination is 10 megohms to a high accuracy. Small corrections must be made because of lead resistance and resistor self-heating that are not common in the series and parallel cases. We feel that these corrections, which are about 1 ppm, may be made with good accuracy. The low temperature coefficient of these resistors aids greatly in the self-heating problem.

We have calibrated this system three times over a period of 6 months. We found that the average change of the $1 K/$step decade resistors is 1 ppm compared with the $1 K$ standard for these calibrations. We hope that with this system, the error in doublet measurements from resistor uncertainties will be essentially negligible even for wide doublets.

The second part of my talk will deal with some measurements in the mass region ruthenium to xenon. A paper reporting these measurements

Table 1. Mass Doublets

Doublet*	Mass Difference u	Error**	Doublet*	Mass Difference u	Error**
C_7H_{12}—^{96}Ru	0.186 304 6	38	C_9H_{10}—^{118}Sn	0.176 644 6	71
C_7H_{14}—^{98}Ru	0.204 263 5	29	C_9H_{11}—^{119}Sn	0.182 777 6	72
C_7H_{15}—^{99}Ru	0.211 442 8	30	C_9H_{12}—^{120}Sn	0.191 709 0	112
C_7H_{16}—^{100}Ru	0.220 983 8	37	$C_8H_{12}N$—^{122}Sn	0.193 541 4	80
C_8H_5—^{101}Ru	0.133 549 5	22	$C_7{}^{13}CH_{13}N$—^{124}Sn	0.202 885 6	83
C_8H_6—^{102}Ru	0.142 604 8	32	^{117}Sn—^{116}Sn	1.001 219	11
C_8H_8—^{104}Ru	0.157 171 5	34	^{118}Sn—^{117}Sn	0.998 662	11
^{99}Ru—^{98}Ru	1.000 652	11	^{119}Sn—^{118}Sn	1.001 709	12
^{100}Ru—^{99}Ru	0.998 282	11	^{120}Sn—^{119}Sn	0.998 887	11
^{101}Ru—^{100}Ru	1.001 368	11	^{124}Sn—^{122}Sn	2.001 838	22
^{102}Ru—^{101}Ru	0.998 767	11	C_9H_{13}—^{121}Sb	0.197 910 5	37
C_9H_8—^{116}Sn	0.160 860 7	84	$C_8H_{13}N$—^{123}Sb	0.200 580 0	33
C_9H_9—^{117}Sn	0.167 485 5	127			

* C, H and N refer to ^{12}C, ^{1}H and ^{14}N.

** Throughout this work, the errors refer to the last figure of the particular result.

has been submitted to *Physical Review*. Instead of trying to report on all of these measurements, I will talk in detail only about three elements, for which there are other mass spectroscopic results. I will then conclude by showing some separation and pairing energy graphs determined from the complete set of measurements. The elements to be considered in detail are ruthenium, tin and antimony. DEMIRKHANOV and his co-workers[2, 3] have made measurements on each of these elements. DUCKWORTH and his co-workers[4, 5] have recently published results of measurements on tin and antimony isotopes.

Table 1 shows the doublet values that we have measured for a number of isotopes of these elements. Note that we have measured hydrocarbon doublets and also the wider isotopic doublets. The hydrocarbon results are used to calculate masses employing the values of the secondary standard masses previously measured at Minnesota[6].

Before we consider these masses in detail, I would like to discuss the error analysis used to determine doublet errors.

In the past we have established a total error[6] by considering three sources of error:

1. Statistical error derived from the variation of the set of original measurements. In most cases, 10 runs were made on each doublet over a period of some days. The instrument was refocused between each pair of runs.

2. Resistance errors for the decade resistors and the helipot.

3. Resistor error for the large resistor R.

As an example of this, the contributing errors for the ^{121}Sb doublet were 2.6 μu, 1.8 μu, and 2.0 μu respectively. These are combined by taking the square root of the sum of the squares of the individual errors to yield a final error of 3.9 μu. For wide doublets, on the other hand, the statistical error is small compared with the resistor errors.

In the present work, we have attempted to measure both hydrocarbon doublets and, where possible, isotopic doublets. With doublets of both types, the masses in question are over-determined. This over-determination allows us to perform a weighted least squares adjustment of the masses, similar in a small way to that performed by Professor MATTAUCH and his co-workers[7-9].

Adjustments of this sort allow one to test the errors originally applied to the input doublets. This test was performed by Professor MATTAUCH on some over-determined doublet data made in the light mass region at

Minnesota. He measured the correctness of the original error assignment by computing the external to internal error ratio given in equ. 4.

$$\left[\frac{R_e}{R_i}\right] = \left(\frac{\sum\left(\frac{V_K}{\sigma_K}\right)^2}{N-n}\right)^{1/2}. \tag{4}$$

σ_K is the original error on the Kth doublet.

V_K = difference between the adjusted quantity and the unadjusted quantity.

N = number of input results.

n = number of independent output results.

The value of R_e/R_i should be 1 if the original errors were assigned correctly. Professor MATTAUCH found that the result was 2.65 instead of 1. The conclusion was reached that the errors on these Minnesota doublets were underestimated by 2.65[8]. Unfortunately, many of the measurements made at Minnesota above $A = 30$ are single measurements, and the resulting mass is not over-determined. For this reason, tests of the error assignment to doublets in the heavier region could not be made. When using these doublets, Professor MATTAUCH made the reasonable assumption that all Minnesota errors should be multiplied by 2.65.

Tests of the internal consistency of a number of the present measurements have been made. We have performed weighted least squares fits using two different choices of errors: 1. errors derived in the manner previously described using both the statistical error and all resistor errors and 2. errors due only to the statistical variation of the results omitting any resistor errors. Using only errors derived from statistical variation reduces the total error on isotopic doublets and thus greatly increases their weight. An example of the input data for these two adjustments is found in Table 2.

Table 2. Input Data for Least Squares Fit

Isotope	Mass u	Total Error	Statistical Error
^{86}Sr	85.909 280 5	43	27
^{87}Sr	86.908 878 9	38	29
^{88}Sr	87.905 636 7	50	42
^{87}Sr—^{86}Sr	0.999 618 1	113	13
^{88}Sr—^{87}Sr	0.996 739 6	114	18

Using the results of the least squares fit, we have calculated R_e/R_i for each element. These results are listed in Table 3. Because of the small number of entries in each group, one cannot infer a great deal from these values. It is, however, worthwhile to point out that the ratio for both sorts of errors has a large variation from element to element. The values for strontium and tin are certainly, in a statistical sense, significantly greater than for the results with statistical errors alone.

Table 3. External to Internal Error

Element	(R_e/R_i) total error	(R_e/R_i) statistical error	N	n
Ge	0.71	1.66	9	5
Br......	0.08	0.31	3	2
Kr	0.00	0.00	5	3
Sr......	1.87	3.56	5	3
Zr......	0.49	1.21	5	3
Mo	0.58	1.39	9	5
Ru	0.12	1.16	9	5
Cd	0.39	1.57	9	5
Sn	0.94	2.14	13	7

It is probably more appropriate to consider a group of elements rather than just one element at a time. I chose to divide the elements into two groups, one from germanium to molybdenum and the second from ruthenium to tin. These results are shown in Table 4. One may conclude that of the

Table 4. Values of the External to Internal Error by Groups

	(R_e/R_i) total error	(R_e/R_i) statistical error	$N-n$	(R_e/R_i) expect.
Group I	0.67	1.77	15	1 ± 0.18
Group II	0.65	1.75	17	1 ± 0.17

two ways of calculating the error, the method using the statistical error with resistor errors appears to yield a total error that is perhaps a little large. On the other hand, using a statistical error alone results in an underestimation of the error by roughly a factor of 2. If one chooses to multiply the purely statistical errors by the R_e/R_i value, then on the average, the error is about the same size as the original total error. Also, the adjusted masses in either case are in good agreement. We therefore choose to use the method of error calculation that we have formely employed, using both statistical and resistor errors. It should be pointed out that using the more precise resistor network will result in lower total resistor errors, and the choice of which of the two errors we will use will have to be studied further.

Let us now consider the mass results for ruthenium, tin and antimony. The results for ruthenium and tin are the least squares adjusted values while the antimony results come directly from the measured doublets. Table 5 shows these results.

Table 5. Present Masses

Isotope	Mass u	Error	Isotope	Mass u	Error
96Ru	95.907592	4	116Sn	115.901737	6
98Ru	97.905282	4	117Sn	116.902944	8
99Ru	98.905928	4	118Sn	117.901601	6
100Ru	99.904210	5	119Sn	118.903298	6
101Ru	100.905574	2	120Sn	119.902186	9
102Ru	101.904343	3	122Sn	121.903428	8
104Ru	103.905426	4	124Sn	123.905264	9
			121Sb	120.903811	4
			123Sb	122.904214	4

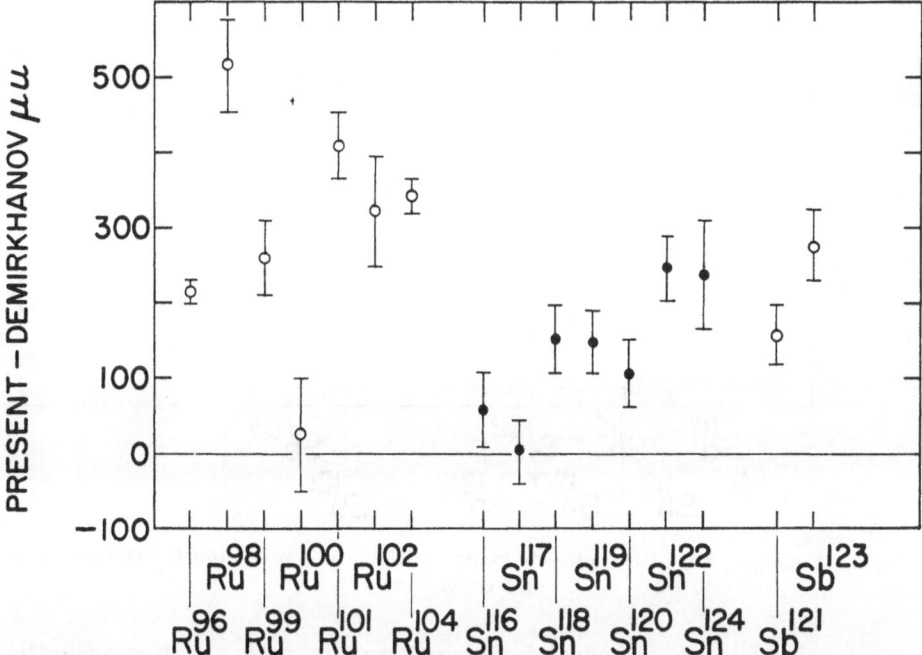

Fig. 2. A comparison of the present masses with masses measured by DEMIRKHANOV and co-workers

We may compare these results with the results of other mass spectroscopic measurements. In order to make this comparison, some doublets were converted to the 12C scale. Masses were then calculated using the same secondary standard masses as the present work. The results of DEMIRKHANOV and co-workers[2, 3] are compared with the present results in Fig. 2. We see that the masses from DEMIRKHANOV are always smaller than the present masses. There is, however, no systematic magnitude of difference. In one case the agreement would be considered reasonable. In the others there is disagreement.

Fig. 3 compares the present work with results from Duckworth and co-workers[4, 5] for tin and antimony. In these cases, the errors of the Minnesota results are somewhat larger than those of Duckworth. We note that in each case, Duckworth's masses are larger than the Minnesota results, but at most by 23 $\mu\mu$. I would conclude that the agreement is reasonable for all cases.

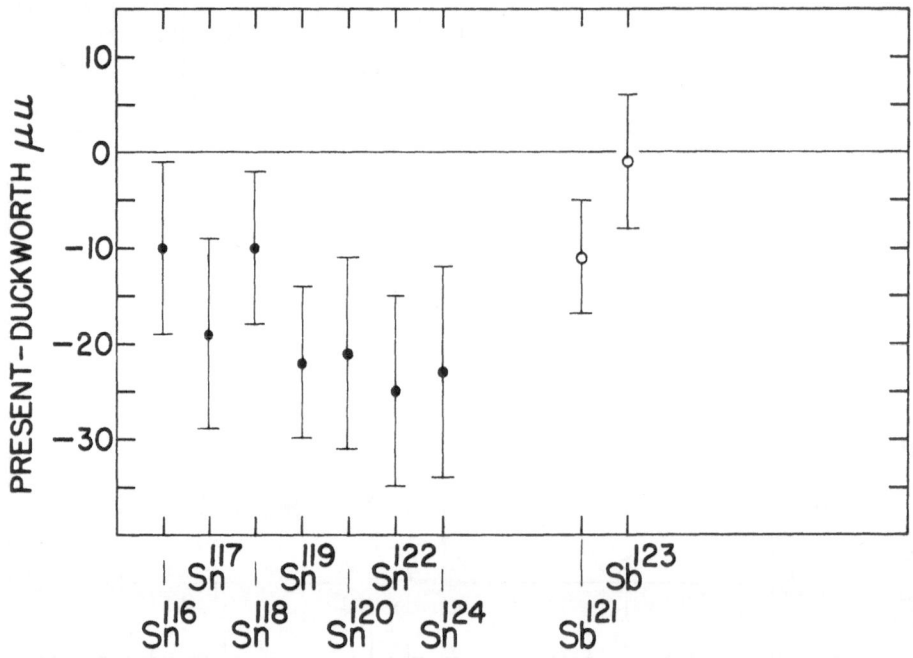

Fig. 3. A comparison of the present masses with masses measured by Duckworth and co-workers

Lastly, let me compare the present results with the less accurate results of Halsted[10] made with a small instrument at Minnesota in 1952. These are shown in Fig. 4. The agreement is good in four of the seven comparisons.

One can also compare differences calculated from the present measurements with differences calculated from reaction energies[11], see Table 6. I have restricted myself here to comparisons that involve reactions linking two isotopes of the same element. In some of these cases, the reaction energy has a rather large error compared with the mass result. The differences ^{101}Ru$-^{100}$Ru and ^{123}Sb$-^{121}$Sb yield particularly good agreement. The excellent agreement between the present tin results and the (d, p) and (d, t) results of PATELL and COHEN[12] lends more confidence to the present tin measurements.

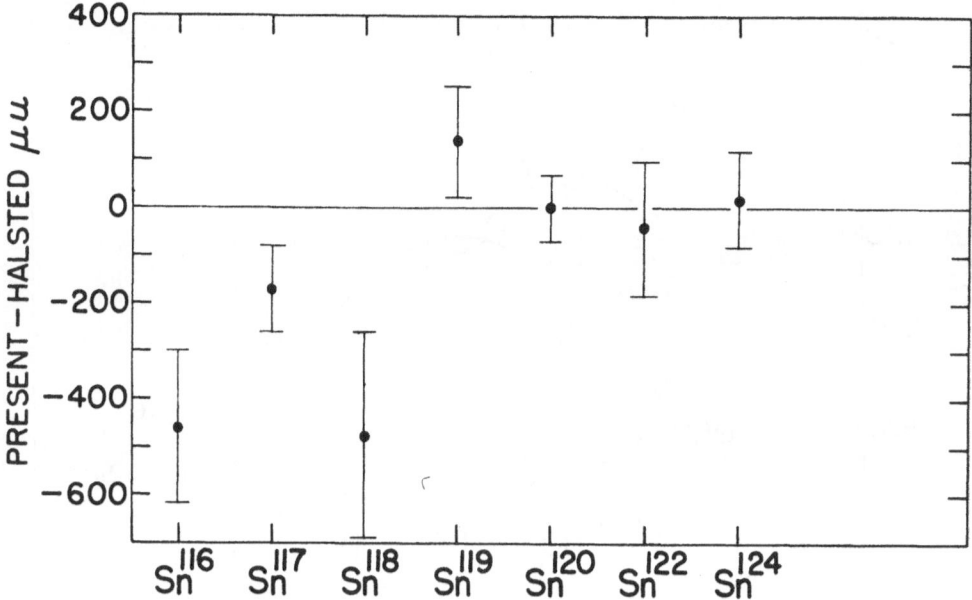

Fig. 4. A comparison of the present masses with masses measured by HALSTED

Table 6. Some Reaction Comparisons

Mass Difference	Reaction Reference	Reaction Difference	Present Difference
^{101}Ru—^{100}Ru	* (γ, n)	1.001096 ± 215	1.001136 ± 6
^{102}Ru—^{101}Ru	* (d, p)	0.999599 ± 60	0.998769 ± 4
^{117}Sn—^{116}Sn	**	1.001192 ± 35	1.001207 ± 10
^{118}Sn—^{117}Sn	**	0.998643 ± 35	0.998657 ± 10
	* (γ, n)	0.998896 ± 200	
^{119}Sn—^{118}Sn	**	1.001702 ± 35	1.001697 ± 8
	* (γ, n)	1.002063 ± 150	
^{120}Sn—^{119}Sn	**	0.998884 ± 35	0.998888 ± 11
^{122}Sn—^{120}Sn	**	2.001251 ± 70	2.001242 ± 12
^{124}Sn—^{122}Sn	**	2.001823 ± 70	2.001836 ± 12
^{123}Sn—^{121}Sn	* (γ, n) (n, γ)	2.000406 ± 54	2.000403 ± 6

 * See reference 11.
** See reference 12.

The complete set of stable atomic masses has been used to calculate a number of radioactive masses. The resulting mass table including both radioactive and stable masses has been used to study nuclear binding systematics in the region near $Z = 50$. Fig. 5 shows a plot of average nuclear binding energy per nucleon for stable nuclei. As in the past, we

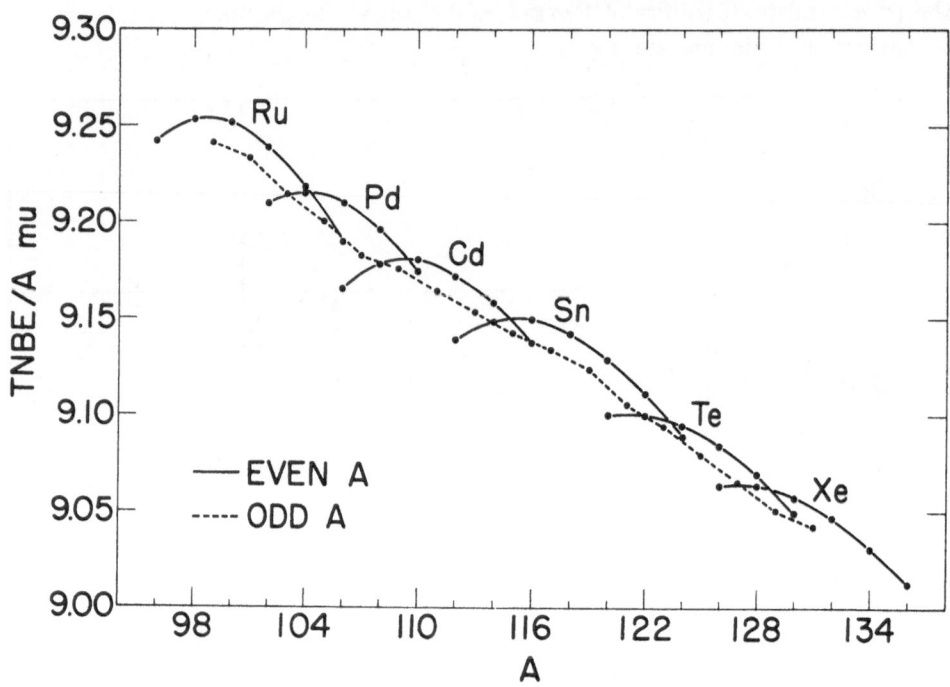

Fig. 5.　The average nuclear binding energy per nucleon

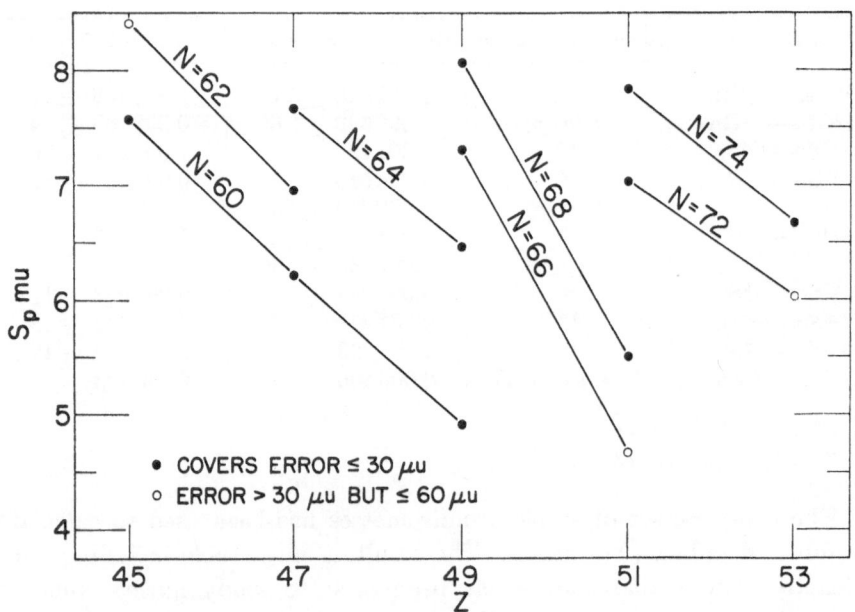

Fig. 6.　The proton separation energy for odd Z nuclei with even N

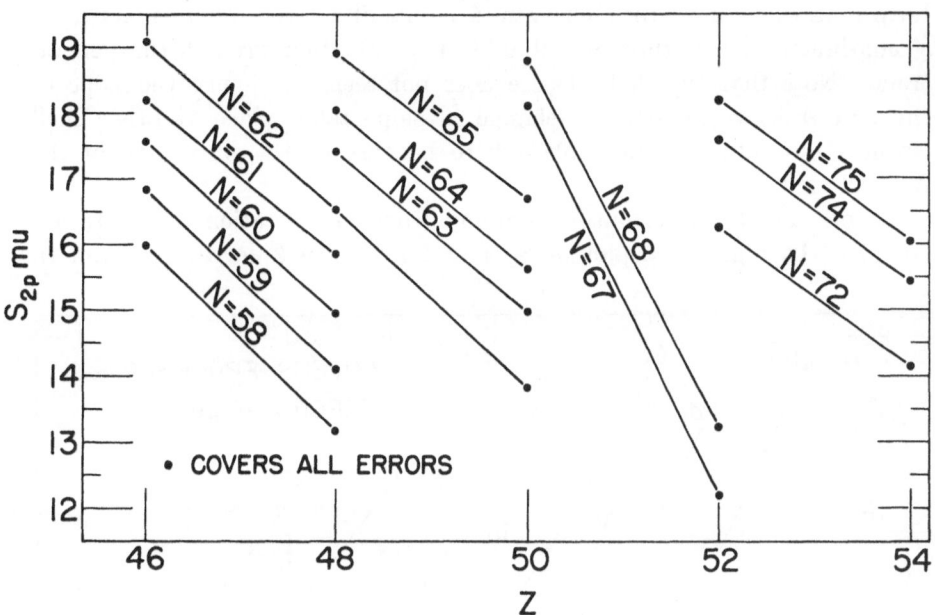

Fig. 7. The binding energy of the last two protons

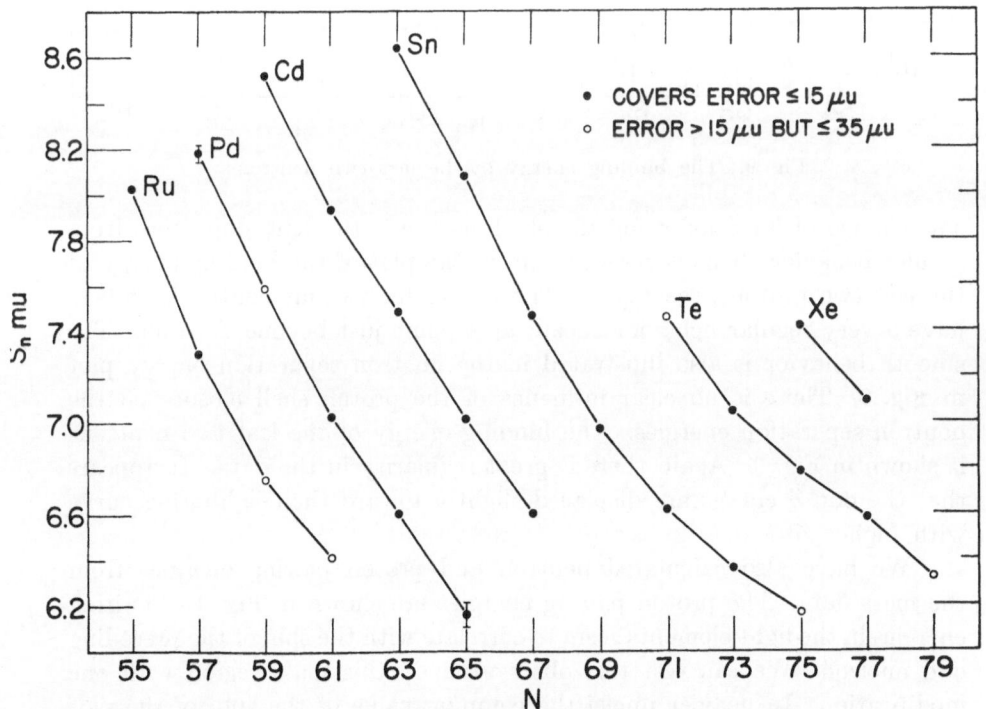

Fig. 8. The neutron separation energy for odd N nuclei

form parabolic curves through even A points of a given element and draw a continuous curve through all odd A points to represent the general trend. Note that the shell closure does not seem to change the slope of the odd A curve greatly. A change in slope takes place at other shell closures but may not take place here because of the proximity of the $N = 82$ shell.

More detailed information is found from consideration of separation energies. In Fig. 6 is a plot of S_p as a function of Z for even N nuclei.

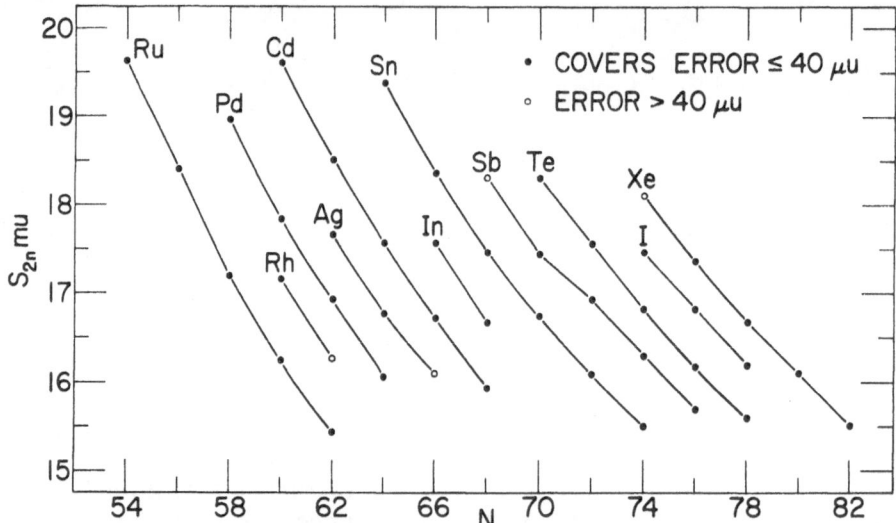

Fig. 9. The binding energy of the last two neutrons

The change of level following the shell closure is the only departure from regular behavior. This is shown again in the plot of the binding energy of the last two protons, see Fig. 7. The curves for various neutron numbers have a very regular behavior except at a point just beyond $N = 50$. The smooth behavior is also illustrated in the neutron separation energy plot in Fig. 8. There is no clear influence of the proton shell closure on the neutron separation energies. The binding energy of the last two neutrons is shown in Fig. 9. Again there is great regularity in the data. It appears that the odd Z curves are displaced slightly toward the neighboring curve with higher Z.

We have also calculated neutron and proton pairing energies from the mass data. The proton pairing energies are shown in Fig. 10. Pairing energies in the light elements seem to correlate with the spin of the preceding odd nucleon. We can test this observation in this mass region with one modification. In heavier nuclei the common value of the spin of the pair of nucleons in question may be the value of the preceding odd nucleon

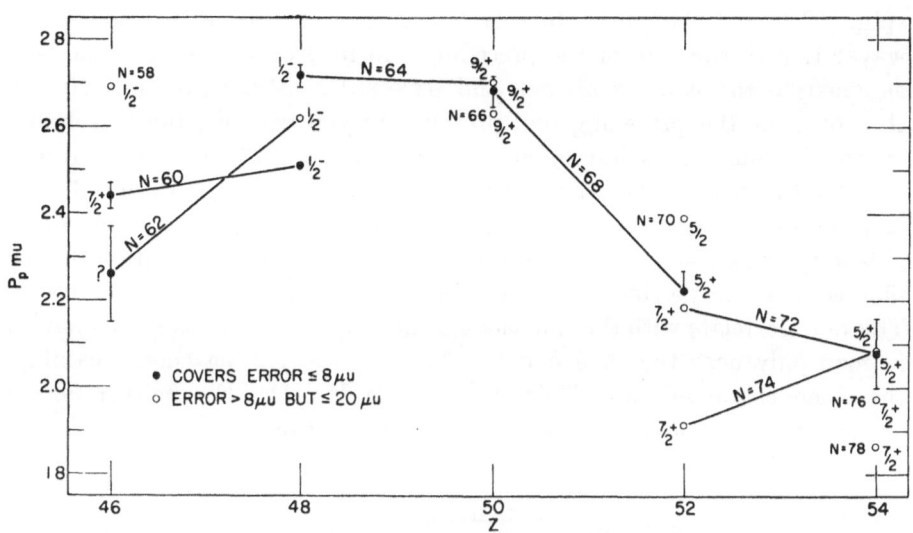

Fig. 10. The proton pairing energy

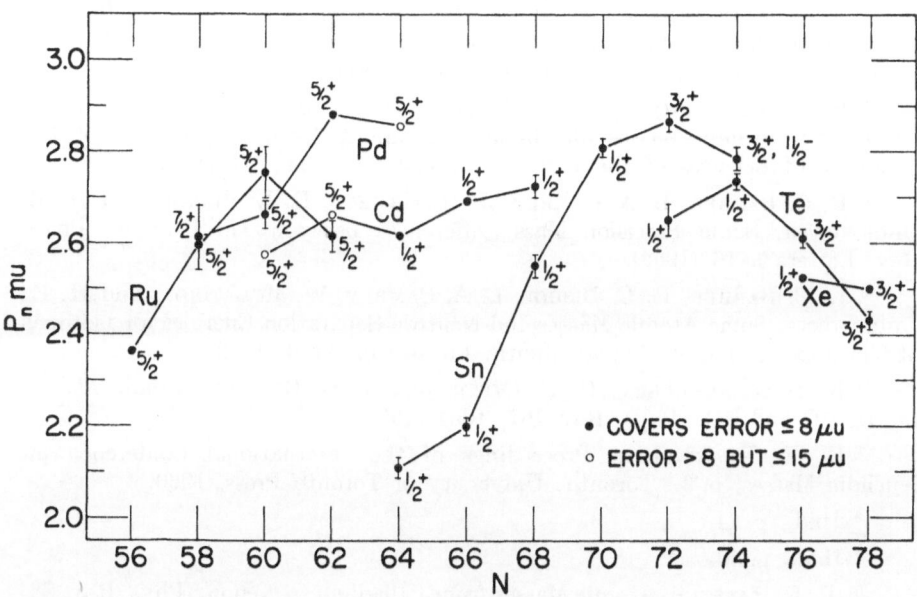

Fig. 11. The neutron pairing energy

in some cases, while in others the common spin of the pair may be a
higher value than the spin of the preceding odd nucleon. This can take
place if we assume that the pairing energy increases with the common
value of the spin of the pair. In some cases, it is then energetically possible
to add an odd nucleon to a zero spin core in a low spin state, but when
adding a second nucleon, the pair will be formed in a state with higher

spin. We have attempted to correlate pairing energy with spin in two ways: 1. with the spin of the preceding odd nucleon and 2. with the spin assigned to the pair by Mayer and Jensen[13]. Shown in Fig. 10 are the spins for the preceding odd proton. As you can see, the correlation is true in some cases but opposite in others. According to Mayer and Jensen the spin of the pair will be 9/2, 1/2, 9/2, 5/2 and 5/2 for the set of proton numbers, from 46 through 54 respectively. Again the correlation is true in some cases but not true in others. The one clear feature of this plot is that the pairing energy decreases as the shell closure is crossed. This does correlate with the spin change and may mean that this explanation is good only near the shell closure. The neutron pairing energy exhibits the same characteristics. This is shown in Fig. 11. For neither choice of spin assignment is there a consistent correlation.

References

[1] H. Hintenberger, Nuclear Masses and Their Determination, p. 185. London: Pergamon Press. 1957.

[2] R. A. Demirkhanov, V. V. Dorokhov, and M. I. Dzkuya, Isotopic Masses and Binding Energies of Nuclei in the Region from Strontium to Ruthenium. Soviet Physics JETP 13, 1104 (1961).

[3] R. A. Demirkhanov, T. I. Gutkin, O. A. Samadashvili, and I. K. Karpenko, Atomic Masses of Tin and Antimony Isotopes. Bulletin of the Academy of Sciences of the USSR, Physical Series 25, 871 (1961).

[4] R. C. Barber, L. A. Cambey, J. H. Ormrod, R. L. Bishop, and H. E. Duckworth, Some Precision Mass Differences between Tin Isotopes. Phys. Rev. Letters 9, 16 (1962).

[5] R. C. Barber, R. L. Bishop, L. A. Cambey, W. McLatchie, and H. E. Duckworth, Some Atomic Masses and Neutron Separation Energies for Isotopes of Tin and Antimony. Canad. Journ. Physics 40, 1496 (1962).

[6] K. S. Quisenberry, C. F. Giese, and J. L. Benson, Atomic Masses of ^{1}H, ^{12}C and ^{32}S. Phys. Rev. 107, 1664 (1957).

[7] H. E. Duckworth, Proceedings of the International Conference on Nuclidic Masses, p. 3. Toronto: University of Toronto Press. 1960.

[8] Ibid., p. 24.

[9] Ibid., p. 39.

[10] R. E. Halsted, Atomic Masses from Palladium to Xenon. Phys. Rev. 88, 666 (1952).

[11] K. Way, A. Artna, W. B. Ewbank, G. H. Fuller, N. B. Gove, M. J. Martin, and H. Ogata, Nuclear Data Sheets, Vol. 5. Washington, D. C.: National Academy of Sciences — National Research Council. 1962.

[12] R. K. Patell, and B. L. Cohen, Q-Value Measurements at the University of Pittsburgh; this conference.

[13] M. G. Mayer, and J. H. D. Jensen, Elementary Theory of Nuclear Shell Structure, p. 74. New York: John Wiley and Sons, Inc. 1955.

Discussion

N. ZELDES: I would like to ask what is your definition of pairing energy?

W. H. JOHNSON: It is the difference in two separation energies, as an example $S_n(N\,Z) - S_n(N-1, Z)$ with N even.

B. HOGG: You measure resolution from the full width at half maximum, others measure resolution at the base of the peaks. It would be nice if some agreement could be made to standardize the way in which resolution is computed for the different mass spectrometers.

W. H. JOHNSON: We used this definition of resolution because it is very easy for us to measure. It turns out that our peaks are closely triangular, so that all one needs to do to convert from one resolution figure to another is to multiply by 2. If you multiply what we call our resolution, one part in 100,000 by 2, then we sould have a base resolution of one part in 50,000.

U. VON ZAHN: I kind of disagree with your error assignment with respect to the hydrogen mass unit calibration. It seems to me that there are only two ways to calibrate your mass scale: Either you trust your standard R-calibration or you use the hydrogen mass unit calibration. What you did is calculate the mass doublets by means of the hydrogen calibration. Applying correction factors to the standard R-calibration means discarding the R-calibration completely. However, you told us that you still assign errors to your mass doublets which partially stem from the standard R-calibration. This looks to me a bit inconsistent. I would rather prefer to substitute this error part by a pure statistical error coming from the hydrogen mass unit calibration which you repeat over and over again anyway.

W. H. JOHNSON: This is quite true. I think if we were going to do this again, the way you suggest is probably the method we would use. What we are really doing in applying this correction is correcting for all linear changes. A change in R is really only one of the linear changes that can occur. Some other linear changes can be caused by contact potentials on the electrostatic analyzer electrodes, thermal EMF's or leakage currents. I am sure that a change in the value of R was not the main cause of incorrect mass units, but rather it was caused by some other mechanism. The corrections were large during and following the time we were measuring selenium and tellurium. I feel that these materials were at least partly responsible for the departures.

If we do our error analysis as you suggest, we still would have to include an error in ΔR in the determination of the value of K. The error in ΔR is about half the size of the error that we claim for R in a typical wide doublet. This will still contribute a large part of the total error in wide doublets and would not reduce the total error greatly.

I believe that the two choices of error analysis that I took to study this situation are the extremes. In these extremes, the masses and errors, after applying a consistency factor, do not really change a great deal. I would hope that if we used your method we would be somewhere between these extremes.

U. VON ZAHN: I agree on that.

A. MCNISH: I am delighted to hear about these very precise measurements. At the National Bureau of Standards we have been concerned for some time about relationships between our electric standards and nuclear mass data. Since the Faraday constant is numerically identical with the conversion factor between electron volts and atomic mass units, the results reported are of special interest with respect to the physical constants, the precision reported being comparable with that obtained in direct experiments on the Faraday. The precision required in the calibration of electric resistors in these experiments approaches what is achievable in a national standardizing laboratory. I hope that after this conference, close cooperation may be established. Perhaps we may be able to help you and you may be able to help us in solving our mutual problems.

K. WAY: Was palladium included in your measurements? When will your results be given to a waiting world?

W. H. JOHNSON: Yes, we have measured all isotopes from gallium to xenon now, except the lowest stable isotope of xenon. We submitted papers reporting the measurements in June. When they are accepted, we will send out preprints.

D. M. VAN PATTER: This comment is directed to Dr. RIES, and relates to the slide showing the proton binding energy S_p in the region of $Z = 40$. At the recent Gordon Research Conference on Nuclear Chemistry, EICHLER (Oak Ridge) reviewed various evidence concerning existence of shell (and sub-shell) closures and concluded that there was some evidence for a sub-shell at 38. Your results for S_p and S_{2p} provide evidence for a proton sub-shell closure at $Z = 38$ rather than at 40, according to the change in slope for the neutron binding energy S_n which you observe between $N = 50$ and $N = 52$ corresponding to the shell closure at $N = 50$.

R. R. RIES: I would like to thank Dr. VAN PATTER for pointing that out. What Dr. VAN PATTER has said is absolutely correct. It was a misinterpretation on my part. I had a fixation from previous literature on a proposed sub-shell at 40 and when I saw the bump I was rather elated and left the matter at that. What the data really do show is exactly what Dr. VAN PATTER has pointed out, that the bump occurs and change in slope is due to the values at 38; I did not point that out correctly.

W. H. JOHNSON: I think that we are not very firm on any of these sub-shell effects. I think it is going to take more analysis and more data before we can really show that there is something anomalous in this region around 40.

E. BREITENBERGER: It makes me sad to see those R_e/R_i figures. At the risk of becoming still more unpopular, I should like to urge that such outmoded tools be abandoned. It is perfectly adequate to give the various values of χ^2 together with the numbers of the degrees of freedom. Exact tests can then be carried out by means of a χ^2-table. One can also make a detailed analysis of variance by means of an F-table. If a very pedantic check on the consistency of the data is required, one can further refine the variance analysis; see for instance reference 14 of my paper. None of this can be done with the R_e/R_i ratio; it is not a practical quantity and has not been tabulated.

W. H. JOHNSON: This is really our first attempt at doing anything of this sort. In the future we will try to refine things.

V. DOROKHOV: In the report of Dr. RIES, the values of doublets and respective errors are given. Will you please explain how can it be that the errors on mass values are less than the errors on the doublets?

W. H. JOHNSON: This reduction in error is the result of the weighted least squares fit. For a given pair of neighboring stable isotopes we measure a hydrocarbon doublet for each isotope and the mass difference between the pair of isotopes. These three measurements allow us to quote errors on the two isotopic masses that are somewhat smaller than any of the errors for the individual measurements.

V. DOROKHOV: Would you tell me how you check for the possible presence of systematic errors? Do you compare only with other measurements or do you try to check somehow with your apparatus, so to say, through an internal check?

W. H. JOHNSON: The internal check that we are using now is the one that I describe here where we measure hydrocarbon units, hydrocarbon doublets, and also, independently, isotope-isotope doublets, to see how they compare. Let me say that we are very happy that you are measuring in the same region we are, because this is a much better check, I think, than the things we can do alone. If instruments are quite different, I think the chance of you having the same systematic error that we do is perhaps quite a bit less than if we were just to do a doublet several ways. We are glad to see the activity of mass measurement at Hamilton and also from your group.

V. DOROKHOV: This problem excites us because the comparison of these data from the point of view of systematic error seems to be the basic problem in mass spectroscopy at present. My last question is do you think that by improvement of the precision of measurements, the values of the neutron and proton separation energy and other quantities tend to become more linear.

W. H. JOHNSON: It certainly seems that way from the data. If you look at the plots that we made with earlier data with less precision, the plots were not nearly as smooth. It looks as if this is due to the large experimental error. We were quite surprised at the smoothness of the curves that we see with the new data. It may be that as we measure better, separation energies will generally get smoother. I am not sure this is the case for pairing energies, however.

V. DOROKHOV: We find approximately the same thing.

J. W. DEWDNEY: My question is related to an earlier one asked by Dr. VON ZAHN and is related to the use of measurements of the hydrogen mass to do something which is essentially a re-calibration of resistors. Can you tell us something about the short term variation of your single hydrogen differences?

W. H. JOHNSON: The short term effect is well within the sort of statistical errors that Dr. RIES showed, approximately a few parts per million.

N. ZELDES: I would like to make two remarks. One concerns the linear trends in the S_n and S_{2n} lines just mentioned by Dr. DOROKHOV. We also noticed over two years ago that these always improved when new, more accurate, measurements become available. So much so, indeed, that I am now a priori inclined to reject data strongly violating these smooth trends. In particular, S_n always decreases with N when N increases by two, except at the onset of the large deformations in the rare earths, so the one point where it increases in bromine, shown by Dr. RIES, must probably be wrong.

The second remark suggests a possible check for systematic errors in the mass doublets. One can combine a whole group of doublets in one least-squares adjustment, treating H, C, N and O as unknowns as well. The values obtained for these fundamental masses can then be compared with the values from the

Mattauch-Wapstra tables. We used this method to check your measurements in the rare earths region presented at the Hamilton Conference.

W. H. Johnson: I think the smoothness is going to give us another means of detecting problems in the measurements. Another possibility is extrapolation of these lines out in either direction, to either higher or lower neutron or proton number. I think that with lines that are as smooth as this, the extrapolation to the next point is probably as good as the mass formula extrapolation. In regard to your second comment, it would be interesting to try this type of data analysis with the more accurate results that we now have. We will certainly consider this method.

Recent Mass Values Obtained at McMaster University*

By

R. C. Barber, R. L. Bishop, L. A. Cambey**, H. E. Duckworth,
J. D. Macdougall, W. McLatchie, J. H. Ormrod, and P. van Rookhuyzen

McMaster University, Hamilton, Ontario, Canada

With 14 Figures

In reporting the recent work done at McMaster University, we are including the first detailed description of the mass spectrometer as well as a discussion of the mass difference values that have been obtained.

A. Description of the Mass Spectrometer

1. Geometry

The mass spectrometer employs a first-order direction-plus-velocity focusing arrangement consisting of a 90° radial electrostatic analyser followed by a 180° uniform magnetic field as indicated in Fig. 1. Both analysers are used symmetrically so that the magnification in each is unity, and the radii of curvature, r_e and r_m, are equal. The radii are approximately 2.7 meters (9 ft). The total length of the ion path is approximately 15 meters.

The various slits indicated in the figure have the widths given and perform the following functions. S_1 and S_5 are the principal and collector slits, respectively. S_2 defines the angular extent of the ion beam such that the half angular spread, α, is 8×10^{-4} radians. S_3 is used to limit the height of the ion bean that enters the magnetic analyser. This ensures that the magnetic field encountered by the transmitted beam possesses adequate uniformity over the height of the beam. S_4 determines the range of energies which can be transmitted by the instrument.

The second order image aberrations are given by

$$y_B = r_m \left(B_{11} \alpha^2 + B_{12} \alpha \beta + B_{22} \beta^2 \right),$$

* This work was supported by the U. S. Air Force through the Air Force Office of Scientific Research of the Air Research and Development Command, and by the National Research Council of Canada.

** Now with Nuclide Analysis Associates, State College, Pennsylvania, USA.

25 a

where the symbols are those used by Hintenberger and König[1]. The coefficients have the values

$$B_{11} = 0.702, \qquad B_{12} = -0.365, \qquad B_{22} = -2.12.$$

Thus, although the instrument possesses no second-order focusing, the aberrations corresponding to the second order terms are unusually small.

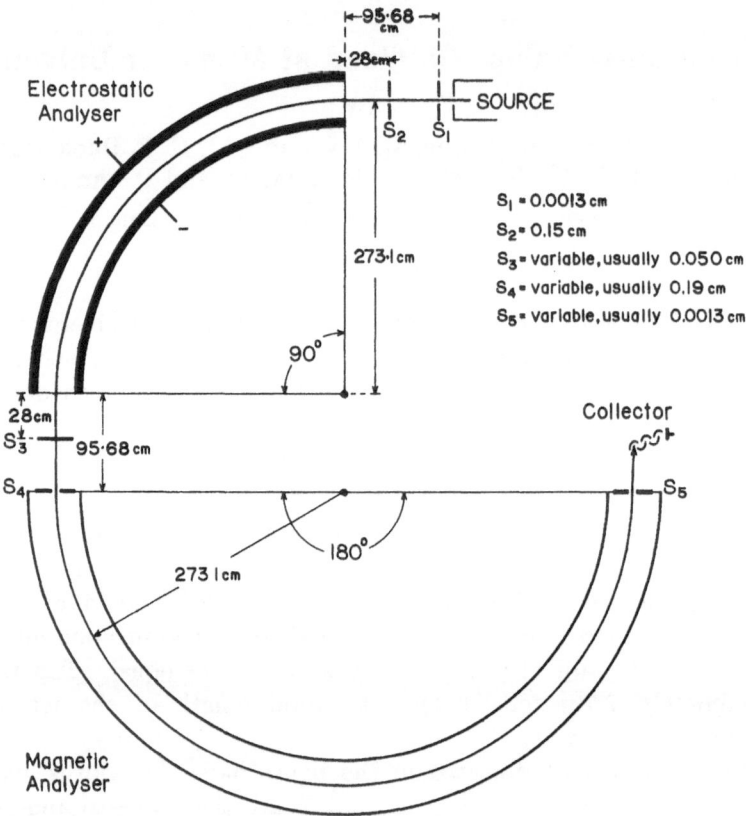

Fig. 1. Geometry of the mass spectrometer

Indeed, the combined effect of the $\alpha\beta$ and β^2 terms appears to be even smaller than in the several instruments that provide partial second-order direction focusing (that is, $B_{11} = 0$).

2. Vacuum System

A somewhat more detailed diagram of the apparatus is given in Fig. 2.

The source region is evacuated by a 90 l/sec oil diffusion pump (1). The remaining five pumping stations (2, 3, 4, 5, 6) are identical. Each consists of a 400 l/sec oil diffusion pump, water-cooled baffle, liquid-air trap and pneumatically-operated gate valve. An automatic protection system closes the valves in the event of electrical or water failure or accidental deterioration of the vacuum.

Two large vacuum gate valves (V_1 and V_2) permit the entire apparatus to be sub-divided into three separate chambers.

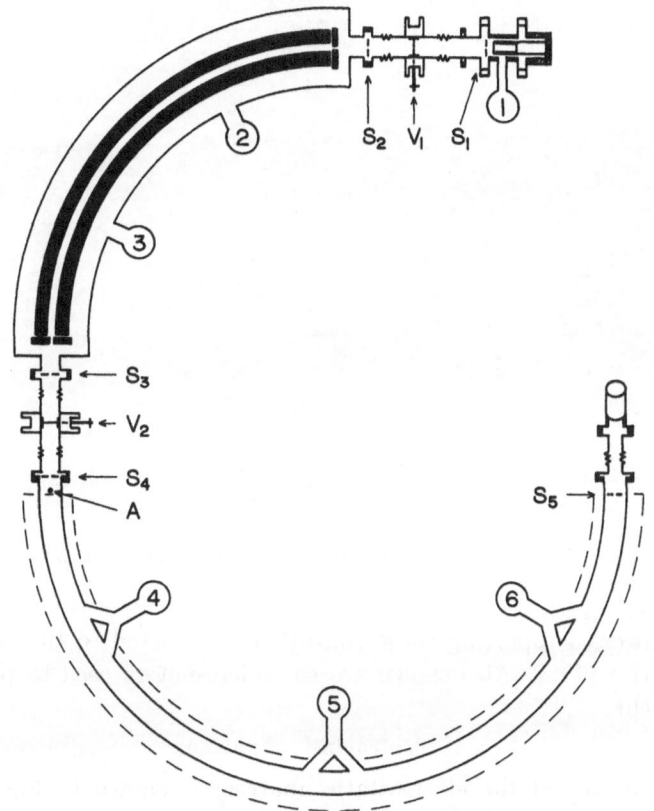

Fig. 2. Diagram of the mass spectrometer showing vacuum system

During typical operation of the instrument, the pressure in the electrostatic analyser is less than 4×10^{-6} torr, while that in the magnetic analyser is less than 5×10^{-7} torr.

3. Ion Source

The source which has been used to date is a conventional electron bombardment source, fitted with either a gas leak or some suitable form of oven. The source potential (~ 30 kV) is derived from a commercial unit (Beta Electric type 2069) and has a short term stability of one part in 30,000. It has been found, by decreasing the width of S_4, that the energy spread amongst the ions emerging from the source is less than 3 volts, corresponding to $\beta \leqslant 5 \times 10^{-5}$.

Fig. 3 is a photograph of the source end of the spectrometer. The controls for the adjustment of the source position along the ion beam (l_e') and in the two directions perpendicular to the ion beam are clearly shown.

Fig. 3. Photograph of source end of instrument showing controls for adjustment of source's position

The framework supporting these controls is connected to the electrostatic analyser base plate. Also shown are the micrometers used to position the principal slit.

4. Electrostatic Analyser

The interior of the electrostatic analyser is shown in Fig. 4.

Each of the two plates of the cyclindrical condenser consists of eleven gold-plated iron blocks which are carefully fitted together to form an accurate, continuous surface. Each block is supported by three 1.588 cm (0.625 in) thick alumina insulators which rest in turn on the steel base plate. The surfaces between which the field is established were positioned relative to the inner edge of the base plate so that the gap was 2.540 $\pm$ $\pm$ 0.007 cm (1.000 $\pm$ 0.003 in). The plates are 10.2 cm (4.0 in) high and roughly 7.5 cm (3 in) thick. Grounded diaphragms, positioned in accordance with HERZOG's theory[2], terminate the electric field at the physical boundaries of the condenser.

Fig. 5 is a wide angle photograph of the entire instrument. The vacuum housing of the electrostatic analyser, seen in the right side of the picture, is constructed of 1.27 cm (0.50 in) welded aluminum and is sealed to the steel base plate by a neoprene gasket.

Fig. 4. Interior of electrostatic analyser

Fig. 5. Wide angle photograph of entire mass spectrometer

Three ball bearing supports, one of which is shown in Fig. 5, provide for easy movement of the analyser about the point (A in Fig. 2) at which the ions enter the magnetic field. This arrangement makes it possible to vary the angle of entry, ε_m', to the magnetic field without altering any of the parameters of the electrostatic analyser.

5. Magnetic Analyser

The 180° magnetic field consists of 28 equal angle sectors, each of which is in itself, a "C" electromagnet. The pole pieces are 17.8 cm (7.0 in) wide and have a nominal separation of 2.057 cm (0.810 in). The field strength may be adjusted from sector to sector by means of trimming rheostats in series with each coil. The variation of the field within a given sector may be adjusted by tilting the upper pole piece of the sector.

Fig. 6. Collector end of instrument

A numerical calculation[3] was performed to determine the effect of non-uniformities on the focusing properties of the magnet. The non-uniformities considered were those in which the field varies along the ion path, and those in which it varies radially.

It was found that random variations of 1% from sector to sector would not destroy the double focus although its position might be shifted slightly. However, radial variations in the field of the order of 0.01% per cm could shift the direction and velocity foci by differing amounts so that a double focus would not be formed.

In our instrument, the velocity focus is most sensitive to radial non-uniformities in the region of 45° from the entrance boundary while the direction focus is most sensitive to those in the region of 90° from the entrance boundary. During the initial focusing it was necessary to adjust certain of the magnet pole pieces accordingly.

A soft iron diaphragm was introduced at the entrance boundary of the magnetic field to reduce the extent of the fringing field. The fringing field was measured with a Hall generator and the effective boundary found to be 0.99 cm (0.39 in) outside the physical boundary.

The magnet current is regulated by a combination D. C plus A. C. amplifier circuit described by GARWIN et al.[4] and modified slightly for this application. Long term stability is about one part in 10^5 and appears to depend primarily on the stability of the reference voltage. The short term regulation is better than one part in 10^6.

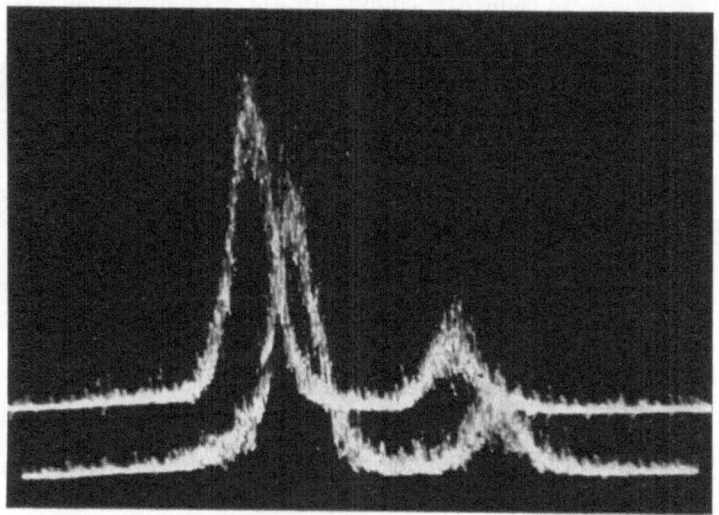

Fig. 7. Focusing test: direction focus only

6. Ion Detection

Fig. 6 shows the collector end of the magnetic analyser.

On passing through the collector slit, the ion beam is detected by a 14 stage Allen type electron multiplier with a gain of about 10^6. The multiplier output is amplified and viewed on an oscilloscope screen. When the sawtooth voltage from this oscilloscope is used to modulate the electrostatic analyser voltage, the ion beam is swept across the collector slit and a peak is seen on the oscilloscope screen.

7. Focusing Procedure

In the initial focusing, the principal slit was set at its theoretical position with respect to the electrostatic analyser. The distance between the electrostatic analyser and the magnet was also set at its theoretical value. With the electrostatic analyser thus set up, the value of ε_m' was varied in order to bring, in turn, the direction and velocity foci to the collector.

The velocity and direction foci are identified as follows. If, on every second sweep of the oscilloscope, the source potential is changed by 15 volts, lack of velocity focusing results in a peak being displaced with respect to itself. Coincidence occurs for some value of ε_m'. The direction focus is characterized by a very marked increase in the sharpness of the peaks as ε_m' passes through its optimum value. The relative positions of the two foci may be deduced from the geometry.

Fig. 7 and Fig. 8 are photographs of the oscilloscope screen taken when these tests were being carried out. In Fig. 7 the instrument is adjusted

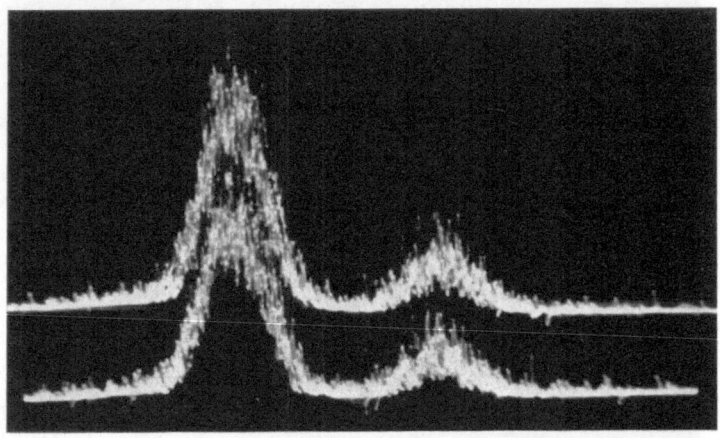

Fig. 8. Focusing test: velocity focus only

for direction focusing without velocity focusing. In Fig. 8 the reverse is true. Alternate sweeps are displaced with respect to each other to facilitate viewing.

When ε_m' is varied, the field necessary to bring ions to the same stage of the trace on the oscilloscope screen is a maximum for normal entry. If the final direction focus occurs at this same angle, this is an indication that the direction focus of the electrostatic analyser is formed at the entrance boundary of the magnetic field. Having thus established that the electrostatic analyser was focusing properly, the magnet pole pieces were adjusted in the manner described above to bring the two foci into coincidence.

Thereafter, the routine focusing operation consists of varying ε_m' to determine the relative positions of direction and velocity foci, and then making any small adjustment of l_e' that is necessary to bring the direction and velocity foci into coincidence.

Fig. 9 shows the doublet of Figs. 7 and 8 after these adjustments have been completed. The mass separation of this doublet is 1/45,000, indicating that the resolution on this occasion was approximately 1/100,000.

Among the most recent doublets studied is one for which the separation is 1/126,000 and one for which the separation is only 1/130,000. In these cases it was necessary to have a resolution of about 1/200,000. This figure was attained regularly for the period needed by reducing the width of the collector slit. The actual resolution under these conditions still agreed with the theoretical value.

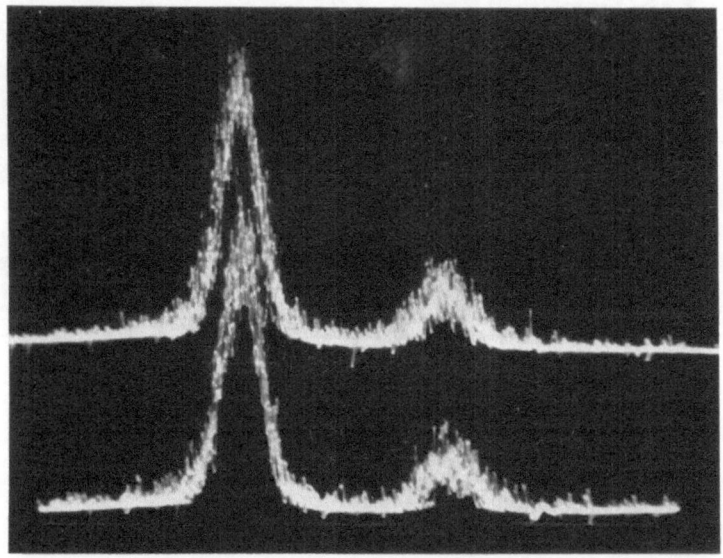

Fig. 9. Focusing test: double focus

8. Peak Matching

The spacing of a doublet is measured in the following way. Let us suppose that, for a given magnetic field and electrostatic analyser voltage V, ions of mass M are arriving at the collector. If, then, on every second sweep of the oscilloscope, the voltage is changed an amount ΔV to $V + \Delta V$, a peak due to ions of mass $M - \Delta M$ will be brought to the same stage of the oscilloscope sweep. At the same time, the source potential must also be switched by an amount ΔV_a on alternate sweeps. The matching procedure is illustrated in Fig. 10. There are, of course, two ways of doing this: the lighter ion may be matched to the heavier one or vice versa.

When the doublet spacing is close, the error in measuring V is larger than the magnitude of ΔV. Accordingly the mass difference for either of these two cases is given by

$$\Delta M = M \frac{\Delta V}{V}.$$

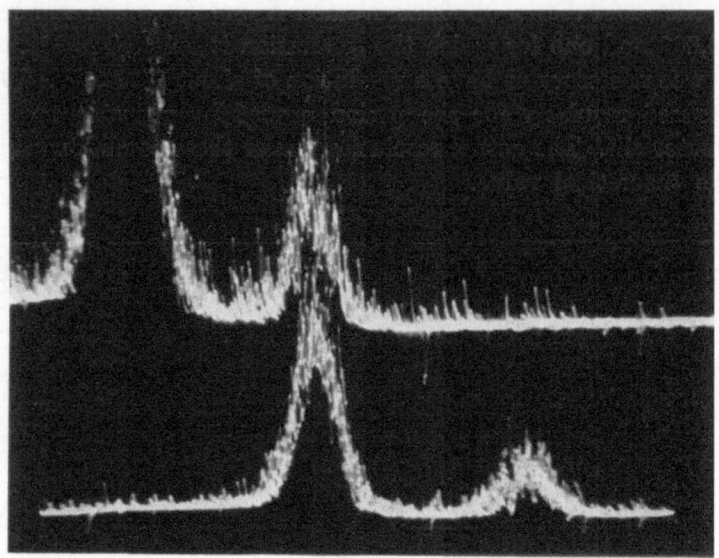

Fig. 10. Matched doublet

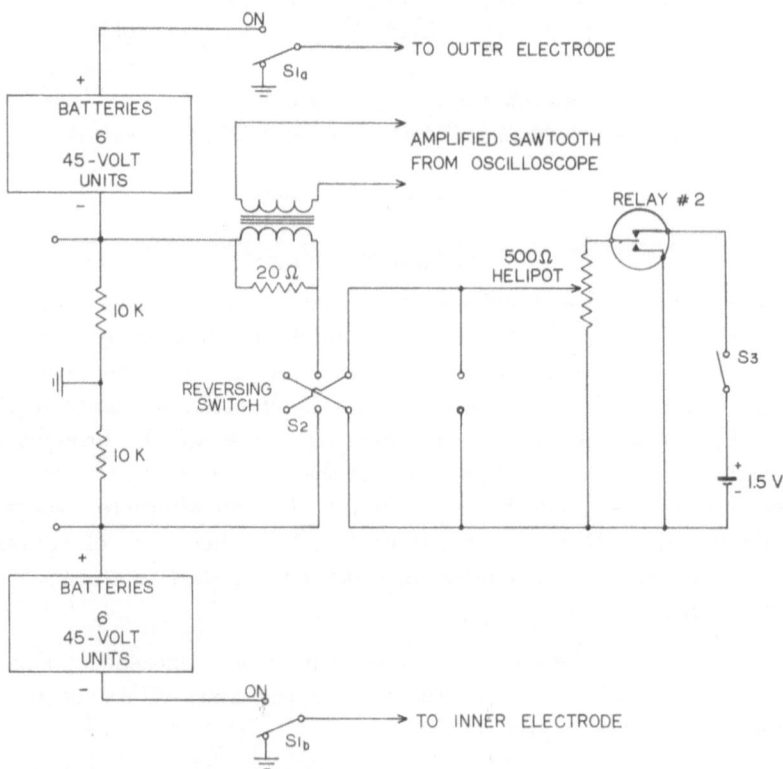

Fig. 11. Circuit to provide electrostatic analyser potential

Considering the two methods of matching, the two directions of sweeping the ion beam, and the fact that one trace on the oscilloscope may be made to appear either above or below the other, we see that there are 8 possible matching situations. The 8 values so obtained constitute one "run" and the mean of these values is the value of the run. By adopting this procedure we believe that we have eliminated the possibility of significant errors arising in the switching mechanism or in the operator's judgment of the matched condition.

The circuit which provides the voltage to the electrostatic analyser plates is given in Fig. 11. V is supplied by twelve 45 volt batteries which are stable to one part in 10^4 during a day and to better than one part in 10^6 over two or three minutes. A differential voltmeter is used to measure V with an accuracy of 0.042%. A potentiometer with an accuracy of 0.015% is used to measure ΔV.

9. Calculation of Errors

Twenty to thirty runs are made in order to determine a doublet spacing, and from these values the probable error is computed in the usual way. This probable error is then combined with the errors (given above) in V and ΔV in order to arrive at the final probable error. It should be noted that this procedure treats the errors ascribed to the voltage measurements as if they were 50% confidence limits. We understand these values to represent limits of error rather than 50% confidence limits. Because of this, we should expect the final error associated with a doublet to be somewhat conservative.

10. Examples of Consistency of Runs
a) Least Consistent Results

In Table 1 are given the results for the various runs on the doublet $^{121}Sb - ^{12}C^{35}Cl^{37}Cl_2$. The results for this doublet are presented here as they are much less consistent than those obtained for any other doublet so far.

The values are grouped according to the day on which the run was performed. Below each group is given the mean of the group. The mean of all 23 values is given near the bottom of the table. We also present the data in the form of a histogram and note that they appear to follow a normal distribution.

The probable error representing the reproducibility of the results was calculated and is so identified. When this probable error was combined with the error associated with the voltage measurements, the final probable error was obtained.

The statement that the values are the least consistent that have been encountered so far is borne out by the fact that the probable error

26*

Table 1. Least Consistent Set of Runs
^{121}Sb—$^{12}C^{35}Cl^{37}Cl_2$ (mu)

	July 31/62 [1]	Aug. 1/62 [2]	Aug. 2/62 [3]
	3.1212	3.1626	3.1138
	3.1542		3.1598
Mean ...	3.1377	3.1626	3.1368

	Aug. 3/62 [4]	Aug. 7/62 [5]	Aug. 9/62 [6]
	3.1778	3.2074	3.1475
	3.1959	3.1849	3.1598
	3.1593		3.1499
	3.1784		3.1717
	3.1701		3.1749
	3.1577		3.1438
	3.1607		
	3.1615		
	3.1625		
	3.1523		
Mean ...	3.1676	3.1962	3.1579

Mean of 23 runs: 3.1621 mu.
Probable error (reproducibility): 2.9 μu.
Total Spread of results: 93 μu.
Final probable error: 3.2 μu.
Value quoted: 3.162 ± 3 mu.

```
                                      6
                                  4       6
                                  4   4   6
                          6       4   4   4
                          6       3   4   4
            3   1     6   1   2   4   5   4   5
            |         |       |       |       |
          3.12     3.14    3.16    3.18    3.20
```

representing the reproducibility corresponds to locating one peak with respect to the other to only one part in 4.2×10^7 of the mass at which it occurs. In the next worst case, the corresponding figure is one part in 8.0×10^7.

The value quoted by us for this doublet[5] is given at the bottom of the table. This result was confirmed by a closed loop formed by it and three other doublets. The value of the loop was $0 \pm 5 \mu$u.

b) Most Consistent Results

We present now in Table 2 the most consistent set of results that we have obtained, namely, for the doublet $^{123}Sb^{35}Cl$—$^{121}Sb^{37}Cl$. Although these are the most consistent results, the values for several other doublets are almost as good.

Table 2. Most Consistent Set of Runs
$^{123}Sb^{35}Cl$—$^{121}Sb^{37}Cl$ (mu)

May 3/62 [1]	May 4/62 [2]	May 7/62 [3]	May 10/62 [4]
3.34559	3.34393	3.34909	3.34037
3.34333	3.34205	3.33831	3.34775
3.33782	3.34441	3.34040	3.34120
3.34457	3.35088	3.34855	3.34350
3.33867	3.34028	3.34239	3.34431
3.34095	3.34718	3.34022	3.34086
Mean... 3.34182	3.34479	3.34316	3.34300

Mean of runs: 3.34319 mu.
Probable error (reproducibility): 0.49 μu.
Total Spread of results: 13 μu.
Final probable error: 1.6 μu.
Value quoted: 3.3432 $\pm$ 16 mu.

```
                    4
                    4
                    4
                    4
                    4
                    3
                    3
                    3
                    2
                    2
                    2 4
                    2 3
                  3 1 3
                  1 1 2
                  1 1 1 2
                  |   |
                3.34 3.35
```

Again the results are grouped according to the days on which the runs were performed with the mean value for each day at the bottom of the column. In the histogram of the results, the mass scale has been expanded by a factor of two over that used for the histogram shown in Table 1.

For this doublet, the probable error associated with the reproducibility of the results corresponds to locating one peak with respect to another to within one part in 3.2×10^8 of the mass at which the doublet occurs. At a resolving power of 1/100,000 this corresponds to matching the peaks to 1/3200 of the peak width, and represents the most favourable operation of the instrument in its present form.

11. Pressure Effect

A doublet of the type discussed in 10 (b) above is peculiar in that the two members of the doublet are chemically identical. Because of this, the two ion groups emerge from the source with the same distribution in angle and energy. The amount, ΔV_a, that the source potential should be changed in switching from one doublet member to the other is calculated from the doublet spacing and set accurately using a voltmeter. This procedure will eliminate the effect of second order terms on the doublet spacing. For this reason, results from this type of doublet should be particularly reliable.

In a doublet formed by two chemically dissimilar ion species, it may be that, for a given source potential, the two ion groups possess different energies after passing through the electrostatic analyser. This may be verified in our instrument because an intermediate direction focus is formed and because the energy resolving power of the electrostatic analyser is high. It has been confirmed that, for the doublet $^{121}Sb - ^{12}C^{35}Cl^{37}Cl_2$, this energy difference can be as large as 6 eV. It was further noted that this energy difference arose, at least in part, from a high pressure in the source region. When the pressure was increased beyond 7×10^{-4} torr, the energy difference increased rapidly. When the work on this doublet was being done, the source pressure was kept sufficiently low that the energy difference between doublet members was always less than one electron volt.

This pressure effect has been suspected for some time[6] and it now appears that the earlier conjectures concerning it were correct.

B. Results to Date

1. New Mass Differences

The mass spectrometer has been in regular operation since March, 1962; that is, for about sixteen months. During this period 35 doublets have been studied involving isotopes of titanium, zinc, germanium, selenium, zirconium, molybdenum, cadmium, tin, antimony, tellurium, and mercury. The mass differences for the majority of these doublets have been published[5, 7-9] and the remainder are shown in Table 3. The errors range from 1 to 5 keV, but are usually about 2 keV.

Most of the mass differences are of the type

$$\Delta M = \pm (^{A+2}G \ X \ ^{35}Cl - ^{A}G \ X \ ^{37}Cl),$$

where G is the symbol for the metallic atom and X is a constituent (which may or may not be present) common to both members. Thus, by use of the known mass difference $^{37}Cl - ^{35}Cl = 1.9970414 \pm 36$ u, as given in the 1961 mass table of KÖNIG, MATTAUCH, and WAPSTRA[10], the mass difference $^{A+2}G - ^{A}G$ may be computed. Further, this difference may be expressed in the convenient form

$$^{A+2}G - ^{A}G = 2 \ n - S_{2n},$$

where $2 \ n$ represents the mass of two neutrons and S_{2n} is the so-called *double neutron separation energy*, or the energy in MeV required to remove the last two neutrons from the heavier nuclide. Unfortunately, the uncertainty in the $^{37}Cl-^{35}Cl$ mass difference is roughly double that of the doublet spacings, with the result that the precision of the double neutron separation energies is diluted to 4 keV. This error will, of course,

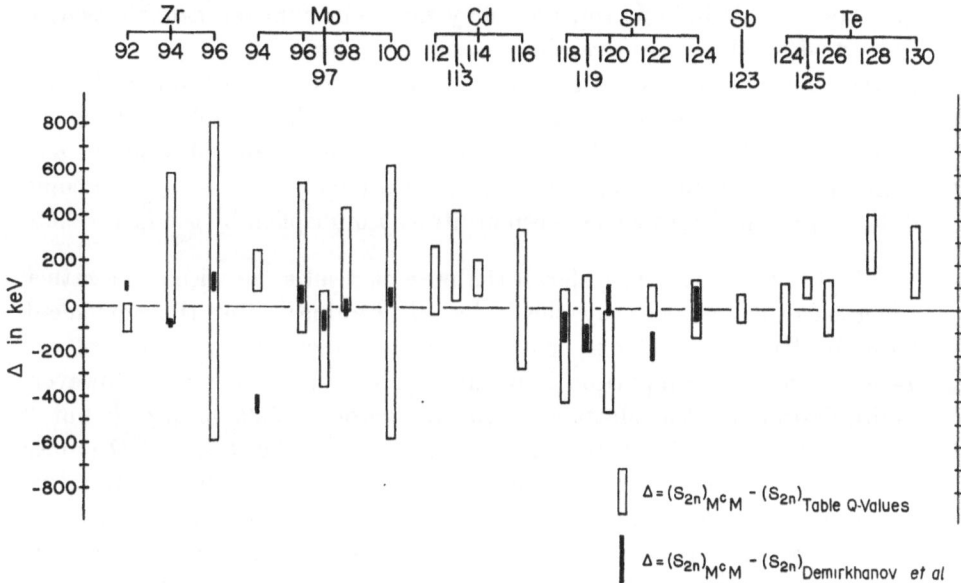

Fig. 12. Comparison of McMaster values with existing data

Table 3. New Doublet Spacings (on scale $^{12}C = 12$ u)

Doublet	Spacing in μu	$M/\varDelta M$	Intensity ratio of doublet members
$^{47}Ti^{35}Cl^{37}Cl-^{49}Ti^{35}Cl_2$	946.4 ± 1.1	126000	0.46
$^{46}Ti^{35}Cl^{37}Cl-^{48}Ti^{35}Cl_2$	1726.8 ± 1.1	68000	0.035
$^{68}Zn^{35}Cl-^{66}Zn^{37}Cl$	1757.9 ± 1.0	59000	2.04
$^{70}Zn^{35}Cl-^{68}Zn^{37}Cl$	3429.5 ± 1.5	31000	0.10
$^{74}Ge^{35}Cl-^{72}Ge^{37}Cl$	2047.5 ± 1.1	53000	4.11
$^{76}Ge^{35}Cl-^{74}Ge^{37}Cl$	3175.7 ± 1.5	35000	0.66
$^{80}Se^{35}Cl-^{78}Se^{37}Cl$	2164.8 ± 1.4	53000	6.54
$^{94}Mo^{16}O_2{}^{35}Cl-^{92}Mo^{16}O_2{}^{37}Cl$	1234 ± 2	130000	1.75
$^{96}Mo^{16}O_2{}^{35}Cl-^{94}Mo^{16}O_2{}^{37}Cl$	2359 ± 2	69000	5.54
$^{97}Mo^{16}O_2{}^{35}Cl-^{95}Mo^{16}O_2{}^{37}Cl$	3138 ± 2	52000	1.83
$^{98}Mo^{16}O_2{}^{35}Cl-^{96}Mo^{16}O_2{}^{37}Cl$	3690 ± 2	45000	4.40
$^{100}Mo^{16}O_2{}^{35}Cl-^{98}Mo^{16}O_2{}^{37}Cl$	5019 ± 2	33000	1.24
$^{110}Cd^{35}Cl-^{108}Cd^{37}Cl$	1764 ± 5	82000	0.47
$^{112}Cd^{35}Cl-^{110}Cd^{37}Cl$	2701 ± 2	54000	5.94
$^{113}Cd^{35}Cl-^{111}Cd^{37}Cl$	3174 ± 2	47000	2.94
$^{114}Cd^{35}Cl-^{112}Cd^{37}Cl$	3547 ± 2	43000	3.67
$^{116}Cd^{35}Cl-^{114}Cd^{37}Cl$	4353 ± 2	35000	0.80
$^{202}Hg^{35}Cl-^{200}Hg^{37}Cl$	5271 ± 3	45000	3.98

be automatically reduced when a better value is obtained for the $^{37}Cl - {}^{35}Cl$ mass difference. As they are, however, the values of S_{2n} are considerably more accurate than any that have been reported to date for most of the elements studied. This is especially true for the elements zirconium, molybdenum, cadmium, tin, antimony and tellurium, as can be seen in Fig. 12 where the present results are compared both with the values given in the Table of Q-Values[11] and with those deduced from the recent mass spectroscopic results of DEMIRKHANOV et al.[12, 13]. The Table of Q-Values, it will be recalled, is based on the Mass Table of KÖNIG, MATTAUCH, and WAPSTRA[10] which, in this mass region, is based largely on mass spectroscopic data whose stated errors have been arbitrarily multiplied by the factor 2.65.

In 16 of the 23 values shown the present results fall within the rather generous errors allowed by the Table of Q-Values. But the agreement with the Russian results is not nearly as satisfactory. In only 3 of the 14 cases where a comparison is possible do the errors overlap. However, multiplication by the Mattauch-Wapstra factor of 2.65 brings all but 3 into the fold. This fact and the agreement with the Table of Q-Values in 70% of the cases, suggests that the factor 2.65 is unnecessarily large. Perhaps Professors MATTAUCH and WAPSTRA and their colleagues will take a somewhat less incredulous attitude toward stated mass spectroscopic errors in preparing their next Mass Table.

Table 4. Some Double Neutron Separation Energies for Ti, Zn, Ge and Se (in MeV)

Nuclide	McMaster	From Table of Q-Values	$M - Q$ (keV)
^{48}Ti	20.507 ± 4	20.507 ± 5	0 ± 6
^{49}Ti	19.780 ± 4	19.769 ± 8	11 ± 9
^{68}Zn	17.261 ± 4	17.244 ± 12	17 ± 13
^{70}Zn	15.704 ± 4	15.693 ± 17	11 ± 17
^{74}Ge	16.991 ± 4	16.690 ± 70	301 ± 70
^{76}Ge	15.941 ± 4	16.950 ± 90	— 9 ± 90
^{80}Se	16.882 ± 4	16.921 ± 45	— 35 ± 45
^{202}Hg	13.989 ± 5	14.013 ± 22	— 24 ± 22

The new mass differences involving titanium, zinc, germanium, selenium and mercury provide certain useful spot checks of the reliability of the Mass Table in those regions, as is shown in Table 4. Except for the glaring discord in the case of ^{74}Ge, the two sets of values are in satisfactory agreement. This is especially gratifying for the titanium, zinc and mercury examples, where the stated accuracy in the Table of Q-Values is relatively high.

2. Systematics of Double Neutron Separation Energies between the 50 and 82 Neutron Shells

Inasmuch as our experiments yield double neutron separation energies directly, as described above, we shall give special emphasis to this type of quantity. In Fig. 13 we have plotted (as a function of N) the double

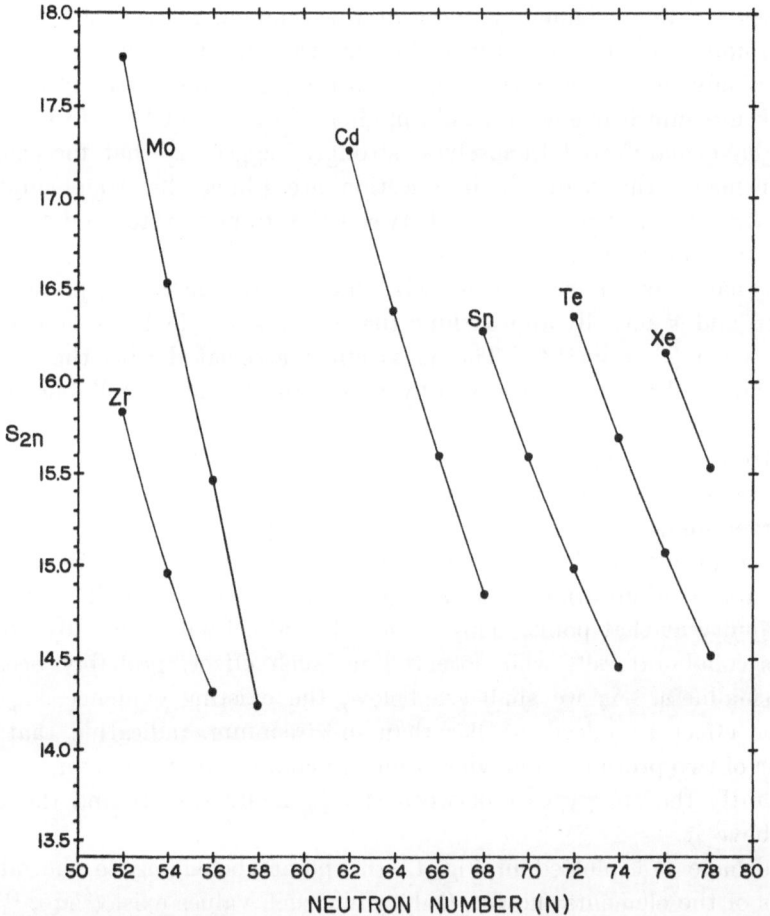

Fig. 13. Double neutron separation energies for even-A nuclides plotted as a function of neutron number. Points belonging to the same element are joined

neutron separation energies for ^{92}Zr, ^{94}Zr, ^{96}Zr, ^{94}Mo, ^{96}Mo, ^{98}Mo, ^{100}Mo, ^{110}Cd, ^{112}Cd, ^{114}Cd, ^{116}Cd, ^{118}Sn, ^{120}Sn, ^{122}Sn, ^{124}Sn, ^{124}Te, ^{126}Te, ^{128}Te and ^{130}Te as obtained in our laboratory, and for ^{130}Xe and ^{132}Xe as obtained at the University of Minnesota[14]. The two last-named, with errors of 5 and 6 keV, respectively, are the only other published S_{2n} data for even-A nuclides in this region whose accuracies are comparable to our own. The dots representing the several values of S_{2n} are actually about three times greater than the error limits, but they are drawn large for visibility.

The curves obtained by connecting the points belonging to the same elements display negative slopes, indicating that the double neutron separation energy decreases with neutron number. This decrease is almost entirely caused by the increasing neutron excess, an effect which forms the basis for the asymmetry term in the semi-empirical mass formula. Furthermore, for the elements cadmium, tin and tellurium the curves are strikingly smooth, slightly concave to the right, and virtually parallel, as the semi-empirical mass formula predicts they should be. No sub-shell effects have manifested themselves, strongly suggesting that for each of these elements the neutrons in question are filling the same sub-shell and/or the energy gaps are small between the ground states and the first excited neutron states.

The curve for zirconium is likewise smooth. It is more sharply curved, however, and is actually approaching the end of a sub-shell, the 56 neutron configuration ($d_{5/2}^6$) in ^{96}Zr. The mass effect associated with the closure of this sub-shell has been observed by B. L. Cohen and his collaborators[15]. Unfortunately, we are not able to extend the zirconium curve beyond 56 neutrons, inasmuch as ^{96}Zr is the heaviest stable isotope of zirconium. The effect will be seen below, however, when our data are looked at from a different angle.

As mentioned earlier, the curve for molybdenum passes through the 56 neutron configuration and does in fact display a small, but real, discontinuity at that point. This element has also been studied by Cohen and his collaborators[16], who observed no such effect, probably because of its smallness. As we shall see below, the existing evidence suggests that the effect is indeed smaller than in zirconium, indicating that the addition of two protons (in moving from zirconium to molybdenum) reduces significantly the energy gap between the $d_{5/2}$ neutron state and the ones next above it.

We have not plotted in Fig. 13 the points belonging to the odd-A isotopes of the elements shown. Only a few such values exist (^{97}Mo, ^{113}Cd, ^{119}Sn, ^{125}Te, and ^{131}Xe), and they are too few to establish any pattern. They do lie, however, very close to the even-A curves for their respective elements.

3. Systematics of Double Neutron Separation Energies for Families of Isodiapheres

In Fig. 14 we have plotted the same data as in Fig. 13 but, instead of connecting points belonging to the same element, we have connected points for nuclides with a constant neutron excess, Δ. The curves are, thus, isodiapheric curves.

In general, S_{2n} increases as these isodiapheric curves move toward heavier nuclides, in agreement with the predictions of the semi-empirical mass formula. It is again the asymmetry term in the mass formula which is primarily responsible for this trend. This term, which has the effect of decreasing the binding force, contains the mass number (A) in the denominator, with the result that for constant neutron excess its absolute magnitude decreases with increasing A.

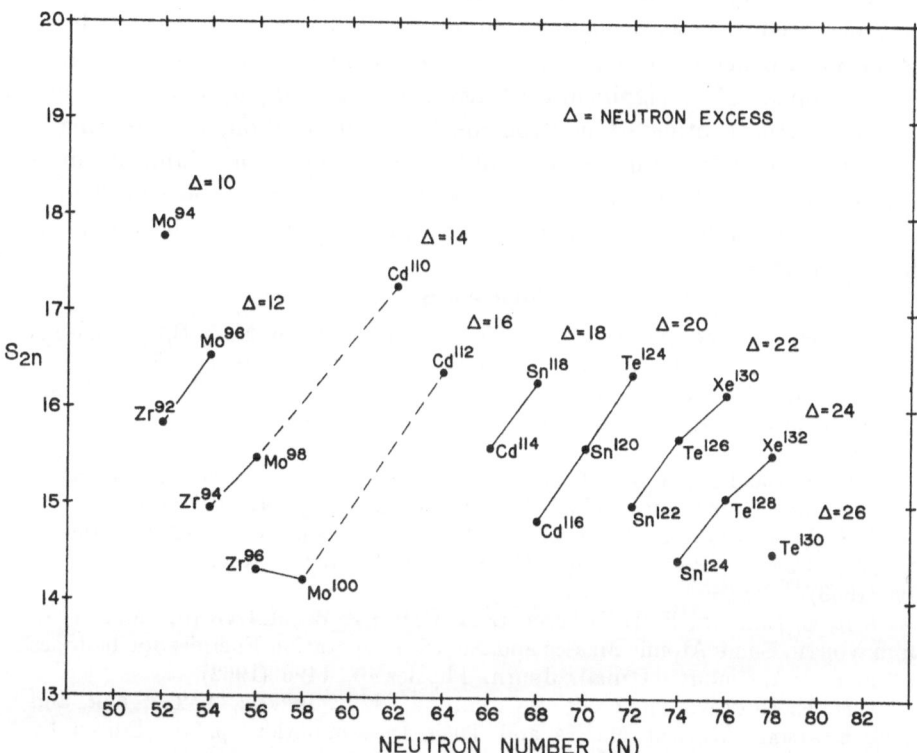

Fig. 14. Double neutron separation for even A nuclides plotted as a function of neutron number. Points for nuclides with constant neutron excess, $\varDelta$, are joined

The 56 neutron effect in ^{96}Zr is clearly seen in Fig. 14 as a dislocation in the curve passing through ^{96}Zr, ^{100}Mo and ^{112}Cd. The corresponding effect in ^{98}Mo should appear as a dislocation at 56 neutrons in the curve passing through ^{94}Zr, ^{98}Mo and ^{110}Cd. Unfortunately, the lack of S_{2n} values for ^{102}Ru and ^{106}Pd prevents the effect from being seen. It would appear, however, from the existing points that the effect is indeed smaller than in ^{96}Zr.

4. Single Neutron Separation Energies

We realize, of course, that our data are also useful in the calculation of single neutron separation energies. Some such calculations have been made in our published work. For the purposes of this Conference, however, it seemed best to leave this subject to Dr. Katherine Way, who knows a great deal more about it than we do. With this in mind, we communicated all but our most recent data to her prior to the Conference and, as you have heard, she has already made some use of them in her paper.

5. Summary and Future Plans

In summary, we have confirmed the 56 neutron effect in ^{96}Zr, we have observed a smaller, but real, 56 neutron effect in ^{98}Mo, but we have found no evidence for other sub-shells. In particular, we have eliminated the possibility of a significant 64 neutron effect ($d_{5/2}^6 \, g_{7/2}^8$) in ^{112}Cd but not, of course, in other 64 neutron nuclides. In addition, we have made a few spot checks involving isotopes of titanium, zinc, germanium, selenium and mercury, and the results are in general agreement with the Mass Table.

In the coming year we hope to extend our work into the region of the rare earths.

References

1 H. Hintenberger, and L. A. König, Advances in Mass Spectrometry, p. 16. Edited by J. D. Waldron. London: Pergamon Press. 1959.

2 R. Herzog, Ablenkung von Kathoden- und Kanalstrahlen am Rande eines Kondensators, dessen Streufeld durch eine Blende begrenzt ist. Z. Physik 97, 596 (1935).

3 L. A. Cambey, J. H. Ormrod, and R. C. Barber, Focusing Properties of Non-uniform Magnetic Fields. Canad. Journ. Physics 42, 103 (1964).

4 R. L. Garwin, D. Hutchinson, S. Penman, and G. Shapiro, Efficient Precision Current Regulator for High-Power Magnets. Rev. Sci. Instr. 40, 105 (1959).

5 R. C. Barber, R. L. Bishop, L. A. Cambey, W. McLatchie, and H. E. Duckworth, Some Atomic Masses and Neutron Separation Energies for Isotopes of Tin and Antimony. Canad. Journ. Physics 40, 1496 (1962).

6 H. E. Duckworth, G. R. Bainbridge, N. R. Isenor, J. T. Kerr, and P. C. Eastman, Nuclear Masses and Their Determination, p. 75. Edited by H. Hintenberger. London: Pergamon Press. 1957.

7 R. C. Barber, R. L. Bishop, W. McLatchie, P. van Rookhuyzen, and H. E. Duckworth, Neutron Separation Energies for Zirconium and the ^{96}Zr—^{96}Nb and ^{94}Zr—^{94}Nb Decay Energies. Canad. Journ. Physics 41, 696 (1963).

8 R. C. Barber, W. McLatchie, R. L. Bishop, P. van Rookhuyzen, and H. E. Duckworth, Neutron Separation Energies for Tellurium and the ^{130}I—^{130}Te and ^{123}Te—^{123}Sb Decay Energies. Canad. Journ. Physics. 41, 1482 (1963).

9 R. L. Bishop, R. C. Barber, W. McLatchie, J. D. Macdougall, P. van Rookhuyzen, and H. E. Duckworth, Systematics Governing Double Neutron Separation Energies Between 50 and 82 Neutrons. Canad. Journ. Physics 41, 1532 (1963).

10 L. A. König, J. H. E. Mattauch, and A. H. Wapstra, 1961 Nuclidic Mass Table. Nuclear Phys. 31, 18 (1962).

11 L. A. König, J. H. E. Mattauch, and A. H. Wapstra, (1961) Nuclear Data Tables, Part 2. Edited by Katherine Way (National Academy of Sciences, National Research Council, Washington 25, D. C.).

[12] R. A. Demirkhanov, V. V. Dorokhov, and M. I. Dzkuya, Isotope Masses and Binding Energy of Nuclei in the Region from Strontium to Ruthenium. J. Exptl. Theoret. Phys. (USSR) 40, 1572 (1961).

[13] R. A. Demirkhanov, T. I. Gutkin, O. A. Samadashvili, and I. K. Karpenko, Atomic Masses of Tin and Antimony Isotopes. Izvest. Akad. Nauk. SSSR, Ser. Fiz. 25, 871 (1961).

[14] R. R. Ries, R. A. Damerow, and W. H. Johnson, Preceedings of the International Conference on Nuclidic Masses, p. 446. Edited by H. E. Duckworth. Toronto: University of Toronto Press. 1960.

[15] I. Talmi, Effective Interactions and Coupling Schemes in Nuclei. Revs. Modern Phys. 34, 704 (1962).

[16] B. L. Cohen, and O. Chubinsky, Stripping-Reaction Studies in the Zr and Mo Isotopes. Bull. Amer. Phys. Soc. 8, 376, U 7 (1963).

Discussion

A. H. Wapstra: I would like to add just two remarks. The first is about the chlorine mass difference. Indeed it's true that the chlorine mass difference by the Minnesota group brings in a much better agreement with the other data as Professor Duckworth said. Here you see just our trouble. We have here the Minnesota doublet and Brookhaven doublet for ^{37}Cl. The two are different by 16 keV. The errors assigned to the doublets were 1.0 and 1.2 keV respectively. So here we are in a spot. Second, Prof. Duckworth showed us a very nice slide for zirconium where his data agrees well with the Minnesota group and then he made the remark about 2.65. I'm going to reverse the procedure and quote another value from the antimony measurements which is the one that Professor Duckworth said is the best they measured. Here the difference between the Minnesota and the Hamilton measurement is 21 keV and the combined error is 5 keV. So here we have to use a correction factor of 4. Also the difference between ^{121}Sb and ^{119}Sn was measured and that difference was 28 keV in the same direction. So if we add them we have a difference of about 50 keV with a combined error of about 9 keV. This is the worst case, but you see that we have to be a little careful if we want to make somewhat consistent adjustments.

E. Breitenberger: I should like to join the Mattauch forces. The so-called consistency factor is merely a stopgap to compensate for lack of precise meaning in many error statements made by experimentalists. It has only two minor weaknesses. Firstly, the factor may have a value different from the one you think is the best, and if you put it at 2.65 rather than 2.50 or 2.80 you suppress a possible component of variability which is afterwards missing from the errors of the adjusted masses. These errors are, therefore, subject to a microscopic downward bias. Secondly, when you handle an enormous mass of data, there must be a couple of discordant subsets owing to sheer chance. So you might occasionally apply a consistency factor although the data are in fact good. Within these limitations, Mattauch's procedure is the most sensible I can think of. The alternative would be an analysis of variance of truly fabulous complexity.

W. H. Johnson: May I direct a question to the Mattauch-Wapstra group? We are a little unclear where the factor 2.65 comes from. I would like to have a short discussion of its source. And second, do the doublets that we've measured now bear on this number? Will this number change now with the addition of these over-determined doublets that we have measured?

J. MATTAUCH: Professor WAPSTRA will certainly give you the answer to your first question. The factor 2.65 stems of course from the amount of over-determined doublets which we had at that time from your laboratory. We extended it to the others, because we didn't have any over-determination there. Please don't say that in the next adjustment we will stay forever with this factor 2.65. The more over-determinations we have the more we are able to find out what we should do.

W. H. JOHNSON: I think that it is our fault that we have not given you over-determined numbers. In the future, we will try to correct this and measure as many doublets as we can including the same mass.

A. H. WAPSTRA: I can quote the data numerically, because I just happen to have them here. At that time we had a set of 36 doublets in 12 parameters, so that over-determination was about a factor of 3; and there we got a factor of 2.65 for this group of Minnesota doublets. I will be glad to give you the precise data, it is in Nuclear Physics 25 (1961), p. 183. I should also say that the factor of 2.65 was used because SMITH himself made adjustments of this sort. He made two adjustments, one of 14 doublets in 3 unknowns, where he found the factor 3.58, and another one of 39 doublets in 8 unknowns, where he found a factor 2.35. You see, so the factor 2.65 was something like an average of the three most important sets of doublets we had at that time. I completely agree that it was somewhat arbitrary to apply the factor to all other doublets that we used. But on the other hand, I should say we did not have any reason to suspect that they were better than these measurements that had been determined with very high precision.

W. H. JOHNSON: I quite agree with your last statement. It is our fault that we haven't provided you with any other results that will bear on this. If I understand you correctly, your number of degrees of freedom is 24, so that the doublets that we are presented today have about equal weight.

A. H. WAPSTRA: I would very much like to agree, but unfortunately I cannot, because if you look closely at the data as they are now, then your only check on your doublets are the one mass unit differences. They have much larger errors than the direct mass doublets. This means that if you are going to do a consistency check here, then you check essentially only your mass doublets of one mass unit. But the things that are important for mass calculation are unfortunately the other doublets with only two exceptions. In germanium you have quite a lot of very nice doublets. I should say that I am very happy with the germanium doublets because there you have measured doublets using germanium hydrides. That is quite a good set. This remark is also not valid for the tin isotopes, but there, unfortunately, for the reverse reason. Your normal doublets have larger errors so that the check here is really a good check but not of very good data, that is to say, relative to your other data. I still think that your data were very good, much better than any before; but compared with your other data the tin results were somewhat less good.

V. DOROKHOV: Does not Prof. WAPSTRA think that one will be obliged to increase the consistency coefficient (2.65) and not to reduce it, since the errors have been reduced considerably while the discrepancies between some masses have remained rather considerable?

A. H. WAPSTRA: I have no definite answer since I have not yet personally compared all new doublets with one another and with reaction energies. My provisional impression is that the consistency factor may very well be lower than before.

Mass Differences of Doublets Involving H, D, C, N and O

By

Koreichi Ogata and **Satoru Matsumoto**

Department of Physics, Faculty of Science, Osaka University, Nakanoshima, Osaka, Japan

(Presented by K. OGATA)

With 3 Figures

The reconstructing work on our large machine, damaged by a high tide-flood in 1961, has not yet been finished completely. However, the degree of damage to our Bainbridge-Jordan type mass spectrograph was

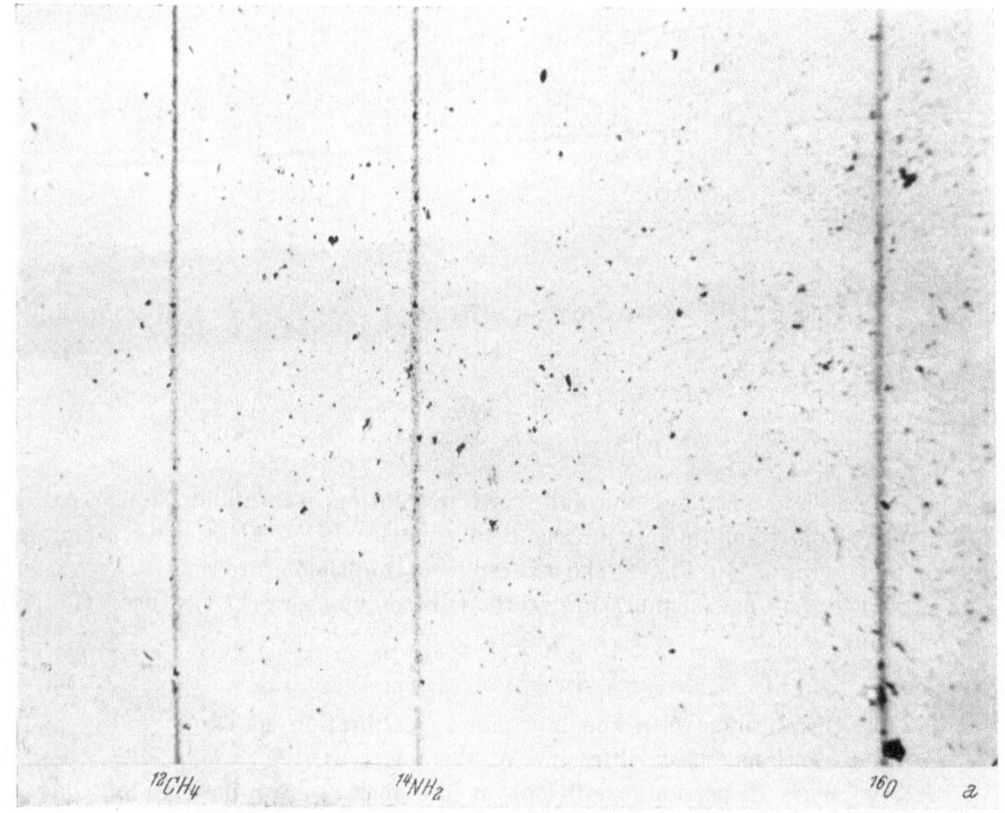

Fig. 1. Multiplet. a) $A = 16$: $^{12}CH_4$—$^{14}NH_2$—^{16}O

not so serious and we were able again to start our mass measuring work last fall. Here I would like to present some data obtained with it.

The essential parts of the apparatus are quite the same as previously reported[1]. However, with the apparatus presently used, an ion source of the Finkelstein type and an electrostatic cylinder lens as the accelerating system is used. In the present work, we used the main slit with a width

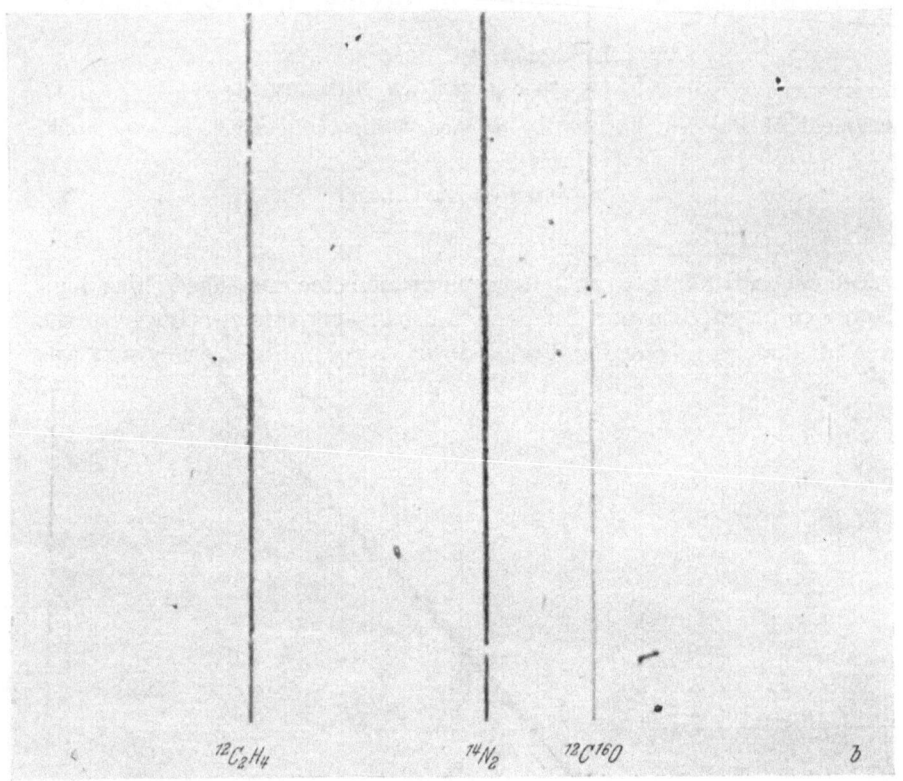

Fig. 1. Multiplet. b) $A = 28$: $^{12}C_2H_4 - ^{14}N_2 - ^{12}C^{16}O$

of several microns, and the full-width resolution was about 30,000 to 40,000. This resolution is sufficient to resolve $^{12}CH - ^{13}C$ at mass range up to C_4-group. In Fig. 1, the examples of multiplets are shown.

As for the mass calibration method, in previous work[2], we used the following relation

$$\varDelta y/\gamma = K(1 + \alpha_2 \gamma^2), \tag{1}$$

where

$\varDelta y$: the distance between the line pair of calibration series,

γ: the fractional mass difference of the line pair,

K: the mass dispersion coefficient at the mass center position of the line pair.

With this method for determining K at one position on photographic plate, it was necessary to have at least four calibration line components with the same mass difference between the two adjacent-lines. In order to determine the dispersion coefficients K over the wide range on the photographic plate, the calibration series must be shifted as small a distance as possible. Then, as a result, the whole spectrogram thus obtained

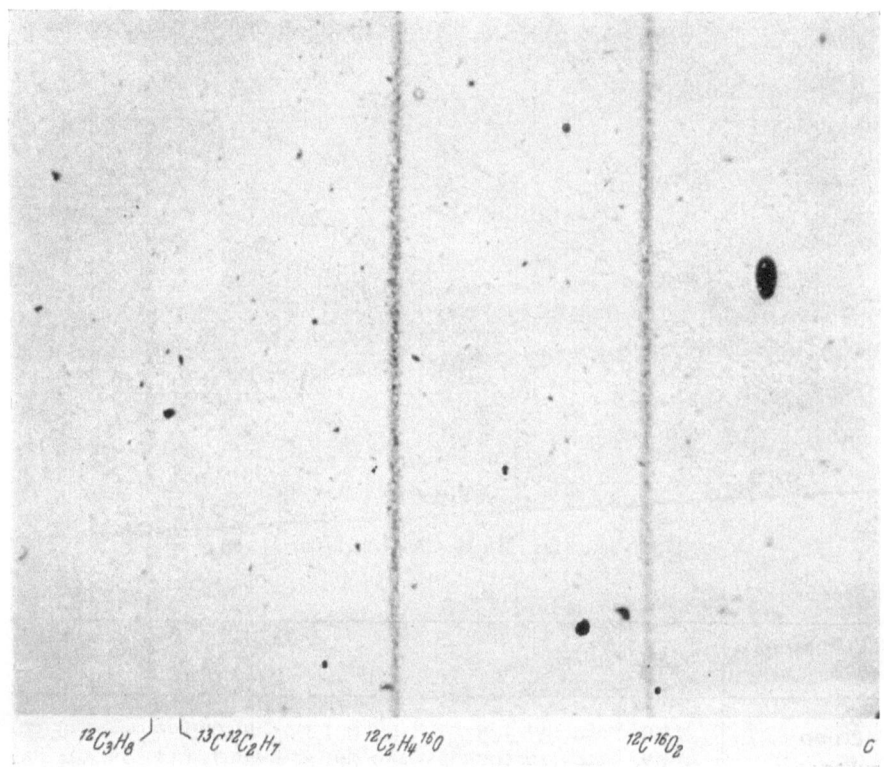

Fig. 1. Multiplet. c) $A = 44$: $^{12}C_3H_8$—$^{13}C^{12}C_2H_7$—$^{12}C_2H_4$—$^{12}C^{16}O_2$

may be very complicated, and it is difficult to assign a mass to each line. In addition, we have to find the mass center position of each line pair. Therefore, such a mass calibration method is not only troublesome but also subject to mistakes.

In order to avoid such a laborious work, we adopted a simpler method for mass calibration. That is, one calibration series, say C_3- or C_4-group, is photographed with an appropriated field strength, then the following empirical relation may be assumed between each line position y and the corresponding mass M of the calibration series for the sake of convenience of calculation.

$$y = a_0 + a_1(M/M_0) + a_2(M/M_0)^2 + a_3(M/M_0)^3, \qquad (2)$$

where M_0 is a mass close to the center mass of the calibration series. The coefficients a_i can be determined with a least squares calculation from M and the position of each line measured. M can be known with sufficient

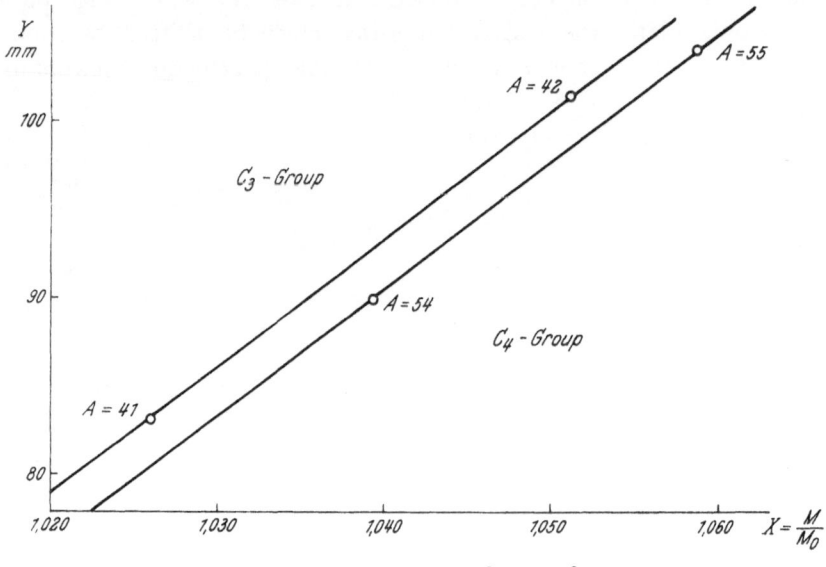

$$y = a_0 + a_1 x + a_2 x^2 + a_3 x^3$$

Fig. 2. Dispersion curve

Table 1a, 1b. Mass Calibration Data

Table 1a. a_i

Calibration series	M/M_0	a_0	a_1	a_2	a_3
C$_3$-group	$M/40$	— 475.136,77	446.061,13	28.038,95	68.796,67
C$_4$-group	$M/52$	— 401.106,56	228.891,37	233.718,17	0.604,47

Table 1b

A	Positions of C$_4$-group lines y (mm)		K (mm)		
	Measured	Calculated from a_i of C$_4$-group	Calculated from		
			C$_3$-group	C$_4$-group	Previous method*
55	103.922,2	103.925,5	768.422,6	768.446,2	768.260,3
54	89.950,0	89.946,8	744.677,1	744.886,3	744.787,6
53	76.145,5	76.145,1	721.484,0	721.681,8	721.779,1
52	62.519,9	62.520,3	698.816,7	698.832,7	698.812,5
51	49.074,3	49.072,5	676.383,2	676.338,8	676.371,4

* Ref. 2.

accuracy when we use hydrocarbon fragments as the calibration series. The dispersion coefficient K at the position of mass M can be calculated from the following equation using a_i obtained in the above.

$$K = M \frac{dy}{dM} = (M/M_0)[a_1 + 2a_2(M/M_0) + 3a_3(M/M_0)^2]. \qquad (3)$$

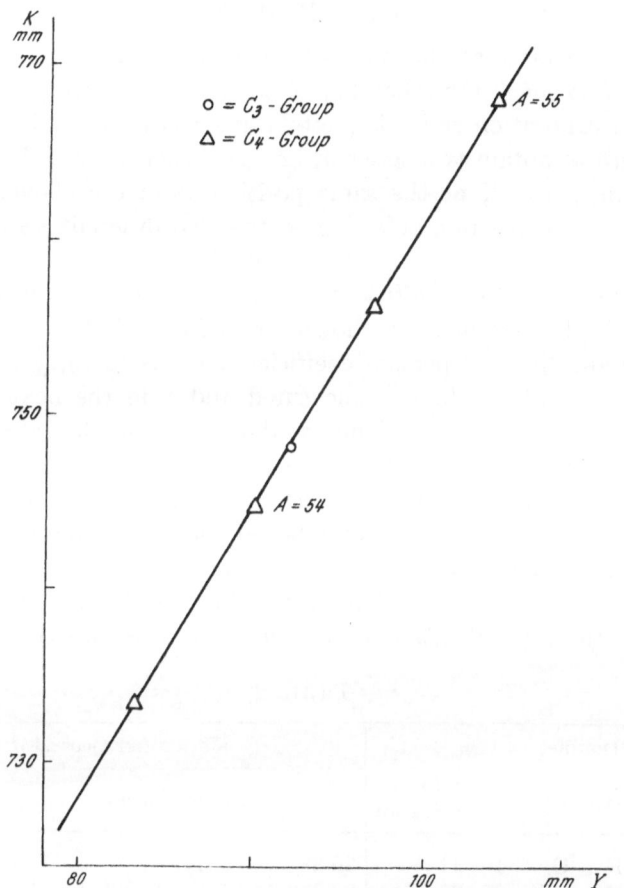

Fig. 3. The curve of mass dispersion coefficient

In this case, of course, the value of mass M to be used must be that referring to the calibration series used above.

For determining the mass difference of a doublet to be measured, we first have to obtain the dispersion coefficient at the mass center position of the doublet. We take two virtual lines which are located sufficiently close to the doublet, with the center of doublet in between. The mass values M_1 and M_2 of these virtual lines are estimated referring to the calibration series, and then by inserting these M_1 and M_2 into the equations (2) and (3) we can obtain the position y_1 and y_2 of the

27*

virtual lines and the values of K_1 and K_2 at the positions y_1 and y_2. Using these y_1, y_2 and K_1, K_2, the mass dispersion coefficient K at the mass center position of the doublet can be calculated with linear interpolation. The mass difference of the doublet can be determined from the K and the distance Δy of doublet components with the formula

$$\Delta M = M \cdot \Delta y / K, \qquad (4)$$

where M is the mass at the mass center of the doublet.

In order to check the reliability of the mass calibration method above described, a calibration series is photographed twice changing the electric field strength to obtain two series of images. Then we use the other series and determined the K at the same position as in the above. The value of K at the same position referring to the two different series of images can be compared.

In Table 1 (a), the coefficients a_i in the dispersion curve, equ. (2), for the C_3- and C_4-groups are shown. And in Table 1 (b), we show line positions y and mass dispersion coefficients K: as for y, y in the second row is that of C_4-lines directly measured and y in the next row is that calculated with the above mentioned method; as for K, the values calculated by using C_3- and C_4-groups and by the previous method[2] are shown. The agreement between K thus obtained is good within 0.03% accuracy. Therefore, one calibration series may be sufficient for determining the mass dispersion coefficients over a wide range of the photographic plate. In Fig. 2, a part of the dispersion curves for the C_3- and C_4-groups are shown, and the mass dispersion coefficient curve is also shown in Fig. 3.

Table 2

Doublet				Mass differences ΔM (μu)		
	Measured	A	No. of measure	Present mean		Previous*
CH_4—O	CH_4—O	16	41 (V)	$36,381.6_2 \pm 2.4_7$		
	C_2H_4—CO	28	99 (X)	$36,383.9_3 \pm 1.8$	$36,383.2_2 \pm 1.0_9$	$36,379.1 \pm 1.2$
	C_2H_4O—CO_2	44	30 (IV)	$36,381.3_0 \pm 2.2_9$		
CH_2—N	CH_2—N	14	34 (IV)	$12,576.8_0 \pm 1.7_3$		
	C_2H_4—N_2	28	66 (VII)	$12.577.8_5 \pm 3.0_5$	$12,577.4_7 \pm 1.1_5$	$12,575.5 \pm 2.9$
NH_2—O	NH_2—O	16	29 (III)	$23,811.2_3 \pm 3.2_5$		
	NH_3—OH	17	29 (III)	$23,811.0_0 \pm 3.0_9$	$23,811.1_2 \pm 2.0_2$	
N_2—CO	N_2—CO	28	53 (V)	$11,227.4_3 \pm 1.0_6$		
H_2—D	H_2O—DO	18	9 (I)	$1,553.1 \pm 10.3$		

* Ref. 2.

The mass differences of doublets thus obtained are shown in Table 2. As a calibration series we use most appropriate hydrocarbon fragments for each case, and the mass of H is assumed to be 1.007825 exactly. The error quoted in the table is the probable error for the measurement of the doublet component distance.

Assuming the mass of H to be 1.00782522^3, the masses of D, ^{14}N and ^{16}O can be calculated from the doublet mass differences in the Table 2. The results obtained are shown in Table 3.

Table 3

$$^1H \; .. \; + \; 7,825.22: \longrightarrow$$
$$(^2D \; .. \; + \; 14,097.3 \; \pm \; 10.3)$$
$$^{14}N \; .. \; + \; 3,072.6_9 \; \pm \; 2.2_6 \; (\mu u)$$
$$^{16}O \; .. \; - \; 5,084.1_3 \; \pm \; 2.2_6$$

In the present work, we use only the photographic method for mass determination. We are now preparing the peak matching method, so in future I hope we shall be able to do mass work with both methods.

References

[1] K. OGATA, and H. MATSUDA, Masses of light Atoms. Phys. Rev. 89, 27 (1953).

[2] K. OGATA, H. MATSUDA, and S. MATSUMOTO, Mass Measurement Work at Osaka University. Proc. Int. Conf. "Nuclidic Masses", Hamilton, 1960, p. 474.

[3] L. A. KÖNIG, J. H. E. MATTAUCH, and A. H. WAPSTRA, 1961 Nuclidic Mass Table. Nuclear Phys. 31, 18 (1962).

Discussion

J. MATTAUCH: This offers another example, Professor JOHNSON, of what we did. You might have noticed that the difference between the CH_4—O doublet and the sum of CH_2—N and NH_2—O doublet is 5.4 μu and the stated error is 1 or 2 μu. What would you do if you used this?

W. H. JOHNSON: I'm not making least squares tables. And I further should say that I don't propose to, even in the far future.

I certainly understand your problem and I am in agreement with what you have done. Let me make that very clear. The factor of 2.65 applied to our other doublets is perfectly acceptable as far as I am concerned, up to now. But now I think we have provided you with some data that you can use and perhaps the factor can be different.

I think one of the things that bothers me is that your factor of 2.65 always goes back to Minnesota values, yet it is applied to all mass spectrometer errors. I think it would certainly be more palatable to me if you just said mass spectrometer errors are wrong by about a factor of $2^1/_2$.

J. MATTAUCH: We have tried to make such consistency tests also, for instance, with Demirkhanov's doublets, and we got a factor somewhat over 3. In some cases, and as Dr. WAPSTRA mentioned already, SMITH has provided us with consistency factors himself. And then perhaps you forget another thing. In 1960, there was a discrepancy between reaction values and mass doublet, and what is more important, the Q value people estimated errors very conservatively, whereas the mass spectroscopists did it really as it should be done. They measured the doublet 20 or 25 times and computed the accidental error and perhaps added something for systematic errors or errors in resistance. It would have been very unfair to the Q-value people if we had accepted doublet errors directly. We should have perhaps reduced the errors of the Q-values.

A. O. Nier: By dividing their errors by 2.65. I'm sure the rest of you will understand this has been a standing joke between our laboratories for a number of years, since the 1956 conference, as a matter of fact.

N. Zeldes: I also want to put in a protest against putting all mass spectroscopic doublets together and multiplying them by 2.65 for the sole reason that this was the Minnesota's consistency factor for the lightest nuclei. It would be more fair to treat each group of doublets measured at the same laboratory over the same period separately, finding its consistency factor, and using it in assigning weights to that group.

A. H. Wapstra: This is exactly what we tried. When we made our 1961 adjustment, we used a very fine set of doublets measured by Demirkhanov and Dorokhov. We made such a pre-adjustment and I can quote you the result. We had 40 doublets in 16 unknowns, a quite respectable degree of over-determination. The factor we found was 2.42. We thought it did not really make any difference if we used 2.4 or 2.6.

N. Zeldes: That is correct with that data but I am sure that in the newer data the factor will have a lower value.

R. R. Ries: May I just make a request, that is by now, quite obvious. It seems that there is now definitely a need for a better and more reliable value for the ^{37}Cl—^{35}Cl mass difference. It is to be hoped that both mass spectroscopic labs as well as nuclear reaction people will soon give new values.

A Direct Determination of the ^{3}H—^{3}He Mass Difference*

By

P. E. Moreland Jr. and **K. T. Bainbridge**

Harvard University, Cambridge, Mass.

(Presented by K. T. BAINBRIDGE)

With 2 Figures

Abstract

The ^{3}H$-^3$He mass difference has been measured directly utilizing the high resolving power obtainable with the Harvard double-focusing mass spectrometer. The mass difference of this doublet was found to be $19.83 \pm 0.18 \,\mu$u (^{12}C $= 12$). Results obtained for the doublets HD$-^3$H, HD$-^3$He and H$_2-$D are respectively $5{,}877.22 \pm 0.69$, $5{,}896.84 \pm 0.42$ and $1{,}547.77 \pm 0.28 \,\mu$u.

Introduction

The initial mass measurements with the large spectrometer have been made on the lightest nuclides which afford close doublets less dependent on an accurate knowledge of the dispersion of the instrument. Among these doublets ^{3}H$-^3$He was chosen for study because of its interest with respect to the rest mass of the neutrino when accurate experimental rather than extrapolated ^{3}H beta-ray energy end points are achieved[1]. Also the ^{3}H$-^3$He mass difference expressed in energy units should be equal to the energy of the extrapolated end point of the ^{3}H beta-rays which has been measured by several methods since 1949[2].

No prior direct measurement of the ^{3}H$-^3$He mass differences has been reported but FRIEDMAN and SMITH[3] measured ^{3}H and ^{3}He using the doublets D$_2-^3$HH, ^{3}HH$-^4$He, and HD$-^3$He.

In this report on the mass differences ^{3}H$-^3$He, results were also obtained for HD$-^3$H, HD$-^3$He as a check measurement, and for H$_2-$D.

The Harvard spectrometer has been described earlier at different stages of its construction and initial operation[4, 5]. A number of

---

* This work supported by the Office of Naval Research.
** Present address: Argonne National Laboratory, Argonne, Illinois.

modifications for improved performance had been made by the time of these measurements in the spring of 1962. It was found that the battery supplies used for the magnet and electrostatic analyzer were sufficiently stable without regulation for use with a "peak-matching"[6] technique. We are indebted to Dr. J. W. DEWDNEY for the beam modulation and display circuit in use then and currently also. With this scheme, the ion beam is modulated sinusoidally by an A. C. field provided by a coil in the gap of the magnet[7]. The modulation frequency is normally derived from the line directly, but a Hewlett Packard function generator synchronized to the line can also supply 60 cycles and even and odd subharmonic sinusoidal currents to the sweep coil. Near or at the end of each modulation cycle, the acceleration and electrostatic analyzer voltages are switched by an amount sufficient to cause members of the doublet to follow identical trajectories and be focused at the image slit with precisely the same value of magnetic field on alternate cycles. This coincidence is accurately determined by observing the mass peaks in the ion detector output on an oscilloscope whose horizontal sweep is proportional to the modulation current. The high speed switching is accomplished with mercury contact relays driven with a Strobotron light link.

The peak display circuit includes provision for matching the amplitude of the doublet peaks and for separating the peaks vertically. In addition, one peak can be superposed on a high frequency square wave, appearing as two traces symmetrically displaced from the other peak on the display. In order to avoid display of the return trace and any switching transients present, the oscilloscope is blanked during half or a greater fraction of each cycle.

The measurement of the fractional mass difference $\Delta M/M_L$* consists of measuring the fractional change in potential required to produce a peak match. This comes about from the requirement that $\Delta M/M_L = \Delta V/V_H$, imposed by the fact that the ions must have the same momentum at peak match. By virtue of the energy focusing property of the instrument, the acceleration voltage change need not be as highly precise. The accuracy of the mass measurement thus depends on the precision with which the fractional potential change at the energy analyzer is known. This potential is derived from a calibrated precision voltage divider. For wide doublets the dispersion must be calibrated by the use of hydrocarbons and the ^{12}C standard.

Measurements

Over one hundred peak matchings were made of each of the four doublets over a period of days. Fig. 1 shows the general appearance of

---

* Subscripts L and H refer to the low and high mass member of the doublets, respectively.

TOP: He³, H³ PEAKS
RESOLVED. BOTTOM: PEAK
MATCH OF H³ - He³.

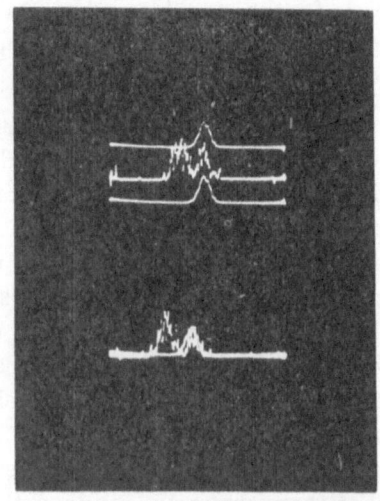

TOP: PEAK MATCH
OF HD-H³, HD PEAK SPLIT.
BOTTOM: PEAK MATCH OF
HD-H³, PEAK SPLITTING NOT
USED.

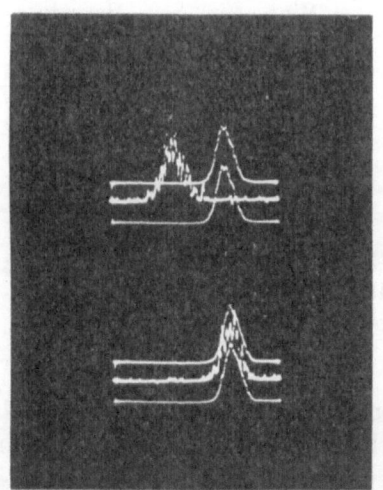

TOP: H₂-D PEAKS
DISPLACED 1 PART IN 10⁵.
BOTTOM: PEAK MATCH OF
H₂-D, H₂ PEAK SPLIT.

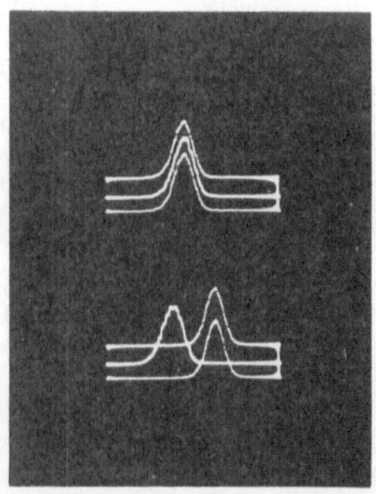

TOP: PEAK MATCH
OF D₂-He⁴, D₂ PEAK SPLIT.
BOTTOM: He⁴ PEAK DISPLA-
CED 1 PART IN 10⁵ FROM
D₂ PEAKS.

Fig. 1

the display at peak match. The atomic ion beams produced by our electron
bombardment sources were rather weak, resulting in poor signal to noise

ratio for these doublets. Using a base resolving power of 150,000 to 165,000, the $^3H-^3He$ doublet, with a separation of about 1 part in 150,000, is just resolved, as shown in Fig. 1.

A relativistic correction factor $(1 + 2\,T/M_0\,C^2)$, where T is the kinetic energy of the ions, was applied to the average of the calibrated voltage divider readings to obtain final values of $\Delta M/M_L$. At an acceleration potential of about 80 kV, this correction is nearly equal to the quoted errors for these doublets. The mass differences for the four doublets are shown in Table 1. The errors listed were obtained by adding an estimated limit of error for the voltage divider calibration to an external error[8], R_{ext}, obtained by applying a consistency factor[9] to the standard error of the readings. Since the consistency factors ranged from 3 to 11, there are errors much larger than the standard error of the readings would indicate. However, after consistency factor adjustment under the assumption that these errors are random, the final R_{ext} is still of the same magnitude as the calibration error.

Table 1. Doublet Mass Differences in μu
$^{12}C = 12$

Doublet	BNL	Minn.	Harvard
^3H—^3He	$20.03 \pm .21$[a]	—	$19.83 \pm .18$
HD—^3H	—	—	$5877.22 \pm .69$
HD—^3He	$5898.03 \pm .07$[a]	—	$5896.84 \pm .42$
H_2—D	$1548.78 \pm .18$[a]	$1547.2 \pm .4$[c]	$1547.77 \pm .28$
	$1548.08 \pm .10$[b]		
(HD—^3He)—(HD—^3H)	—	—	$19.62 \pm .81$

[a] FRIEDMAN, and SMITH, Phys. Rev. **109**, 2214 (1958).
[b] L. G. SMITH, Phys. Rev. **111**, 1606 (1958).
[c] QUISENBERRY, SCOLMAN, and NIER, Phys. Rev. **102**, 1071 (1956).

Previous mass spectrometer measurements of these doublets from the Brookhaven National Laboratory and the University of Minnesota are shown in Table 1. The mass synchrometer results are slightly larger in every case than the mass spectrometer values.

The mass excesses for D, ^3H, and ^3He were obtained by combining the doublet data of Table 1 with the ^1H mass value obtained by KÖNIG et al.[10] These values are shown in Table 2.

The previous measurements of the ^3H extrapolated end point energy together with the proper energy difference $^3H-^3He$ are presented in Fig. 2. Agreement is found between the two mass difference measurements of FRIEDMAN and SMITH and ourselves, and also with the work of PORTER

Table 2. Mass Excess in μu
$$^{12}C = 12$$

Isotope	BNL	Minn.	Harvard
H—1	7825.00 ± .10[a]	7824.7 ± .2[c]	7825.22 ± .08[e]
D—2	14101.90 ± .10[a]	14102.2 ± .5[d]	14102.67 ± .32
³H—3	16048.19 ± .38[b]	—	16050.74 ± .79
³He—3	16028.15 ± .22[b]	—	16030.98 ± .54

[a] L. G. SMITH, Phys. Rev. 111, 1606 (1958).
[b] L. FRIEDMAN, and L. G. SMITH, Phys. Rev. 109, 2214 (1958).
[c] QUISENBERRY, GIESE, and BENSON, Phys. Rev. 107, 1664 (1957).
[d] GIESE, and BENSON, Phys. Rev. 110, 712 (1958).
[e] KÖNIG, MATTAUCH, and WAPSTRA, Nuclear Phys. 31, 18 (1962).

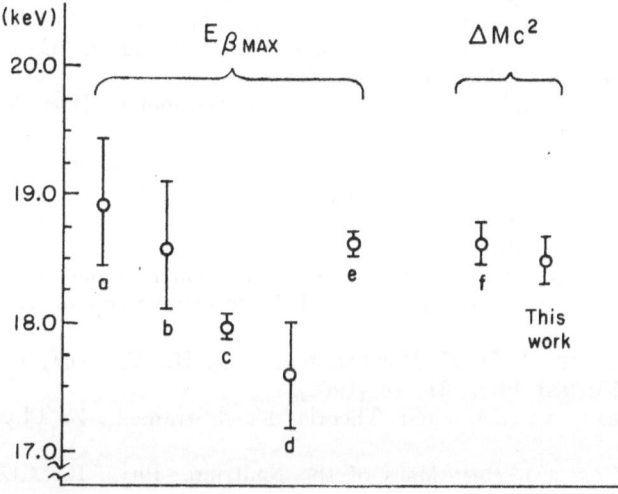

a. G. L. Hanna and B. Pontecorvo, PR 75, 983 (1949).
b. S. C. Curran, et al., PR 76, 835 (1949).
c. L. M. Langer and R. D. Moffat, PR 88, 689 (1952).
d. D. R. Hamilton, et al., PR 92, 1521 (1953).
e. F. T. Porter, PR 115, 450 (1959)
f. L. Friedman and L. G. Smith, PR 109, 2214 L (1958)

Fig. 2

using an iron-free beta-ray spectrometer, and with the earlier work of CURRAN et al. who used a gas proportional counter.

The equivalence of the extrapolated end point data and the mass difference data precludes information about the rest mass of the neutrino which must come from the shape of the beta-ray distribution near the end point, or from the combination of the total energy available for decay and an observed maximum beta-ray energy[11, 12]. It is worth remarking that the total energy change between the ground states of ^{3}H and ^{3}He is available directly and unequivocally from the doublet mass difference.

References

[1] K. T. Bainbridge, Proceedings of the 7th Solvay Congress in Chemistry, p. 69. Brussels: R. Stoops. 1947. — M. Deutsch, and O. Koefeld-Hansen, Experimental Nuclear Physics, Vol. III, p. 569, Ed. E. Segrè. New York: John Wiley and Sons, Inc. 1959.

[2] See references Fig. 2, and R. W. King, Table of Total Beta-Disintegration Energies. Revs. Modern Phys. 26, 327 (1954).

[3] Lewis Friedman, and Lincoln G. Smith, Mass Difference T—^{3}He and the Mass of the Neutrino. Phys. Rev. 109, 2214 (1958).

[4] T. L. Collins, and K. T. Bainbridge, Nuclear Masses and Their Determination, p. 213, Ed. H. Hintenberger. London: Pergamon Press. 1957.

[5] K. T. Bainbridge, and Parker E. Moreland Jr., Proceedings of the International Conference on Nuclidic Masses, p. 460, Ed. H. E. Duckworth. Toronto: University of Toronto Press. 1960.

[6] L. G. Smith, and C. C. Damm, Mass Synchrometer. Rev. Sci. Instr. 27, 638 (1956).

[7] C. F. Giese, and T. L. Collins, Mass Doublet Measurements Using a Modulated Magnetic Field. Phys. Rev. 96, 823 A (1954).

[8] Raymond T. Birge, The Calculation of Errors by the Method of Least Squares. Phys. Rev. 40, 207 (1932).

[9] J. Mattauch, Proceedings of the International Conference on Nuclidic Masses, p. 14, Ed. H. E. Duckworth. Toronto: University of Toronto Press. 1960.

[10] L. A. König, J. H. E. Mattauch, and A. H. Wapstra, 1961 Nuclidic Mass Table. Nuclear Phys. 31, 18 (1962).

[11] E. Fermi, Versuch einer Theorie der β-Strahlen. Z. Physik 88, 161 (1934).

[12] J. J. Sakurai, Rest Mass of the Neutrino. Phys. Rev. Letters 1, 40 (1958).

Discussion

H. H. Staub: I would like to ask Dr. Bainbridge why he excluded the low E_β value of Popov, obtained calorimetrically from his table? Although the shape of the spectrum enters critically, it seems to me that we should know it sufficiently well for such a simple β-transition.

K. T. Bainbridge: The excellent calorimetric measurements of Popov et al. in 1958 and of Jenks et al. in 1950 were omitted because of their dependence on the theoretical shape. Calorimetric measurements are very insensitive to small but finite values of the neutrino mass and so probably provide a good figure for the total energy release if one accepts the theoretical shape for zero neutrino mass.

B. Hogg: Can you give an estimate of the neutrino mass from your doublet measurement?

K. T. Bainbridge: No, it is not possible. There are really no experimental beta-ray end point values except the low value from Langer's graph. Also our error value is probably too large, larger than we had originally hoped to attain.

W. H. JOHNSON: There was a column of numbers on your slide showing the error analysis. Were these the number of runs that were made?

K. T. BAINBRIDGE: The auxiliary numbers which I wrote out were the numbers of individual doublet comparisons for each doublet. The numbers ran from 576 measurements, for ^{3}H—^{3}He, to 178 for the H$_2$—D comparison.

A. O. NIER: What plans do you have to extend the work to higher masses?

K. T. BAINBRIDGE: We plan to work on the mass difference ^{37}Cl—^{35}Cl, and in June obtained some DCl, with the plan of making a measurement of the chlorine difference compared to D.

Nuclidic Masses and Binding Energies of Nuclei from Samarium to Lutetium

By

R. A. Demirkhanov, V. V. Dorokhov and **M. I. Dzkuya**

Physical Technical Institute of the Georgian Academy of Sciences, Sukhumi 2,
USSR

(Presented by V. V. DOROKHOV)

With 8 Figures

There is a great scarcity of experimental data on the precise measurement of nuclidic masses in the range of rare earth elements. The survey paper[1] on the Nuclidic Masses gives a critical review of all the data available up to 1960 and contains averages, calculated by the method of the least squares, of all the values obtained from both nuclear data and mass-spectrometrical measurements. In actual fact, however, the great majority of nuclidic masses reported in that paper have been calculated on the basis of one report[2], since no other mass-spectrometrical data were available at that time, while available nuclear data in this range of masses were very scarce and only fragmentary by nature.

The most comprehensive study covering this range of masses is the report by V. BHANOT, W. JOHNSON JR. and A. NIER[2]. The report contains measurement results of a great number of "isotopic" doublets ranging from gadolinium to gold. In order to determine the absolute value of nuclidic masses it was necessary to make use of "additives" in which the mass of the ion to be measured was directly related (through the doublet) to the mass of an organic compound consisting of H, C, N, and O atoms. Such "additives" were used for practically all the elements measured. Therefore an independently measured sub-standard was introduced into each isotopic series of every element, so that all the masses of a given series were determined in relation to the respective sub-standard (on the basis of "isotopic" doublets).

However, the mass-spectrometer used for the measurement of masses was not capable of resolving the ^{13}C-containing fragment from the basic fragments of the organic compounds. Therefore the authors of the report cited above were obliged to choose organic compounds the fragments of which did not contain the "satellite"-^{13}C. The experimental value obtained was corrected for the presence of a minute quantity of an impurity fragment containing ^{13}C. The authors state that in the range concerned, for at least one mass number of practically every element, hydrocarbon ions were detected containing minute concentrations of ^{13}C.

In spite of this, the authors of the report were in some cases obliged to use the ^{13}C-containing fragments (e. g. for the doublets of ^{159}Tb, ^{176}Yb, ^{185}Re, ^{191}Yr and others) as sub-standards and, in fact, corrections were introduced in cases where the mass-spectrometric data were not "consistent" with the results obtained from some other sources (nuclear reactions, alpha- and beta-decay, etc.). As the authors themselves admit, the correction was rather an arbitrary one and it was introduced in order to reduce arbitrary adjustments to a minimum.

Thus, out of 23 basic doublets, 3 were definitely discarded and 12 were corrected by a value equal to or even exceeding the experimental error. After having tried this type of adjustment the authors have suggested a value of about 150 to 200 micro-Mass Units (μu) as a mean error for the whole range of masses measured. The authors state that these errors (degree of accuracy) may be considered standard errors; that the maximum limit of error may reach the triple value of the standard error, and that errors above 1000 μu cannot be tolerated.

Judging by the magnitude of possible error, as indicated by the authors of the report, and taking into consideration the fact that they have themselves discarded 3 basic doublets as inconsistent with the nuclear data, it may be concluded that the authors are not sure of the magnitude and of the accuracy of the correction that they introduce in order to account for the presence of a minute impurity of a ^{13}C-containing fragment in the spectrum of the organic matter. By fixing at 1000 μu the maximum value of an admissible error the authors introduce a limitation for the calculation of the binding energy of the nucleus, of the binding energies of the last neutrons and of the last protons, of the pairing energies of neutrons and protons, as well as of other nuclear parameters.

The purpose of the present research was to obtain more precise values of the nuclidic masses in the range of rare earths and to verify the calculation of the energy parameters of the nuclei (B_n, B_p, P_n and others) in the beginning of the second region of deformed nuclei ($A \simeq 155$).

The masses were measured by means of a double-focusing mass-spectrograph[3]. The resolution capacity of the apparatus, equal to some 50 000–70 000, has made it possible to resolve completely the ^{13}C-containing impurity fragments, and therefore no corrections were needed to account for any ^{13}C-quantities. On the other hand, an adequate resolution capacity of the apparatus allowed an unrestricted use of any organic compound. Moreover, in cases of a sufficient intensity of a ^{13}C-containing fragment the latter was used as one of the doublets for measurement of the nuclidic mass of the unknown isotope.

For each isotope the nuclidic mass was computed on the basis of several independent doublets of various composition. In a number of cases doublets formed by A Cl-type ions coupled with mercury isotopes (A = rare earth isotope) were used. For the measurements chlorides of rare earths elements derived from the oxides were used. The impurities present in the materials did not exceed 0.1%.

The use of several doublets for measuring the mass of an unknown isotope has permitted verification of the "internal" compatibility and consistency of the results. Moreover, a quality control of the results was achieved by measuring the so-called "isotopic" doublets, i. e. doublets formed by two neighbouring isotopes of the same element, and sometimes even of elements differing from each other by only one or two mass units.

Final values of the nuclidic masses were calculated by taking into account the "weight" of the measurement. It is to be noted that the mass value of each measured isotope did neither directly nor indirectly depend on the measurement of other masses.

The scattering (dispersion) due to the measuring apparatus was determined in the majority of cases by using the fragments of organic compounds differing only by one hydrogen mass unit. For the whole range of measured masses this procedure ensures a relative accuracy amounting to some $\sim 10^{-5}$ of the scattering (dispersion) of the results. The average relative accuracy (limit of error) attained in the present research, when taking into account the "internal" consistency of the results, amounts to $\dfrac{\varDelta M}{M} = 2$ to 5×10^{-7}, which corresponds to some 30 to 80 keV. The approximations assigned to the mass values are equal to the mean statistical errors calculated in the standard manner.

The masses of ^{1}H, ^{13}C, ^{14}N, ^{16}O, ^{35}Cl, ^{37}Cl and mercury isotopic were used as sub-standards, in accordance with the suggestions contained in the paper[1]. All the results given in the present report were calculated for the scale of ^{12}C = 12.000000.

The numerical values of $\varDelta X$ (doublet) were obtained by the method described in the paper[3].

A total of 140 doublets were measured on the basis of which it was possible to calculate the masses of 41 stable isotopes from lutetium to samarium.

Mass Measurements

Lutetium. For the measurement of the mass of lutetium, fragments of anthracene ($C_{14}H_{10}$, $m = 178$) with masses of 175 ($C_{14}H_7$) and of 176

($C_{14}H_8$) were used as standards. The mass 175 gave a triplet formed by ^{175}Lu ions and $C_{14}H_7$ and $C_{13}^{13}CH$ anthracene fragments. The mass values of lutetium thus obtained were verified by measurement of the "isotopic" doublet $^{176}Lu-^{175}Lu$.

Ytterbium. The nuclidic mass of ytterbium was measured from both the doublets with various organic compounds and "isotopic" doublets differing by one or two mass units from one another. For the formation of doublets on the nuclidic masses of ytterbium ($m = 176$ to $m = 168$) and for the determination of the dispersion coefficient organic compounds of anthracene ($C_{14}H_{10}$, $m = 178$) α-nitroso-β-naphtol ($C_{10}H_7O_2N$, $m = 173$), hydrazo-benzol ($C_{12}H_{12}N_2$, $m = 184$) and their fragments were used.

Thulium. Diphenylamine ($C_{12}H_{11}N$, $m = 169$) was used as standard. This organic compound gives fragments allowing a precise determination of the dispersion. In addition, the value of the nuclidic mass of ^{169}Tu was verified by measuring a doublet formed by $^{169}Tu^{35}Cl$ and the ^{204}Hg mercury isotope.

Erbium. Erbium ions were obtained by evaporation of erbium chloride ($ErCl_3$). Hydrazo-benzol ($C_{12}H_{12}N_2$, $m = 184$), fluorene ($C_{13}H_{10}$, $m = 166$), carbazol ($C_{12}H_9N$, $m = 167$) and their fragments were used as organic sub-standards.

In measuring nuclidic masses of erbium a very thorough verification of the compatibility and consistency of the results was carried out. Each erbium isotope was measured from several doublets formed with sub-standards of various organic compositions and obtained by different methods. All the "isotopic" doublets were measured. Some nuclidic masses of erbium (^{164}Er, ^{166}Er and ^{167}Er) were directly related, by means of chlorides (of $^{164}Er^{35}Cl$-type), to some mercury nuclidic masses.

The following observation deserves a special mention: when hydrazo-benzol ($C_{12}H_{12}N_2$, $m = 184$) was used for measuring the erbium nuclidic masses, the fragments of this compound, with respective masses of 166 ($C_{12}H_8N$) and of 167 ($C_{12}H_9N$), were used for the formation of the doublet pairs; by their organic composition these fragments corresponded to carbazol ($C_{12}H_9N$, $m = 167$) and to its fragment ($C_{12}H_8N$, $m = 166$) obtained by elimination of one hydrogen mass unit. Thus, molecular ions, other ions originated from fragments after a loss of 1 to 22 mass units, as well as mercury masses known from other experiments, were used as sub-standard masses.

In no case were any deviations of the mass values observed which depended on the choice of sub-standard used. The same is true for the "isotopic" doublets. For this reason all the measured doublets were used for the calculation of masses, irrespective of the method applied for obtaining the sub-standards.

Holmium. The nuclidic masses of ^{165}Ho were measured on the basis of five different doublets. The fragments of fluorene ($C_{13}H_{10}$, $m = 166$) and of carbazol ($C_{12}H_9N$, $m = 167$) on the mass of 165 were used as organic sub-standards.

Dysprosium. Such organic compounds as fluorene ($C_{13}H_{10}$, $m = 166$), benzal-methyl-ethyl-ketone ($C_{11}H_{12}O$, $m = 160$), naphto-chinon ($C_{10}H_6O_2$, $m = 158$), dipyridile ($C_{10}H_8N_2$, $m = 156$) and their fragments were used for measuring dysprosium nuclidic masses. Besides, some statistical data on the measurements of masses of dysprosium-chlorides were obtained by a direct comparison with mercury nuclidic masses. During the process of measuring these masses the doublets $^{164}Dy^{35}Cl - {}^{162}Dy^{37}Cl$ and $^{163}Dy^{35}Cl - {}^{161}Dy^{37}Cl$ were obtained and studied; a resolution capacity of about $\sim 45\,000$ was required to distinguish between these doublets. The lines of these doublets were completely resolved, and this confirms the accuracy of our estimate that the mean resolution capacity of our mass-spectrograph amounts to $\sim 60\,000$.

Terbium. A fragment of naphto-chinon ($C_{10}H_6O_2$, $m = 158$) containing a carbon isotope $^{13}C - C_9{}^{13}CH_6O_2$ and ions of associative origin $C_{10}H_7O_2$ was used as an organic sub-standard for measuring the mass of ^{159}Tb. The mass values of ^{159}Tb obtained by these two different methods have shown a rather good agreement.

Gadolinium. Phenyl-cyclohexane ($C_{12}H_{16}$, $m = 160$), naphto-chinon ($C_{10}H_6O_2$, $m = 158$), dipyridile ($C_{10}H_8N_2$, $m = 156$), acenaphthene ($C_{12}H_{10}$ $m = 154$), and their fragments were used for the measurement of the gadolinium nuclidic masses. The internal consistency of measurements was verified for the majority of isotopes.

Europium. Fragments of organic compounds of acenaphthene ($C_{12}H_{10}$, $m = 154$) and of camphor ($C_{10}H_{16}O$, $m = 152$) were used for the measurement of the masses of ^{153}Eu and ^{151}Eu. The dispersion constants were calculated on the basis of the same compounds. In addition, isotopic doublets formed by europium and samarium isotopes differing by 1 or 2 mass units were measured separately.

Samarium. Organic compounds of acenaphthene ($C_{12}H_{10}$, $m = 154$), dimethyl-amino-benzaldehyde ($C_9H_{11}NO$, $m = 149$), cinnamic acid ($C_9H_8O_2$, $m = 148$), phthalemide ($C_8H_5NO_2$, $m = 147$), naphthol ($C_{10}H_8O$, $m = 144$) and naphthyl-amine ($C_{10}H_9N$, $m = 143$) were used for measurements of nuclidic masses of samarium.

The great number of organic compounds used for the measurement of nuclidic masses of samarium shows that there was ample opportunity, in this case, to verify the "internal" compatibility and consistency of the results. The quality control of the measurements was also made by means of isotopic doublets and by forming a closed cycle of all the samarium isotopes (except ^{144}Sm).

Comparative Analysis of Measured Masses

The values of the doublets, the masses deduced therefrom and the mean values of nuclidic masses of lutetium, ytterbium, thulium, erbium, holmium, dysprosium, terbium, gadolinium, europium and samarium are given in Table 1. The same Table contains the nuclidic masses as reported

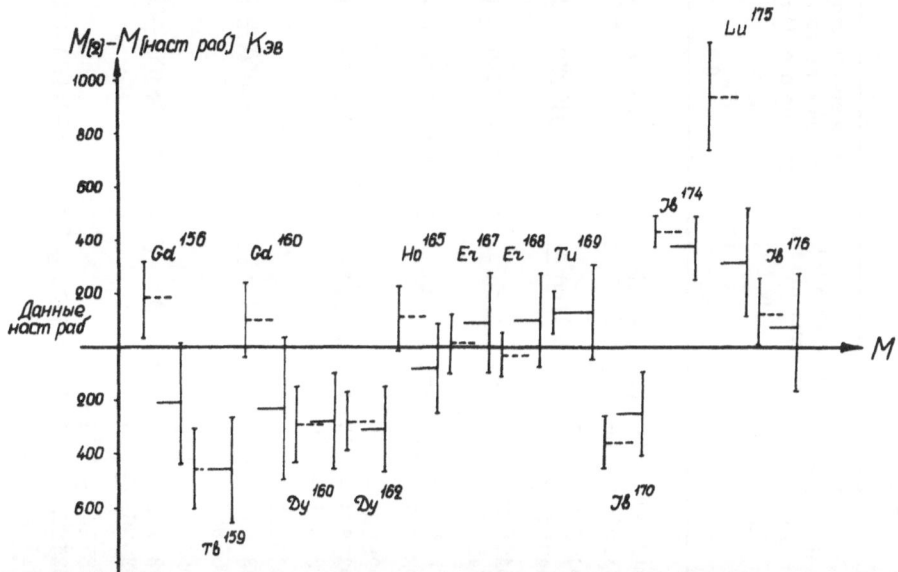

Fig. 1. Differences in absolute values of the nuclidic masses as given in paper[2] and as obtained in the present research. Ordinates: $M_{(2)} - M_{\text{(present research)}}$, in keV. Abscissae: Results according to the present research

in the paper[2] and in the survey paper[1]. It is to be noted that in view of the great scarcity of experimental results in this range of masses, the mass values suggested in [1] are, with a few exceptions, based mainly on the data contained in the paper[2]. Therefore, the results contained in both papers are largely interdependent. Corrections to the mass numbers were introduced in the survey paper[1] only in those cases where it was impossible to reconcile the results as given in the paper[2] with other nuclear data (and, in particular, with the Q-values for beta- and alpha-decays and with the data of nuclear reactions). This applies, in particular, to the gadolinium range, in respect of which the authors of the paper[1] have disregarded the corrections proposed in the paper[2].

As will be seen from Table 1*, in some cases no complete "internal consistency" was reached within the limits of the statistical errors for

---

* The values of the doublets and oft he masses for lutetium, ytterbium and thulium in the scale of $^{16}O = 16.000000$ are given in reference 4.

Table 1*

Mass Number / Atomic Number	Doublet	Value of ΔM in mu	Value of the nuclidic mass of the isotope u	Average value of the nuclidic mass u	Mass spectroscopical data according to reference[2] (1960) u	Data according to reference[1] (1962) u
1	2	3	4	5	6	7
176/71	$C_{14}H_8$—^{176}Lu ^{176}Lu—^{175}Lu	119.962 ± 49 1001.982 ± 60	175.942639 ± 50 175.942579 ± 60	175.942620 ± 30	175.942850 ± 150	175.942740 ± 80
175/71	$C_{14}H_7$—^{175}Lu $C_{13}{}^{13}CH_6$—^{175}Lu	114.121 ± 37 109.763 ± 36	174.940655 ± 37 174.940542 ± 36	174.940600 ± 40	174.940920 ± 160	174.940890 ± 160
176/70	$C_{14}H_8$—^{176}Yb $C_{13}H_6N$—^{176}Yb	119.980 ± 46 107.189 ± 109	175.942622 ± 46 175.942837 ± 109	175.942650 ± 90	175.942730 ± 130	175.942740 ± 130
174/70	$C_{14}H_6$—^{174}Yb ^{176}Yb—^{174}Yb	108.308 ± 38 2003.997 ± 50	173.938643 ± 38 173.938657 ± 100	173.938640 ± 20	173.939020 ± 100	173.933020 ± 140
173/70	$C_{14}H_5$—^{173}Yb $C_{10}H_7O_2N$—^{173}Yb ^{174}Yb—^{173}Yb	101.030 ± 67 109.807 ± 63 1000.704 ± 50	172.938096 ± 67 172.937874 ± 63 172.937941 ± 50	172.937960 ± 50	172.938370 ± 110	172.938300 ± 160
172/70	$C_{10}H_6O_2N$—^{172}Yb ^{172}Yb—^{171}Yb ^{173}Yb—^{172}Yb	103.558 ± 60 999.947 ± 226 1001.974 ± 118	171.936297 ± 60 171.936240 ± 230 171.935985 ± 127	171.936240 ± 70	171.936520 ± 120	171.936530 ± 160
171/70	$C_{10}H_7ON_2$—^{171}Yb $C_{11}{}^{13}CH_{12}N$—^{171}Yb ^{171}Yb—^{170}Yb	119.638 ± 270 164.143 ± 80 1001.215 ± 60	170.936202 ± 270 170.936188 ± 80 170.936367 ± 65	170.936290 ± 40	170.936460 ± 120	170.936460 ± 170
170/70	$C_{12}H_{12}N$—^{170}Yb $C_{11}H_8ON$—^{170}Yb $C_{11}{}^{13}CH_{11}N$—^{170}Yb	161.831 ± 43 125.366 ± 150 157.324 ± 210	169.935146 ± 43 169.935225 ± 150 169.935182 ± 210	169.935150 ± 30	169.934900 ± 130	169.934880 ± 140

	Doublet					
168/70	$C_{12}H_{10}N$—^{168}Yb	147.008 ± 103	167.934318 ± 103	167.934300 ± 30	167.933910 ± 150	167.933900 ± 270
	^{170}Yb—^{168}Yb	2000.907 ± 200	167.934245 ± 200			
169/69	$C_{12}H_{11}N$—^{169}Tu	154.923 ± 61	168.934229 ± 62	168.934220 ± 20	168.934350 ± 150	168.934350 ± 260
	^{204}Hg—$^{169}Tu^{35}Cl$	70.424 ± 98	168.934203 ± 100			
170/68	$C_{12}H_{12}N^{**}$—^{170}Er	161.205 ± 70	169.935772 ± 70	169.935780 ± 20	169.935580 ± 180	169.935510 ± 300
	^{170}Er—^{168}Er	2003.453 ± 70	169.935809 ± 60			
168/68	$C_{11}^{13}CH_9N$—^{168}Er	144.524 ± 29	167.932332 ± 30	167.932360 ± 30	167.932460 ± 150	167.932380 ± 160
	$C_{12}H_{10}N^{**}$—^{168}Er	148.884 ± 44	167.932442 ± 45			
	^{168}Er—^{167}Er	1000.284 ± 31	167.932324 ± 50			
167/68	$C_{12}H_9N$—^{167}Er	141.480 ± 27	166.932021 ± 28	166.932040 ± 30	166.932140 ± 160	166.932050 ± 160
	$C_{12}H_9N^{**}$—^{167}Er	141.516 ± 52	166.931985 ± 52			
	$C_{13}H_{11}$—^{167}Er	153.838 ± 130	166.932239 ± 130			
	^{202}Hg—$^{167}Er^{35}Cl$	69.735 ± 60	166.932041 ± 60			
	^{204}Hg—$^{167}Er^{37}Cl$	75.432 ± 61	166.932155 ± 65			
	^{167}Er—^{166}Er	1001.722 ± 31	166.932059 ± 55			
166/68	$C_{12}H_8N$—^{166}Er	135.376 ± 29	165.930300 ± 30	165.930340 ± 40	165.930400 ± 160	165.930400 ± 170
	$C_{12}H_8N^{**}$—^{166}Er	135.415 ± 59	165.930261 ± 60			
	$C_{13}H_{10}$—^{166}Er	147.741 ± 58	165.930509 ± 60			
	^{201}Hg—$^{166}Er^{35}Cl$	71.186 ± 35	165.930275 ± 40			
164/68	$C_{12}H_6N$—^{164}Er	120.876 ± 39	163.929150 ± 40	163.929150 ± 20	163.929240 ± 160	163.929290 ± 250
	^{199}Hg—$^{164}Er^{35}Cl$	70.310 ± 81	163.929093 ± 84			
	$^{164}Er^{35}Cl$—$^{164}Er^{37}Cl$	2001.105 ± 78	163.929232 ± 90			
	^{166}Er—^{164}Er	2001.214 ± 46	163.929123 ± 65			

* Errors (degree of approximation) in the values of the doublets and of the nuclidic masses refer to the last decimals.

** Ions of dissociative origin which have lost a fragment the mass of which exceeds 10 mass units.

(Table 1, continued)

Mass Number Atomic Number	Doublet	Value of ΔM in mu	Value of the nuclidic mass of the isotope u	Average value of the nuclidic mass u	Mass spectroscopical data according to reference[2] (1960) u	Data according to reference[1] (1962) u
1	2	3	4	5	6	7
162/68	$C_{12}H_4N - {}^{162}Er$	105.591 ± 74	161.928784 ± 75	161.928650 ± 50	161.928730 ± 180	161.928780 ± 350
	$C_{13}H_6 - {}^{162}Er$	118.426 ± 170	161.928525 ± 170			
	${}^{164}Er - {}^{162}Er$	2000.556 ± 48	161.928590 ± 56			
165/67	$C_{13}H_9 - {}^{165}Ho$	140.043 ± 29	164.930384 ± 30	164.930350 ± 20	164.930260 ± 150	164.930300 ± 250
	$C_{12}H_7N - {}^{165}Ho$	127.537 ± 28	164.930314 ± 28			
	$C_{11}{}^{13}CH_6N - {}^{165}Ho$	122.967 ± 53	164.930413 ± 53			
	${}^{200}Hg - {}^{165}Ho{}^{35}Cl$	69.116 ± 33	163.930374 ± 35			
	${}^{202}Hg - {}^{165}Ho{}^{37}Cl$	74.474 ± 54	164.930240 ± 60			
164/66	$C_{13}H_8 - {}^{164}Dy$	133.320 ± 38	163.929280 ± 38	163.929270 ± 20	163.928774 ± 160	163.928830 ± 240
	$C_{12}{}^{13}CH_7 - {}^{164}Dy$	128.920 ± 34	163.929209 ± 34			
	${}^{201}Hg - {}^{164}Dy{}^{37}Cl$	75.086 ± 42	163.929333 ± 46			
	${}^{199}Hg - {}^{164}Dy{}^{35}Cl$	70.087 ± 31	163.929315 ± 37			
163/66	$C_{13}H_7 - {}^{163}Dy$	125.906 ± 36	162.928869 ± 36	162.928850 ± 20	162.928352 ± 160	162.928370 ± 210
	${}^{164}Dy - {}^{163}Dy$	1000.446 ± 28	162.928830 ± 35			
	${}^{200}Hg - {}^{163}Dy{}^{37}Cl$	73.527 ± 42	162.928921 ± 46			
	${}^{198}Hg - {}^{163}Dy{}^{35}Cl$	68.979 ± 37	162.928936 ± 42			
	${}^{164}Dy{}^{37}Cl - {}^{163}Dy{}^{37}Cl$	1000.392 ± 48	162.928884 ± 52			
	${}^{163}Dy{}^{37}Cl - {}^{164}Dy{}^{35}Cl$	996.637 ± 54	162.928871 ± 58			
	${}^{164}Dy{}^{35}Cl - {}^{163}Dy{}^{35}Cl$	1000.540 ± 25	162.928736 ± 32			

162/66	$C_{13}H_6$—^{162}Dy	120.115 ± 19	161.926835 ± 19			
	^{163}Dy—^{162}Dy	1001.985 ± 38	161.926867 ± 43			
	^{199}Hg—$^{162}Dy^{37}Cl$	75.661 ± 41	161.926699 ± 46			
	$^{164}Dy^{35}Cl$—$^{162}Dy^{37}Cl$	5.589 ± 19	161.926645 ± 30	161.926760 ± 30	161.926451 ± 150	161.926470 ± 190
	$^{163}Dy^{37}Cl$—$^{162}Dy^{37}Cl$	1002.174 ± 40	161.926718 ± 45			
	$^{162}Dy^{37}Cl$—$^{163}Dy^{35}Cl$	994.931 ± 42	161.926741 ± 47			
	$^{163}Dy^{35}Cl$—$^{162}Dy^{35}Cl$	1002.164 ± 35	161.926688 ± 42			
161/66	$C_{13}H_5$—^{161}Dy	112.246 ± 25	160.926879 ± 25			
	^{162}Dy—^{161}Dy	1000.022 ± 40	160.926741 ± 47			
	^{198}Hg—$^{161}Dy^{37}Cl$	74.132 ± 56	160.926741 ± 58			
	$^{163}Dy^{35}Cl$—$^{161}Dy^{37}Cl$	5.204 ± 62	160.926606 ± 66	160.926750 ± 30	160.926619 ± 160	160.926600 ± 210
	$^{162}Dy^{37}Cl$—$^{161}Dy^{37}Cl$	1000.151 ± 70	160.926612 ± 74			
	$^{162}Dy^{37}Cl$—$^{161}Dy^{35}Cl$	1000.078 ± 23	160.926685 ± 33			
	$^{161}Dy^{37}Cl$—$^{162}Dy^{35}Cl$	996.916 ± 67	160.926637 ± 72			
	$^{164}Dy^{35}Cl$—$^{161}Dy^{37}Cl$	1005.610 ± 48	160.926624 ± 52			
160/66	$C_{12}H_{16}$—^{160}Dy	200.047 ± 68	159.925156 ± 68	159.925160 ± 70	159.924882 ± 160	159.924830 ± 190
158/66	$C_{10}H_6O_2$—^{158}Dy	112.868 ± 98	157.923913 ± 98	157.923910 ± 100	157.924016 ± 180	157.923960 ± 330
156/66	$C_{10}H_8N_2$—^{156}Dy	145.131 ± 96	155.923619 ± 117	155.923620 ± 120	155.923814 ± 200	155.923760 ± 420
159/65	$C_9{}^{13}CH_6O_2$—^{159}Tb	114.842 ± 50	158.925292 ± 50	158.925350 ± 40	158.924880 ± 150	158.924950 ± 250
	$C_{10}H_7O_2$***—^{159}Tb	119.238 ± 25	158.925368 ± 25			
160/64	$C_{12}H_{16}$—^{160}Gd	198.152 ± 52	159.927051 ± 52	159.927050 ± 50	159.926820 ± 220	159.927120 ± 190
158/64	$C_{10}H_6O_2$—^{158}Gd	112.444 ± 33	157.924337 ± 33	157.924320 ± 20	157.923840 ± 210	157.924100 ± 180
	^{158}Gd—^{157}Gd	1000.392 ± 48	157.924290 ± 50			
157/64	$C_{10}H_5O_2$—^{157}Gd	105.080 ± 62	156.923876 ± 62			
	$C_{10}H_6N_2$—^{157}Gd	152.722 ± 62	156.923854 ± 62			
	$C_9{}^{13}CH_8N_2$—^{157}Gd	148.171 ± 65	156.923934 ± 65	156.923900 ± 20	156.923690 ± 200	156.923940 ± 180
	^{157}Gd—^{156}Gd	1001.863 ± 60	156.923930 ± 60			

*** Ions of associative origin which have gained one or two hydrogen masses.

(Table 1, continued)

Mass Number Atomic Number	Doublet	Value of ΔM in mu	Value of the nuclidic mass of the isotope u	Average value of the nuclidic mass u	Mass spectroscopical data according to reference[2] (1960) u	Data according to reference[1] (1962) u
		3	4	5	6	7
1	2	3	4	5	6	7
156/64	$C_{10}H_8N_2$—^{156}Gd	146.661 ± 38	155.922089 ± 38	155.922060 ± 20	155.921850 ± 200	155.922100 ± 180
	$C_{10}{}^{13}C_2H_{10}$—^{156}Gd	162.814 ± 55	155.922146 ± 55			
	$C_{12}H_{12}$***—^{156}Gd	171.923 ± 44	155.921979 ± 44			
	$C_{11}{}^{13}CH_{11}$***—^{156}Gd	167.384 ± 43	155.922041 ± 43			
	^{156}Gd—^{155}Gd	999.416 ± 33	155.922096 ± 34			
155/64	$C_{10}H_7N_2$—^{155}Gd	138.213 ± 38	154.922712 ± 38	154.922680 ± 30	154.922320 ± 200	154.922590 ± 180
	$C_{11}{}^{13}CH_{10}$—^{155}Gd	158.921 ± 42	154.922685 ± 42			
	$C_{10}{}^{13}C_2H_9$—^{155}Gd	154.449 ± 140	154.922686 ± 140			
	$C_{12}H_{11}$***—^{155}Gd	163.527 ± 72	154.922550 ± 72			
154/64	$C_{12}H_{10}$—^{154}Gd	157.149 ± 40	153.921103 ± 40	153.921150 ± 30	153.920540 ± 200	153.920720 ± 190
	$C_{10}H_6N_2$—^{154}Gd	131.977 ± 243	153.921123 ± 243			
	$C_{11}{}^{13}CH_9$—^{154}Gd	152.550 ± 106	153.921231 ± 106			
	$C_{10}{}^{13}C_2H_8$—^{154}Gd	148.030 ± 94	153.921280 ± 94			
	^{155}Gd—^{154}Gd	1001.480 ± 60	153.921200 ± 72			
152/64	$C_{12}H_8$—^{152}Gd	142.870 ± 50	151.919732 ± 50	151.919740 ± 20	151.919430 ± 170	151.919530 ± 120
	^{154}Gd—^{152}Gd	2001.401 ± 50	151.919751 ± 56			
153/63	$C_{12}H_9$—^{153}Eu	149.103 ± 18	152.921324 ± 18	152.921360 ± 50	152.920980 ± 250	152.920860 ± 180
	$C_{11}{}^{13}CH_8$—^{153}Eu	144.606 ± 30	152.921350 ± 30			
	$C_9{}^{13}CH_{16}O$—^{153}Eu	201.934 ± 38	152.921538 ± 38			

	Reaction					
151/63	$C_{12}H_7$——^{151}Eu	134.920 ± 37	150.919856 ± 37	150.919830 ± 20	150.919550 ± 130	150.919630 ± 180
	$C_{10}H_{15}O$——^{151}Eu	192.487 ± 66	150.919806 ± 66			
	^{153}Eu——^{151}Eu	2001.567 ± 33	150.919793 ± 58			
154/62	$C_{12}H_{10}$——^{154}Sm	155.830 ± 29	153.922422 ± 29	153.922450 ± 20	153.921940 ± 150	153.922010 ± 280
	^{154}Sm——^{153}Eu	1001.082 ± 42	153.922442 ± 64			
	^{154}Sm——^{152}Sm	2002.664 ± 43	153.922519 ± 50			
152/62	$C_{12}H_8$——^{152}Sm	142.764 ± 32	151.919838 ± 32	151.919850 ± 30	151.919370 ± 150	151.919490 ± 120
	^{152}Sm——^{151}Eu	1000.095 ± 47	151.919927 ± 52			
	^{153}Eu——^{152}Sm	1001.544 ± 42	151.919816 ± 64			
150/62	$C_{12}H_6$——^{150}Sm	129.805 ± 140	149.917146 ± 140	149.917320 ± 60	149.916910 ± 80	149.917010 ± 130
	$C_8{}^{13}CH_{11}NO$——^{150}Sm	170.029 ± 25	149.917392 ± 25			
	$C_9H_{12}NO^{**}$——^{150}Sm	174.612 ± 47	149.917280 ± 47			
	^{151}Eu——^{150}Sm	1002.796 ± 62	149.917036 ± 65			
	^{152}Sm——^{150}Sm	2002.563 ± 31	149.917292 ± 40			
149/62	$C_9H_{11}NO$——^{149}Sm	166.820 ± 33	148.917246 ± 33	148.917200 ± 20	148.916810 ± 100	148.916930 ± 130
	$C_8{}^{13}CH_{10}NO$——^{149}Sm	162.408 ± 46	148.917188 ± 46			
	$C_8{}^{13}CH_8O_2$——^{149}Sm	138.597 ± 29	148.917189 ± 29			
	^{150}Sm——^{149}Sm	1000.149 ± 30	148.917171 ± 61			
	^{149}Sm——^{148}Sm	1002.282 ± 31	148.917150 ± 52			
148/62	$C_9H_{10}NO$——^{148}Sm	161.275 ± 31	147.914966 ± 31	147.914920 ± 40	147.914430 ± 100	147.914560 ± 130
	$C_9H_8O_2$——^{148}Sm	137.540 ± 26	147.914891 ± 26			
	$C_8{}^{13}CH_7O_2$——^{148}Sm	133.028 ± 57	147.914932 ± 57			
	^{150}Sm——^{148}Sm	2002.429 ± 51	147.914891 ± 73			
147/62	$C_9H_7O_2$——^{147}Sm	129.703 ± 17	146.914903 ± 17	146.914880 ± 30	146.914490 ± 80	146.914620 ± 50
	$C_8H_5NO_2$——^{147}Sm	117.197 ± 40	146.914833 ± 40			
	^{148}Sm——^{147}Sm	1000.110 ± 44	146.914758 ± 61			
	^{149}Sm——^{147}Sm	2002.317 ± 58	146.914880 ± 60			
144/62	$C_{10}H_8O$——^{144}Sm	145.448 ± 51	143.912068 ± 51	143.911980 ± 60	143.911660 ± 90	143.911650 ± 240
	$C_9{}^{13}CH_9N$——^{144}Sm	164.955 ± 46	143.911900 ± 46			

measurements. At the present time investigations are going on in order to find the causes of such deviations from the mean value.

A comparison of the mass values as measured in the present research work (see Table 1) with the respective "adjusted" values in the paper[2] shows that there are no significant discrepancies between the values of the measured masses. The greatest discrepancies (~ 0.4 mu) are within the limits of the maximum error of measurement proposed by the authors of the paper[2]. In many cases the results obtained in both investigations remain within or slightly exceed the aggregate measurement error.

Fig. 1 shows the differences, in absolute values, between the nuclidic masses as given in the paper[2] and the results based on the present research for cases where both the direct measurements and the "adjusted" values are available. The results of the present research are taken as a base line. The horizontal dashed lines correspond to the mass values obtained from the doublets with the organic compounds and the full horizontal lines correspond to the "adjusted" values. The vertical lines give aggregate errors in the respective differences.

The major part of corrections introduced by the authors of the paper[2] in their experimental results enhance the consistency of their data with the results of the present research (see, for instance, ^{165}Ho, ^{170}Yb, ^{176}Lu and others). In a number of cases, however (^{174}Yb, Lu^{175}) the correction should, in our opinion, be larger.

The only serious contradiction is an unjustified and "arbitrary" correction for the base masses of ^{156}Gd and ^{160}Gd by 0.390 and 0.330 mu respectively. This correction was necessary because of the nuclear data which have indicated that the differences Gd$-$Dy and Gd$-$Tb obtained in the paper[2] were not correct. For this reason the authors of the paper[2] have corrected the values for gadolinium. However, both an analysis of the available data and the results of the present research have shown that it was also necessary to correct approximately by the same amount (~ 0.350 mu) the masses of Dy and Tb. The authors of the paper[1] have not accepted the corrections introduced for the masses of gadolinium and have assumed other values which agree well with the results of the present research. The values of nuclidic masses of terbium and dysprosium in the paper[1] have also been corrected, so as to bring them into better conformity with the results of the present investigation; we consider, however, that these corrections are still insufficient.

A cyclical analysis of the results obtained was also made in the course of the present research. The main idea of this examination was to compare masses obtained from several independent doublets with a neighbouring mass (differing from it by only 1 or 2 u) obtained in the same manner. The difference between these masses must be equal to the independently measured "isotopic" doublet (within the limits of the experimental errors).

Such a verification was carried out for all the elements having more than one isotope. In such a closed cycle errors at any one of the stages would lead to a discrepancy in the mass values obtained by different methods, and this discrepancy would by far exceed the aggregate error for the cycle. Moreover, the value of any mass comprised in this cycle is indirectly confirmed by the measurement of each doublet when determining any nuclidic mass comprised in the cycle. For the purpose of checking such a cycle, the nuclidic masses are calculated by means of "isotopic" doublets (one of the isotopes being taken for a basis), and then the degree of accuracy (error) obtained by these calculations is compared with the aggregate error resulting for the closed cycle concerned. Such cycles were formed for Lu, Yb, Er, Dy, Gd, Eu and Sm. In none of these cycles did the internal error exceed the aggregate error occured in the measurement of the cycle concerned.

As an illustration of an analysis by means of a closed cycle we shall describe the treatment of data obtained for the nuclidic masses of erbium.

Verification of Cycles on the Basis of Erbium

The mass of isotope ^{162}Er is taken for a basis. By using the "isotopic" doublets this mass serves for the determination of masses of all the others erbium isotopes. When passing from one mass to another a comparison is made of the mass value obtained from several independent doublets (their arithmetic mean) with the mass value obtained by means of "isotopic" doublets, related to each other. The results of calculations are given in Table 2.

As can be seen from the above Table 2, the masses obtained in different ways differ by quantities falling within the admissible error, and this is true for all the masses. The final aggregate error exceeds the mass differences by almost twice their values. This result may be considered as fully satisfactory.

Table 2

A	Nuclidic mass u	Deviation from the mean arithmetic value mu	Error for the cycle mu
164	163.929211 $\pm$ 110	0.066	0.120
166	165.930425 $\pm$ 130	0.089	0.140
167	166.932147 $\pm$ 140	0.059	0.150
168	167.932431 $\pm$ 150	0.044	0.160
170	169.935884 $\pm$ 170	0.112	0.200

A similar verification was carried out for all the elements listed above. For none of them did a deviation in a given cycle exceed the aggregate

Table 3. The Values of Mass Differences are given in Mass Units

Mass differences	A-results of the present research (1962)	B-results from the paper[2] (1960)	Nuclear data	$\Delta = A - B$, in 10^{-3} u*
1	2	3	4	5
176Lu—175Lu	1.002020 ± 60	1.001940 ± 60		0.080 ± 0.120
176Lu—176Hf**	0.001300 ± 170	0.001060 ± 200	0.001120 ± 30 (1)	0.240 ± 0,370
176Yb—174Yb	2.004010 ± 100	2.003700 ± 90		0.310 ± 0,190
174Yb—173Yb	1.000680 ± 60	1.000660 ± 60		0.020 ± 0,120
173Yb—172Yb	1.001720 ± 90	1.001860 ± 60	1.001670 ± 90 (1)	— 0.140 ± 0,150
172Yb—171Yb	0.999950 ± 90	1.000080 ± 60		— 0.130 ± 0,130
171Yb—170Yb	1.001140 ± 50	1.001570 ± 60		— 0.420 ± 0,110
170Yb—168Yb	2.000850 ± 50	2.000980 ± 90		— 0.130 ± 0,140
174Hf—170Yb	4.005300 ± 120	4.005400 ± 200	4.005350 ± 220	— 0.100 ± 0,320
170Er—168Er	2.003420 ± 40	2.003120 ± 95		0.300 ± 0,140
168Er—167Er	1.000320 ± 50	1.000340 ± 60	1.000310 ± 40 (1)	— 0.020 ± 0,110
167Er—166Er	1.001700 ± 60	1.001740 ± 60	1.001610 ± 90 (1)	— 0.040 ± 0,120
166Er—164Er	2.001190 ± 50	2.001160 ± 90		0.030 ± 0,140
166Er—165Ho	0.999990 ± 50	1.000140 ± 200	1.000020 ± 420 (1)	— 0.150 ± 0,250
164Er—162Er	2.000500 ± 60	2.000510 ± 95		— 0.010 ± 0,160
165Ho—164Er	1.001200 ± 30	1.001020 ± 200	1.000990 ± 70 (1)	0.180 ± 0,230
164Dy—163Dy	1.000420 ± 40	1.000430 ± 60		— 0.060 ± 0,100
164Dy—162Dy	2.002510 ± 50	2.002340 ± 100		0.170 ± 0,150
163Dy—162Dy	1.002090 ± 50	1.001950 ± 60	1.001865 ± 120 (1)	0.140 ± 0,110
162Dy—161Dy	1.000010 ± 60	0.999890 ± 60		0.120 ± 0,120
162Dy—160Dy	2.001600 ± 80	2.001560 ± 100		0.040 ± 0,180
161Dy—160Dy	1.001590 ± 80	1.001780 ± 60		— 0.190 ± 0,140
160Dy—159Tb	0.999810 ± 90	0.999990 ± 180	0.999725 ± 430 (1)	— 0.180 ± 0,270
160Dy—158Dy	2.001250 ± 130	2.000870 ± 100		0.380 ± 0,230
158Dy—156Dy	2.000290 ± 160	2.000200 ± 100		0.090 ± 0,260

158Gd—157Gd	1.000420 ± 30	1.000160 ± 45	1.001890 ± 430 (6) 1.000760 ± 60 (7) 1.000110 ± 10 (1)	0.260 ± 0,080
157Gd—156Gd	1.001840 ± 30	1.001840 ± 45	1.001800 ± 120 (1) 1.000300 ± 60 (7)	0.000 ± 0,080
156Gd—155Gd	0.999380 ± 40	0.999580 ± 60	0.999510 ± 10 (1)	—0.200 ± 0,100
155Gd—154Gd	1.001530 ± 50	1.001830 ± 60		—0.300 ± 0,110
154Gd—152Gd	2.001410 ± 40	2.001110 ± 90		0.300 ± 0,130
160Gd—158Gd	2.002730 ± 70	2.002980 ± 90		—0.250 ± 0,160
153Eu—151Eu	2.001530 ± 70	2.001430 ± 380		0.100 ± 0,450
153Eu—152Sm	1.001510 ± 80	1.001610 ± 400	1.001360 ± 170 (1)	—0.100 ± 0,480
152Sm—151Eu	1.000020 ± 50	0.999820 ± 280		0.200 ± 0,330
151Eu—150Sm	1.002510 ± 80	1.002640 ± 220		—0.130 ± 0,300
153Eu—152Gd	1.001620 ± 70	1.001550 ± 420	1.001330 ± 150 (1)	0.070 ± 0,490
152Gd—148Sm	4.004870 ± 60	4.005009 ± 260	4.004940 ± 30 (1)	—0.140 ± 0,320
152Sm—152Gd	0.000110 ± 50	—0.000060 ± 320	—0.000050 ± 20 (1)	0.170 ± 0,370
150Sm—149Sm	1.000120 ± 80	1.000100 ± 180	1.001565 ± 320 (6) 1.000165 ± 60 (6) 1.000065 ± 30 (7)	0 020 ± 0,260

* The error in the difference was assumed to be equal to the sum of component errors.
** The nuclidic mass of ^{176}Hf was determined on the basis of the isotopic doublet ^{177}Hf—^{176}Hf, see ref. 11.

error for the respective cycle. Thus, a verification of the cycle on the basis of lutetium established that the relation between the mass differences and the internal error of the cycle was: 0.045 ± 0.085 mu; for ytterbium it was: 0.250 ± 0.430 mu; for dysprosium*: 0.130 ± 0.130 mu; for gadolinium: 0.026 ± 0.120 mu; for europium: 0.010 ± 0.170 mu; and for samarium: 0.210 ± 0.170 mu.

A comparison of absolute values of masses as established in the present research study with the values given in the papers[1] and [2] shows that the corrections of the mass values obtained from the doublets with the organic substances[2] will, in the majority of cases, lead to a higher degree of consistency[4] between the results of the present research and the values given in the paper[2]. The same trend is apparent in most of the cases described in the paper[1] after the results in the paper[2] have been corrected (^{176}Lu, ^{175}Lu, ^{173}Yb, ^{168}Er, ^{160}Gd and many others).

However, it is interesting to compare not the absolute values of nuclidic masses, but the differences between neighbouring masses, as the majority of experimental values given in the paper[2] is precisely in this form ("isotopic" doublets). Table 3 contains both the values of experimental mass differences and the differences obtained from averaged values. In a number of cases it was possible to verify the nuclidic mass differences by means of nuclear reactions; such data are figuring in the 4th column of the Table 3. Nuclear data were used only in such cases where with their help it was possible to establish a correlation between two stable isotopes. A comparison of the mass differences obtained by two, and in some cases even by three completely unrelated methods shows that there are no significant discrepancies in these values. The maximum discrepancies amount to $\sim 300\ \mu$u. Taking into consideration the fact that the errors in the compared values are statistical errors, and that their aggregate value amounts to some 100 to 200 μu, the degree of consistency may be considered as fully satisfactory. If the error adjustment factor is assumed to be equal to 2.65 (according to a suggestion of the authors of the paper[1]), then the deviation would not, even for a single value, exceed the aggregate error of the measurements.

Two particular cases stand out from a comparison of mass-spectrometric results and nuclear data; it is worthwhile to consider them in more detail.

1. *Mass difference* $^{156}Gd—^{155}Gd$. Mass-spectroscopical measurement of this difference in the paper[2] (neglecting the corrections for standard values of the masses of ^{156}Gd and ^{160}Gd) is in good agreement with the results of the present research. The mean value of this difference is equal

---

* The dysprosium cycle contained 4 isotopes: ^{161}Dy, ^{162}Dy, ^{163}Dy and ^{164}Dy.

to 0.999476 ± 60 u. This value was obtained from two independent measurements of an "isotopic" doublet and from the results of the two papers on the basis of averaged values of the masses of ^{156}Gd and ^{155}Gd.

According to the references[7] and [8] the nuclear value for this mass differences is equal to 1.0003000 ± 60 u. The difference is 824 ± 100 μu, i. e. it exceeds almost by 8 times the aggregate error of the measurements. The mass-spectroscopical results, which are in good agreement, within the limits of permissible error, are supported by a great number of repeated measurements of the masses of ^{156}Gd and ^{155}Gd; moreover, the measurements were carried out by means of two different equipments. An analysis of all the values proposed for this mass difference shows that preference should be given to the mass spectroscopical results; it must therefore be admitted that the Q-value for the reaction ^{155}Gd$(n, \gamma)^{156}$Gd contains an error of approximately 800 μu. The results of the paper[1] support this conclusion.

2. *Mass difference* $^{158}Gd - ^{157}Gd$. Table 3 shows the evolution of the Q-value for the reaction ^{157}Gd$(n, \gamma)^{158}$Gd. The value of Q has changed by some 1800 μu. The results of the paper[2] differ from those of the present research by some 250 μu which exceeds by three times the aggregate error of measurements. The last value of Q for this reaction and the values of gadolinium masses calculated therefrom[1] agree, within the limits of experimental errors, with the results of the present research.

An analysis of absolute values of the nuclidic masses and of their differences leads to the conclusion that the most significant discrepancies in the mass values are observed for the isotopes: ^{174}Yb, ^{173}Yb, ^{168}Yb, ^{159}Tb, ^{158}Gd and ^{154}Gd.

The nuclear reactions data show that the arbitrary corrections for the masses of gadolinium in the paper[2] are unjustifiable. Without regard to the present research, in an independent way, some corrections were introduced in the paper[1] which bring into a satisfactory agreement the mass values proposed therein with the gadolinium mass values obtained by measurement in the present research study.

The discrepancies for the ytterbium masses remain rather significant (they exceed by 2 or 3 times the aggregate error of the measurements). In the survey paper[1] the mass values for ytterbium were accepted without changes as compared with the paper[2].

From a comparison of the results of the present research with the data of BHANOT et al.[2] one can draw one common conclusion: the most significant discrepancies occur in those cases where organic compounds were used for the measurement of nuclidic masses. In the present research the major part of nuclidic masses was measured by means of the "isotopic" doublets. Such doublets permit calculation of the mass differences differing only by 1 or 2 mass units. In order to obtain an absolute value of the mass it is, however, necessary, at least for some of the nuclei, to compare an unknown mass to the ^{12}C-standard, or to H, ^{14}N and ^{16}O sub-standard values.

It is therefore natural to suspect that the cause of discrepancies in absolute values of the nuclidic masses lies with the determination of the "basic" mass-doublets obtained by using organic compounds. This seems to be confirmed by the fact that the isotopic differences given in both sources under consideration are in rather good agreement.

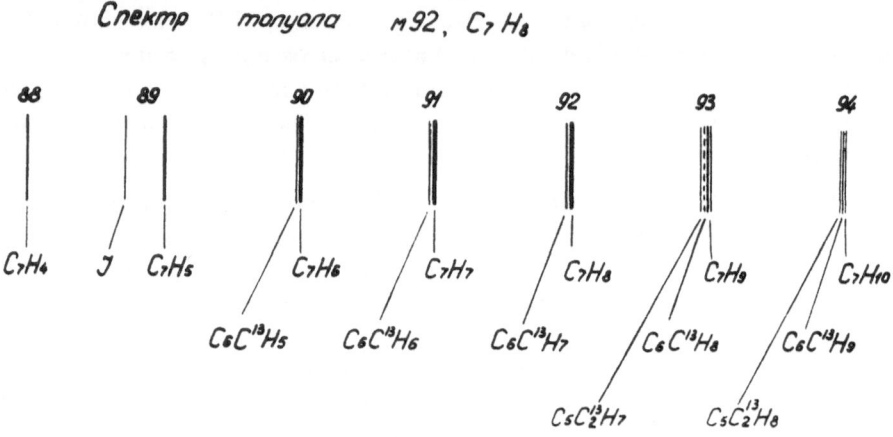

Fig. 2. Toluol spectrum

The principal causes of discrepancies lie, as it seems to us, with the use, in the paper[2], of a mass-spectrometer having a poor resolution capacity ($M/\Delta M = 14\,000$). With such resolution capacity the most "dangerous" impurity (trace component), capable of causing systematic errors, is the one that contains the ^{13}C-isotope; in this range of masses ($A \approx 170$) a resolution capacity of some $40\,000$ to $45\,000$ is necessary to resolve the ^{13}C-isotope.

A comparison of the mass values derived from the doublets obtained by the use of organic compounds shows that a major part of discrepancies between the results of the present investigation and the results of Bhanot et al.[2] can be explained by the presence, in their case, of unresolved impurities.

Working experience with heavy organic compounds in our laboratory shows that in a plasma ion source, besides the doublets $^{12}CH - {}^{13}C$ on any mass, and besides the fragments of the organic compound introduced, the ions of an associative origin are present in the spectrum; these ions are of the type: $A + H$, $A + H_2$ and there are even ions of the type: $A + CH_m$, where $m = 0, 1, 2, 3, 4$ etc.; A is molecular ion of the organic compound introduced in the source. In the channel beam tube the ions of a sufficient intensity of the type $A + H$, $A + H_2$ and others were also observed[5]. A spectrum of Toluol (C_7H_8, $m = 92$) is reproduced for illustration in Fig. 2. Similar spectra were obtained in all other mass ranges.

As can be seen from Fig. 2, the mass spectrum obtained is rather a complicated one. The doublets $^{12}CH - ^{13}C$ are present on the masses of 91, 92, 93 and 94; moreover, the ions of the $A = H$ and $A + H_2$ type of a comparatively high intensity are present on the masses of 93 and 94. Where the intensity of ions containing the ^{13}C-isotope is sufficiently high, the mass which is by one unit greater can contain type $^{12}C_5{}^{13}C_2H_7$-ions, i. e. the ions containing two atoms of the ^{13}C-isotope and accordingly a double doublet $^{12}CH - ^{13}C$ on the mass under consideration.

The presence of ions of associative origin together with ions containing ^{13}C and $^{13}C_2$ can lead to systematic errors, if an apparatus having an insufficient resolution capacity is used; this situation occurs when a comparison is made of BHANOT et al. data with the results of the present investigation.

Thus, in particular, a discrepancy for the mass of ^{186}Os can be explained by the presence of an unresolved admixture of an ion of associative origin $C_{12}H_{28}N$ (dihexylamine $C_{12}H_{27}N$, $m = 185$ was used) together with the basic series $C_{11}{}^{13}CH_{27}N$ used for measurement of the ^{186}Os-mass. The presence of this unresolved impurity of admixture will shift the centre of the line towards the heavier masses. The value of the doublet will appear to be higher, and the value of the ^{186}Os will appear to be correspondingly lower (by ~ 0.6 mu). That is exactly what is observed when a comparison is made with the data of the present research[10].

A similar effect is observed for the measurement of the masses of ^{186}W, ^{190}Os and ^{191}Yr.

The discrepancy for the ^{175}Lu-mass can be explained by the presence of an intensive line containing $^{13}C - C_{12}{}^{13}CH_{18}$ (hydrocarbon $C_{13}H_{19}$, $m = 175$ was used in this case). The error is so significant (~ 1 mu) that the authors have discarded this doublet. The discrepancy for the ^{176}Hf-mass supports the validity of this explanation; in fact, if the organic compound used ($C_{13}H_{19}$, $m = 175$) gives a fragment containing a large amount of the isotope ^{13}C ($C_{12}{}^{13}CH_{18}$), then this fragment, in its turn, will give a fragment $C_{11}{}^{13}C_2H_{18}$, the presence of which can explain the discrepancy for the ^{176}Hf-mass.

It is to be noted that the relative quantity of the fragment containing the ^{13}C-isotope can vary very considerably, depending on the nature of the organic compound used, on its properties and on the ion spectrum that it gives. There are some cases where the ^{13}C-containing ion has a greater intensity than the ion not containing ^{13}C; but reverse situations

can also occur. Therefore a constant correction for the presence of the
^{13}C-containing ions is, to a certain extent, an arbitrary one and can also
lead to a systematic error in the values of the measured masses.

As can be seen from the Toluol spectrum shown above (Fig. 2), it
may occur, in case of a sufficiently high intensity of fragments of the
molecular ion, that besides the $C_n{}^{13}CH_m$-type ions some $C_n{}^{13}C_2H_m$-type
lines may also be observed, and their intensity may be comparable to that
of the other lines.

If the mass spectrometric lines of this type are not resolved, this
may lead to an under-estimation of the doublet's value and to a corresponding
over-estimation of the mass value (^{176}Hf).

For a safe use of organic compounds as sub-standards for measurement
of the masses, it is therefore necessary to use mass-spectrometers having
such a resolution capacity as to be able to resolve the doublet $^{12}CH-^{13}C$
(≈ 4.5 mu) in the range of masses to be measured.

Binding Energies of Nuclei

Based on the masses of stable isotopes as measured in the present
investigation the masses and the binding energies of 64 radioisotopes were
calculated. These data were used for calculating the binding energy per
nucleon E/A (for the stable isotopes), the binding energy of the last
neutron B_n, the binding energy of the last proton B_p, the binding energy
of the two last neutrons B_{2n}, and the pairing energy of neutrons P_n
and protons P_p. The results of these calculations are given in Table 4.

For calculating the masses of radioisotopes on the basis of nuclear
reactions and of alpha- and beta-transformations the values were taken
from the references[1, 6-9]. Where possible the determination of the mass
of a radioactive nucleus was made in several independent ways. Some
70 Q-values were used for the calculation of nuclidic masses of radioisotopes.
Seven values were discarded as inconsistent with the results of other
reactions and with the mass spectroscopic results. No error limits (degree
of accuracy) are given for binding energies of the nuclei (column 5), but
in no case should they have exceeded 150 keV. The binding energies of
the nuclei for the masses of which no error limits (degree of accuracy) are
given in Table 4 were calculated on the basis of systematic variations of
the parameters B_n, B_p, B_{2n}, P_n and P_p. It is not possible to make a
direct calculation of the masses and of the binding energies for these
nuclei, since the nuclear reactions connecting an unknown mass and a
known mass have only an upper or a lower limit of the Q-value. In the
majority of cases the errors for the values B_n, B_p and B_{2n} did not
exceed 200 keV and the errors for the values P_n and P_p did not exceed
300 keV.

Table 4

Isotope	Z	N	Nuclidic mass u	Binding energy MeV	B_n	B_p	B_{2n}	P_n	P_p
							MeV		
1	2	3	4	5	6	7	8	9	10
^{143}Sm	62	81	142.913620 ± 80	1186.1	—	—	—	—	—
^{144}Sm*	62	82	143.911980 ± 60	1195.7	9.6	—	—	—	—
^{145}Sm	62	83	144.913300 ± 200	1202.6	6.9	—	16.5	—	—
^{146}Sm	62	84	145.913200 ± 100	1210.7	8.1	—	15.0	1.2	—
^{147}Sm*	62	85	146.914880 ± 30	1217.2	6.5	—	14.6	—	—
^{148}Sm*	62	86	147.914920 ± 40	1225.3	8.1	—	14.6	1.6	—
^{149}Sm*	62	87	148.917200 ± 20	1231.2	5.9	—	14.0	—	—
^{150}Sm*	62	88	149.917320 ± 60	1239.2	8.0	—	13.9	2.1	—
^{151}Sm	62	89	150.919940 ± 30	1244.8	5.6	—	13.6	—	—
^{152}Sm*	62	90	151.919850 ± 30	1253.0	8.2	—	13.8	2.6	—
^{153}Sm	62	91	152.922230 ± 60	1258.8	5.8	—	14.0	—	—
^{154}Sm*	62	92	153.922450 ± 20	1266.7	7.9	—	13.7	2.1	—
^{155}Sm	62	93	154.924990 ± 300	1272.4	5.7	—	13.6	—	—
^{156}Sm	62	94	155.925670 ± 80	1279.8	7.4	—	13.1	1.7	—
^{150}Eu	63	87	149.919815 ± 130	1236.1	—	—	—	—	—
^{151}Eu*	63	88	150.919830 ± 20	1244.1	8.0	4.9	14.2	—	—
^{152}Eu	63	89	151.921850 ± 100	1250.3	6.2	5.5	14.2	—	—
^{153}Eu*	63	90	152.921360 ± 50	1258.9	8.6	5.9	14.8	2.4	—
^{154}Eu	63	91	153.923290 ± 30	1265.1	6.2	6.3	14.8	—	—
^{155}Eu	63	92	154.922940 ± 30	1273.5	8.4	6.8	14.6	2.2	—
^{156}Eu	63	93	155.924700 ± 30	1280.0	6.5	7.6	14.9	—	—
^{157}Eu	63	94	156.925270 ± 60	1287.5	7.5	7.7	14.0	1.0	—
^{150}Gd	64	86	149.918665 ± 140	1236.4	—	6.5	—	—	1.9
^{151}Gd	64	87	150.920120	1243.1	6.7	7.0	—	—	2.1
^{152}Gd*	64	88	151.919740 ± 20	1251.5	8.4	7.5	15.3	1.5	—
^{153}Gd	64	89	152.921657 ± 60	1257.8	6.3	7.4	14.7	—	—
^{154}Gd*	64	90	153.921150 ± 30	1266.3	8.5	7.4	14.8	2.2	—
^{155}Gd*	64	91	154.922680 ± 30	1273.0	6.7	7.9	15.2	—	—
^{156}Gd*	64	92	155.922060 ± 20	1281.6	8.6	8.1	15.3	1.9	—
^{157}Gd*	64	93	156.923900 ± 20	1288.0	6.4	8.0	15.0	—	—
^{158}Gd*	64	94	157.924320 ± 20	1295.7	7.7	8.2	14.1	1.3	—
^{159}Gd	64	95	158.926372 ± 50	1301.8	6.1	—	13.8	—	—
^{160}Gd*	64	96	159.927050 ± 50	1309.3	7.5	—	13.6	1.4	—
^{161}Gd	64	97	160.929475 ± 90	1315.1	5.8	—	13.3	—	—
^{153}Tb	65	88	152.922106	1256.6	—	5.1	—	—	—
^{154}Tb	65	89	153.923470	1263.4	6.8	5.6	—	—	—
^{155}Tb	65	90	154.922903	1272.0	8.6	5.7	15.4	1.6	—
^{156}Tb	65	91	155.924482	1278.6	6.6	5.6	15.2	—	—
^{157}Tb	65	92	156.923807	1287.3	8.7	5.7	15.3	1.9	—
^{158}Tb	65	93	157.925437 ± 60	1293.8	6.5	5.8	15.2	—	—
^{159}Tb*	65	94	158.925350 ± 40	1302.0	8.2	6.3	14.7	1.7	—
^{160}Tb	65	95	159.927060 ± 100	1308.5	6.5	6.7	14.7	—	—

* Nuclidic masses as measured according to the present research.

(*Table 4, continued*)

Isotope	Z	N	Nuclidic mass u	Binding energy MeV	B_n	B_p	B_{2n}	P_n	P_p
					MeV				
1	2	3	4	5	6	7	8	9	10
^{161}Tb	65	96	160.927 325 ± 30	1316.3	7.8	7.0	14.3	1.3	—
^{156}Dy*	66	90	155.923 620 ± 120	1278.6	—	6.6	—	—	1.3
^{157}Dy	66	91	156.924 732	1285.6	7.0	7.0	—	—	1.4
^{158}Dy*	66	92	157.923 910 ± 100	1294.5	8.9	7.2	15.9	1.9	1.5
^{159}Dy	66	93	158.925 402 ± 60	1301.2	6.7	7.4	15.5	—	1.6
^{160}Dy*	66	94	159.925 160 ± 70	1309.5	8.3	7.5	15.0	1.6	1.2
^{161}Dy*	66	95	160.926 750 ± 30	1316.1	6.6	7.6	14.9	—	0.9
^{162}Dy*	66	96	161.926 760 ± 30	1324.1	8.0	7.8	14.6	1.4	0.8
^{163}Dy*	66	97	162.928 850 ± 20	1330.2	6.1	—	14.1	—	—
^{164}Dy*	66	98	163.929 270 ± 20	1337.9	7.7	—	13.8	1.6	—
^{165}Dy	66	99	164.931 749 ± 30	1343.7	5.8	—	13.5	—	—
^{166}Dy	66	100	165.932 855 ± 100	1350.7	7.0	—	12.8	1.2	—
^{160}Ho	67	93	159.928 477	1305.6	—	4.4	—	—	—
^{161}Ho	67	94	160.927 909	1314.2	8.6	4.7	—	—	—
^{162}Ho	67	95	161.929 150 ± 100	1321.1	6.9	5.0	15.5	—	—
^{163}Ho	67	96	162.928 861 ± 30	1329.4	8.3	5.3	15.2	1.4	—
^{164}Ho	67	97	163.930 320 ± 100	1336.2	6.8	6.1	15.1	—	—
^{165}Ho*	67	98	164.930 350 ± 20	1344.2	8.0	6.3	14.8	1.2	—
^{166}Ho	67	99	165.932 335 ± 100	1350.4	6.2	6.7	14.2	—	—
^{167}Ho	67	100	166.933 117 ± 120	1357.8	7.4	7.1	13.6	1.2	—
^{161}Er	68	93	160.929 646	1311.8	—	6.2	—	—	1.8
^{162}Er*	68	94	161.928 650 ± 50	1320.8	9.0	6.6	—	—	1.9
^{163}Er	68	95	162.929 799	1327.8	7.0	6.7	16.0	—	1.7
^{164}Er*	68	96	163.929 150 ± 20	1336.5	8.7	7.1	15.7	1.7	1.8
^{165}Er	68	97	164.930 489	1343.3	6.8	7.1	15.5	—	1.1
^{166}Er*	68	98	165.930 340 ± 40	1351.5	8.2	7.3	15.0	1.4	1.0
^{167}Er*	68	99	166.932 040 ± 30	1358.0	6.5	7.6	14.7	—	0.9
^{168}Er*	68	100	167.932 360 ± 30	1365.8	7.8	8.0	14.3	1.2	0.9
^{169}Er	68	101	168.934 581 ± 20	1371.8	6.0	—	13.8	—	—
^{170}Er*	68	102	169.935 780 ± 20	1378.7	6.9	—	12.9	0.9	—
^{171}Er	68	103	170.937 989 ± 50	1384.7	6.0	—	12.9	—	—
^{172}Er	68	104	171.939 241 ± 100	1391.6	6.9	—	12.9	0.9	—
^{164}Tu	69	95	163.933 007 ± 40	1332.1	—	4.3	—	—	—
^{165}Tu	69	96	164.932 010	1341.1	9.0	4.6	—	—	—
^{166}Tu	69	97	165.932 917 ± 130	1348.3	7.2	5.0	16.2	—	—
^{167}Tu	69	98	166.932 700	1356.6	8.3	5.1	15.5	1.1	—
^{168}Tu	69	99	167.934 220 ± 80	1363.2	6.6	5.2	14.9	—	—
^{169}Tu*	69	100	168.934 220 ± 20	1371.3	8.1	5.5	14.7	1.5	—
^{170}Tu	69	101	169.936 043 ± 30	1377.7	6.4	5.9	14.5	—	—
^{171}Tu	69	102	170.936 399 ± 45	1385.4	7.7	6.7	14.1	1.3	—
^{172}Tu	69	103	171.938 261 ± 80	1391.8	6.4	7.1	14.1	—	—
^{167}Yb	70	97	166.934 759	1353.9	—	5.6	—	—	0.6
^{168}Yb*	70	98	167.934 300 ± 30	1362.4	8.5	5.8	—	—	0.7
^{169}Yb	70	99	168.935 448	1369.4	7.0	6.2	15.5	—	1.0

(Table 4, continued)

Isotope	Z	N	Nuclidic mass u	Binding energy MeV	B_n	B_p	B_{2n}	P_n	P_p
							MeV		
1	2	3	4	5	6	7	8	9	10
^{170}Yb*	70	100	169.935150 $\pm$ 30	1377.7	8.3	6.4	15.3	1.3	0.9
^{171}Yb*	70	101	170.936290 $\pm$ 40	1384.8	7.2	7.1	15.4	—	1.2
^{172}Yb*	70	102	171.936240 $\pm$ 70	1392.9	8.0	7.5	15.2	0.8	0.8
^{173}Yb*	70	103	172.937960 $\pm$ 50	1399.3	6.4	7.5	14.5	—	0.4
^{174}Yb*	70	104	173.938640 $\pm$ 20	1406.8	7.5	—	13.9	1.1	—
^{175}Yb	70	105	174.941101 $\pm$ 40	1412.6	5.8	—	13.3	—	—
^{176}Yb*	70	106	175.942650 $\pm$ 90	1419.2	6.6	—	12.4	0.8	—
^{177}Yb	70	107	176.945263 $\pm$ 170	1424.8	5.6	—	12.2	—	—
^{170}Lu	71	99	169.938852 $\pm$ 120	1373.5	—	4.1	—	—	—
^{171}Lu	71	100	170.938305	1382.1	8.6	4.4	—	—	—
^{172}Lu	71	101	171.939240	1389.3	7.2	4.5	15.8	—	—
^{173}Lu	71	102	172.938702 $\pm$ 60	1397.9	8.5	5.0	15.7	1.4	—
^{174}Lu	71	103	173.940332 $\pm$ 80	1404.4	6.6	5.1	15.1	—	—
^{175}Lu*	71	104	174.940600 $\pm$ 40	1412.2	7.8	5.4	14.3	1.3	—
^{176}Lu*	71	105	175.942620 $\pm$ 30	1418.4	6.2	5.8	14.0	—	—
^{177}Lu	71	106	176.943773 $\pm$ 150	1425.4	6.9	6.2	13.2	0.8	—

Table 5*

Designation of the isobar	Value according to the present research (1963)	Value according to paper[2] (1960)	Value according to paper[1] (1962)
1	2	3	4
$_{62}$Sm$^{152}_{90}$—$_{64}$Gd$^{152}_{88}$	0.110 $\pm$ 0.040	— 0.060 $\pm$ 0.240	— 0.040 $\pm$ 0.180
$_{62}$Sm$^{154}_{92}$—$_{64}$Gd$^{154}_{90}$	1.300 $\pm$ 0.040	1.400 $\pm$ 0.250	1.290 $\pm$ 0.340
$_{66}$Dy$^{156}_{90}$—$_{64}$Gd$^{156}_{92}$	1.560 $\pm$ 0.140	1.960 $\pm$ 0.300	1.660 $\pm$ 0.470
$_{64}$Gd$^{158}_{91}$—$_{66}$Dy$^{158}_{92}$	0.410 $\pm$ 0.120	— 0.180 $\pm$ 0.280	0.140 $\pm$ 0.380
$_{64}$Gd$^{160}_{96}$—$_{66}$Dy$^{160}_{91}$	1.890 $\pm$ 0.120	1.940 $\pm$ 0.300	2.290 $\pm$ 0.270
$_{68}$Er$^{162}_{91}$—$_{66}$Dy$^{162}_{96}$	1.890 $\pm$ 0.080	2.280 $\pm$ 0.250	2.310 $\pm$ 0.400
$_{66}$Dy$^{164}_{98}$—$_{68}$Er$^{164}_{96}$	0.270 $\pm$ 0.040	— 0.470 $\pm$ 0.280	— 0.460 $\pm$ 0.350
$_{70}$Yb$^{168}_{98}$—$_{68}$Er$^{168}_{100}$	1.940 $\pm$ 0.060	1.450 $\pm$ 0.220	1.520 $\pm$ 0.320
$_{68}$Er$^{170}_{102}$—$_{70}$Yb$^{170}_{100}$	0.630 $\pm$ 0.050	0.680 $\pm$ 0.240	0.630 $\pm$ 0.330
$_{72}$Hf$^{174}_{102}$—$_{70}$Yb$^{174}_{104}$	1.800 $\pm$ 0.130	1.280 $\pm$ 0.200	1.240 $\pm$ 0.220
$_{70}$Yb$^{176}_{106}$—$_{72}$Hf$^{176}_{104}$	1.330 $\pm$ 0.240	0.940 $\pm$ 0.180	1.090 $\pm$ 0.160
$_{70}$Yb$^{176}_{106}$—$_{71}$Lu$^{176}_{105}$	0.030 $\pm$ 0.120	— 0.120 $\pm$ 0.200	0.000 $\pm$ 0.160
$_{71}$Lu$^{176}_{105}$—$_{72}$Hf$^{176}_{104}$	1.300 $\pm$ 0.180	1.060 $\pm$ 0.200	1.090 $\pm$ 0.160

* The values of the differences and the error limits (degree of accuracy) are given in mu.

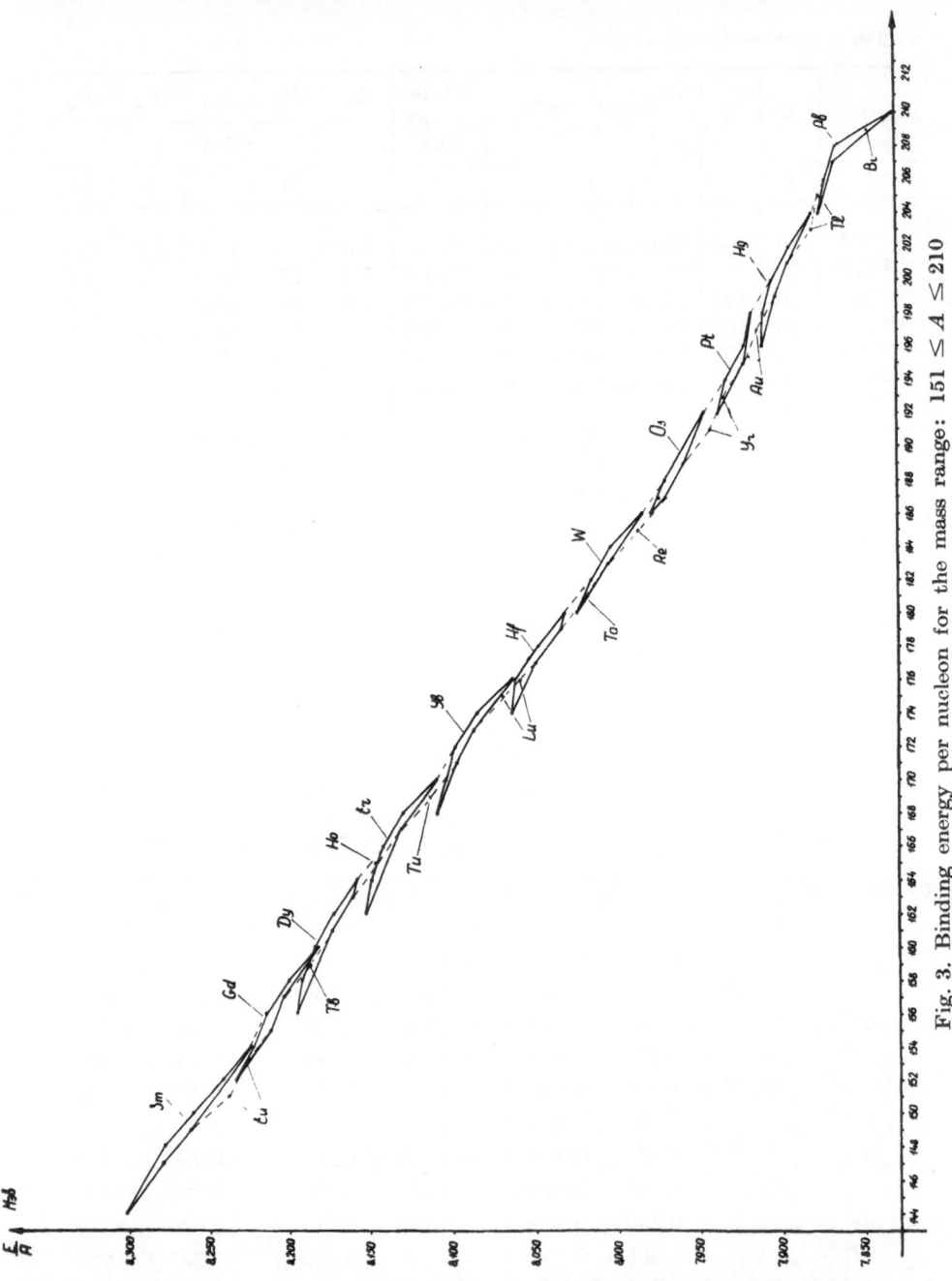

Fig. 3. Binding energy per nucleon for the mass range: $151 \leq A \leq 210$

The experimental results on the nuclidic masses as obtained in the present research study provide a possibility of calculating the mass differences for 11 isobaric pairs. In Table 5 the values of the differences so obtained are compared with the results given in the paper[2]. As can be seen from this table, generally speaking, there is an agreement between the values, within the limits of experimental errors. In some cases, however, a deviation in the values of the mass differences for isobaric nuclei exceeds the aggregate measurement error by a factor of 1.5 and even of 2.0 (^{164}Dy$-^{164}$Er). It was to be expected, since the sub-standards used in the paper[2] are sometimes higher and at other times lower than the values obtained for the same isotope in the present investigation (see Fig. 1).

Column 4 of Table 5 contains the values of the isobar pairs taken from the paper[1]. The data of the present investigation agree much better with the results obtained in the paper[1] than with those of the paper[2]; this is particularly well apparent for the isobars which include gadolinium. For other isobars the agreement is also much better. A significant discrepancy exists only for 2 isobars pairs ($m = 164$ and $m = 174$). We believe that these discrepancies are due to the wrong values of the nuclidic masses of ^{164}Dy and ^{174}Yb given in the paper[2] and maintained practically without any changes in the paper[1]. This conclusion is supported by a number of other considerations discussed above.

The binding energies of the isobar nuclei in which 2 protons were replaced by 2 neutrons have, in the majority of cases, a rather high value; for the two isobar pairs, however, (^{160}Gd$-^{160}$Dy and ^{176}Yb$-^{176}$Hf) this trend is broken; nevertheless, the deviations are practically within the limits of the measurement errors. An analysis of the values of the isobar pairs for lighter nuclei ($84 < A < 104$) shows that in this range the deviations are more significant. The cause of this phenomenon is examined with sufficient detail in the paper[12] for the isobar pair $_{71}$Lu$_{105}^{176}-_{72}$Hf$_{104}^{176}$ the negative value of the difference of the binding energies is due to the fact that the comparison is made between the nuclei of different parities (odd-odd and even-even).

The value of the binding energy per nucleon gives a general characteristic of the binding energy between the nuclei. Fig. 3 shows the binding energy per nucleon (for the stable isotopes) in the mass range measured. In order to complete the picture, the binding energy per nucleon is also given for the range of $174 < M < 210$ the curves of which are plotted on the basis of nuclidic masses measured in the previous investigations[10, 11]. The binding energies of nuclei having the same Z are connected by full lines. The binding energies of nuclei of elements having odd Z are connected by a dash-dotted line to the odd "A" of the elements having an even Z. The errors on the values E/A do not exceed 0.8 keV. No sharp bents are observed in the measured range which would correspond to a filling of the shell with regard to protons or neutrons (see $A = 208$). The general aspect of the curve is rather a smooth one. The beginning and the end of distorted nuclei ($A \approx 155$ and $A \approx 185$) do not appear on this curve. An analysis of the binding energies of the last neutron, of a pair of neutrons, of the last proton and of the pairing energies of neutrons and protons made it possible to determine the Q-value for the majority of nuclear reactions in those cases, where only the upper or the lower limit of the energy of a given reaction is known. This, in turn, allowed determination of the masses and the binding energies of 14 radioisotopes figuring in Table 4*. In three cases we were obliged to introduce (for

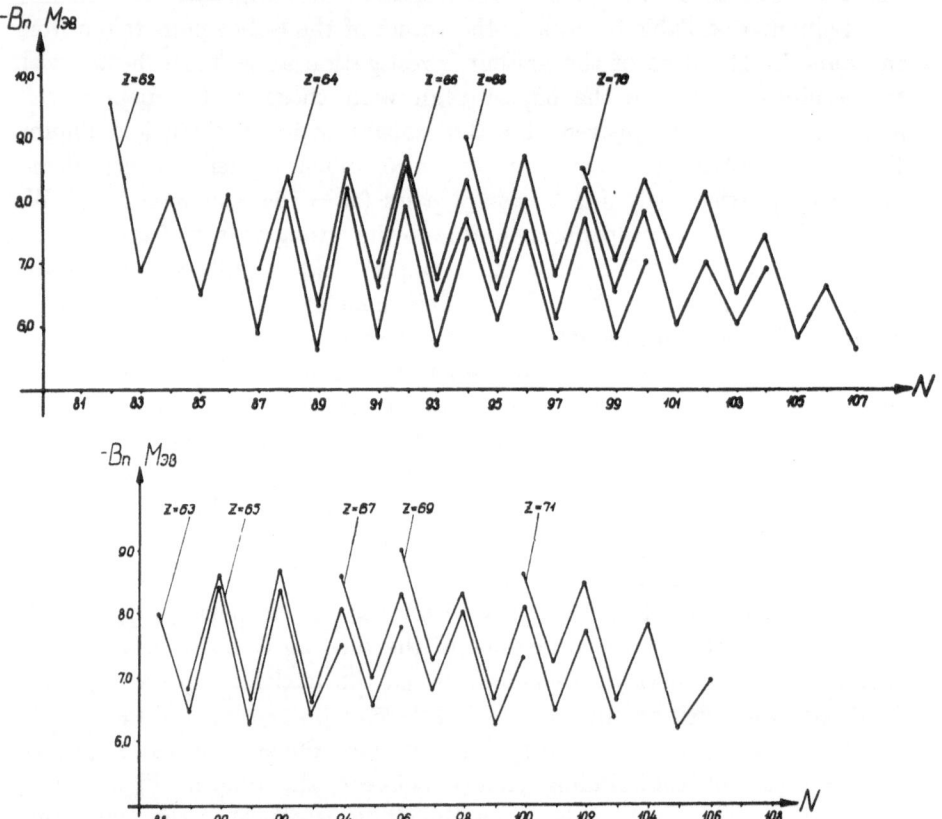

Fig. 4. Binding energy of the last neutron as a function of the number of neutrons N. 4a: for even Z; 4b for odd Z; Ordinates: B_n in MeV. Abscissae: Number of neutrons N

^{159}Dy, ^{164}Tu and ^{170}Tu) a correction of about 0.250 mu to the Q-values for the given nuclear reactions. Otherwise the regularity of the parameters B_n, B_p, B_{2n} and the others would be greatly disturbed.

In Fig. 4 the binding energy of the last neutron is plotted against the number of neutrons, for even and odd values of Z. As can be seen from this figure, B_n tends to decrease slightly with the increasing number of neutrons. This is apparent for both the even and the odd values of N.

---

* No limits of errors (degree of accuracy) are given for the masses of these isotopes.

The difference of B_n values for the even and the odd values of N is clearly apparent from the "saw-tooth" shape of the curve $B_n = f(N)$. For $N = $ constant the binding energy of the last neutron increases with the increasing Z. The downward trend of B_n with the increasing N is

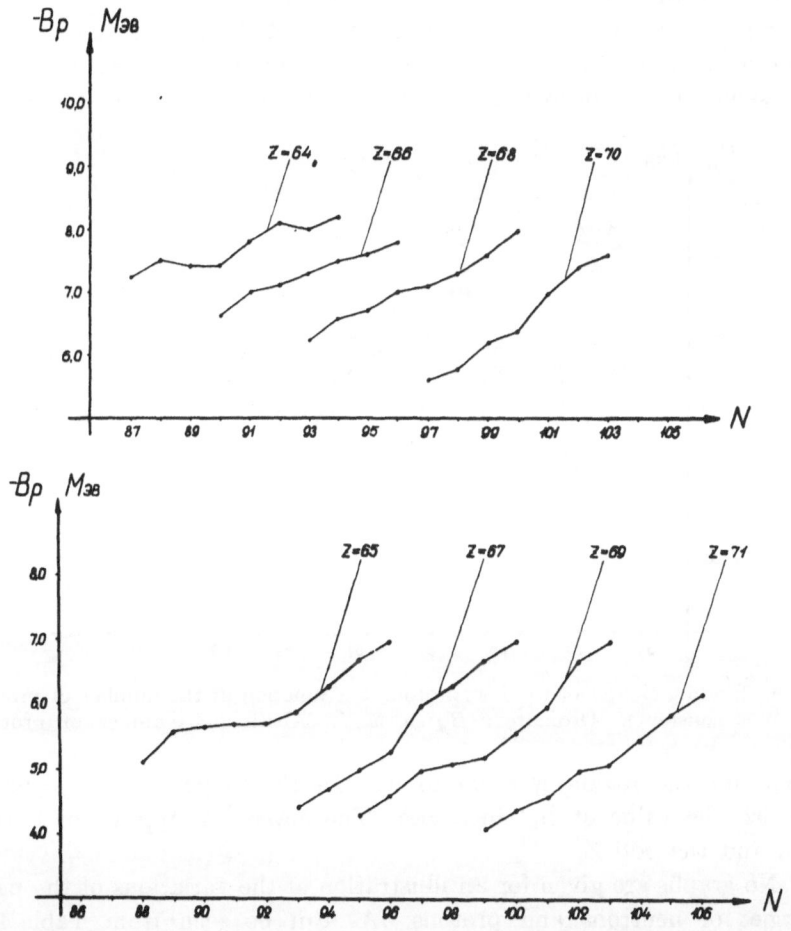

Fig. 5. Binding energy of the last proton as a function of the number of neutrons N. 5a: for even Z; 5b: for odd Z; Ordinates: B_p in MeV. Abscissae: Number of neutrons N

broken only for $N = 90$ where the B_n values are somewhat greater than those of the neighbouring isotopes which contain an even number of neutrons. This increase, although it falls within the limits of experimental errors, violates the downward trend of B_n with the increasing number of neutrons.

Fig. 5 shows the variations of $B_p = f(N)$ for $Z = $ constant. For both even and odd Z the binding energy of the last proton increases with the increasing number of neutrons. For $N = $ constant B_p reaches its maximum value for the nuclei with the least number of protons. In Fig. 6

the energies of the last proton are plotted against the number of protons. For the even protons the values of B_p are always greater than those for the neighbouring odd protons. For Z = constant the value of B_p increases with the increasing N. For both the even and the odd Z the value of B_p decreases with the increasing Z (for N = constant).

Fig. 7 gives the curves of the variations of the binding energies of the two last neutrons as a function of the number of neutrons for Z = constant. The value of B_{2n} decreases rather monotonically in the whole range,

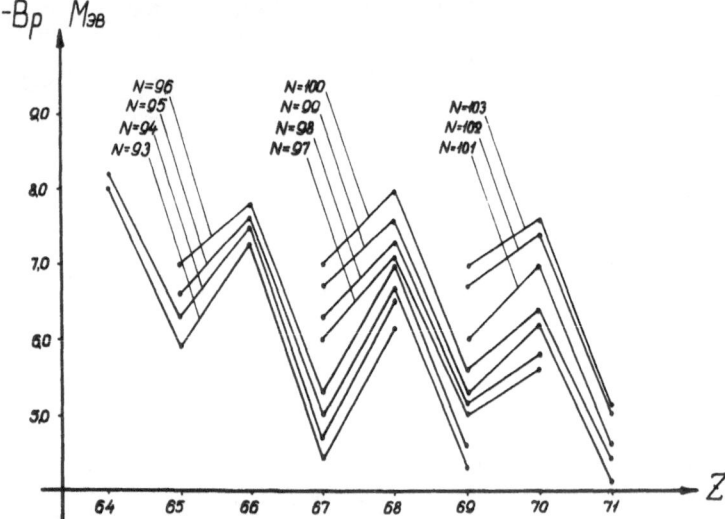

Fig. 6. Binding energy of the last proton as a function of the number of protons Z (for N = constant). Ordinates: B_p in MeV. Abscissae: Number of protons Z

except for the region $N = 90$ to 92. In this region, for $N = 90$ and $N = 92$, the value of B_{2n} increases. The maximum appear for both the even and the odd Z.

No graphs are given for an illustration of the variations of the pairing energies of neutrons and protons. As can be seen from Table 4, for $N = 90$ and $N = 92$ somewhat higher values of the pairing energies of neutrons are observed for some nuclei (^{155}Eu, ^{154}Gd, ^{157}Tb and ^{158}Dy).

All the aforesaid parameters vary comparatively monotonically, except for the region $N = 90$ to 92[2, 13]. Some irregulatities are observed in the region in question which distort the general trend of the variations of all these parameters.

It seems that a slight irregularity in the evolution of the B_n-values, a slight increase of the value of the pairing energy of neutrons and sharply apparent maxima of B_{2n}-values, for $N = 90$ and $N = 92$, are attributable to the structural changes of the nuclei and to their transition from a spherical to a deformed shape.

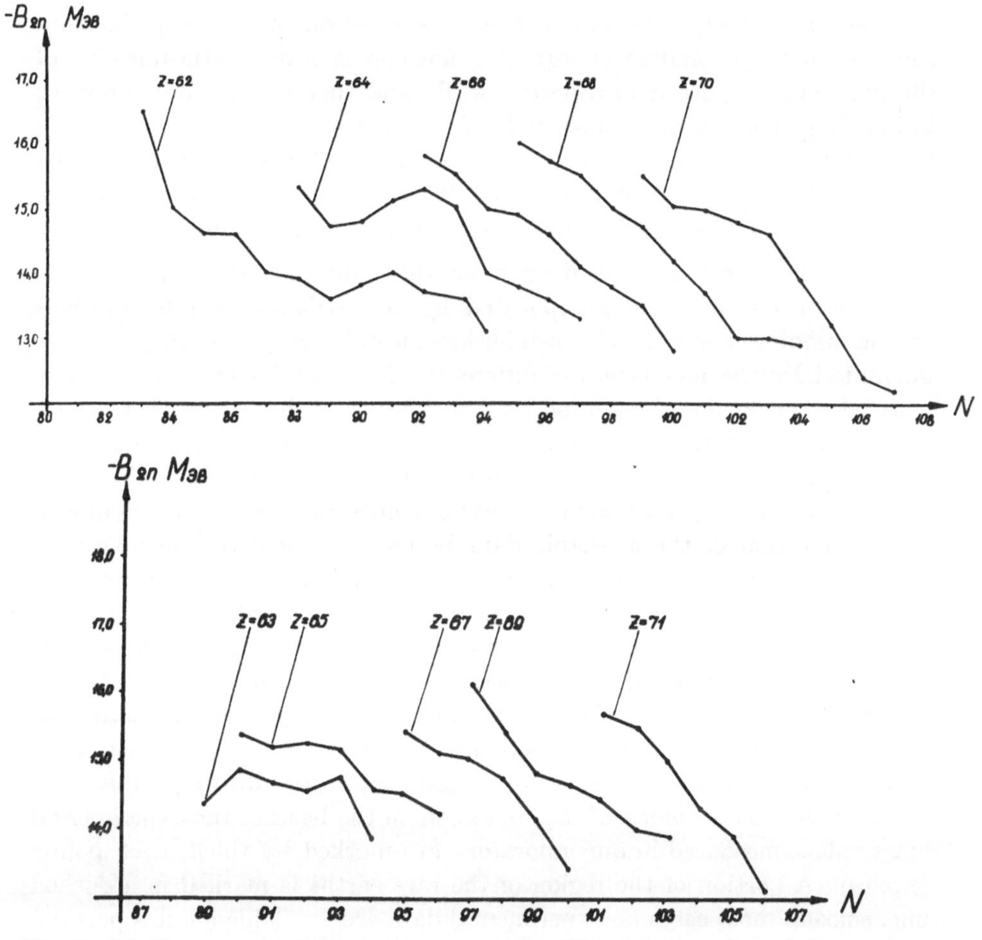

Fig. 7. Binding energy of the two last neutrons (B_{2n}) as a function of the number of neutrons N. 7a: for even Z; 7b: for odd Z

Systematization of the Binding Energies of the Two Last Neutrons

In confronting the experimental mass values and the binding energies it is of interest to follow the variations of the binding energies of the last neutrons (B_n) and of the last protons (B_p) and particularly of the binding energies of the last pair of neutrons (B_{2n}) and of the last pair of protons (B_{2p}) as a function of the mass number A.

It has been long ago observed[14, 15] that if one plots any one of these values against the values of A, then the points corresponding to the nuclei with the same values of N and Z fall approximately on straight lines, giving thus a parallelogrammic network. This parallelogrammic network may be considered as a systematization of the binding energies of the last neutrons and of the last protons; this network is somewhat similar to the systematization of the energies of the beta-decay. The linear relationship between these energetical parameters as a function of A is explained by the fact that the binding energy of a nucleon is a quadratic function of the number of nucleons in different shells, and therefore the differences of the binding energies (the values of B_n, B_p, B_{2n} etc.) must vary linearly with the number of nuclei. An analysis of the systematization of the binding energy of a nucleon itself is a rather tricky problem because of the fact that this binding energy is a quadratic function of the number of nuclei.

It is most convenient to examine the values of B_{2n}, since in this case the situation is not complicated by the even-and-odd fluctuations of the binding energy. The establishment of a B_{2n}-network is further facilitated by the fact that the difference of masses is very often known with a higher degree of precision than the values of the masses themselves, even when these latter are known. A comparatively large number of nuclear reactions is available fór which the Q-values are known, and there are also many mass spectroscopical measurements of the isotope doublets. A confrontation of the available data becomes particularly meaningful if the experimental points are plotted on a semi-empirical net covering a large number of nuclides. Such a graph gives the possibility of checking some doubtful values and of determining in which regions it is possible to estimate the unmeasured values by extrapolation or interpolation. Fig. 8 represents a network of the binding energies of the two last neutrons (B_{2n}) for a wide range of nuclides $80 < A < 252$. The data taken from references[1, 2, 7, 16, 17, 18] were used for drawing this network graph. The values of B_{2n} obtained on the basis of the experimental mass values measured in our laboratory are marked by thick black points (circles). A portion of the region of the rare earths is marked in a dashed line, since in most cases no experimental data are available for this portion.

The "magic" numbers $N = 50, 82, 126, 152$ and $Z = 50, 82$ are sufficiently sharply apparent in Fig. 8. At the intercepts with the "magic" numbers the B_{2n}-values decrease sharply. The lines connecting equal N or equal Z in different sections are straight lines. In the range of $N = 52$ to 80 a saturation of the binding energy B_{2n} is observed which manifests itself by a reduction of $\dfrac{\delta B_{2n}}{\delta Z}$ with the increasing Z and correspondingly by a reduction of $\dfrac{\delta B_{2n}}{\delta N}$ with the increasing N.

Moreover, it is possible to see that the points B_{2n} which correspond to the nuclei differing by an entire number of alpha-particles are also situated on straight lines, but the inclination of these lines is different. (A similar observation was made in the paper[16] with respect to the energies of beta-decay.)

On the whole, the variations of B_{2n} with increasing A give evidence of a fine structure which cannot be directly related to the recognized

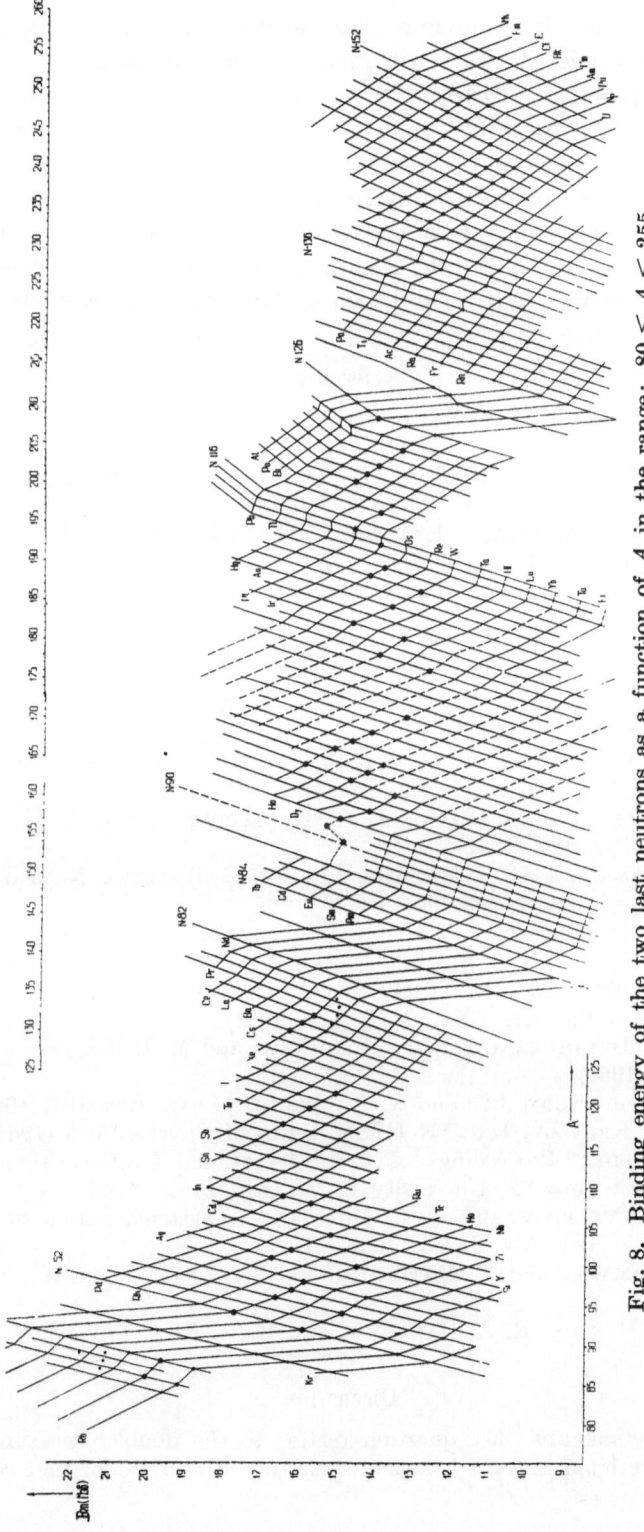

Fig. 8. Binding energy of the two last neutrons as a function of A in the range: $80 \leqq A \leqq 255$

"magic" numbers. It is obvious that as new experimental data become available the whole situation will gradually become clearer.

For the time being the data of the present systematization can be used for the determination of the unknown mass values, in the first place by means of an interpolation. In this connection it may be said that the degree of accuracy of the estimated value will not exceed ± 200 keV.

The authors are happy to express their thanks to E. E. Baroni and V. M. Soifer for providing the organic compounds and for preparing the chlorides of the rare earths, and also to Mrs. G. A. Dorokhova for her effective assistance.

References

[1] L. A. König, J. H. E. Mattauch, and A. H. Wapstra, Nuclear Phys. 31, 18 (1962).

[2] V. B. Bhanot, W. H. Johnson, and A. O. Nier, Phys. Rev. 120, 235 (1960).

[3] R. A. Demirkhanov, T. I. Gutkin, V. V. Dorokhov, and A. D. Rudenko, Atomic Energy (USSR) 2, 21 (1956).

[4] R. A. Demirkhanov, V. V. Dorokhov, and M. I. Dzkuya, DAN USSR 146, No 1, 72 (1962).

[5] R. A. Demirkhanov, T. I. Gutkin, O. A. Samadashvili, and I. D. Karpenko, Izd. AN USSR, vol. XXV, No 7, 871 (1961).

[6] R. W. King, D. M. van Patter, and W. Whaling, Revs. Modern Phys. 26, 327, 402 (1954).

[7] D. M. van Patter, W. Whaling, and L. J. Lidofsky, Revs. Modern Phys. 29, 757, 773 (1957).

[8] F. Everling, L. A. König, J. H. E. Mattauch, and A. H. Wapstra, Nuclear Phys. 25, 177 (1961).

[9] L. A. König, J. H. E. Mattauch, and A. H. Wapstra, Nuclear Phys. 28, 1 (1961).

[10] R. A. Demirkhanov, T. I. Gutkin, and V. V. Dorokhov, JETPh 37, vol. 5 (II).

[11] R. A. Demirkhanov, T. I. Gutkin, and V. V. Dorokhov, Izv. AN USSR, series of Physics, XXV, No 1, 124 (1961).

[12] R. A. Demirkhanov, V. V. Dorokhov, and M. I. Dzkuya, JETPh 40, vol. 6, 1572 (1961).

[13] W. H. Johnson Jr., and V. B. Bhanot, Phys. Rev. 107, 1669 (1957).

[14] V. A. Kravzov, Izv. AN USSR, series of Physics 18, 5 (1954).

[15] N. Zeldes, Proceedings of the International Conference on Nuclidic Masses, p. 151. Toronto: University of Toronto Press. 1960.

[16] B. S. Djelepov, and G. F. Dranitzina, Systematisation of the beta-decay energies. Izd. AN USSR, 1960, M.

[17] M. Yamada, and Z. Matumoto, J. Phys. Soc. Japan 16, No 8, 1529 (1961).

[18] J. L. Benson, R. A. Damerow, and R. R. Ries, Phys. Rev. 113, 1105 (1959).

Discussion

M. Higatsberger: My question relates to the doublet measurements of the different erbium isotopes by using cycles of ^{12}C and ^{13}C organic compounds for determining one isotopic mass. From the small inconsistencies there one can draw some conclusions about the sort of systematic errors that are involved. What systematic errors are characteristic for your machine?

V. V. DOROKHOV: As a matter of principle there may be some misfunction of the double focussing and, moreover, we use the dissociation doublets which have a certain velocity spread. There are some other factors also. We are trying to eliminate these sources of errors by using completely different doublets and by arranging of cycling with a minimum internal error.

H. E. DUCKWORTH: What is the value of β, the velocity spread amongst the ions, in your mass spectrograph.

V. V. DOROKHOV: This factor has not been taken into account in our work, since the spread of ion energies is of the order of magnitude of 2 eV, and the value of acceleration intensity of the order of 50 keV.

K. WAY: Have your results been published?

V. V. DOROKHOV: Our results have been partially published in the Soviet literature, but the major part presented at this conference has not yet been published, since these are quite recent data.

A. O. NIER: I am very pleased to see this new work. The more different laboratories that work in difficult regions such as this will make it possible to ultimately arrive at more accurate values. We at Minnesota are very much pleased to see you working in this region because this is going to help everyone to arrive at the very best values.

Before closing this session, Dr. VON ZAHN has a very brief communication which he would like to give that has relevance to this whole technique of measurement which we have been talking about today.

U. VON ZAHN: With higher and higher precision of mass measurements the resistor calibration turns out to be one of the serious problems in the whole experiment. Therefore, I would like to mention a new technique of voltage measurement which avoids the whole resistor calibration completely. The general idea may stem from TRISCHKA but as far as I know has never been applied to mass spectrometry.

This technique uses a small molecular beam resonance apparatus with an electric C-field [see G. GRÄFF, W. PAUL, and C. SCHLIER, Zeitschrift für Physik 153, 38 (1958)]. Having a suitable diatomic molecule within this C-field, we may induce rf transitions between different energy levels split by the Stark effect. From the transition frequency one can calculate the electric field strength in the C-field to an extreme precision.

To make a long story short, we hope to measure changes in a voltage of 1000 volts to about a tenth of a millivolt which is $1/10$ ppm. This method promises to have three advantages:

1. Neither a standard R nor a variable ΔR calibration is necessary.

2. Exact measurement of the voltage ratio simultaneously with the mass determination is possible.

3. There is almost no limit in the relative voltage jump. Hence, measurement of doublets separated by many mass units seems possible.

As a first step in using this technique, we in Bonn will check our resistor calibration against the molecular beam apparatus and first results of this test should come before the end of this year.

E. Breitenberger: You need a reasonably large splitting in relation to the line width. I presume you use a first-order effect. Could you give us an indication where you find suitable transitions and in what molecule?

U. von Zahn: A suitable molecule might be thallium fluoride, TlF. Transitions are of the type $\Delta J = 0$ and $\Delta m_J = \pm 1$ (with J the rotational quantum number). The first order Stark effect is averaged out by the rotation of the molecule, hence, a quadratic Stark effect is measured. Assuming an electric field strength of 2 kV/cm the transition occurs at about 400 Mcps., the line width is expected to be less than 8 kcps (= full width at half height). Localizing the line to 1% of their width at half height, therefore, yields $\Delta v/v = 2 \times 10^{-7}$ or because of the quadratic effect, $\Delta V/V = 1 \times 10^{-7}$.

W. H. Johnson: This method could be used as an absolute standard because apparently the constant ahead of the quadratic in E is a term which may be calculated. One should point out that for measurements such as we have, we are not interested in absolute standards but just $\Delta V/V$. It would turn out that many of the errors and difficulties in the experiment are eliminated because we are not interested in absolute measurements.

K. T. Bainbridge: Your idea is very interesting. At this time, I might feel I was jumping from the frying pan into the fire to adopt an entirely new approach because of contact differences of potential which may run into millivolts.

W. H. Johnson: The thing that I've been thinking about since I heard of this technique a couple of weeks ago in Bonn was the possibility of being able to run both of these experiments in the same electric field. If this would be the case then the contact potentials would be eliminated.

R. Elliot: It has been suggested by Dr. von Zahn that if resistance calibration errors can be reduced by the molecular beam technique then unit mass differences could be used with more confidence for mass measurement. However, the figures of Dr. Ries this morning for the variation of the measured hydrogen mass difference indicated that this variation may become important if unit mass differences are used. Correction could, of course, be made if the dispersion constant does not change and I would like to ask Dr. Ries how rapidly the dispersion constant varies.

R. R. Ries: As regards the hydrogen mass unit measurements, let me say that formerly, no data at Minnesota was considered acceptable unless we measured the hydrogen mass to be, on the average, equal to the accepted value. When this did not occur, we re-calibrated the resistors, or baked the analyser plates, or removed and cleaned the plates. Because of some of the semi-conducting materials which were measured, we had difficulty getting precise hydrogen mass unit measurements, so we adopted the dispersion constant correction. It appears that the mass unit measurements give long-term, constant, but incorrect values. This correction appears to be a linear effect according to our preliminary studies. Applying one of the solutions mentioned above, seems to give a new and different, but fixed value.

U. von Zahn: I like to emphasize the advantage of the molecular beam technique which is the lack of any calibration for the standard R and the variable R. However, it does not take care of contact potentials and surface charges in the electrostatic analyzer as does partially the hydrogen calibration.

Summation of the Conference

By

A. H. Wapstra

Instituut Voor Kernphysisch Onderzoek Ooster Ringdijk 18, Amsterdam, Holland

I would like to start by making a list of the various topics of the conference. First, are the atomic constants. The next points we have treated are the initial mass data: mass spectroscopic data and Q-values. The last point will include also alpha and beta decay energies. Then, we consider the combination of these values in order to get a consistent set of atomic masses. The fifth point is trying to represent these data by a mass formula, and then the sixth point, treated for the first time in a mass conference, is looking at excited states. They also have masses, and the difference in masses between different excited states is very interesting. And then the last point is considering the statistical methods used. This last point is of interest for all the other points so I had better start with that.

The methods of least squares is known to give a best value if the initial data are known to be gaussian distributed. But of course they are not, mostly, as was realized very well by DuMond and Cohen. In their treatment for getting best sets of atomic values, they mention that the method is also applicable under simpler assumptions. If a result $A \pm \sigma$ may be considered to mean that there is a distribution $N(x)$ for the thing measured with the properties that

$$A = \int x\, N(x)\, dx / \int N(x)\, dx \qquad (1)$$

and that

$$A^2 + \sigma^2 = \int x^2\, N(x)\, dx / \int N(x)\, dx, \qquad (2)$$

then you can indeed prove that the best value is given by the method of least squares. But there is still another "if": the "best value" means here the optimum value of a linear combination of the separate values given by the experiments. I do not feel at all sure whether this is always true; in fact I doubt it seriously for counting statistics (Poisson distributions). And then, I am not at all convinced that there exists any exact way of determining the precision σ in the results of our measurements. It is true, there exists no better way of treating experimental data than using the method of least squares on data weighted with (inverse square) probable errors determined as conscientiously as possible. But personally I always feel that this method is mainly a convention, and that one should not insist on an exactness in its application higher than warranted by this state of affairs.

This remark may be considered as a note by the contributions of BREITENBERGER, THIELE and COHEN. But as to the last one, I would like to make a few more remarks about the atomic constants. I very much enjoyed seeing how beautifully our measuring methods develop. I was especially enchanted about the contribution of HUGHES and BAILEY showing what high energy physics may give to this subject. I never realized before that making measurements on mu mesons already gives values which are very nearly significant for adjustments such as COHEN and DuMOND are making.

Let us now consider our atomic masses again. Here I want to make a small remark about what an atomic mass is. The mass is the number of protons times the mass of a hydrogen atom, plus the number of neutrons times the mass of the neutron, plus the nuclear binding energy, plus the binding energy of electrons. The order of magnitudes of the terms is the following: The mass of the hydrogen atom is 931 MeV and that of the neutron is about 1 MeV higher. The nuclear binding energy is of the order of 8 MeV per nucleon, and BREITENBERGER states that the electron binding energy is

$$0.8 \left(\frac{Z}{100} \right)^{2.5} \text{MeV}.$$

Here you can already see that the electron binding energy is not at all negligible. Many masses have been measured with a precision of the order of 20 keV or better and the electron binding can be as large as about 800 keV. Is now this binding energy of the electrons included in our mass measurements or not? It is certainly in the mass spectroscopic measurements. Let us take an example: a mass doublet $C_7H_6{}^+ - {}^{90}Zr^+ = \Delta$. What we want to know is the mass of the zirconium atom. Assume we know the masses of the C atom and of an H atom. The mass of that zirconium atom now equals seven times the mass of the C atom plus six times of the hydrogen atom minus Δ, but also plus the difference in the electron binding energy in the $C_6H_7{}^+$ ion and seven C atoms and 6 H atoms, and minus the difference in the electron binding energy of a zirconium atom and a zirconium ion. Up to now, we did not take into account these electron binding energy differences; can we justify this? I would estimate that the molecular electron binding energy term in $C_7H_6{}^+$ should certainly be less than

0.1 keV and the ionization energy in Zr even far less. You have seen that the precision of that mass doublet is of the order of 5 or 10 keV, thus we find that the electron binding energies are taken well enough into account.

How is it now in nuclear reactions? Let's look at the worst example: alpha decay, for instance ^{210}Po going over by alpha particle emission to ^{206}Pb. One normally uses the equation

$$M(^{210}\text{Po}) = M(^{206}\text{Pb}) + M(\alpha) + Q_\alpha.$$

Now we can ask, is this correct? Here a polonium atom changes into a lead atom, and if you again use that formula for the electron binding energy from BREITENBERGER, then we find easily that the electron binding energy for polonium is higher than that in lead by about 30 keV; so this looks quite serious. We perhaps neglect amounts of energy of 30 keV per alpha particle for alpha decay chains of some ten members. But it is not so serious as it appears because in reality that 30 keV cannot be lost, and it has been demonstrated that by far the larger part of this energy is just added to the energy of the alpha particle. Thus Q_α is again the alpha decay energy of the polonium atom — it is not the alpha decay energy of the bare polonium nucleus. The same argument probably applies to all reaction energies: here too, we need not take the electron binding energy separately into account. It will be done automatically with a precision of perhaps half a kilovolt at most.

One more remark about the electron binding energy. BREITENBERGER remarked that in mass formulae there is no provision for the electron binding energy. You saw that this energy is not at all negligible, it is as large as about 1 MeV for high Z. Do we have to add some term to the mass formulae for the electron binding energy? I think that this is not the case because in the mass formula we have already a term for the Coulomb energy of nucleus itself which is proportional to $Z^2/A^{1/3}$. For the sake of fast computation $A^{1/3}$ is about proportional to $Z^{1/3}$, so this Coulomb term is approximately proportional to $Z^{5/3}$, whereas the electron binding energy should be proportional to $Z^{2.5}$. Now the question is how large errors do we make by omitting the last term and slightly changing the constant in the nuclear Coulomb term. The following table shows this difference for various Z values if made zero at $Z = 0$ and $Z = 80$:

$$Z = 0 \quad \ldots\ldots\ldots \quad 0 \text{ keV}$$
$$Z = 20 \quad \ldots\ldots\ldots \quad 31 \quad ,,$$
$$Z = 40 \quad \ldots\ldots\ldots \quad 63 \quad ,,$$
$$Z = 60 \quad \ldots\ldots\ldots \quad 60 \quad ,,$$
$$Z = 80 \quad \ldots\ldots\ldots \quad 0 \quad ,,$$

I think in reality the trouble is even less. Of course, it means that the Coulomb term will now not represent the charge of the nucleus alone:

its constant there will be a little bit falsified. In that respect, BREITEN-BERGER is entirely correct. On the other hand, as an atomic mass formula the result will be quite acceptable.

Talking about atomic mass formulas, I would like to say that in several respects the situation here looks quite a lot better then before. Earlier we always had the idea that one should not only adjust the mass formula to major shells, needing about 30 constants, but that in addition one might have to introduce new constants for every sub-shell for high accuracy. I still don't know whether every theorist agrees, but we have heard in the first session that some theorists now think that new constants for every sub-shell will not be necessary. We have also seen many separation energy curves which really seem to show that this is correct. So many people have shown these beautiful curves that I will not mention them here separately. I am impressed by the smoothness in all differences in the nuclear masses. On the other hand the mass formula may theoretically look in better shape than before, in this respect, but may still be more complex in others. Thus, Dr. WAY showed me a letter from YAMADA who says that he has obtained some indications that the compressibility of nuclear matter might be much larger than had been thought before, and this might be of considerable influence in the nuclear mass formula.

I had a few misgivings about the introduction of excited states in this conference — after all, atomic mass considerations refer almost exclusively to ground states. This session of our conference, however, demonstrated that the total energy of different special states varies very smoothly with the mass number. The ground state masses form the envelope of a number of these graphs, and if you want to know details about how the ground states vary then it might be quite apt to study excited states too. Now that I have said this, I should perhaps add that I had some more misgivings about this conference before it started. I was somewhat afraid that this conference was a little too early — that there was not enough new in this conference. But the last two days I have become convinced that this conference has been quite timely. We have seen quite a lot of very important new material, to start with the beautiful new measurements by Minnesota and Hamilton. These measurements are really extremely good, and I am very impressed by the degree of consistency reached here. But there have been more new results. We just now heard the talk of DOROKHOV in which he mentioned quite a lot of new mass spectroscopic data in the region of rare earths. The precision of atomic masses in the rare earth region is improved considerably with these new results. I haven't yet studied them closely but the errors here probably will reach a value of about 30 keV or less. And for the masses in the region of Uranium we have the new reaction measurements reported here by MIDDLETON and ALLEN.

There have been more improvements, for instance, SPERDUTO's talk giving the recalibrating of all the wonderful measurements at M. I. T. Personally, this will spare me quite a lot of work. I am also very satisfied about the measurements of Dr. COHEN. I haven't checked them very well, but I hope they will check several mass doublets in the region of the light rare earths. If you remember the discussions about the consistency factor 2.65, you will understand the importance of such checks.

I would like to add one or two remarks about the outlook in the future. We will soon have a mass table which is good everywhere within some 30 keV or so for the main isotopes of every element. Now we can ask ourselves—does it make sense to go on trying to improve the precision? I think this is a question which should be considered because, after all, if we decide that after some time it would not be worth while I think several of us could stop and go on to something more useful. But I don't think that this is the case. First we still don't have a very good check of the mass-energy relation by EINSTEIN. In the first day of the conference, I told you that now the polonium-alpha decay energy has come up so much that we now arrive at a slight discrepancy on the other side. If we look now, this affects the mass energy relation: taking everything at face value, then you should say that the mass energy relation is off by about 1 part in two thousand or so, with a precision of about half the difference. Now there are other ways of checking this relation, somewhat more accurate, that disprove such a difference. On the other hand, we should realize that there is still a little question here: DICKE has reported a possible explanation for the discrepancy from masses as derived from nuclear reaction measurements and masses as derived from mass doublets. Thus, investigating the relation between energy and mass will be worth while.

I should like to insist that we should not trust too much the final statistical error in any adjustment of a great number of nuclear reaction measurements to a great number of mass doublets. I feel quite convinced that any kind of measurement will contain some systematic error, and therefore one should not have too much confidence in the reduction in errors that you get from a big adjustment. If one wants to make an experiment in this direction, then one should try to find a place where we could compare just a few mass doublets with a few reaction measurements, both measured with the best precision that we can get. Since the last 5 years, mass doublets have been measured with very high accuracy, something like 1 part in 10^5 or so in cases, but if you look at reaction energies their precision is much less. It's normally of the order of one part in a thousand or at most one part in 5000 or so. I'm curious to see whether Dr. BUECHNER will live up to his remark that he can improve his precision by an order of magnitude—I think it would be very desirable in this respect.

I cannot see immediate use for improvement by one order of magnitude in the methods of measuring mass doublets. I would prefer remeasurement of some nuclides with the present precision, for instance in order to eliminate the discrepancy between ^{27}Al and the older data. This is not meant to discourage mass spectroscopists to go on. It may be that if we gained say three orders of magnitude in precision, it might be that we come in a region where we get some information on the electron binding energies, which might be important. I would point to intensity measurements in mass spectroscopes which also have become important in a way that probably in the beginning nobody would have thought possible. It might very well be that if we improve the mass doublet measurements another 3 or 4 orders of magnitude, we could again get a tool of very high power in molecular physics.

Subject Index

Manzsche Buchdruckerei, Wien IX